International Symposium on Earth Sciences: History, Contemporary Issues and Prospects (ESHCIP 2019)

IOP Conference Series: Earth and Environmental Science
Volume 350

Moscow, Russia
28 March 2019

ISBN: 978-1-7138-0146-7
ISSN: 1755-1307

Printed from e-media with permission by:

Curran Associates, Inc.
57 Morehouse Lane
Red Hook, NY 12571

Some format issues inherent in the e-media version may also appear in this print version.

No changes have been made to the content of these proceedings. There may be changes to pagination and minor adjustments for aesthetics.

Printed with permission by Curran Associates, Inc. (2020)

Additional copies of this publication are available from:

Curran Associates, Inc.
57 Morehouse Lane
Red Hook, NY 12571 USA
Phone: 845-758-0400
Fax: 845-758-2633
Email: curran@proceedings.com
Web: www.proceedings.com

TABLE OF CONTENTS

HISTORICAL ASPECTS OF EARTH SCIENCES

GEOCOLOGY, GREEN POWER ENGINEERING AND ENVIRONMENTALLY FRIENDLY SOLUTIONS

CARTOGRAPHY, GEOINFORMATICS AND LAND MANAGEMENT

IOP Conf. Series: Earth and Environmental Science **350** (2019) 011001 doi:10.1088/1755-1315/350/1/011001

PREFACE

The International Symposium "Earth sciences: history, contemporary issues and prospects" was successfully held in Moscow, Russia, on March 28, 2019. The organizer of the symposium was S. I. Vavilov Institute for the History of Science and Technology RAS, Moscow. Today, the research work in IHST RAS is conducted in various fields such as the history of natural sciences, the history of technology and technical sciences, the science of science, and other. Moreover, IHST RAS is one of the leading centers of international cooperation in the field of historical and scientific and historical and technical research work.

The key goal of ESHCIP 2019 was to heighten the interest to the issues of the history as well as the current state of Earth sciences. Especial attention was paid to the ecological and environmental issues. Researchers from all across Russia came to the symposium to share the results of their work and discuss the prospects of Earth sciences. The day of the symposium became the day of intense learning and certainly gave all the participants the impetus to further research.

More than 90 papers were submitted on geology and geophysics, geoecology, green power engineering, land use and monitoring of lands, cartography and geoinformatics. All the submitted manuscripts were carefully reviewed by the editors and the invited reviewers. The most valuable papers on the topical issues of Earth sciences were selected for publication. We hope that this volume will not only deliver to the readers an overview of the latest research results in Earth sciences, but also give an opportunity to learn more about the Russian academic community of scientists and educators.

We would like to express our deep gratitude to the members of the ESHCIP 2019 organizing committee as well as to all the participants. Moreover, we would like to thank the reviewers and editors for their careful work and great contribution to preparing of the ESHCIP 2019 proceedings.

IOP Conf. Series: Earth and Environmental Science **350** (2019) 011002 doi:10.1088/1755-1315/350/1/011002

Peer review statement

All papers published in this volume of *IOP Conference Series: Earth and Environmental Science* have been peer reviewed through processes administered by the proceedings Editors. Reviews were conducted by expert referees to the professional and scientific standards expected of a proceedings journal published by IOP Publishing.

Exhibiting nature, modernising agriculture: ecology and technology at the All-Russian Agricultural Exhibition, 1923

O Yu Elina

S.I. Vavilov Institute of the History of Science and Technology, 14, Baltiyskaya st., Moscow, 125315, Russia

E-mail: olga.elina25@gmail.com

Abstract. This paper examines the history of the First All-Russian Agricultural and Handicraft Industrial Exhibition, which took place in Moscow during August-September of 1923. The study aims to reconstruct the environmental and agronomic contexts of the expositions presented at the Exhibition. The objectives of the study included analysing the contents, forms and slogans of these expositions, as well as identifying the ambitions of the Bolshevik leaders to publicly demonstrate the achievements of agrarian science and the results of cataloguing the natural environment and resources in the USSR. It is shown that the primary goal of the Exhibition was to convince the domestic public, especially peasants, in the advantages of the science-based "Sovietisation" of the countryside, along with modernisation of the backward Russian agriculture. The paper argues that the Exhibition became a tool for promoting various natural resources and biodiversity of the different regions of the USSR. No less important was the presentation of agronomy and agricultural technologies as practical tools for modernising agriculture.

1. Introduction

This article is a part of an interdisciplinary research project which deals with the interaction between science, technology and art at the First All-Russian Agricultural and Handicraft Industrial Exhibition (further – Exhibition). Alluded to the ideas of the Revolutionary renovation of Russia, the Exhibition was inspired by the Bolshevik leaders to promote and implement the scientific and technological modernisation into agriculture. Besides, vast and diverse natural environment and resources of the different regions of the country were on display. The primary focus of this article is analysing the Exhibition in terms of boundaries and intersections of environmental and technocratic contexts of its expositions. In particular, the problem under the question is exhibiting of natural diversity as an independent scientific and humanitarian value, on the one hand, and as a resource and means of modernisation of agriculture, on the other hand. The other goal is tracing the possibilities of art to reflect these complex interactions.

2. Materials and methods

This paper is based first and foremost on the documents of the Chief Committee of the Exhibition from the Russian State Archive of the Economy (RGAE). Therefore, the materials of this collection, most of that are cited for the first time, constitute the source plot of the article. Further, as the critical approach towards the study of the ecological and technological contexts of the Exhibition lays in the interdisciplinary realm, a number of methods of historical and socio-cultural research were used in this study, including not only the classic problem-historical approach, but also the methods of art history, sociology, historiography, and cultural anthropology.

3. Context and tasks of the Exhibition

The organisation of the Exhibition began in the spring of 1922, during an awkward moment of the country: the civil war had just ended; people were struggling with the ravages of the famine of 1921–22; the peasantry was in turmoil, still remembering *prodrazvyorstka* (the surplus appropriation system) [1]. Nevertheless, the Soviet leadership sought to demonstrate publicly "revolutionary achievements and plans" in the main branch of the economy of then Russia – agriculture, at the same time getting ready to present the results of scientific cataloguing of natural environment and resources. This kind of united exhibitions hadn't been previously held: traditionally, expositions were supposed to be divided according to the topic (agricultural, industrial, environmental, ethnographic, etc.). The project of the Exhibition, which passed two rounds of selection (open and closed) belonged to A.V. Shchusev; I.V. Zholtovsky became the chief architect; V.K. Oltarzhevsky was responsible for the construction as a deputy chief architect. The Exhibition occupied a vast territory of more than 100 hectares (the zone on both sides of the Crimean Bridge which is the present-day Gorky Park and Museon, and the Neskuchny Garden); 225 buildings were erected, 150 pavilions operated. All this was done in a record short time – just ten months! The Exhibition was opened on the 19th of August with a fiery speech from Klara Tsetkin as well as speeches from members of the All-Union Central Executive Committee, the Council of People's Commissars. Vladimir Lenin visited the Exhibition during his last visit to Moscow, on October, 19. More than 200 scientific institutions became exhibitors: institutes of the Academy of Sciences, the Moscow Timiryazev Agricultural Academy, experimental stations of People's Commissariat of Agriculture (Narkomzem), the State Institute of Experimental Agronomy, the All-Ukrainian Scientific Committee, the Anikovsky Biological Station, the Main Botanical Garden, etc. [2]. The Exhibition is known primarily as the triumph of the Russian architectural and artistic avant-garde [3]. Several examples of the co-creativity of people of science and the best Moscow artists in the embodiment of the ideas and concepts will be considered in the following section.

The overall goal set for the organisers was to promote the Leninist plan of cooperation and the "linkup between the urban and the peasantry". People entering the Exhibition were met by a massive pyramid of planted flowers, on the one side of which there was a portrait of Lenin. On the other side his long saying was placed, "Our goal is to restore the linkup, to prove to the peasant that we begin with the thing that he understands; that we can help him; that the communists are helping the ravaged little peasant at the moment of this hardship." [2, op. 1, d. 129b, p. 6] The target of the Exhibition was also the global audience: foreign institutions were invited to participate as exhibitors, thus confirming the innovative intentions of the Soviets.

Consequently, the Exhibition had several tasks. One of them was to convince the public, first and foremost peasants, of the advantage of the scientific and technical modernisation of agriculture, by demonstrating new machines, technologies and agricultural science data. Meanwhile, it was modernising the very carrier of agrarian practices – the peasant. So, the Exhibition contained agitation and propaganda aim. Another task was to demonstrate the "natural wealth of the country" by combining a wide range of geographic, geological, botanical, zoological, ethnographic and anthropological materials. Besides, the global audience was in focus as well: more than 600 foreign firms, companies and organisations, both state and private, were invited to participate as exhibitors and potential investors.

It should be noted that the Exhibition was held during the New Economic Policy (NEP) period with its elements of capitalist economic management. Besides, during the setting up of the Exhibition, the republics in the USSR were united, so the geography and themes of the expositions were significantly expanded. Therefore, in addition to thematic pavilions, the republic's pavilions appeared at the Exhibition: Azerbaijan, Ukraine, Belarus', Armenia, Kirgizia (Kirrepublic, as it was named at the time), Turkestan, etc. [4].

4. Environmental issues at the Exhibition: expositions of the Scientific and Educational Department

According to the organisers' idea, the Exhibition was supposed to be opened by the introductory exposition of the Scientific and Educational Department, containing a vivid overview of the natural resources and biodiversity of various regions of the USSR. It was distributed over several pavilions; the leading one was the so-called Main House.

In the Main House, the central section of the Department – "Nature of Russia" – was located. Its objective was "to show the environment in which agriculture develops, as well as the natural resources and the forces which it relies on" [5, p. 10]. Visitors were greeted with all sorts of maps, charts, graphs, etc. Most of these exhibits were prepared for the Exhibition. Thus, a new geological map was presented (scale: 60 versts, about 60 km (39, 77 miles), a soil map (size: 100 versts (66, 29 miles), maps of groundwater and artesian waters, minerals of European Russia which had agricultural importance. Different maps of climatic regions, diagrams and graphs reflecting the distribution of various elements in the marine area were shown in the subsections "Climate" and "Sea". The attention was drawn to the soil profile along the meridian through European Russia in the subsection "Soils" in addition to the corresponding cartograms. The sections "Vegetation" and "Animal World" presented the relevant botanical-geographical and zoogeographical maps and profiles. To explain the graphic exhibits, short essays were prepared in two variants: complete scientific version (5–15 p.) and a popular one (1–1.5 p.) [2, op. 1, d. 3, p. 4].

The large and almost independent section "Population" contained a lot of various historical and ethnological exhibits ("a tribal map of the USSR", map of the ancient settlement of the Russian Plain, etc.). These were followed by the characteristics of the population of the main climatic zones using models, slides, photographic and film materials. A title painting (oil on canvas) was presented for each zone, giving an artistic concept of the ethnological appearance of an area [2, op. 1, d. 129b, p. 29]. Living ethnographic exhibits, like a parade of the peoples of the USSR, were also presented. Not only dwellings and utensils of small nationalities of the USSR were collected for the Exhibition, but also so-called "exhibitors-residents" were invited (most often peasants); the Central Exhibition Committee and local committees provided the transportation of their houses with their families, pets, belongings, and also sponsored their stay at the exhibition [2, op. 1, d. 13, p. 8]. The underlying theme of a special "Forest" pavilion with the most precious exhibits also adjoined to the section "Nature of Russia". The pediment of this pavilion was painted by A.A. Ekster [3, p. 76].

The exposition of the Scientific and Educational Department pursued the primary aim of demonstrating natural and human resources as a material base and means of modernising agriculture. However, equally important was the presentation of natural diversity as an independent asset. It comes as no surprise that the "environmental protection" unit was prepared with the participation of a wide range of specialists from central and regional academic institutions (institutes of the Academy of Sciences, among them Astrophysical, the Main Botanical Garden, nature preserves and others). The environmental theme of the local level was also presented at the expositions of many experimental stations (more than 110) [5, p. 19–24].

As an example, the author will take the botany section (Muratov Base) of the Shatilov Agricultural Experiment Station, which was headed by a geobotanist, a student of S.G. Navashin, V.N. Durnovo [6]. The Base prepared the data for the exposition of plant diversity (a booklet, later a comprehensive volume of 600 pages) of the Oryol province which the station belonged to. The materials on nature reserve Galichya Gora, located on the Eastern slope of the Central Russian Upland, bordering the Oka-Don Lowland, in the North-Don Relic Botanical Region, were also collected for the Exhibition. The Galichya Gora project was successfully conducted by V. Khitrovo. He managed to resolve the issue of the protected-area status of this natural monument. The reserve combined unique botanical and geographical characteristics of relic historical landscape and flora, not typical for Central Russia. Nowadays the reserve is registered in the Guinness World Records as the smallest in the world. The expositions of the Muratov Base were located in different places, often separately from other sections of the Shatilov Station. Thus, publications about the Base took place among other editions of the Shatilov Station in its display case in the Pavilion of Experimental Stations. The exhibits on meadow typology, weed vegetation, clover polymorphism (plant and soil samples, photographic materials, maps, etc.) were demonstrated in the Land Reclamation Pavilion in the display cases of the Oryol, Bryansk and Ryazan provinces, the Bryansk forestry and the Orel seed control station. The works on clover received the Exhibition Medal; the department received an award from the local authorities: in the spirit of the time, they were given a horse for travelling [7].

5. Technological context of the Exhibition: agronomy and machinery

The Shatilov Station was undoubtedly invited to the Exhibition not only to demonstrate the results of the Muratov Base. There were numerous varieties of oats, rye, clover, and other crops among the achievements of the experimental station. Also, the activities of the Shatilov Gossemkul'tura (the State Seeds Cultivation, or the State Management of Grain Culture), a specialised scientific and practical institution that carried out mass reproduction of elite varietal seeds, were presented. Visual exhibits (sheaves of breeding varieties, the model of a factory for cleaning high-grade grains, etc.) and numerous charts, diagrams, scientific publications and popular science booklets were demonstrated by the Shatilov station at the Exhibition. For example, a booklet "Shatilov Gossemkul'tura", which was beautifully designed in the avant-garde style, explained the advantages of the system of mass reproduction of high-quality grain, which included the peasantry cooperation as one of the steps [6].

The demonstration of technocratic and environmental expositions in the windows of the Shatilov station reflected the non-confrontational, concurrent development of these trends in the activity of large experimental stations that characterises the mid-1920s. These trends wholly corresponded to the situation in the field of environmental management: still peaceful division of the departmental responsibility of the Narkomzem (experimental stations) and the People's Commissariat of Education, or the Narkompros (its Committee for Nature Protection and a subdivision of the Nature Conservancy). Furthermore, the conflict between the Narkomzem and the Narkompros on the reassignment of reserves would occur only in a year and would be lost by the Narkomzem. Thus, the Exhibition slogans reflecting these polar trends are also significant: "Exterminated nature monuments cannot be resurrected!" (Department of Nature Conservation) [2, op. 1. D. 34b, p. 35]; "The knowledge of nature makes it possible to subordinate its forces!" (Narkomzem) [2, op. 1. D. 34b, p. 30].

However, the underlying message of the Exhibition was symbolised by numerous "mild" slogans about agrarian modernisation with the primacy of scientific knowledge in this process. For example, "Only a close union of labour and science will conquer vast areas of wild lands"; "If you know nature, its hardships are not so terrible" (Nature Section) [2, op. 1. D. 34b, p. 30]; "Do not cut down the forest for no reason, so as not to fight for the scraps" (Forest Department) [2, op. 1. D. 34b, p. 40]; "Knowledge is capital, without which agriculture cannot be improved" (Department of Experimental Stations) [2, op. 1. D. 34b, p. 34]; "No knowledge – no bread" (Department of Agricultural Assistance) [2, op. 1. D. 34b, p. 35]; "Observation of natural phenomena facilitates the work of a peasant" (Narkompros) [2, op. 1. D. 34b, p. 35]; "The Museum of the native land is a living book about the riches of the country" (Museum Department of the Narkompros) [2, op. 1. D. 34b, p. 35]; "Electricity is needed in the place, where a peasant uses agricultural machinery" (Department of Agricultural Electrification) [2, op. 1. D. 34b, p. 35]; "There is no life without water" (Department of Land Reclamation); "Our raw material is processed by foreigners – let's learn to do it ourselves" (Processing Department) [2, op. 1. D. 34b, p. 40–41].

The expositions within the framework of what we call the "modernisation" trend were numerous and variable. For example, many agro-industrial syndicates that arose during the NEP period, such as the All-Russian Tobacco Syndicate, introduced innovative (for that time) automation of processing of raw materials and production of finished goods. By the way, the production line was presented at the Exhibition in the famous pavilion "Makhorka" of K.S. Melnikov.

In addition, as an example, I will tell you about irrigation projects meant to increment agricultural arable land. They were presented in the Pavilion of Turkestan by architect F.O. Schechtel (in addition to the Land Reclamation Pavilion). Thus, the project was performed (1920) as part of the unimplemented irrigation program of the Golodnaya (Hungry) Steppe Irtur under the supervision of Schechtel along with the exposition "The Golodnaya Steppe". It included dams, administrative, public, residential buildings containing signs of the architectural heritage of Central Asia. According to the architect's plan, the project should have clearly shown that giant hydraulic structures and settlements are a phenomenon which is closely connected with the culture of the peoples inhabiting this land [8]. However, uncertainty with land ownership – land in several vital areas belonged to public associations – turned out to be an obstacle for a successful implementation.

As for the construct of a peasant, in the framework of the Exhibition we also see the harmony of the present and the future. Peasant's "Present day" was portraited by the already mentioned "exhibitors-residents" with a transferred fragment of their habitat. As well as by numerous participants of ethnographic processions, demonstrations, festivals (labour, culture, arts and crafts, etc.) that took place in the "Old Village" exposition. "The Future" was represented by "New Village" exposition with modern electrified dwellings, public buildings, as well as agricultural technology, innovative materials and land cultivation. "New peasant" was positioned as a master of the machinery, electricity and the innovative methods of reclamation, selection, agrochemicals, etc. [10] "The Future" was illustrated by the avant-garde architecture of the pavilions, agitation posters, etc. Those forms should have been demonstrated by artists, sculptors and architects of the Russian avant-garde: A.V. Shchusev, A.A. Ekster, V.M. Mukhina, K.S. Melnikov and others. An influential art propaganda machine worked together with the writers such as D. Bedny, V.V Mayakovsky, M.A. Bulgakov and others.

6. Conclusion

The First All-Russian Agricultural and Handicraft Industrial Exhibition became a unique demonstration of the advancement of modernised technologies in agriculture while maintaining an expressed environmental trend. Combination of the scientific and technological expositions presented at the Exhibition provided future exhibitions, both central and local, with the valuation methodology. More than 1.5 million visitors saw the Exhibition in less than two months. Public interest was fueled by a combination of scientific and technological exhibits, on the one hand, and entertainments and public cultural events, on the other.

References

[1] Roberts P C 1970 War Communism: A Re-Examination *Slavic Rev.* **29(2)** 238–61

[2] *Hereinafter, the documents of the Chief Committee of the Exhibition* RGAE F 480 Op 1 Dd 3, 8, 28, 34б 40б, 94a, 95б 122a, 129б, 145, 156, 157, 179, 181; Op 13 D 31

[3] Ryazanov I V 1976 *The Art of the Soviet Exposition Ensemble 1917–1970* (Moscow: Soviet Painter)

[4] Chayanov S K 1923 Report of the member of the Presidium of the CCE on the Section of Agronomy on 6 March 1923 *Bull. of the CCE* **2** 2–16

[5] *Collection of programs of the departments of the First Agricultural and Handicraft Industrial Exhibition of the USSR 1923* (Moscow: Publishing department of the CCE)

[6] Elina O Yu 2015 Pyotr Lisitsyn and his Shatilov Gossemkul'tura as a Project of State Seeds Cultivation *Studies in the History of Biology* [in Russian – Istoriko-biologicheskie issledovanija] **7(4)** 8–39

[7] Khitrovo V N 1928 Historical essay on the Muratov Base – Botanical Department of the Shatilov Station *Proceedings of the North-Chernozem Regional Agricultural Experiment Station* **1** pp 67–80

[8] Rizenkampf G K 1921 *Irrigation problems of Turkestan* (Moscow: VSNH)

[9] Elina O Yu 2002 Planting Seeds for the Revolution: The Rise of Russian Agricultural Science 1860–1920 *Sci. in Context* **15(2)** 209–37

[10] Fitzpatrick S 1993 Ascribing Class: The Construction of Social Identity in Soviet Russia *J. of Modern History* **65(12)** 745–70

The beginning of comprehensive studies of the nature and population of Ciscaucasia in the XVIII century

I.A. Kerimov[1,2,3], Z.Sh. Gagaeva[3,4], U.T. Gayrabekov[3,4] and S.V. Badaev[4]

[1]S.I. Vavilov Institute for the History of Science and Technology RAS, 14, Baltiyskaya st., 125315, Moscow, Russia
[2]M.D. Millionshchikov Grozny State Oil Technical University, 100, Kh. Isaev Ave., 364061, Grozny, Russia
[3]Academy of Sciences of Chechen Republic, 13, M. Esambaev Ave., 364024, Grozny, Russia
[4]Chechen State University, 32, A. Sheripov st., 364024, Grozny, Russia

E-mail: zsh_gagaeva@mail.ru

Abstract. The paper covers the history of geographical studies of Ciscaucasia carried out by academic expeditions of the 18th century. Large-scale expeditionary activities of this period marked the beginning of comprehensive studies of the nature and economy of the northern part of the Caucasus, made it possible to collect geographic information on the territory. New information contributed to the development of new research areas, formed a basis for the development of new textbooks and cartographic materials. The information about the Pre-Caucasus enriched science. It has a scientific value.

1. Introduction

The second half of the XVIII century is a bright page in the history of geographical research of the Caucasus. It is associated with activities of academic expeditions. It marked a new era in the development of geography. Expeditionary activitieswere aimed at studying remote territories of the Russian Empire and synthesizing information about its nature and population. The idea of expeditionary studies in the Russian Empire belonged to M.V. Lomonosov (1711-1765). European education shaped his views.Hewanted to see Russia an enlightened and developed European country. St. Petersburg Academy established relations with Germar states and had great prestige in Europe [1].

Prior to the second half of the XVIII century, there was only scattered informationon the territories of the Russian Empire. Most of data were obtained by travelers. Military departments played a crucial role in collecting asnd accumulating information about the Caucasus. Diverse natural conditions of the Caucasus, the multinational composition of the population, traditions, religion, unique ethnocultural values, trade exchange between nations wereresearch areas. All this determined the most important task of the expeditions: implementation of the socio-political order, collection of information about the natural wealth of Russia, especially its north-eastern and southern territories. Activities of academic expeditions are covered in many literary sources and archives [2, 3]. However, there are many "white spots" which can be of scientific interest for modern scientists.

2. Materials and methods
Research materials are literary and archival sources. Historical research methods, complex scientific analysis were used as research methods.

3. Results
Geographical information about the Caucasus has been accumulated since antiquity. The routes of European travelers of the XIII-XV centuries passed through the Caucasus. A. Nikitin (15th century) was one of the first Russian travelers who collected information about his journey, including the Caucasus. His route to India covered the Caspian coast of the Caucasus, Persia. However, detailed geographical information about the Caucasus was obtained thanks to the activities of 18th century academic expeditions. The wide coverage of issues, the scope and content of the research program had no analogues in the scientific world.

In 1768-1774, the Imperial Academy of Sciences sent five expeditions, three of which were from Orenburg (P.S. Pallas (1741-1811), I.I. Lepekhin (1740-1802), J.P. Falk (1727-1774) and two ones were from Astrakhan (S.G. Gmelin (1744-1774) and J.A. Güldenstädt (1745-1781). The main goal of these expeditions was to collect information about the nature and population of the little-studied territories of Russia. Uniform instructions, plans c and routes were developed for the expiditions [4–6]. The expeditions had to collect information on destroyed monuments, ruins, barrows, graves, tools and other material evidence that could be attributed to archaeological monuments. At the same time, special attention was paid to coins and their origin [2]. "They should make every effort to spread their knowledge and increase it, so that all memorable things such as animals, birds, fish, insects, and will be described" [7]. According to the instructions, the leaders of the expedition were obliged to send their reports to the Academy.

Noting the socio-political significance of the expeditions, M.A. Polievktov (1935) writes: "The expedition of the Academy of Sciences serving economic interests and aspirations of the nobility of feudal Russia of the second half of the XVIII century ... examined the southeast of European Russia and a number of adjacent areas. Gmelin and Güldenstädt captured the entire Caucasus" [3, p. 12].

Expeditionary teams headed by J.A. Güldenstädt and S.G. Gmelin were the first ones sent to the Caucasus (the Astrakhan academic expedition of 1768-1774). Young J.A. Güldenstädt was instructed to examine vast territories of the south-east of the European part of Russia. The Caucasus route of J.A. Güldenstädt ran through Kizlyar, Kabarda, Ossetia, Georgia [8]. He was entrusted with "... a survey of those areas of the Caucasus where in the XVI-XVII centuries Moscow State carried out surveys but which were not covered by the Russian Caucasus studies of the 13th century" [3, p. 123]. There are detailed descriptions of Güldenstädt route through Novgorod, Porkhov, Old Russa, Ostashkov, Rzhev, Mozhaisk, Moscow, Tula, Yelets, Voronezh, Tsaritsyn and Astrakhan. Güldenstädt went to Kizlyar, examined its surroundings, banks of the Kambilee, Sunzha, Aksai, Koisu and other rivers. He conducted surveys of oil fields and mineral springs [3].

Complex studies of the route of the expedition headed by Güldenstädt (1768-1775) are of great value from a scientific point of view, because factual materials were collected. Numerous reports, diaries and cartographic materials collected by J.A. Güldenstädt are stored in St. Petersburg branch of the archive of the Russian Academy of Sciences [9, 10, etc.] and published in a number of papers [11–13]. The results of his work formed a basis of many works which contain detailed descriptions of his travels [3, p. 125]. One of the works is highly appreciated by the scientific community: "... For the Caucasus, mainly Northern and Reisen Georgia, J.A. Güldenstädt's work is crucial. It covers all aspects of natural science (especially botany and zoology), provides information on local productive activities (agriculture, silkworm breeding, mining, oil, manufacturing, etc.) and trade, ethnography and history of the region; dictionaries of Georgian, Mingrelian, Svan, Circassian, Chechen and other languages and the Caucasian adverb are given .. "[3, p. 125].The information about the geographical location of the Caucasus Mountains and their general natural conditions, the Terek river and its tributaries, Caucasian villages and peoples (statistical data on the number, customs, customs, religion, etc.), industry and the economy, etc. have not yet lost its scientific significance. This was written by K. German (1809) in the preface to the Geographical and Statistical Description of Georgia and the Caucasus: "Güldenstädt's journey through Russia and the Caucasus Mountains contains precious data..." [11].

The route of the expedition headed by S.G. Gmelin passed through Astrakhan, the Terek estuary, Derbent and Baku, the coastal part of the Caspian Sea. They visited Shamakhi, Salyan, Iran. From 1771-1773, they explored the entire coast of the Caspian Sea. S.G. Gmelin had to study the Caspian region. S.G. Gmelin began his journey from Astrakhan, visited "... northeastern Ciscaucasia, western and eastern shores of the Caspian Sea ..." [3, p. 116]. He had to study areas of the Caucasus which attracted attention of Western European and Russian travelers [3, p. 116]. The program developed by the Academy of Sciences was intensive. Gmelin's research results formed a basis of his five-volume description of his journey. Detailed information on natural history, ethnographic descriptions of various ethnic groups of the Northern Ciscaucasia were obtained. Gmelin described Astrakhan and Astrakhan territory, Mozdok, Kizlyar, Derbent; flora, fauna, political relations, history of the region, historical monuments [3, p. 116]. Descriptions are accompanied by illustrations and maps. Gmelin's handwriting and other materials are stored in the St. Petersburg ARAS [14–16, etc.]. The most important results of his research (valuable scientific materials, observations, records and travel diaries) were processed and published by Güldenstädt and Pallas in 1777-1806 [17].

The expedition headed by P.S. Pallas was crucial. They did a lot in the field of geography, geology, paleontology, botany, history, archeology, economics, philology. Pallas is known as a scientist and naturalist. "They had to examine all areas of the" Russian Empire "; this expedition focused on the south-eastern sector of European Russia, the Urals, the Caspian region, the Caucasus and part of the Black Sea. A significant contribution to the study of this part of the Caucasus was made by the second expedition (1793-1794 (to the Astrakhan region) [3, p. 156]. Throughout the journey, Pallas kept detailed records. The factual material collected became a basis for his famous work "Journey to different places of the Russian state" which was a reference book for many generations of scientists [17]. Pallas paid significant attention to the reliability of information and considered this to be the main feature of journey descriptions. Pallas made it possible to raise the level of knowledge in various scientific fields. He studied water, soil, farming, diseases and treatment methods, beekeeping, sericulture, and cattle breeding. He studied the mineral wealth and mineral waters, arts, crafts, crafts, etc. P.S. Pallas collected all sorts of information about customs and beliefs of the peoples [17]. P.S. Pallas headed the scientific expedition of J.A. Güldenstädt after his tragic death in 1781.For three years, P.S. Pallas worked on his manuscripts and diaries. The result of this work was the two-volume book "Journey to Russia and the Caucasus Mountains (1787-1791) (J.A. Güldenstädt) published in Germany and translated into Russian in 2002 [18].

The Caucasian route of J.P. Falk (1768-1775) passed through Astrakhan and Kizlyar. He investigated the Terek greenhouses and Kizlyar waters [17]. There is not much information about Falk in the domestic and foreign historical literature. During the journey to the Russian Empire, he collected a lot of factual materials that are stored in the St. Petersburg branch of ARAS [6]. Falk's records were processed by his contemporaries who tried to translate Falk's texts. Moreover, the recordings were made in three languages. Materials on the flora are of great importance.

J.P. Falk tried to make the most accurate descriptions of nature (plants, animals), life and activities of the population. Many entries were made in three foreign languages. Materials collected by J.P. Falk were being processed for many years. The main results were published in the first quarter of the XIX century [19].

4. Conclusion

Academic expeditions of the eighteenth century opened an era of integrated geographical research. Valuable factual materials on the nature, ethnography, folklore, religion, traditions and customs of the peoples of the Caucasus were collected and systematized.

Detailed information on the natural and cultural wealth of the Caucasus was collected. Drawings, descriptions, plans of archaeological monuments were created. This was part of the geographical description of the territories of the Russian Empire [20]. Of particular interest were field materials on natural resources of the North Caucasus [9]. The results of the expeditions were used in scientific publications and cartographic materials on the Caucasus. Information collected by academic expeditions enriched geographical information about poorly explored territories of the Caucasus. The latest information on the territories of Russia allowed Russia to strengthen its political influence in the world. This paper continued a series of works on the history of geographical studies of Ciscaucasia whose results were published in a number of papers [21, 22, etc.].

5. Acknowledgments

The work was financially supported by Russian Foundation for Basic Research (RFBR), project number 17-33-00038 a1.

References

[1] Osipov V I 1993 *Scientific heritage of P S Pallas Letters 1768–1771* (St. Petersburg)
[2] Gnucheva V F 1940 *Materials for the history of expeditions of the Academy of Sciences in the XVIII and XIX centuries: chronological reviews and description of archival materials* (*Proceedings of the Archive* iss 4) (Moscow; Leningrad: Academy of Sciences of the USSR)
[3] Polievktov M A 1935 *European travelers XIII–XVIII centuries across the Caucasus* (Tiflis)
[4] *St. Petersburg Branch of the Archive of the Rus. Academy of Sci.* **3(32)** p 1
[5] *St. Petersburg Branch of the Archive of the Rus. Academy of Sci.* **3(33)** p 1
[6] *St. Petersburg Branch of the Archive of the Rus. Academy of Sci.* **3(35)** p 11
[7] *St. Petersburg Branch of the Archive of the Rus. Academy of Sci.* **3(23–6)**
[8] Kopelevich Yu H 1997 *Johann Anton Güldenstädt 1745–1781* (Moscow: Nauka)
[9] *St. Petersburg Branch of the Archive of the Rus. Academy of Sci.* **3(33)** p 5
[10] *St. Petersburg Branch of the Archive of the Rus. Academy of Sci.* **3(33)** p 7
[11] *Geographical and statistical description of Georgia and the Caucasus from the trip of academician J.A. Güldenstädt through Russia and the Caucasus Mountains in 1770, 71, 72 and 73 respectively* 1809 (St. Petersburg)
[12] Güldenstädt J A 1779 Geographical, historical and statistical news about the new frontier line of the Russian Empire between the Terek and the Sea of Azov *Historical and Geographical Monthly in 1779* (St. Petersburg)
[13] Güldenstädt J A 2002 *Journey through the Caucasus in 1770–1773* (St. Petersburg: Petersburg Oriental Studies)
[14] *St. Petersburg Branch of the Archive of the Rus. Academy of Sci.* **3(34)** p 1
[15] *St. Petersburg Branch of the Archive of the Rus. Academy of Sci.* **3(34)** p 16
[16] *St. Petersburg Branch of the Archive of the Rus. Academy of Sci.* **3(34)** p 29
[17] Aleksandrovskaya O A, Shirokova V A, Romanova O S and Ozerov N A 2011 *Lomonosov and the 18th century academic expeditions* (Moscow: RTSoft)
[18] Tkachenko D S 2015 Expeditions of the Academy of Sciences in the formation of the image of the Caucasus of the Russian Empire border XVIII–XIX centuries *History: facts and symbols* [in Russian – Istoriya: Facty i simvoly] **3** 82–94
[19] Travel Notes of Academician Falk *Full Collection of Travel Scholars* (St. Petersburg) vol VI 1824; vol VII 1825
[20] Kolesnikova M E 2010 At the root of Russian Caucasian studies: Academic expeditions to the North Caucasus in the second half of the 18th – early 20th centuries *VIET* **1** 55–62
[21] Kerimov I A, Gagaeva Z Sh, Badaev S V, Akhmatkhanov R S and Chimaeva Kh R 2018 Academic expeditions of the XVIII century and nature studies of the North Caucasus *Bull. of the Academy of Sci. of the Chechen Republic* **3(40)** 54–57 DOI: 10.25744/ vestnik.2018.40.3.009
[22] Kerimov I A and Gagaeva Z Sh 2018 Academic expeditions of the XVIII century in the Caucasus: traveler scientists (to the 250th anniversary of the physical expeditions of the Academy of Sciences) *Grozny Natural Sci. Bull.* **3 6(14)** 112–8

Analysis of trends in annual time series of precipitation in the Republic of Bashkortostan, Russian Federation

V M Gaponov[1], A N Elizaryev[1,3], S G Aksenov[1] and A Longobardi[2]

[1]Department of Fire Safety, Ufa State Aviation Technical University, 12, K. Marx Street, Ufa, 450000,The Republic of Bashkortostan, Russia
[2]Department of Civil Engineering, University of Salerno, Via Giovanni Paolo II, 132 - 84084 - Fisciano (SA), Italy
[3]Water Problems Institute, Russian Academy of Sciences, Moscow, 3, Gubkina st., 119333, Russia

E-mail: elizariev@mail.ru

Abstract. Nowadays, the problem of climate change receives a wide resonance throughout the world. According to meteorological observations in Bashkortostan over the past 30 years, annual precipitation has increased. The occurrence of natural hazards is directly related to climate change, the number of which has been steadily increasing in recent years. If in the period from 1998 till 2005, their number in Bashkiria ranged from 5 to 11, then in 2010 it was recordedas 37, and as 34 in 2016. In this article, the object of study is the Republic of Bashkortostan, its territory is 142,947 km^2. Available data consists of annual time series of precipitation for 34 stations, they were analyzed to identify potential trends and their significance. As a first step, precipitation data was checked to assess the quality of the data, to find potential points of change in the time series. For this purpose, parametric (The Student's t -test) and non-parametric (Pettitt's test) statistical criteria were applied. Then, a statistical test was used to determine the significance of linear trends (the Mann-Kendall trend test and the Sen's test). Thus, the main purposes of the article are to detect points of change, as well as, to check, with a help of parametric and non-parametric tests, annual trends for precipitation data.

1. Introduction

Nowadays, the problem of climate change is receiving a wide resonance throughout the world. This is confirmed in the articles of many authors [1-4]. Moreover, climate change in recent decades has been called unprecedented. Actually, the analysis of regular meteorological measurement data impartially confirms the increase of average global air temperature over the past century. According to the Hydrometeorological Center of Russia, the growth rate of average annual air temperature in Russia is 2.5 times higher than the growth rate of global temperature. Significant changes have occurred in precipitation. In general, on the territory of Russia, in the period 1976–2016, annual rainfall is growing by 2.1% in every 10 years. Spring precipitations are especially fast growing, 5.9% growth per 10 years (in Eastern Siberia it is up to 15–20% per 10 years). Some decrease in annual sums was noted in most of the European territory of Russia, in the Southern Urals, and in the Volga Federal District [5].

According to meteorological observations in Bashkortostan over the past 30 years, annual precipitation has increased. With an average level for the region of 499 millimeters per year, over the last decade it was 502 mm. Several researches prove this [6, 7]. The occurrence of natural hazards is directly related to climate change, the number of which has been steadily increasing in recent years. If in the period from 1998 to 2005, their number in Bashkortostan ranged from 5 to 11, in 2010 it was noted 37, and in 2016 - 34 [8]. According to the IPCC, higher temperatures and increase in precipitation can raise the frequency of fires [9].

The most important indicators of climate are air temperature and rainfall [10].

In this article, the object of research is the Republic of Bashkortostan, its area is 142,947 km^2. Available data consists of annual time series of precipitation for the 34 stations, analyzed to identify potential trends and their significance.

As a first step, precipitation data was checked to assess the quality of the data, to find potential points of change in the time series. For this purpose, parametric (The Student's t -test) and non-parametric (Pettitt's test) statistical criteria were applied. Subsequently, a statistical test was used to determine the significance of linear trends (the Mann-Kendall trend test and the Sen's test). These tests were widely used in different countries, such as Italy, Iraq, China, India etc. Their effectiveness has been proven by many authors [11-16].

Thus, the main objectives of the article are to detect points of change, as well as to check, using parametric and non-parametric tests, trends on an annual basis for precipitation data.

2. Materials and methods

There are 34 weather stations in the Republic of Bashkortostan, Russia.Bashkortostan belongs to a temperate climate zone with an Atlantic-continental climate. The climate is rather humid, the winter is moderately severe, the summer is warm. Thus, the average annual air temperature in the central and southwestern regions of the republic is +2, + 3°, and in the mountainous and northeastern regions 0, + 1°, respectively. The average temperature of the warmest month of July is 17-19°, in mountainous areas is 16°, the absolute maximum in the territory of the republic is 38-41°. The average temperature of the coldest month of January is -14, -16°, the absolute minimum is -50, -52°. In some cold years there are severe frosts for a long time. For example, in January 1969 during 21-23 days the minimum temperature was below -30°, within 12-15 days of them below -35°, within 3-8 days below -40°. The summer heat is increasing due to the neighborhood in the south of the republic witharid steppe territories of Orenburg and Chelyabinsk regions, as well as in Kazakhstan. Most of the precipitation (40-60%) falls during 3 summer months. Average annual rainfall is 400-550 mm, in mountainous areas 500-600 mm. Summertime is more typical for thunderstorms. The diversity of the climate is greatly affected by the relief. Ridges of the Ural Mountains, which stretched from north to south,create a sharp difference in the climatic conditions on the western and eastern hillsides. [5]

Annual precipitation data were collected from the Bashkir Territorial Administration for Hydrometeorology and Environmental Monitoring (BashHMC) and from the All-Russian Research Institute of Hydrometeorological Information - the World Data Center (ARSRIHMI-WDC). The database includes annual records at 34 meteorological stations in the Republic of Bashkortostan from 1936 to 2009.

In order to detect change points, we used two tests: the parametric The Student's t –test [17, 18] and the non-parametric Pettitt's test [19]. The non-parametric Mann-Kendall trend test [20-23] and The Sen's test [24, 25] were used to detect the linear trend in time.

3.Change point detection methodology

As a preliminary step in any homogeneity analysis, it is very useful to make a schedule of time series in linear scale (Figure 1). A visual inspection of the site(territory) may reveal the presence of noticeable changes in the time series, which can be further investigated by statistical procedures. So, in some stations (for example, Askino) we immediately detect the presence of missing data. Failure in data collection and, thus, missed data have mainly occurred during the period 1940–1950, that is the period of the Second World War and immediately after it. Even if the missing data are not related to heterogeneity, there may be points of change during this period.

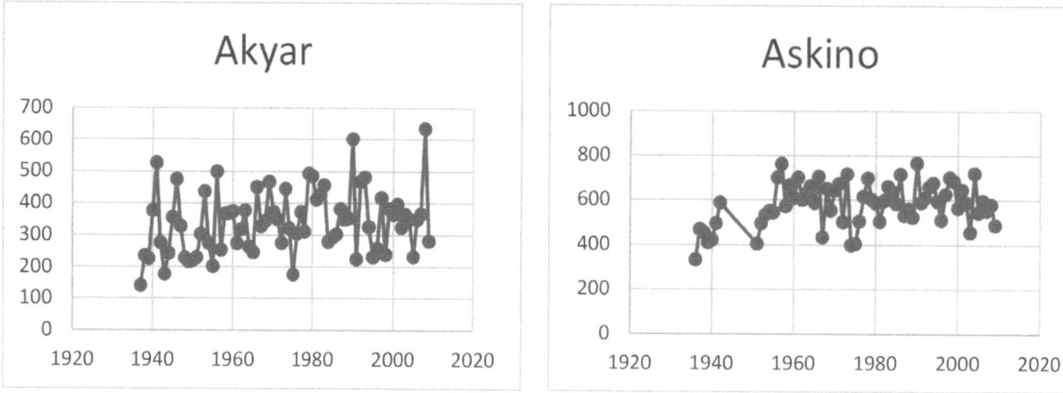

Figure 1. Graphs of time series in a linear scale as seen in the example of Akyar and Askino stations.

A series of homogeneous climate data is defined as a series in which changes are caused only by changes in weather and climate [26]. Climate factors do not really affect on most of the long-term climate series: changes in instruments, station locations, station environments, etc. They make climate data unrepresentative of temporal climate variability. Heterogeneities lead either to sharp discontinuities or to a gradual bias in the data, which can be detected using statistical tests. Actually, a large number of approaches have been proposed to detect change points, and when we use them to the same series, they can actually lead to conflicting conclusions due to different climatic elements relating to the time series under research [27]. For this reason, it would be advisable to apply a number of methods to detect points of change and compare. [28]

In order to detect change points, we used two tests: the parametric Student's t –test [17, 18] and the non-parametric Pettitt's test [19].

3.1 Results of the detection of change pointsanalysis

The parametric Student's t –test and the non-parametric Pettitt'stestwere applied to detect discontinuities in the data set on long-term annual precipitation at 34 stations of the Republic of Bashkortostan from 1936 to 2009. For each meteorological station and for each test, the corresponding results are shown in Figure 2, for the significance level $\alpha = 5\%$.

From the article, it can be seen that the various approaches applied to long-term annual precipitation data show different results. Letus pretendthat time series can be defined as non-uniform if two tests indicate this. Further, only homogeneous precipitation time series will be analyzed additionally to identify trends. The Student's t –test indicates that a change point has been detected in all time series, but the probability of its occurrence is very small ($p = 0.0001$). Based on this, we will assume that the time series in which Pettitt's test indicates a positive result are homogeneous. In this way, we can see, that at 16 meteorological stations, time series are non-uniform. Accordingly, they are excluded from the following trend detection analysis.

4 Trend Analysis

Trendin meteorological variables in the series of observations during the period under consideration [29]. We tested 18 meteorological stations, in which time series are homogeneous, to detect a linear trend in time. For this, the non-parametric Mann-Kendall trend test [20-23] and The Sen's test [24, 25] were used.

4.1 Results of trend detection analysis

The tests, which we discussed above, were applied to detect a linear trend from precipitation data of 18 meteorological stations in the Republic of Bashkortostan for the period from 1936 to 2009, in which the time series are homogeneous. Whereashalf of the time series turned out to be heterogeneous, it is difficult to talk about the general picture of changes in precipitation in the Republic of Bashkortostan. But in most cases, Sen's slope is positive, which indicates an increase in annual precipitation amounts.

IOP Conf. Series: Earth and Environmental Science **350** (2019) 012003 doi:10.1088/1755-1315/350/1/012003

4 among 18 meteorological stations have a negative trend, but it is not significant. For illustrative purposes, a map was made (Figure 2). Based on the map data, made on the basis of the results of the Mann-Kendall test and the Sen test, it can be seen that most stations have a positive trend. The maximum positive trend was found at the Dyurtyuli meteorological station.

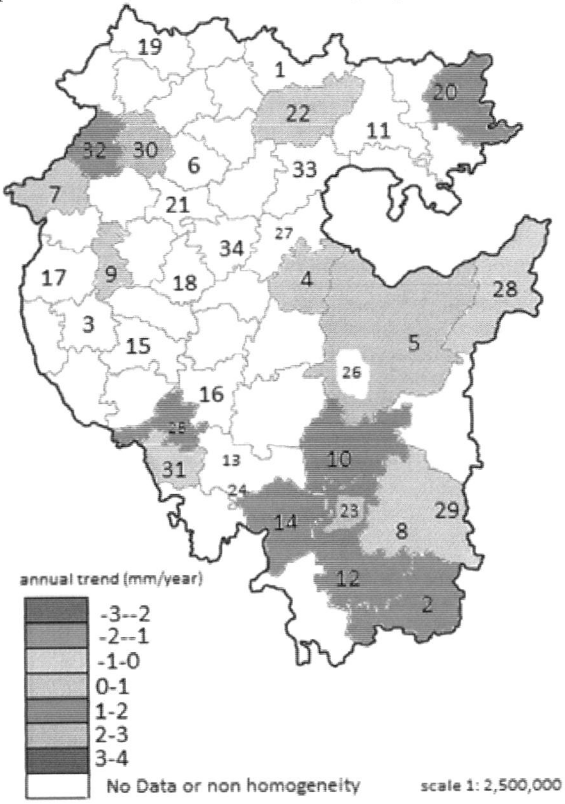

№, Station name	Pettitt's test	The Student's t-test
1 Askino	n	
2 Akyar	h	
3 Aksakovo	n	
4 Arkhangelskoe	h	
5 Beloretsk	h	
6 Birsk	n	
7 Bakaly	h	
8 Baymak	h	
9 Buzdyak	h	
10 Bashgoszapovednik	h	
11 Duvan	n	
12 Zilair	h	
13 Meleuz	n	
14 Mrakovo	h	n
15 Rayevsky	n	
16 Sterlitamak	n	
17 Tuymazy	n	
18 Chishmy	n	
19 Yanaul	n	
20 Emashi	h	
21 Kushnarenkovo	n	
22 Karaidel	h	
23 Kananikolskoe	h	
24 Kumertau	n	
25 Sterlibashevo	h	
26 Toucan	n	
27 Ulu-Telyak	n	
28 Uchaly	h	
29 Sibay	h	
30 Dyurtyuli	h	
31 Fedorovka	h	
32 Verkhneyarkeevo	h	
33 Pavlovka	n	
34 Ufa-Dema	n	

n - nonhomogeneity h - homogeneity

Figure 2. Map of average annual precipitation decrease (negative trend,green scale) or increase (positive trend, blue scale) inmillimeters per year.

5. Conclusion

The main purpose of this research was to analyze the time series of precipitation, to identify potential trends and assess their significance over the wide (about 142,947 km²) territory of the Republic of Bashkortostan and over a long time interval (1936-2009). To this purpose, climate data on precipitation were preliminarily checked for data uniformity using parametric The Student's t –test and non-parametric Pettitt's test. It was found, that the time series of 16 meteorological stations are heterogeneous. Later on, only homogeneous time series (18 meteorological stations) were tested for trend detection on an annual scale, affected using non-parametric tests: the Mann Kendall test and the Sen test. According to the study, it was found that on an annual scale, the study area is experiencing a positive trend, even though in the eastern part, minor negative trends were observed.

The growth of precipitation represents a serious threat to the Republic of Bashkortostan, this leads to an increase in the probability of emergencies. An increased wind, a sharp drop of air temperature, and a thunderstorm accompany precipitation, thus greatly complicating the situation. They may cause injury and death.

6. Acknowledgments
The presented research was financially supported by the Russian Science Foundation (grant no. 17-77-30006).

References
[1] Kokorin A 2005 *Climate Change: A Review of the State of Scientific Knowledge of Anthropogenic Climate Change RREC, GOF, WWF* (Russia) p 20
[2] Cook J, Nuccitelli D, Green S, Richardson M, Winkler B, Painting R, Way R, Jacobs P and Skuce A 2013 Quantifying the consensus on anthropogenic global warming in the scientific literature *Environmental research letters* **8** 1-7
[3] Asadi Zarch M A, Sivakumar B, Malekinezhad H and Sharma A 2017 Future aridity under conditions of global climate change *Journal of Hydrology* **554** 451–469
[4] Shiqin Xu, Zhongbo Yu, Chuanguo Yang, Xibin Ji and Ke Zhang 2019 Trends in evapotranspiration and their responses to climate change and vegetation greening over the upper reaches of the Yellow River Basin *Agricultural and Forest Meteorology* **263** 118-129
[5] *Bashkir Hydrometeorology and Environmental Monitoring Department. Climate change and its effects* 2015 Available at: http://www.meteorb.ru/
[6] Afanasev I, Volkova T and Elizaryev A 2014 Analysis of interpolation methods for the Republic of Bashkortostan, Russian Federation *WSEAS transactions on environment and development* pp 405-416
[7] Golovina A V 2012 *Evaluation of the modern variability of the stock R. Belay (Republic of Bashkortostan) under the influence of natural and anthropogenic factors,* Abstract of candidate dissertation (Kazan) p 25
[8] *RBC-Ufa study: How climate change affects Bashkiria* 2017 Available at: https://ufa.rbc.ru/ufa/29/06/2017/5954ddb39a7947defd54412b
[9] *IPCC Working group III fourth assessment report, Summary for Policymakers* 2007
[10] Myakisheva N V 2008 *The climatic system of the Earth. A manual for masters. Direction "Hydrometeorology". Specialization "Rational using and protection of water resources"* (St. Petersburg: RSHU Publishers) p 95
[11] Longobardi A, Buttafuoco G, Caloiero T and Coscarelli R 2016 Spatial and temporal distribution of precipitation in a Mediterranean area (Southern Italy) *Environmental earth sciences* **75(3)** 189
[12] Yavuz Selim Güçlü 2018 Multiple Şen-innovative trend analyzers and partial Mann-Kendall test *Journal of Hydrology* **566** 685–704
[13] Prateek Verma and Sanjay Kumar Ghosh 2019 Trend analysis of Climatic Research Unit temperature for Gangotri glacier, India *Dynamics of Atmospheres and Oceans* **85** 83–97
[14] Omar M A, Mahmood Agha, S Cağatay Bağçacı and Nermin Şarlak 2017 Homogeneity Analysis of Precipitation Series in North Iraq *IOSR Journal of Applied Geology and Geophysics (IOSR-JAGG)* **5(3) II May-June** 57-63
[15] Zhigao Zhou, Lunche Wang, Aiwen Lin, Ming Zhang and Zigeng Niu 2018 Innovative trend analysis of solar radiation in China during 1962-2015 *Renewable Energy* 675-689
[16] Bihrat Onoz and Mehmetcik Bayazit 2003 The Power of Statistical Tests for Trend Detection *Turkish J. Eng. Env. Sci.* **27** 247-251
[17] Hald A 1952 *Statistical Theory with Engineering Applications* (Wiley: New York)
[18] Panofsky H A and Brier G W 1968 *Pennsylvania State University, University Park* p 224
[19] Pettitt A N 1979 A non-parametric approach to the change-point problem *Appl. Statist.* **28(2)** 126-135
[20] Mann H B 1945 Nonparametric tests against trend *Econometrica* **13** 245-259
[21] Kendall M 1975 *Multivariate Analysis* (London: Charles Griffin &Company)
[22] Hirsch R M, Slack J R and Smith R A 1982 Techniques of trend analysis for monthly water quality data *Water Resources Research* 107-121
[23] Hirsch R M and Slack J R 1984 A nonparametric trend test *Water Resources Research* 727-732

[24] Sen P K 1968 Estimates of Kendall's tau *J. Am. Stat. Ass.* ***63*** 1379-1389
[25] Gilbert 1987 *Statistical Methods for Environmental Pollution Monitoring* (New York: John Wiley & Sons, Inc.)
[26] Conrad V and Pollak C 1950 *Methods in Climatology* (Cambridge: Harvard University Press)
[27] Reeves J, Chen J, Wang X L, Lund R and Lu Q Q 2007 A review and comparison of climate control data for climate data *J. Appl. Meteorol. Clim.* **46** 900-915
[28] Longobardi A and Villani P 2010 Trend analysis in the Mediterranean area *Int. J. Climatol.* **30** 1538–1546
[29] Gruza G and Rankova E 2003 The climate of Russia: warming continues *Science and life* [in Russian – Nauka i zhizn] **11**

National importance of surveying and ancient linear measures in the history of development of Russia

T V Ilyushina

Moscow State University of Geodesy and Cartography, 4, Gorokhovsky Lane, Moscow, 105064, Russia

E-mail: tilyushina@yandex.ru

Abstract. Issues of rational use of land resources are urgent contemporary problem. It is considered possible to solve it by analyzing centuries-old experience of history of land surveying, geodetic studies and mapping. The purpose of the study is to show the national importance of land surveying and ancient linear-measuring of boundary measures in the history of the state. The tasks of the article are connected with the analysis of types of geodetic surveys, linear measures and instruments during land surveying in different historical periods of development of the state. Historical and archival materials, ancient geodetic tools and maps were used as materials for the study. The study revealed that the development of surveying, the system of linear measurements and geodetic instrumentality had national importance to establish the boundaries of possessions, registration of quantity and quality of the land funds, protection of property rights, and termination of land disputes, valuation of land and establishment on the basis of equitable taxation. The main provisions of this study can add to the history of land relations in Russia and contribute to the further development and improvement of the state system of land surveying.

1. Introduction

The emergence of the first linear measurement of land property boundaries in the times of Ancient Rus and its principalities in the 10th – 15th centuries. History of state formation is closely connected with the development of land. The establishment and consolidation of state and private boundaries, the determination of the quantity and quality of land, the protection of ownership rights and the cessation of land disputes, the evaluation of land and the establishment of fair taxation on this basis were the main tasks on the way to the development of the state and the use of land resources. And here there was a need for measuring boundary works, which would guarantee the determination of the location of the boundaries, the description and evaluation of property and recognition of property rights. Measuring instruments were used for the works, and many of them not only became a part of history, but also significantly influenced the course of its development. It is known that since ancient times the first "earth-separating" tools were the hands and feet of a man, the length of his body, steps. But the first artificial measuring instrument, apparently, was a stick or a pole. If all measuring poles are compared with one and the same pole, then, this one pole becomes a standard. Therefore, the system of ancient measures of land gradually created a set of samples — verst, sazhen, arsheen, step, and others. Measures became an integral part of life, firmly established in everyday life, in traditions and foundations, and had state significance. A verst was originally defined as the distance from one turn of the plow to the other during plowing.

The verst was mentioned in the written sources of the 11th century. One verst consisted of 1000 sazhens in the time of Tsar Alexei Mikhailovich. One verst was equal to 500 sazhens at Peter the Great time (1066.8 m). A verst was also called a milestone on the road. The value of the verst was changed several times depending on the number and size of sazhens which were included in the verst. One of the main laws – the Code of 1649 established "Mezhevaya (Boundary) verst" – which included one thousand sazhens, and was equal to two versts (2.16 km). The "Boundary verst" was used as a boundary measure, usually in determining the land around large cities, and on the outskirts of Russia, especially in Siberia, and to measure distances between settlements. For measuring the distances in the European part of Russia, 500 sazhen's verst was used. Long distances, in Eastern Siberia, were measured in days of travel. The Mezhevaya verst was gradually ousted in the 18th century, and in the 19th century the only verst which remained was a "track" verst, equal to 500 sazhens [1].

Sazhen was one of the most common measures of length in Ancient Russia. There were more than ten differently defined sazhens - different in purpose and size. Sazhen as an ancient measure of length was mentioned by Nestor in 1017. As a result of the measurements, the topographers got value of the sazhen – 151.4 cm. The results of the measurements of temples and Russian folk measures matched this value. Sazhen's dimensional ropes and folding wooden "skladeny" tools existed in the measurement of distances and in construction. The sazhens were used for measuring until the introduction of the metric system.

The basic unit of linear measures was arsheen. The arsheen was one-third of the sazhen. It was possible that initially, the arsheens were determined by the length of human step and was the basis for sazhens and versts. There were several arsheens – printed (pechatny), branded, official (kazennyi), folding and correct yardstick. Russian arsheen was a ruler with applied scale for measurement, and contained 16 vershoks or 28 inches. Three arsheens were equal to 7 feet (500 sazhens are equal to one verst). This measure was widely used not only in measuring works but was very popular in folk's economic and cultural life.

In Moscow Rus, arpent (Dessiatina) was a universal measure of land in the state and possessive lands; in the peasant ones it was replaced by "quarters", which were half its size and were divided into even smaller units. There were some local measure units: in the Great Novgorod— "korobya" (1 kor. = 2 quarters of arable land), to the Dvina's area — "ropes". The next units - "vut" and "sokha"(plow), Novgorod "sokha" (plow), "soshka" and "obzha" were used to collect the dues from the peasants and state taxes [3].

2. Materials and methods

Historical and archival materials, ancient geodetic instruments and maps of different Museum collections were used as materials for the study. This includes instrumental collections of Educational and historical center of Moscow state University of Geodesy and Cartography, of Moscow State Polytechnical Museum, and of Moscow Museum of the company "Geostroyizyskaniya" (ZAOGSI). Geodetic instrument making today is undergoing a new stage of its development. The growing need for surveying instruments allows creating new models, developing new tools and technologies of topographic, boundary and cadastral surveys. The historical method of research allowed studying the texts of primary sources and finding material evidence for the study, as well as presenting the events of different historical periods in Russia. A historical-comparative method has a wide cognitive capabilities, reveals the essence of the study and the importance of land surveying and ancient linear-measuring measures in the history of the state.

3. The beginning of Russian Chronicles refers to the 11th century.

Back in those days the division of lands did not require complex measurements. It was enough to measure the lanes of land of equal width, and denote them with bordering signs-small pits, stakes, stones or furrows. Natural or man-made signs were used to indicate the boundaries. Rivers, streams, ravines, swamps, forests, trees, stones could all be part of the signs that determined the land boundaries. In addition to the designation of the boundaries were used like- poles, pits, mezniki, signs on trees, fences, etc. There were no rules governing the location and shape of boundary markers. Everything depended on the custom and convenience of setting signs in the area. However, when trees or poles were marked by signs, often pits were dug, in which were placed the coals, stones and bones. On apicultural trees ["borti" (apiculture) the oldest form of beekeeping] and boundary oaks the signs represented various subjects, for example a hand, a skobel (special knife), a scythe, etc. [3] were made by axe.

The first artificial land boundaries - [the fence, lanes of uncultivated land, "mezhniki", "faces", "predely", etc.] appeared on business needs — for protection from wild animals, rapprochement with neighbors in high-value lands: arable lands, meadows, and beekeeping areas. The emergence of fisheries-hunting, cattle breeding and others also caused the appearance of the first signs indicating the places of land use of individuals. Thus, the boundaries of the possessions were formed during of land surveying. Boundary marks could be installed for separating of the land of one owner from another. Land relations were regulated by special boundary legislation. The accuracy of the boundary and its preservation were the cause of quarrels and clashes of owners, and the accuracy of land surveying depended on the applied surveying instruments and methods of the first boundary measurements. The value of land was increased with increasing population, and the boundaries of possessions protected by law and authority became stronger. In articles of the First Russian legislative document - "Russkaya Pravda " the first law regulations for protection of possession according to the rights of estates [4] were raised. The eighth Chapter of "Pravda" stated that every possession had its limits, approved by the government, and their signs were sacred to the people [4].

Methods of land delimitation depended on the structure of ancient Slavic settlements. All convenient lands were either considered the property of the settlement or divided between individual families. Plans of individual parcels were made on the base of the results of inspections, inventories and simple measurements. The lengths of the boundary lines were measured with a measuring rope, the areas of agricultural land and forests were determined. In the "Russian Truth" a guidance on the boundary marks was provided and boundaries that were established by measuring and determining areas of land for taxation. Distribution, modification and supplement of the local boundary laws led to the creation of specific legislation in the principalities (Novgorodskay, Pskovskay and Dvinskaya judicial Charters). There were the articles about the land boundaries, the decisions of disputes about lands and forests in the chapters of "Russian Truth". "Bortnaya" boundary shared part of the forest, "Roleinaya" one – divided arable fields. Boundary (meznoi) oaks marked the property lines, cut down the oak meant to destroy boundary marker [4]. As a result of the policy of creating a united state, there was a need for more accurate land surveying. The land use and accounting of the one could not be implemented without the measurement of land, i.e. geodetic works. There was a need for land surveying (location, description of borders, calculation of areas, issuance of legal documents for ownership, resolution of disputes about lands), as well as information about state borders and natural resources of state territories. In this regard, work begun on the creation of Scribe books (1500), which included descriptions of land and methods of establishing and describing of boundaries, was fixed by certificates (boundary, bills of sale, granted, etc.) [4]. Thus, we can say that in the 10th-15th centuries not only the first important concepts of surveying were developed, but also the first measuring instruments and the first techniques and methods of land measurements were established.

4. The development of the initial system of land surveying in the Moscow State during the scribe mezhevanie (surveying) in 16th – 17th centuries.

Moscow state, worrying about the safety of its borders, created fortifications along them, the descriptions of which already in the 16th century were accompanied by plans of the border lanes. The local system required frequent descriptions, measurements and delineations of estates and patrimonial estates. Plans of separate districts of the Moscow State were based on the results of inventories, inspections of the area and the simplest of measurements until 16th century. Grant of land for service, for Tsar or the Sate, was one of the reasons for the frequent land censuses in the 15th – 16th centuries. Scribe books were used in the 14th – mid 17th centuries, contained legal and financial issues (property valuation, tax rates; indicators of population statistics and land ownership; classification by land quality). Patrimony were endowed for special services that were confirmed by diplomas and borders. The "Book of "Soshnogo" letter" was primordially Russian manual, outlining the methods of measuring the areas, the oldest copy of which referred to 1629. Under Ivan the Terrible was compiled the first Russian guide for Surveying of lands [5]. The Pomestny (Domestic) Prikaz endowed the serving people land, oversaw the Fund of state lands, registered and controlled all measurements on the property, conducted general and private descriptions of land, was a court, developed a new system of taxation (1684). Boundary laws of the Council code of 1649 for the first time defined the basis of state surveying as a consequence of the developed state life, established the order of disputed and indisputable surveying. The Council code determined the punishment for damaging of boundary markers, for illegal appropriation of land and the procedure of allotment of land. Land tenure plans were important legal documents as they were used in court proceedings.

A significant event in the history of Russian land surveying was the total land surveying of 1680-1686 – the legal strengthening of the rights and boundaries of possessions, there were many established boundaries of possessions, determined the quantity and quality of land in the state, and eliminated many land disputes. Land surveying gave the government the basis for further arrangement of ownership and proper taxation.

The practical geometry was based on the determination of distances and areas (areas were measured with a rod and measuring ropes in dessiatinas of 80 sazhens in length and 30 sazhens in width at the Behest of 1684). In the Code of laws (Sudebnik) of 1497 there were several articles on the order of land surveying possessions and regulation of feudal-land relations. The landmark laws of Moscow state provided for punishment for violation of the borders. In the code of laws (Sudebnik) of Ivan III and Ivan IV (1497 and 1550) for the infringement between the princely, boyar and monastic lands were administered, in addition to the monetary penalty, the criminal punishment - the whip; meanwhile, monetary penalty was supposed for violation within the community between peasant lands [6].

The Council code of 1649 established a system of land surveying and determined the size of the sazhens, issued boundary laws that determined the order of land surveying, provided penalties for damage to boundary marks. In cases of requests from the owners to survey their estates the Code prescribed to survey them with old-timer witnesses, dig holes on the borders, set poles, and do all sorts of signs on the lands to exclude disputes. If for some reason the surveyor could not resolve the dispute with his power, the disputed lands were bypassed with the icon ("figurative walking") [4]. Penalty for damage of boundaries, forcible seizure of lands, and the wrong actions of surveyors also defined in the code: who damaged boundaries, and boundary posts; or pits would align, or the land plowed, and a court would prove it, those people should be whipped and be sent to jail for a week. The same punishment was also applied to the peasants who spoiled their boundaries, but without fine [7]. In 1684 a Scribe of land cadaster related to the formation of the Russian state.

5. Change of the main provisions of the subdivision during the development of the Russian state (the Russian Empire) during the reign of Peter the Great and continuation of reforms of the 17th – 18th centuries

One of the main problems of this period was a patchwork of lands. Such land fragmentation devalued land, and had a harmful impact on the economy, prevented the use of advanced crop systems, gave rise to disputes and agricultural unrest, and undermined the legal basis of ownership. With the existence of intermingled property was impossible "to approve the peace of mind of owners" and to set boundaries of ownership. New educational institutions and the publication of books related to surveying like *Geometry that is Land Surveying in Slavic* (1708), *Geometry, Methods of Compass And Ruler* (2nd ed.) were published, and had a significant impact on the future of the land surveying (1709) and *Practical Geometry* (1709) [5]. The survey was carried out by the surveyors of Peter the Great with the preparation of land maps of each county. At this time, accurate instrumental surveys were started to obtain maps of individual territories and the General map of the State with the determination of the coordinates of "astronomical points".The Senate decided in 1718: "The surveyors should study geodesy, and use theodolites and compasses in their works" [2]. Measuring chains and ribbons served as main instruments back in that times. Two method were used for surveying parcels using measuring chains and stakes: the breakdown of the parcel into triangles or parallel lines method [9]. The structure of the geometry allowed them to determine distances and angles (with the measuring chains and the first astrolabes with telescope), differential leveling- (method of water leveling) and areas. The manual of A. F. Kleshnina (1723) set out the mathematical essence of the method of "zasecheck" (used in cases where the mutual location of two points was known and it was necessary to determine the position of the third point, often inaccessible for measuring) and explaining the method of equalizing the points, i.e. eliminating the so-called "triangle of errors" by changing the rhumbs and distances.

Peter the Great declared arsheen as the main unit for measures, supervision over the correctness of trading measures and weights were assigned to the town hall, magistrates, mayors, governors and especially the police.

Figure 1. Normal arsheen 1806

Normal arsheen was arranged as follows: on a massive rectangular base made of copper alloy, two limiters were installed, with polished chalcedony inserts fixed on the inner side. Chalcedony has a higher hardness than steel and does not wear out when inserting metal standard rods. Thus, the "arsheen's standard" was like calibration device, a calibration "bench" for testing of the standards themselves. So, the surveyor after checking the "arsheen's standard" out on the calibration “bench”, felt confident to use it. This tool was equipped with two sliders with scales, which allowed not only to accurately set the length of any rods less than the arsheen, but also to mark the working standard with the division of value in the tenth part of a vershok. Branded metal tapes and measuring chains with lengths of 5 and 10 sazhens with divisions of value were made of iron, steel or copper alloys. The error allowed for measuring chains and steel tapes should not exceed 5 lines (1.27 cm) for each sazhen of length.

With the socio-economic development of the state, land tenure required more precise definitions of boundaries, location and affiliation. Since the Order (Nakaz) of 1684, more than 65 years had passed, and many of the newly appeared possessions remained without approved borders, the old parcels lost their boundaries. These all had very negative impact on the land holdings in the state, giving rise to the appropriation, arguments, fights and murders on the borders. Poorly protected and unknown to the government lands could not bring income to the state. The Domestic Office (Pomestny Prikaz) was transformed into the Patrimonial Board (Votchinnaya Colegia), and many of its functions were lost, which negatively affected the accounting and evaluation of land. However, the accuracy of field measurements based on geometry and the use of surveying instruments were improved. Military interests demanded display terrain and aquatic objects. The government organized surveys to obtain maps of individual territories and the General map of the state.

6. The formation of developed system of land survey at the time of the General survey of the Empire in the second half of the 18th–19th centuries

Catherine II had to solve the problem of insecurity of land relations. For restoring of the order in the area of land ownership to prevent "disputes between owners, lawsuits, fights, and death kills", the General rules was composed and new land survey office and land survey expedition in Senate were opened, and also two provincial land survey offices in Volokolamsk and Serpukhov [10]. By conducting land surveying, the government sought not only to legalize their borders, but also to optimally arrange the type of management, to prevent the ways of violation of borders and the possibility of damage to natural resources. Several important legislative documents were issued in 1766: "The Methods of General Land Surveying" and "The Instruction for the Land Surveying of the All-Empire" which was made on the basis of the General rules. The instruction described how to draw up plans (1:8400) focused on the magnetic Meridian, which were used to draw up maps of counties (from 1:42000 to 1:336000) and atlases of the bordered lands, and detailed techniques and technologies of surveying. [11]. The angles of turn of the boundaries of land ownership for the first time were marked by wooden pillars with coats of arms, in treeless places triangular pits, and in flood areas – barrows were arranged. The Instructions determined the form of boundary marks, the rules of drawing up of plans, boundary books and documents [12]. Professor of astronomy from Oxford University E. Gunter (1581-1626) [13] proposed measuring chains, which were used for surveying, in the beginning of the 17th century.

Figure 2. Measuring chain (England, 1795) the most convenient and advanced design. Ten-per-sazhen links of equal length with planted plaques of different shapes and with a different number of teeth

Measuring chain usually was composed of 70 steel links, where every seven units (i.e., every seventh sazhen), was attached to the copper plaques engraved on them the numbers. Thus, sazhens and feet were counted on the plaques and according to the number of individual links. There were also "centesimal" measuring chains consisting of 100 links of 0.1 of sazen, or hundred-foot (one hundred links in length per foot). Compared with the chain – the measuring steel tape had several advantages: first, it was lighter, and secondly, did not stretch, less sagged in measurements through the pits and ditches, and therefore, gave more accurate results. But the tape was not as strong as the measuring chain and in case of rupture could not be corrected by local means. For measuring of small distances a thin canvas rope – braid – was used of 10 sazhens in length, which had scale divided into sazhens, feet, inches and even smaller parts. The braid was kept in a leather case with a rotating axis in its center, equipped with a folding handle. For ease and convenience of use, it surpassed even the measuring tape, but it was very stretched and did not last long, and for measuring lines in the open areas it was not suitable at all.

Formed in 1779 Konstantinovskiy Land Surveying Institute trained engineers in the field of land management and land surveying. At all times, Russia was an exclusively agricultural power, and land policy issues played a major role in the development of the state. That is why the role of a professional higher school in the boundary survey, legal, geodetic, geographical and cartographic study of land and other natural resources of the state was incredibly important. Education and training of highly qualified specialists, the development of professional Sciences of the geodesic cycle, the dissemination of scientific achievements and knowledge in the field of astronomy, gravimetry, photogrammetry and other Sciences, the improvement of land law, the establishment and protection of state borders – these were the main tasks that were solved as a result of educational activities, researches and expeditions of the Boundary Institute (Mezevoy Institute) [9]. The rules of land surveying, faced with new ways and conditions of economic use of natural lands, encouraged the government to simultaneously address the issues of regulation and management. There was a gradual formation of the system of surveying on a legislative basis with accurate geodetic and cartographic supports. In different provinces of the Empire, the tasks of land surveying coincided, but the priorities at the same time, according to natural and climatic conditions, the availability of natural resources, the development of the economy, the national way of life, were different. Land surveying was still to be involved in the urgent state tasks: the elimination of the patchwork lands; the allocation of land to state peasants; the return of illegal possession of state lands; the delimitation of joint possessions; the provision of land to cities and many others.

References

[1] Dal V I 1880–1882 *Explanatory dictionary of the living great Russian language 1863–1866* in 4 volumes (St. Petersburg; Moscow) 2rd ed
[2] Khrenov L C 1987 *Chronology of domestic geodesy from ancient times to the present day Academy of Sci. of the USSR* (Leningrad)
[3] Ushakov T I 1935 *Explanatory dictionary of the Russian language* in 4 volumes (Moscow)
[4] Herman I E 1914 *History of Russian land survey* 3rd ed (Moscow)
[5] *Geometry of slavensky surveying: Methods of compass and ruler* 1709 (Moscow)
[6] Gerberstein S 1908 *Notes in Moscow Affairs* (St. Petersburg)
[7] Malinovsky F L 1844 *Historical view on land surveying in Russia until 1765* (St. Petersburg)
[8] Tomsinov V A 2012 *Legislation of Tsar Fedor Alekseevich 1676–1682 Legislation of Tsars Ioann Alekseevich and Peter Alekseevich 1682–1696* (Moscow: Zertsalo)
[9] Ilyushina T V 2012 *Cadastre of natural resources of Russia. Essays on history (X – early XX centuries)* (KG Saarbrücken, Germany: Palmarium academic publishing; AV Akademikerverlag GmbH & Co)
[10] Herman I E 1911 *Materials for the history of the General land surveying in Russia* (Moscow)
[11] Malinowski F L 1868 *On the state of the General land surveying in Russia* (St. Petersburg)
[12] Kavelin S P 1914 *Land Surveying and Land Management: theoretical and practical guidance. With drawings and samples of office work* (Moscow: Publication of the legal bookstore Jurisprudence I K Golubev)
[13] Bik A N 1909 *The lower course of geodesy* Part 3, 5th ed (Moscow)

Formation of the xenobiotic profile in anthropogenic systems

V V Gorin, T A Sokolova and N V Khvatysh

State University of Land Management, 15, Kazakov st., Moscow, 105064, Russia

E-mail: Natasha_hv@inbox.ru

Abstract. The article deals with the issue of formation of the xenobiotic profile of a biogeocenosis as a combination of foreign substances in the environment in a form that allows them to enter into chemical and physicochemical interactions with biological objects of the ecosystem. At the initial stage, patterns of distribution of xenobiotics in various natural environments (water, soil, air, and living organisms) and their subsequent effects on the biogeocenosis were determined.

1. Introduction

At the beginning of the twentieth century, a new sphere of the Earth's shell began to form. It is a technosphere which is part of industrial and non-productive human activities with its own laws of material and energy flows. Being part of the biosphere, humans interact with it developing within the limits of physical, chemical, biological and other states of the biosphere. Natural processes are increasingly intertwined with man-made ones, increasing the exchange of matter, energy and information.

At the initial stage, transformation of the natural system takes into account possible transformations of its biosphere framework (biological, biocenotic, ecological, etc.). In the future, changes take place under the influence of both anthropogenic activities and dynamic processes characteristic of open self-organizing natural systems.

Patterns of the processes of entry, distribution, transformation and metabolism of pollutants are complex and rely on environmental information whose core is a live matter in all forms of its manifestation and a human as a socio-biological species. All other variables must be considered in relation to the overall biological nucleus.

2. Methods and materials

Currently, there are no methods describing patterns of xenobiotic behavior in the environment. The available research materials are devoted to environmental problems and methods of analytical chemistry of persistent organic pollutants, including monitoring which generally does not identify patterns of their migration and transformation in various natural objects. This article is a phenomenological description of formation of a xenobiotic profile as one of the most important environmental factors.

3. The xenobiotic profile and man-made migration paths

The xenobiotic profile should be considered as one of the most important environmental factors (along with temperature, light, humidity, trophic conditions, etc.) which can be described by qualitative and quantitative characteristics.

Xenobiotic profiles of the environment developing forming during evolutionary processes belong to natural xenobiotic profiles (XBP). They are different in different regions. The biocenoses in these regions are more or less adapted to the corresponding natural XBP.

The evolution of natural XBE is determined by biogeochemical laws and controlled by the geochemical balance in the processes of biogeochemical cycles of biogenic components of the natural environment.

Determination of the balance of nutrient components of the oceans and land, identification of geochemical processes caused by implementation of large land-reclamation projects, prediction of the impact of production and economic activities on the environment are some problems of migration flows, geochemical balances of territories with different spatial dimensions.

Geochemical development and the balance of the substance of territories are determined by a surface and underground runoff, a biogenic and technogenic migration, a movement of air and water masses of seas and oceans.

At the present stage, migration of the substance with a surface runoff is well studied. For individual river basins, indicators of ionic and solid runoff were determined. For the most studied basins of the underground runoff, it is possible to determine the specific distribution of salt removal or supply over the entire area of the basin; for less studied ones, it is possible to determine the overall balance of solutes.

The atmospheric transfer occurs as a result of general circulation of the atmosphere and local movements of air masses. The main elements of the atmospheric transfer of salts from the ocean are chlorine and sodium, and a number of trace elements (iodine, boron, bromine, lithium). Estimates of the atmospheric transfer of substances differ by tens and hundreds of times due imperfect methods [1, 2].

Technogenic migration is characterized by a direct movement of substances as a result of economic activities. It is understudied, although it is of paramount importance.

The chemical products entering the environment are distributed and transformed in the atmosphere, soil and water bodies; they enter food chains and are accumulated by living organisms. Many eco-toxicants extend over long distances and retain dangerous properties for a long time. The ability of eco-toxicants to spread over long distances depends on their physical, chemical, physicochemical properties (evaporation, volatility, solubility, soil absorbability, resistance to decomposition).

The study of anthropogenic migration is of great importance.

In the geochemical aspect, the anthropogenic impact involves transition of a substance from man-made migration flows into the natural ones. Since products of anthropogenic production that migrate with the anthropogenic flow have different resistance, they are transformed into natural migration flows with different intensity.

To characterize the technogenic geochemical impact, in addition to elements involved in the technogenesis, a number of indicators are used.

One of these indicators is technophilicity of chemical elements (N) [3] which characterizes the degree of utilization of the chemical elements of the lithosphere.

$$N = Q_i/Q_z \tag{1}$$

where Q_i – annual element mining (t), Q_z– clarke.

To assess a possible impact of technogenesis on living organisms, an indicator of destructive activity of elements (A^d) has been introduced. It is the ratio of the weight of an element in the annual production plus its release into the environment from the burning of combustible minerals to the weight of this element in the annual biological production of terrestrial vegetation [4].

The geochemical analysis of toxicity of chemical elements and compounds ensures purity of the atmosphere of urban agglomerations, preserves the natural composition of soils and natural waters. Currently, around 500 thousand compounds are used; 40 thousand compounds are harmful, and 112 thousand are extremely toxic. Technogenic pollution affects ecological systems and is determined by the ability of chemical elements to be included in natural migration cycles.

Technogenic migration occurs in gaseous, liquid and solid forms. Caused by intense human activities, it accelerates the movement of various substances so that the cycles of substances become incomplete. The products of material production involved in the living cycle partially accumulate in the biosphere creating obstacles to natural processes.

The mechanism of specific toxicity is understudied. For ecotoxicology, toxicity of the substance that enters the environment and products of its transformation is important. When contaminated with industrial waste, fuel combustion products, garbage, a mixture of products of complex composition is formed. The total toxic effect of the mixture may differ from the total toxicity of its individual components.

Of particular interest are molecules with bioavailability, i.e. molecules able to interact non-mechanically with living organisms. As a rule, these are compounds that are in a gaseous or liquid state, aqueous solutions; finely dispersed dust (the particle size less than 50 microns); substances entering the body with food.

Some bioavailable compounds are utilized by organisms participating in the processes of their plastic and energy exchange with the environment, i.e. act as habitat resources. Other ones are not used as sources of energy or plastic materials; they modify normal physiological processes. These are xenobiotics.

An important element of the xenobiotic profile is a foreign substance in the organs and tissues of living beings; all of them are consumed by other organisms (they have bioavailability). In contrast, chemicals that are fixed in solid, non-dispersible objects (rocks, solid industrial products, glass, plastic, etc.) do not have bioavailability. They can be considered as potential sources of XBP formation.

Economic activities change the natural XBP of many regions (especially urbanized ones). Chemical substances that are not typical of the environment accumulate in the medium in sufficient quantities to initiate the toxic process in the biocenosis, change the natural XBP and transform into an eco-toxicant (xenobiotic).

Determination of quantitative parameters at which an eco-politanant is transformed into an eco-toxicant is a difficult task. When solving it, it is necessary to take into account that in real conditions, the whole xenobiotic profile of the environment acts on the biocenosis modifying the biological activity of a particular pollutant. Therefore, in different regions, different xenobiotic profiles, different biocenoses, quantitative parameters of transformation of the pollutant into ecotoxicants are different. Assessment of technogenic pollution is used to identify the amount of an element or substances entering the human body according to the scheme: natural objects (soil, water, air) - organisms (plants, animals) - food products (vegetable, animal) - food - human.

Even if according to hygienic standards, environmental pollution is not dangerous, it is advisable to assess the MPC of pollutants and their effects to determine the maximum allowable load taking into account all the pathways of harmful substances.

Environmental studies analyze a) the relationship between accumulation of harmful substances and severity of toxic effects; b) the statistically significant dependence (positive or negative) between the content of xenobiotics in various natural environments and diagnostic substrates; c) the comparison of identified patterns in the distribution of xenobiotics in various natural environments with data on the health status of the population obtained on the basis of medical statistics.

Implementation of this approach is important with respect to priority organic pollutants: pesticides (aldrin, chlordan, DDT, mirex, etc.), industrial substances (PCB, hexabromobiphenyl, chlorine paraffins, etc.), by-products (PCDD, PCDF, PAH), organic metal compounds. Their danger is due to their ability to cumulate; various diseases can develop after exposure to these substances.

Persistent organic pollutants that are part of priority pollutants do not fall into any of the five hazard classes, since most of them have a powerful toxic effect and can have mutagenic, teratogenic and carcinogenic effects on humans and animals [5].

Pollution with persistent organic pollutants (POPs) is caused by their migration between natural environments; all elements of the biosphere are subject to the anthropogenic pressure regardless of its sources [6].

4. Theoretical substantiation of xenobiotic migration in natural environments

Atmospheric air is a nonstationary, inhomogeneous, multiphase, multicomponent system associated with the problems of transboundary transportation of POPs. Entering from various sources, POPs are carried by air currents and spread throughout the globe. Based on the data on POP sources, the model of horizontal transportation and turbulent diffusion of substances in the atmosphere [7] complemented by models of the POP exchange with soil and water surfaces was built. "Dry" deposition of POPs (D) is calculated as a stream on the underlying surface for a certain period of time using the deposition rate W_g:

$$D = (Cx,y.0)\ W_g dt \quad (2)$$

To calculate the amount of POPs that fall out with precipitation (F), the following formula is used:

$$F = C(x,y.0)h\gamma dt \quad (3)$$

The total amount of pollutants. Pollutants entering with atmospheric transfers (Q) are

$$Q = Cx,y.0)\ (W_g + h\gamma)dt \quad (4)$$

Calculations show that atmospheric transportation of POPs is one of the main sources of environmental pollutants.

Natural water is an open-type multiphase heterogeneous system exchanging substances and energy with other media and a biological component. Accumulating POPs, bottom sediments contribute to their removal from water; on the other hand, they are a constant source of water pollution. Hundreds of thousands of tons of POPs are deposited in the bottom sediments of large river deltas, including OCPs, PCBs, PCDDs, PCDFs.

Irrigation waters are a significant source of POPs. Calculation of the removal of chlorine-containing pesticides from agricultural land is a complex multi-dimensional task [8]. The determining factor is a dose of pesticides. The total amount of removal is proportional to the amount of precipitation and the area of agricultural lands.

For small and medium catchments, the average concentration of pesticides in the river in the closing section without taking into account the transformation in the riverbed is calculated by formula

$$C_{cp} = (M_1C_1F_1 + M_2C_2F_2 + \cdots + M_nC_nF_n)/\ M_0F_0, \quad (5)$$

where $M_1 \cdots M_n$ – rainwater flow modules on selected agricultural fields; $C_1 \cdots C_n$– pesticide concentrations in the surface runoff; $F_1 \cdots F_n$– areas of individual agricultural land treated with pesticides; M_0 – rainwater flow module in the river station; F_0 – catchment area [6].

In the total balance of POPs in sludge, the share of metabolites becomes larger than the share of initial compounds.

In the category of global functions, the lithosphere with its surface layers determines a direction and diversity of the soil-forming process and depends on the thin layer of soil that covers. A weathering crust and a sedimentary layer experience effects of soil formation. Due to the fact that the soil does not have mobility characteristics of other natural environments, it is a place for the collection and storage of all types of xenobiotics, including POPs.

When xenobiotics enter the soil, they are subject to complex physico-chemical and biological transformations. Under the influence of some microorganisms, the POPs are oxidized to carbon dioxide and turn into less toxic metabolites. The rate of destruction depends on climatic conditions, the degree of pollution and conditions of reproduction of microorganisms. Unlike other natural objects (water, air) in which self-purification is quick, they are slow. Moreover, lead and mercury compounds accumulate in soils. Substances entering the soil migrate. It is obvious that modeling the space-time behavior of POPs taking into account their persistence, synergism and antagonism is a difficult task. Monitoring of the soils of industrial centers and adjacent territories indicates a certain regularity in the spatial distribution of xenobiotics according to the degree of contamination:

- risk zones - areas with a high level of soil pollution with POPs;
- potential risk areas - areas affected by POP sources;
- areas of POP unloading or degradation - areas around eliminated or modernized sources of POPs, reclaimed land;
- zones not subject to risks - background areas.

Xenobiotic pollution can be assessed only at the qualitative level. When creating a quantitative model of xenobiotic profile formation, it is necessary to take into account that XBP effects on the biocenosis manifest themselves as qualitative, step-wise changes in the state of the system or the mode of its development in response to monotonous and slow changes in parameters. These processes belong to nonlinear dynamical systems.

The theory of "point" dynamic systems is based on differential equations

$$\frac{du_i}{dt} = \frac{1}{\tau_i} F_i(u_1, u_2, \cdots, u_n) \quad (6)$$

where u_i concentration of reactants, $F_i(u_i)$ – nonlinear functions describing the interaction, τ_I- characteristic time of change of variables u_i, i = 1,2 ,, ⋯n.

When specifying a type of function Fi, their solutions are determined by the initial conditions. However, surprises arise when decisions lose their stability with small deviations from the corresponding decision.

Equation (6) describes the development of processes in time rather than in space. However, in these processes, dynamic variables of concentration or a number of living organisms move in space due to diffusion or migration.

In the simplest case, changes in the flow of particles Δu_i for Δt due to space effects is proportional to the gradient of variables

$$\left(\frac{\Delta u_i}{\Delta t}\right)_{\text{пр}} = D\,\mathrm{div}\,j = D\,\mathrm{div}\,\mathrm{grad}\,u_i = D\Delta u_i \quad (7)$$

Equation (6) takes into account spatial effects

$$\frac{\partial u_i}{\partial t} = \frac{1}{\tau_i} F_i(u_1, u_2, \cdots, u_n) + D\Delta u_i \; i = 1,2 ,, \cdots n \quad (8)$$

where D_i – diffusion coefficients.

$$\Delta = \frac{\partial^2}{\partial x^2} + \frac{\partial^2}{\partial y^2} + \frac{\partial^2}{\partial z^2}$$

Equation (8) describes important phenomena of self-organization of systems in which diffusion occurs. Functions $F_i(u_1, u_2, \cdots, u_n)$ describe chemical reactions at x, and variables u_i - concentrations of substances involved in the reaction, values τ_I is a typical time of reactions.

The study of solutions of equation (8) will allow for the analysis of the phases of formation of xenobiotic profiles which may involve fronts, traveling pulses, spiral waves, dissipative structures and other processes in active media.

5. Conclusion

Anthropogenic human activities change the natural xenobiotic profile of territories, especially urbanized ones. Chemicals accumulating in the environment and causing changes in the natural xenobiotic profile act pollutants. A change in the xenobiotic profile may be a result of excessive accumulation of eco-polyutants.

Along with experimental studies of POPs in the atmosphere, water and bottom sediments, soils, plants, biological media and food products, we can trace patterns of migration, transformation and deposition of pollutants into eco-toxicants and assess their impact on the biosphere.

References

[1] Bondarev L G 1974 *Perpetual motion* (Moscow: Mysl)
[2] Bazilevich N I and Rodin L E 1972 *Productivity and circulation of elements in natural and cultural phytocenoses Biological productivity and circulation of chemical elements in plant communities* (Leningrad: Nauka) pp 49–51
[3] Perelman A I 1972 *Geochemistry of elements in the hypergenesis zone* (Moscow: Nedra)
[4] Glazovsky M A 1974 *Technogenic flows of matter in the biosphere Mining and geochemistry of natural ecosystems* (Moscow: Nauka) pp 7–28
[5] Klyuev N A 2000 *Ecological and analytical control of persistent organic pollutants in the environment* (Moscow: James)

[6] Maistrenko V N and Klyuev N A 2004 *Ecological and analytical monitoring of persistent organic pollutants* (Moscow: BINOM, Laboratory of Knowledge)
[7] Dutchchuk S V, Pavlova N K, Shatalov V E and Vulykh N K 2000 *Polychlorinated biphenyls Supertoxicant XXI century* vol 5 (Moscow: VINITI)
[8] Djnigian A S and Vtitr D W 1986 Stream transport and agricultural runoff of pesticide for exposure assessment *Athens Environ Res. Lab.* 104

IOP Conf. Series: Earth and Environmental Science **350** (2019) 012006 doi:10.1088/1755-1315/350/1/012006

State and use of land resources in the Republic of Tuva of the Russian Federation

P V Klyushin, A A Murasheva, G V Lomakin, P P Lepekhin and A V Changaev

State University of Land Use Planning, 15, Kazakova str., Moscow, 105064, Russia

E-mail: klyushinpv@gmail.ru

Abstract. The area of the Republic is 168,604 km^2 (3.28% of the territory of the Siberian Federal district, 0.98% of the area of the Russian Federation - the 20th in terms of area). Tuva basin, stretching more than 400 km from West to East along the Yenisei river, is an open hilly steppe space, located at altitudes from 500 to 1200 m. Mountain ranges, crossing it, divide the basin conditionally into three parts: Chemchik, Ulug-Khem and Turano-Uyuk. Of the total land area, most of the land is classified as forest land (64.5%), agricultural land accounts for 19.9% of the main specialization of livestock (mainly sheep and dairy cattle, goat and horse breeding). The area of farmland is more than 4.5 million hectares. The peculiarity of the pastures of Tuva is the ability to use their considerable part all year round for grazing cattle and livestock. In the conditions of the high-altitude zone, a wide variety of types of territories determines the nature of land use. As we approach the mountain-foothill landscapes, the degree of development of the territory decreases, the share of agricultural land, the share of intensive types of agricultural land use (arable land, intensive crop rotations and so on) decrease. As the height increases, the cadastral value of land plots should be reduced, which is caused by an increase in the possibility of dangerous geomorphological, climatic, neotectonics, hydrological processes and an increase in the complexity of agricultural land use. In this regard, the basis for the formation of agricultural land use in the conditions of high-altitude zones should be the similarity of the types of natural environment.

1. Introduction

Land has always occupied a special place among other resources used in social production, especially in agriculture. The total world reserves of agricultural land are about 5 billion hectares (including 1.4 billion hectares of arable land). The Russian Federation has 9.5% of the world's arable land. Russia has a high supply of land resources. In Russia, there are about 12 hectares of land per 1 person, including 1.3 hectares of agricultural land; among these, 0.8 hectares of arable land, while the world average is 0.14 hectares [1-3].

The geographical position of Tuva has a direct influence on its socio-economic development. In Siberia, large enterprises are located in flat areas, whereas in the Republic all lands are used for production of agricultural production. The population of the Republic according to Rosstat is 324420 people (2019, 78th place in Russia). The population density is 1,92 people/km2 (2019). The urban population is 53.59% (2018), despite the fact that the region has significant land resources for living. It should be noted that more than 2/3 of the population lives in the city of Kyzyl (116983 people, 2019). The rest of the population in the greater part of the 1/3 lives in the Tuva basin (Ulug-Khem coal, Kyzyl, Barun-Khemchikskaya, Dzun-Khemchikskaya, Pius-Khem coal districts) [3-7].

The territory stretches from west to east for 745 km with a width from south to north from 100 to 450 km. The area of the Republic is 168,604 km2 (3.28% of the territory of the Siberian Federal district, 0.98% of the area of the Russian Federation – the 20th in area). The Republic of Tuva is part of 17 municipalities and two cities of Republican subordination of the subject are located in the steppe zone at the junction with the taiga areas of semi-deserts. The second feature of the geographical location is that the territory of the Republic is far from the seas and oceans (Atlantic, Pacific, Arctic oceans) at a distance of 2500 to 3000 km). The third feature is that the Republic is bordered by mountains (especially from Russia).

The relief of Tyva is characterized by the presence of small intermountain basins with a sharply different landscape from the surrounding mountains. Between the ridges and uplands are enclosed basin: the Tuvan steppe, semi-desert UVS Nuur and the taiga-forest and Todzhu Tere-Khol. The average height of the basins is 650-1200 m above sea level. Throughout the year, the Republic is dominated by anticyclonic dry and clear weather, cold in winter and hot in summer. The most pronounced feature of the climate is a sharp continental. Cold winters with little snow, hot summers, low rainfall and high temperature range are characteristic features of Tyva climate. Especially noticeable effect on the climate is the influence of the absolute height of the terrain. There are three vertical climatic zones: low, middle and high mountains (figure 1).

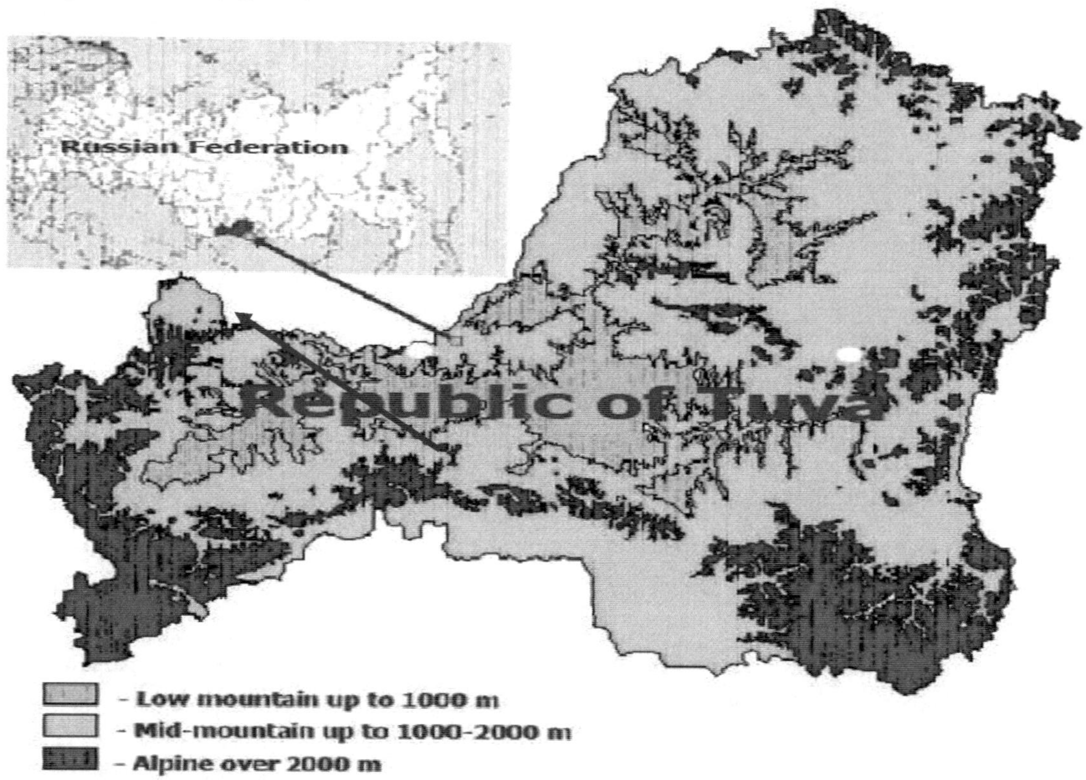

Figure 1. Distribution of the territory of the Republic of Tuva by latitudinal zones

Tuvan basin, stretching more than 400 km from west to east along the Yenisei river, is an open hilly steppe space, located at altitudes from 500 to 1200 m. Mountain ranges, crossing it, divide the basin conditionally into three parts: Chemchik, Ulug-Khem and Turano-Uyuk.

2. Materials and methods

One of the methods of geomorphological research is morphometric analysis, which involves the study of quantitative characteristics of landforms (which include the absolute and relative height of individual landforms, slope angles, exposure of slopes, areas occupied by positive and negative forms, etc.) by special measurements, in particular, using a digital elevation model (DEM). To conduct a vertical zoning of the relief, to highlight the main hypsometric levels that characterize a particular form of relief, allows the hypsometric map.

The use of GIS technologies and digital elevation models can significantly deepen and detail the study of the structure of river valleys, which is necessary in terms of giving them a significant role in ensuring the efficiency of agricultural land use in mountainous areas. In order to study the slope processes, and first of all, their manifestations such as landslides, landslides, mudflows, determine the scale and dynamics of these processes, which have a serious impact on the use of agricultural land in high-altitude zones.

3. Land use efficiency in high-altitude zones

Of the total land area, most of the land is classified as forest land (64.5%), agricultural land accounts for 19.9% (table 1). If the average size of the land plot, given per inhabitant of Russia, accounts for 2.5 hectares of agricultural land in the study area according to statistics in Tuva - 12.66 hectares. The area of agricultural land in all categories of land in 2017 amounted to 3833.2 thousand hectares or 22.7% of the total land fund of the Republic (table 7). The share of non-agricultural land accounts for 13027.2 thousand hectares or 77.3% [5, 8–10].

Table 1. Structure and dynamics of the land fund of the Republic of Tuva from 1990 to 2017, thousands of hectares

Land category	1976	1990	2000	2007	2017
Agricultural land	4623.2	6812.2	2831.6	3164.8	3366.0
Lands of settlements	59.0	61.8	39.7	43.5	45.2
Industrial lands and other special-purpose	28.5	25.7	16.8	16.2	20.1
Lands of specially protected territories and objects	-	336.6	659.1	655.1	655.2
Lands of the forest Fund	11048.9	8486.6	10874.3	10874.6	10874.6
Water Fund lands	46.9	26.2	96.3	96.3	96.3
Reserve lands	1053.9	1111.3	2342.6	2009.9	1803.0
Total for Republic	16860.4	16860.4	16860.4	16860.4	16860.4

The main specialization is animal husbandry (mainly sheep and dairy cattle, goat and horse breeding). The area of farmland is more than 4.5 million hectares. A feature of the pastures of Tuva is the possibility of using a significant part of them throughout the year for grazing small and cattle, the total number of which in Tuva can reach 2 million heads. The peculiarity of natural and climatic conditions of the Republic has historically determined the features of the agricultural sector. Traditionally, the priority is the development of animal husbandry. Cattle, sheep, goats, pigs, horses, poultry are bred in the farms of the Republic.

Analysis of the dynamics of the land fund shows that significant changes occurred from 1990 to 2000 and only since 2000, the dynamics of the land fund has become stable. Agriculture of the Tuvan ASSR (1985-1990) is characterized by land development and, as a result, an increase in the area of agricultural land by 66.8%. At the same time, the creation of Arat and peasant farms was of great importance, and in the process of privatization — the reorganization of collective and state farms. Therefore, the change in the area of agricultural land during this period was insignificant.

Since 1995, negative general economic problems did not allow one to take advantage of the newly created economic structures. In the conditions of unstable production relations, reduction of state support, reduction of consumer purchasing power, growth of non-payments, loss of working capital, as well as increased costs, many agricultural enterprises have become unprofitable, which has led to their liquidation (reorganization) and subsequently reduced the area of agricultural land.

During the observed period (2016-2017) there were no changes in the structure of the land fund. The area of agricultural land in all categories of land in 2017 amounted to 3833.2 thousand hectares or 22.7% of the total land fund of the Republic. The share of non-agricultural land has 13027,2 thousand hectares or 77,3% (table 2).

Table 2. Structure and dynamics of agricultural lands of the Republic of Tuva, thousand hectares

Grounds	1976	1990	2000	2007	2017
Arable	493.1	429.5	155.7	131.4	191.3
Deposit	5.7	1.3	49.6	61.4	147.9
Hayfields	107.5	56.0	53.2	52.1	76.5
Pastures	4016.6	3086.8	2233.2	2209.4	3416.6
Perennial plantings	0.3	0.2	0.2	0.1	0.9
Total for Republic	4623.2	3573.6	2491.9	2454.4	3833.2

Grasslands consisting of agricultural land have always prevailed. During the existence of the large farm, households (aratskiy) of farms were supported by the planned system pastebot (until 1991). However, too much density of cattle and excessive load were allowed, conditions of pasture rotation were not maintained, other measures for restoration of vegetation cover were not carried out. The change in the form of management has only led to a change in the spatial distribution of the pasture load: some pastures are abandoned and not used; others still experience overgrazing.

Hayfields, occupying a small area, spread in narrow ribbons in river valleys and on the slopes of mountain uplands. They represent a highly productive and valuable by the composition of the herbage of the meadows. Their economic use decreased during the analyzed period. It is possible that part of hayfields was transferred to pastures [8].

The main negative process developing in the steppes is desertification, which leads to a decrease in productivity, an increase in the nudity of the earth's surface, a decrease in the reproduction of steppe ecosystems. A number of changes in the parameters of the geographical environment for the duration and consistency allowed one to establish some trends and patterns of development of desertification.

In the conditions of high-altitude zone, a wide variety of types of territories determines the nature of land use. As we approach the mountain-foothill landscapes, the degree of development of the territory decreases, the share of agricultural land, the share of intensive agricultural land use (arable land, intensive crop rotations, etc.) decreases. At the same time, the degree of fainting out of the territory development, which consists in the formation of land use on the basis of individual elements of the relief, types of soils, meadows and pastures, depending on the orographic conditions of the area, is increasing [6, 8].

Accordingly, the forms of agricultural land use are changing. Animal husbandry comes instead of irrigated agriculture in the plains, using pasture resources on the territory of mountain landscapes. Forest-steppe zones of foothill landscapes are characterized by agriculture based on alternation of irrigated and rainfed lands in combination with natural forage lands. Consequently, the formation of agricultural land use in the conditions of high-altitude zone should be based on the similarity of the types of natural environment, which allows us to distinguish the following groups of land uses:

- agricultural land use, characterized by a high level of development of land with a combination of irrigated and rainfed agriculture (based on the use of flat-terraced land forms of relief), involving crops of grain crops, row crops, annual grasses;
- agricultural land use, characterized by a pronounced development of livestock sectors of agriculture

(based on the use of land mountain landscapes), involving the organization of summer pastures and providing seasonal livestock farms lowland areas;

- agricultural land use, characterized by continuous rain-fed agriculture with high ploughing (based on the use of land foothills and low mountains), providing for planting vegetables, potatoes, fruit and berry plantations, etc. This type of agricultural land use involves the management of dairy and beef cattle.

4. Conclusion

The final assessment of lands for agricultural land use in the conditions of the high-altitude zone is derived taking into account the degree of influence of the composition of vegetation and their percentage in the landscape-ecological contour. The predominant type of land use in the landscape-ecological contour (type of pastures when using land for transhumance) takes into account the geographical features of the territory. Thus, the assessment of land depends on the productivity of land, natural biological resources within the boundaries of landscape and ecological contours. Although steppe landscapes in Tuva make up no more than a third of the Republic's area, the region has a very high degree of preservation of steppe ecosystems, occupying one of the first places in Russia. Accordingly, the steppes today occupy large areas, and the complex mountainous terrain and the biological geographic transition character of the region determine their significant diversity.

Thus, in order to improve the cadastral registration of land in the territories of the high-altitude zone, it is proposed to introduce a local indicator – a coefficient that depends on geographical conditions and affects the cadastral value of land in the conditions of high-altitude zone. The locality of this indicator is its distribution in the territory, characterized by processes that complicate agricultural human activity in the conditions of high-altitude zones, and the dependence on these processes. Its introduction will allow one to objectively correlate the cadastral value of land and agroecological parameters of land, taking into account the geographical features of the territory.

The differentiation of the local indicator and the analysis of areas characterized by the possibility of occurrence of such dangerous and negatively affecting agricultural land use as earthquakes, mudflows, avalanches, low temperatures, hurricanes, etc. lead to the following conclusion. As the height increases, the cadastral value of land plots should be reduced, which is caused by an increase in the possibility of dangerous geomorphological, climatic, neotectonic, hydrological processes and an increase in the complexity of agricultural land use.

5. Acknowledgments

The research was carried out within the framework of the State program of the Republic of Tuva "Development of agriculture and regulation of markets of agricultural products, raw materials and food in the Republic of Tuva for 2014-2020" (sub-program "Development of agro-industrial complex").

References

[1] Volkov S N, Shapovalov D A, Klyushin P V, Shirokova V A and Khutorova A O 2017 Solutions of problems in defining indicators of agricultural land within the framework of activities for the implementation of the concept of development monitoring in the Russian Federation *Int. Multidisciplinary Scientific GeoConf. SGEM* **17(52)** 819-828

[2] Shapovalov D A, Klyushin P V, Shirokova V A, Khutorova A O and Savinova S V 2018 Problems and efficiency of land use in the North Caucasian federal district *18th Int. Multidisciplinary Scientific GeoConf. SGEM* **18(5.1)** 667-674

[3] Murzakmatov R T, Koshkarova V L and Farber S K 2014 Ordination of forest types Ulug Khem forest district Tyva in terms of climate and terrain *Siberian forest journal* **3** 107-116

[4] 2017 *Report on the state and use of the lands of the Republic of Tuva in 2017* (Kyzyl: Office of Rosreestr in the Republic of Tuva)

[5] Dubrovsky N G, Namzalov B B and Oorzhak V A 2014 *Steppe, and deposits of Tuva* (Kyzyl: Publishing house RIO Tuvgu)

[6] Zhulanova V N 2012 Evaluation of agricultural land use and fertility of Tuva arable soils *Vestnik of BSAA named after V R Filippov* **1(26)** 46-52
[7] 2000 *Programme of action to combat desertification for the Republic of Tuva* (Abakan)
[8] Zhumaliyeva A S, Seromycin M V and Chistyakov K V 2017 High-Altitude landscape zones and nature use regimes of the ridge At-Bashi *Society. Environment. Developments* **1** 101-108
[9] Zharova T F 2017 Introduction of environmental resource-saving technologies based on soil protection systems of agriculture in the Republic of Tuva *Regional Economics and management: electronic scientific journal. Category: agribusiness and agriculture of the region* [in Russian – Regionalnaja ekonomika i upravlenie] **1-1(49)** 1-11
[10] Christen M, Kowalski J and Bartelt P 2010 RAMMS: Numerical simulation of dense snow avalanches in three-dimensional terrain *Cold Regions Science and Technology* **63(1–2)** 1–14

IOP Conf. Series: Earth and Environmental Science **350** (2019) 012007 doi:10.1088/1755-1315/350/1/012007

Scientific societies and exploration of the territory of the Russian empire

G G Krivosheina

S I Vavilov Institute for the History of Science and Technology of the Russian Academy of Sciences, 14, Baltiiskaia St., Moscow, 125315, Russia

E-mail: krivosheina@gmail.com

Abstract. In the nineteenth century scientific societies being nongovernmental voluntary associations of professional as well as amateur scientists were the leading institutions of Russian science. However, their role in the scientific, social, and cultural life of the country is essentially underestimated. This especially refers to the societies seated outside Petersburg. The present paper discusses the key characteristics that facilitated adaptation of the societies to the cultural realities of the nineteenth century Russia and fostered their research and educational activity. Special focus is given to the contribution of the societies to the exploration of the territory of the country.

1. Introduction

In the eighteenth century Academy of Sciences created by Peter the Great played the dominant role in exploration of the territory of Russia, its natural and human resources. Large-scale expeditions headed by members of the Academy (G.F. Muller, J.G. Gmelin, S.P. Krasheninnikov, P.S. Pallas and others) were sent to different parts of the empire. Materials collected by these expeditions “enriched every branch of natural science from mineralogy and botany to meteorology and local geography, including valuable ethnographic and demographic material and useful information on agricultural techniques, human and animal diseases, and local arts and crafts” [1].

In the nineteenth century the leadership in arrangement of academic expeditions passed to scientific societies, which managed to extend their research activity to the whole territory of the country. Usually the Russian Geographical Society (RGO in Russian initials) is credited to be the main actor in exploration of the territory of Russia. Indeed, this society founded in 1845 and seated in Petersburg was one of the largest and the most prestigious and only few Russian scientific associations could be on par with it. RGO had a number of affiliations all through the country and supported a whole series of large-scale expeditions to the Urals, Siberia, Kamchatka, the Far East, the Caucasus, and so on. In this respect, expeditions of numerous provincial natural history societies, while no less important, were not so impressive and famous and their contribution still remains underestimated. Most of these societies didn’t survive the 1917 revolutions or were dissolved during the first two decades of the Soviet power and, as a consequence, were of little interest to historians of science. Matters were made worse by the fact that Soviet and modern Russian historiography holds to the view that scientific societies as such represent an archaic type of scientific institutions, which do not meet the needs of modern science [2] and deserve serious attention. In recent years, however, history of scientific societies has gained certain attention. New studies not only broaden our knowledge of societies’ activities and their contribution to the progress of Russian science but also make us reconsider our understanding of the history of scientific

IOP Conf. Series: Earth and Environmental Science **350** (2019) 012007 doi:10.1088/1755-1315/350/1/012007

societies and their role in social and cultural developments [3–5]. The present paper deals with those neglected by historians scientific societies and their contribution to natural history exploration of the territory of Russia.

2. Materials and methods

The present paper is based on published sources concerned with activity of various scientific societies that were accumulated in the course of research. The sources fall into two slightly overlapping groups. The first includes minutes of scientific societies (most of them had limited circulation), jubilee publications on societies and biographies of their leaders, materials of naturalists' congresses etc. The sources of this group were used to reconstruct actual scientific contribution of the societies. The second group of sources is represented by publications of more general character (history and sociology of Russian science and scientific institutions, history of Russian civil society, etc.). These were analyzed with regard to their assessment of the role of societies as institutes of science as well as contribution of individual societies to Russian science and culture.

3. General description of Russian scientific societies

The Scientific society is a voluntary association of intellectuals who seek to promote certain academic disciplines or scientific ideas. In Western Europe the first associations of the kind appeared in the seventeenth century: Accademia dei Lincei (Italy, 1603), Academia Naturae Curiosorum (Germany, 1652), the Royal Society of London (Great Britain, 1660), etc. In Russia they, understandably, were established much later. Just as the Russian Academy of Science they were modeled after European institutions and adapted to local contexts. While the first scientific society was founded in Russia as late as in the last third of the eighteenth century, by the middle of the following century they had already occupied the leading position in Russian science, both fundamental and applied, as the most enterprising and efficient scientific institutions and at the beginning of the twentieth century there were registered more than three hundred scientific societies, not to mention numerous medical societies and more than five thousand agricultural association, a certain part of which were involved in scientific research. Rapid increase in number and growth in influence of scientific societies was fostered by a number of factors [6]. Firstly, societies, as distinct from the Academy of Science, were nongovernmental associations and this gave them relative freedom in formulating scientific policy and realizing their research and education programs. Secondly, elected society members included not only academic scientists, but also amateurs – school teachers, military men, physicians, public officers, country gentlemen, who were interested in scientific ideas and natural history observation and research. This was very important, since even in university cities of the period scientific community was thin. Thus, recruitment of amateur scientists was inevitable: it made it possible to realize certain ambitious projects and establish a really working scientific body. Thirdly, philanthropists, who were not involved in research activity but were ready to support societies financially, were also elected full members. And this provided means of existence for the societies and their research and educational programs, equipping expeditions, establishing research stations, educational centers, and research laboratories. As for philanthropists they received honorary awards and titles in exchange to their donations.

Among the nineteenth century Russian scientific associations societies of multidisciplinary type prevailed [6]. Their interests stretched over a wide range of disciplines from zoology, botany, and geology to ethnography, archaeology, and rural economy. Selection of the disciplines varied from society to society and sometimes changed in the course of time. Besides, research work of most societies was geographically confined to the areas of their location (educational district, governorate or even uezd). Practically speaking regional or local multidisciplinary societies were best adapted to the needs and national circumstances of Russian science. Only in Petersburg the situation was quite different. The capital was the place where the Academy of Science resided and where Russian academic and intellectual elite was concentrated, so specialized scientific societies closely cooperating with the Academy were the predominant type. They had been appearing here since the first quarter of the 19th century. The first of them, St. Petersburg Mineralogical Society, was established in 1817. It was followed by the Russian Geographical Society in 1845, Russian Entomological Society in 1859, and later on by some others. Only in the end of the 19th – beginning of the 20th century the serious demand for specialized societies was universally recognized and new societies of this type were being established both in the capital (Russian Anthropological Society, 1888; Russian Botanical Society, 1915; Russian Paleontological Society, 1916, et al.) and provincial towns. By the way, regional dimension was still present, e.g. in addition to the reputable Russian Entomological Society in 1913 Moscow Entomological Society was established.

4. The Free Economic Society

The Free Economic Society (VEO in Russian initials) or to use its full title The Imperial Free Economic Society for Advancement of Agriculture and Economy in Russia must be mentioned in the context of the present paper though formally it belongs to the group of agricultural societies. The point is that in some ways this society predetermined further development of scientific institutions in Russia and served as a model for other societies, both fundamental and applied. VEO was the first voluntary scientific association in Russia. It was founded in Petersburg in 1765 on the initiative of a group of landowners and high-ranking civil servants and with support of the Empress Catherine II. The latter granted her patronage and financial assistance to the society and full-dress uniform of imperial design to its members. Constitution of the Society chartered by the Empress made VEO the first independent public association of tsarist Russia. This was an important political innovation, though its significance has not been fully appreciated yet [3, 4].

Aims of the society were predominantly practical: "Like its Western models, the Free Economic Society was devoted to the improvement of agriculture through the spread of advanced technology, methods of sanitation, and economic knowledge" [7]. Nevertheless, VEO managed to successfully combine applied and fundamental research – a tradition that was followed by most Russian societies. For example, beginning with 1877 VEO was supporting V.V. Dokuchaev's travels to the Black Earth Region and his further studies of black soils (chernozems). As a result Dokuchaev created a schematic pedological map of the Black Belt of European Russia and published in 1883 his fundamental work "Russian Chernozem", in which he laid the basis for a new scientific discipline, soil science [8].

5. Natural history societies

The idea of independent natural history associations was taken by Russian universities, the more so that the 1804 university charters authorized universities to establish learned societies. Since in Russia of that period the only fully functioning university was the Moscow one, while others (in Petersburg, Kazan, and Kharkov) were still nascent, no wonder that the first societies appeared in Moscow (an attempt to organize the Society of Sciences in the Kharkov University in 1812 was not a success). Two Moscow natural history societies were established several months apart in 1805. The first was the Society for Competition of Medical and Physical Sciences (since 1846 – Physical and Medical Society), the second one was the Moscow Society of Naturalists (MOIP in Russian initials).

The Society for Competition is usually referred as medical society but during the first decades of its existence its research were mostly devoted to natural history and largely overlapped with those of MOIP. They included physical and medical description of cities and other locations of Russia; meteorological observations in Moscow and its vicinity; examination of mineral springs of the Moscow governances and some other localities, including the Caucasus; collecting of plants, especially medicinal, of the Moscow area, etc. [9]. Later the society turned mostly to medicine.

MOIP stated its principal mission as the natural-historical exploration of the territory of Russia for the purposes of advancement of agriculture and trade. As the society was based in Moscow it gave special attention to the study of the Moscow Governorate and its natural resources and developed an ambitious program of local research. Though this program was not realized in full measure, some of its results were published in 1837 in the fundamental "Oryctographie du gouvernement de Moscou" by the first director of MOIP J.G. Fischer von Waldheim. This work contained general geographic and economic description of the Moscow Governorate and materials on geological, mineralogical, and paleontological research of the area. However, the work of the society was not limited to the exploration of the Moscow region alone. In the first half of the nineteenth century MOIP actively sponsored numerous expeditions to different areas of the country. Thus, in 1806-1812 members of MOIP travelled to Transbaikalia and Kamchatka (I.I. Redovskii), to the Urals, Altai, and Transbaikalia (F.M. Tauber, G.F. Helm, J.W. Mohr), to the southern and south-eastern governances (F.W. Londes, A.M. Tauscher), to the Caucasus (B.Ch. von Vietinghoff). Unfortunately, the most part of the collections gathered during these expeditions were destroyed in the 1812 Moscow fire. The largest and the best known of MOIP expeditions of the period was that of Russian naturalist G.S. Karelin and young botanist I.P. Kirilov to Altai and Dzungaria in 1839-1841. During their travel Karelin and Kirilov gathered vast collections of plants, insects, and minerals and made geographic description of these far-off and poorly known lands [10]. For MOIP the expedition was both a success and a failure. The society gained wide popularity among Russian and European naturalists but was nearly ruined financially, so that beginning with the 1850s it stopped sponsoring expeditions and its activity was basically confined to regular monthly meetings and publication of the society's proceedings.

These circumstances by no means implied that scientific societies had lost their leading positions in Russian science. In the 1860s a new generation of university societies came on the scene. They were established according to the 1863 university charters and were aimed at regional research of certain educational districts. It should be noted that educational districts were established in the course of the 1803 educational reform of Alexander I and usually included from three (Dorpat Educational District) to fifteen (Caucasian Educational District) governances and oblasts.

The first society of this type was the Society of Friends of Natural Science (since 1867 – Imperial Society of Friends of Natural Science, Anthropology and Ethnography, OLEAE in Russian initials) associated with the Moscow University [4, 11, 12]. However, even earlier, in 1853, the Dorpat Society of Naturalists, aimed at natural-historical exploration of Livonia and neighboring governorates, was established but formally it was not an individual society but an affiliation of the Livonian Welfare Economic Society. OLEAE was founded on October 15, 1863 on the initiative of professor of zoology of the Moscow University A.P. Bogdanov and declared its mission as the "natural-historical exploration of the governorates of the Moscow Educational District and promotion of natural science among the mass of the public" [13]. The district in question was one of the largest and included about ten governorates (Moscow, Vladimir, Kaluga, Kostroma, Tier, Tula, Yaroslavl, Smolensk, Ryazan, Orel, and Nizhniy Novgorod). At first, because of insufficient funds, the society could not cover this whole territory with its research activity and geography of research was limited to Moscow and its nearest vicinity but soon it expanded to more distant uezds of the Moscow governorate and even to the neighboring governorates. OLEAE arranged excursions to collect local fauna and flora and popular public geological excursions (these were headed by the first president of the society, geologist G.E. Shchurovskii), undertook systematic excavation of kurgans for anthropological purposes. To attract more people to the excursions the society issued concise advisories on collecting various natural history objects (various animal groups, especially insects, plants, minerals, paleontological objects and even human skulls and bones). The advisories were addressed not only to OLEAE members but to everyone who was interested in natural science and were widely disseminated. So soon enough OLEAE acquired a lot of correspondents all over the country who sent their collections to the society. Besides, from the start OLEAE was active providing science popularization projects, including large-scale Moscow scientific exhibitions (Ethnographical, 1867; Polytechnical, 1872; Anthropological, 1878) and this made it widely known in Russia and abroad and consequently gave additional opportunities to expand exploration activity. For example, the famous A.P. Fedchenko Turkestan expedition of 1868-1872, the largest of OLEAE expeditions, was organized after the 1867 Ethnographic Exhibition on the initiative of Turkestan Governor-General K.P. von Kaufmann, who was deeply impressed by OLEAE and its exhibition. This was the first time, when the society got an opportunity to move beyond geographical and administrative boundaries outlined in its constitution. Later on it sponsored many expeditions (zoological, botanical and even ethnographical and anthropological) covering with their research a great part of the territory of Russia and initiated some important programs, e.g. the study of freshwater lakes of European Russia, the study of Russian seas (Black, Baltic, White), etc.

Contribution of OLEA to the progress of scientific research and education in Russia was highly appreciated at the First Congress of Russian Naturalists, held in Petersburg on December 28, 1867-January 4, 1868. The Congress solicited the Ministry of National Education that all Russian universities should found scientific societies on the model of OLEAE [14]. This idea was approved by the Minister and during the next few years naturalists' societies were established in the universities of Petersburg (1868), Kiev (1869), Kazan (1869), Kharkov (1869), and Odessa (1870). Later similar societies appeared in the Warsaw (1888) and Tomsk (1889) universities. All the newly organized societies took over the task of exploration of the territories of respective educational districts. These territories differed in size (the largest including the southern part of European Russia and Siberia accrued to the Kazan Society), as well as climate, relief, hydrography, natural resources, and this obviously shaped the main trends of their research work. For example, among the major contributions the Petersburg Society of Naturalist were exploration of the seaboard of the Barents and White Sea, the first large-scale hydrobiological research of the Kola Bay (Barents Sea) on the schooner "Aleksander Kovalevskii", establishing of a number of marine research biological stations in Solovetsky Islands, in the Kola Bay etc. The Kazan Society of Naturalists focused on Povolzhie and the Caspian Sea region and in addition to geological, botanical, and zoological research paid serious attention to the ethnography of local peoples and prehistoric archaeology.

We have to mention also local natural history societies (Society for the Exploration of the Yaroslavl Governorate in Natural History Relation, Ural Society of Friends of Natural Science in Yekaterinburg) and numerous local history societies established in provincial cities and uezd towns. Their resources and possibilities were obviously lesser than those of larger societies but they played a significant role on the Russian scientific scene. They worked in close cooperation with university and specialized natural history societies, as well as with many well-known Russian scientists and vast collections of local fauna, flora, minerals etc. accumulated in local societies contributed greatly to natural history knowledge of the Russian empire.

6. Conclusion

Most of natural history societies survived the World War I, 1917 revolutions, and the Civil War and resumed their activity under the Soviet power, successfully adapting to the new political and social realities, but the "one field of science – one society" reform of the late 1920s put an end to the existence of scientific societies as fully functional scientific institutions. Some of them were closed, some merged with other societies or state scientific institutes and those which survived were deprived of their research facilities such as research stations, laboratories, and institutes. Besides, they were forced to abandon their traditional activities like organization of expeditions (this function was passed to the Academy of Sciences). So the question still remains: were scientific societies really unable to meet the needs of modern science or their dissolution was politically and ideologically biased.

References

[1] Vucinich A 1963 *Science in Russian Culture: A History to 1860* (Stanford, California: Stanford University Press) pp 150-151
[2] Stepanskii A D 1987 *History of Scientific Institutions and Organizations in Prerevolutionary Russia* (Moscow: MGIAI)
[3] Bradley J 2002 Subjects into citizens: Societies, civil society, and autocracy in tsarist Russia *Am. Hist. Rev.* **107** 1094-123
[4] Bradley J 2009 *Voluntary Associations in Tsarist Russia. Science, Patriotism and Civil Society* (Cambridge Massachusetts, London England: Harvard University Press) pp 95-130
[5] Tumanova A S 2011 *Self-Organization of Russian Civil Society in the Last Third of the 18th – early 20th centuries* (Moscow: ROSSPEN)
[6] Krivosheina G G 2018 Entomologists' associations in the 19th century Russia: The story of the

Entomological Committee of the Society of Friends of Natural Science *Rus. Ent. J.* **27** 451-58
[7] Vucinich A *Op. cit.* 163
[8] Snytko V A and Shirokova V A 2017 *V.V. Dokuchaev's Scientific Heritage: Traditions and Development of Ideas. Materials of the International Seminar* (Moscow: Akkolit)
[9] Lipshits S Yu 1940 *Moscow Society of Naturalists during 135 Years of Its Existence* (Moscow) p 87
[10] Varsanofieva V A 1955 *Moscow Society of Naturalists and Its Role in the Progress of National Science* (Moscow: Moscow State University) pp 15-46
[11] Krivosheina G G 2013 'Society of destroyers of natural science' or 'Moscow academy of science'? (To the 150th anniversary of OLEAE) *VIET* **4** 57-71
[12] Krivosheina G G 2014 Long way to the Anthropological Exhibition: The institutionalization of physical anthropology in Russia *Centaurus* **56** 275–304
[13] Senger N K Minutes of the meetings of the Society of the Friends of Natural Science *Izvestiia OLEAE* **3(1)** Suppl. 1-2
[14] Beketov A N 1868 *Proceedings of the First Congress of Russian Naturalists* (St. Petersburg: Tip. Imp. AN) p 62

History of the study of oil resources of the Mesozoic sediments in North Ossetia-Alania and adjacent territories

Zh G Kusova

North Caucasus Mining and Metallurgical Institute (State Technological University), 44 Nikolaev st., Vladikavkaz, 362021, North Ossetia-Alania, Russia

E-mail: quartz7@mail.ru

Abstract. The article analyzes the studies on petroleum resources of the Mesozoic sediments carried out in the XIX-XX and the beginning of the XXI centuries in the Republic of North Ossetia-Alania and the neighboring republics of the North Caucasus. The purpose is to study the history of research carried out to collect available information for qualitative and quantitative assessment of oil and gas resources of the Mesozoic sediments. Only separate materials on this issue were found. Most of the materials were lost and destroyed during the Chechen wars. Combined historical studies of the oil and gas potential and materials from the North Caucasian and central state archives are required. Using the available data, the article provides information on the history of the study of promising areas and deposits of the region, starting from the beginning of the XIX century and up to the beginning of the XXI century. The analysis indicates a high potential of further scientific research on hydrocarbons of the Mesozoic sediments by improving the scientific and methodological base and research methods.

1. Introduction

The history of development of the oil and gas industry includes a wide range of areas (geological prerequisites, prospecting and exploration, assessing hydrocarbon reserves, effective development of promising areas) [1].

Accumulated information can be used to assess the gradual development of the oil and gas industry in the region and the country, outline new areas for prospecting and exploration, and increase hydrocarbon production. The North Caucasus which is of scientific and practical interest for the further development of hydrocarbon resources.

The territory of the North Caucasus is divided into three areas according to the geological structure and oil and gas bearing conditions: Western Ciscaucasia (Krasnodar Territory), Central Ciscaucasia (Stavropol Territory) and Eastern Ciscaucasia (Republic of North Ossetia-Alania, Republic of Ingushetia, Chechen Republic, Republic of Dagestan and South Ossetia, and South Africa and South Africa and South Ossetia-South Ossetia-Alania, Republic of Ingushetia, Chechen Republic, Republic of Dagestan and South Africa and South Africa areas) [2].

In the Eastern Ciscaucasia, oil and gas potential prospects are associated with the Mesozoic sediments including the Jurassic and Cretaceous sediments. The Cretaceous sediments are composed of thick fissured limestone with marl and clay interlayers. Their reservoir properties with a capacity varying from 200 to 1500 m are caused by the development of fractured rocks and leaching. The Cretaceous deposits are promising.

2. Purpose of the study

This paper is devoted to the history of the studies conducted in the RNO-Alania and adjacent territories in order to collect information for assessing prospects of oil and gas resources of the Mesozoic sediments.

3. Materials and methods

The materials used for this research are results of scientific and industrial-production studies of the oil and gas potential of the North Caucasus, in particular, Eastern Ciscaucasia (monographs, articles, literary sources and stock materials).

4. Results

The geological and geophysical materials accumulated over many years of research conducted in the RNO-Alania remains are stored in various funds, organizations, institutions outside the republic. It is necessary to conduct oil and gas potential historical studies in the RNO-Alania.

In Russia, the oil industry began to develop on the Absheron Peninsula and in the North Caucasus.

For the first time in the Caucasus, near the Mozdok fortress, in 1823, the Dubinins (fortress peasants) built a small oil distillation facility to produce transparent liquid called photogen. As a result, they obtained thick dirty and black liquid named "macluzat" which was further transformed into the word "mazut".

The beginning of the development of hydrocarbon resources in the North Caucasus is the end of the XVIII century and the beginning of the XIX century; the beginning of the systematic geological research is the time of the first and subsequent geological surveys conducted by Abih German Wilhelmovich (1862, 1866, 1973, 1876). In 1873, G.V. Abih presented his first systematic ideas about the geological structure of the area and established the Mesozoic (tertiary) age of oil deposits of the North Caucasus with lithological characteristics of individual parts of the section [3].

For the regional petroleum geology, the surveys conducted by A.M. Konshin in 1892, 1896 were crucial. He argued that oil fields in the Eastern Ciscaucasia are confined to three structural zones of the Terek-Bragunsky, Sunzha-Grozny ridges and the Black Mountains which stretch from Vladikavkaz railway to the coast of the Caspian Sea where the formation of significant oil reserves is possible [4].

In 1901, in the Eastern Ciscaucasia, surveys were conducted by the Geological Committee of the Supreme Economic Council in the Terek-Sunzhensk oil and gas region. Their results revealed its prospects.

Since 1915, drilling works began on the southern wing of Voznesensk fold (Mozdok), and the first industrial oil flows were obtained.

In 1917, A.S. Savchenko established an uplift of the Kabardian fold to the east.

In the 1920s, the survey in the Kharbizhin area began. Geologists obtained reliable information about the geological structure. In 1924, Kudryavtsev identified the existence of an oil-bearing structure. In subsequent years, a whole series of seismic studies were conducted in this area. The Mesozoic sediments were discovered. In addition, the reflecting horizons were traced along the Lower Cretaceous and Upper Jurassic sediments.

In the same years, geological surveys were conducted on the territory of the Argudan area by Shvetsov and Alferov who aimed to assess the oil and gas content of tertiary deposits.

In 1925, Shvetsov discovered a new orographic dome-shaped elevation.

In 1925-1927, within the boundaries of the Malo-Kabardinsky Ridge, Orlovsky and Dolitsky carried out geological surveys to specify its tectonic structure. Malo-Kabardinskaya, Zamankulskaya and Kurpskaya anticlinal folds were identified.

A more detailed study of the geological structure and petroleum resources began in the 1920s and 1930s. This period is characterized by intensified geological studies aimed to identify oil and gas areas and discover oil in the Karagan-Chokrak deposits [5].

In 1933, under the leadership of V.I. Dokunin, a gas survey was conducted on the territory of Akhlov. An elevated methane content which coincides with the Sarmatian deposits and a high content of heavy hydrocarbons associated with the Akchagyl contact and Sarmatian deposits were discovered.

In 1937-1938, the trust "Grozneftegeofizika" directed by N.V. Bogdanovich conducted surveys in the western part of the Malo-Kabardinsky Ridge. As a result, they managed to create a geological map.

In 1940, in the central part of the ridge, Harbikh carried out a geological survey and specified the tectonic structure of the Kabardian anticline.

In addition, based on the research results, the following promising areas were identified: Kardin-Zmeiskaya, Zamankulskaya, Korinskaya, Argudanskaya, Verkhne-Kurpskaya, Kharbizhinskaya, Yuzhno-Kharbizhinskaya, Akhlovskaya, Severo-Malgobekskaya.

In 1940, in order to specify the geological structure of Argudan, N.V. Bogdanovich carried out geological surveying covering the area between the Psygan-su and Argudan rivers. The Argudan structure whose tectonic features were studied is an open sloping anticline with wings falling at an angle of 15-17°. The structure is complicated by a series of violations of various amplitudes and orientations. N.V. Bogdanovich suggested a more detailed study of the fold combined with drilling operations.

In 1948, I.M. Krisyuk conducted a geological survey on a scale of 1: 50,000. P.P. Zabarinsky gave a positive assessment of the oil and gas prospects of Maikop, Karagan-Chokrak and Sarmatian-meotic sediments under the lithological or tectonic isolation of layers.

In 1948, the "Malgobekneft" trust drilled over 15 deep wells in Zamankul area to specify the tectonic structure of the Malo-Kabardinskaya fold and assess the oil and gas content of the Karagan-Chokrak deposits of the northern and southern slopes of the ridge. Gas formations were discovered, and water with gas was obtained. However, the presence of oil was not observed [6, 7].

In 1948-1957, the "Malgobekneft" trust directed by I.V. Hans, L.F. Kostrich, B.K. Lotiev, V.L. Mkrtychev, and Yu.M. Zwedel performed structural drilling works. The "Grozneftegeofizika" trust conducted electrogravimetric and seismic studies. The complex structure of the Zamankul anticline arch was revealed, and a series of longitudinal violations that split the fold into a number of separate blocks were found.

In 1952-1955, in order to assess the prospects of petroleum resources, the Malgobekneft trust carried out structural drilling works along five profiles along the valleys of the Fiagdon, Ardon, Ursdon, Dur-Dur, Urukh rivers in the western part of the Terek-Sunzhensk oil and gas region, in the Korzhin-Zmeiskaya area. As a result, stratigraphic sections of deposits of Maikop, Chokrak and Karagan layers were specified.

In 1952-1953, in the Korin Area, researchers of Grozny Oil Institute directed by of P. P. Zabarinsky and B. K. Lotiev carried out reconnaissance work on the Montenegrin Monocline between the Assa and Psygan-su rivers. The results indicated favorable conditions for accumulation of oil and gas. Due to the strong intersection of the relief covered with deciduous forest, it was impossible to conduct areal seismic surveys; therefore, the main method of geological study was drilling of structural-exploratory wells. A single seismic profile was developed along the Ursdon-Saverdon River.

In 1951-1952 and 1956-1957, in the Verkhne-Kurpskaya area, the studies on the geological structure of the Psedakh structural protrusion and the Kharbizhinsky uplift were carried out. In this area, there were no obvious signs of oil deposits, but during the drilling works, the kern with oil signs was lifted. The main result of the structural drilling was new data on the geological structure of the Kharbizhin anticline and a promising positive structure was mapped in the upper tectonic floor.

In the post-war period, targeted geological studies on Mesozoic deposits began in connection with determining the prospects of oil and gas potential, although these prospects were suggested much earlier [8].

In 1957, prospecting drilling works began in the Zamankul area. As a result, industrial oil inflows from Upper Cretaceous limestones were obtained. In 1958-1981, the Upper Cretaceous sediments were studied in more than 25 wells, most of which gave industrial oil inflows with a flow rate of up to 58 tons / day. As a result, the reserves were calculated for the Upper Cretaceous deposit, and the deposit was put into development.

In 1957-1958, in the Argudan Area, researchers of Grozny Oil Institute conducted geological surveys between the Nalchik and Argudan rivers. In 1959, B.K. Lotiev, Yu.A. Sterlenko and B.G. Vasin described the Argudan structure as a tectonic protrusion complicating the Montenegrin monocline. The authors concluded that the pre-Jenomanian, Pre-Chracian and Pre-Apheronian tectonic movements had the most pronounced effects on the development of the Montenegrin Monocline as evidenced by the erosion of the Cenomanian, Upper Maikop, Upper Sarmatian and visible angular disagreement between the layers of the upper conglomerate strata of the Apsheron and Akchagyl longlines and the underlying complex of rocks.

Despite a significant amount of research carried out during these years, the genesis of the fold, its link with the neighboring Serpentine structure, the cnature of facial-lithological variability of the sediment that forms the region remained unsolved, which became the main problem in assessing the oil and gas potential of the area. In 1962, structural-exploratory drilling works were conducted by the Groznefterazvedka Trust and the Sevkavtsvetmetmetrazvedka Trust. The works were carried out despite the fact that in the western part of the protrusion, incompatibility of this section in the Cenozoic and Cretaceous sediments was established and in the eastern part, the area of oil and gas resources was not specified.

In 1962, exploration works were carried out in the Ukhlov Area; a number of exploratory wells were developed in different parts of the Akhlov structure. In one of the wells, an oil fountain was obtained. Oil was produced by wells drilled on the eastern pericline and the southern wing of the Akhlov anticline. Exploration works were carried out to delineate the Upper Cretaceous reservoir and discover oil deposits in the Lower Cretaceous and Upper Jurassic reservoirs.

Oil deposits were discovered in the marl of the Kumsky horizon, in the lower Cretaceous sediments of the Aptian and Valanginian longlines.

It was necessary to summarize the research data. A number of geological organizations began to conduct case studies in different directions. In 1964, A.F. Lopatin identified buried structures favorable for oil and gas accumulation based on the comprehensive analysis of morphometric and gravimetric materials to specify the tectonics of the Argudan bulge.

Analyzing structural and prospecting drilling operations carried out in 1964-1968, the lack of industrial accumulation of oil and gas in Neogene-Paleogene and Cretaceous sediments was established. Since the test was conducted in the open trunk, the prospects for petroleum potential should be associated with the Dotiton complex of sediments which are in structural and tectonic isolation. The oil-bearing deposits whose sandstones have high reservoir properties can be the Lower Cretaceous sediments. The Maikop sediments and carbonate rocks of upper and lower Cretaceous sediments in fault zones can be promising. Geologists were suggested to drill central and eastern deep exploratory wells in tectonically isolated blocks. After unsuccessful drilling of these wells in the late 1970s due to emergency situations, in the early 1990s, drilling operations began, but at the end of 1992, they stopped at 3581 m with a design depth of 5300 m.

In 1963-1965, in the Korinskaya area, the Sevkavtsvetmetrazvedka trust drilled more than 15 wells in order to identify favorable structures containing oil and gas. As a result, the tectonic structure of the Korin Protrusion was studied. In its central part, a terrace-like Mesozoic deposit was discovered.

In 1964, the Grozneftegeofizika trust gave a passport of the Kharbizhin structure to the "Grozneft". The presence of a favorable structure and reservoir rocks indicated the need and feasibility of launching drilling operations in the Kharbizhin area in order to identify the oil and gas content of the Cretaceous sediments.

To accomplish the tasks, a project was drawn. According to it, in 1969, exploratory drilling works began in the area of the Lower and Upper Cretaceous deposits. In the well drilled in 1971, there were no oil in the Barrem and Apt deposits, and the Upper Cretaceous rocks (Maastricht) produced a short-term fountain of oil. The well watered out quickly. Other drilled wells yielded inflows of water which made it difficult to draw final conclusions about the spatial location of the reservoir.

Exploration works were aimed at searching for oil and gas deposits in deep-lying Valanginian-Berrias and Upper Jurassic deposits. To this end, in 1972, step-by-step drilling of wells was carried out on the indicated horizons. Deposits were discovered in the Lower Cretaceous and Upper Jurassic sediments.

Further works were aimed at searching for the deposits in the intersalt and subsalt Jurassic sediments. In 1988, SevKavNIPIneft developed a project for the exploration of the Upper Jurassic intersalt deposits.

In 1971, the Grozneftegeofizika trust discovered a structure in the Cretaceous sediments that was transferred to the Grozneft company. The proximity of the area to the well-known deposits (Zamankul, Harbizhin) indicated the high oil and gas potential of this area.

In the North-Malgobek area, exploratory drilling began in 1971. In 1973, the SevKavNIPIneft institute drafted an oil deposit project for the Upper Cretaceous sediments. The drilling of the first exploratory wells showed that the arch part of the uplift, outlined along the conditional reflecting Upper Cretaceous horizon, is located slightly to the north, and the depth of productive deposits is significantly lower than that determined by the seismic survey results [9].

In 1978, the first operational estimation of reserves of the Upper Cretaceous reservoir was carried out; the volume of reserves was calculated and approved by the State Reserves Committee in 1985.

In 1995, Sevosetinneftegazprom began to restore the wellbore drilled in 1992 to a depth of 3,581 m in the Argudan area. However, due to drilling difficulties and the lack of funds, they decided to test the Upper Cretaceous and Karagan deposits characterized as oil saturated.

In 1997, materials of previous studies of the Korin Area were summarized. Using these materials, an exploration project and a passport of the oil prospecting structure were created for the Upper Jurassic, Lower Cretaceous, Paleogene, Neogene deposits.

In 2000, Sevkavgeologia developed a project of oil and gas exploration and analysis for the Yuzhno-Kharbizhin area which assessed the C_3 oil resources in the Jurassic, Cretaceous and Paleogene sediments. Exploratory wells were designed, but design solutions were not implemented.

5. Conclusion

Despite longstanding research and exploration, the oil potential of the Mesozoic sediments in North Ossetia-Alania and adjacent territories is still very high.

Historical analysis of the exploration of the territory shows that the main ways to improve the efficiency of geological exploration is to improve the scientific and methodological research and exploration base, increase accuracy of seismic mapping methods, study the hydrocarbon resources of North Ossetia Alania and adjacent territories.

References

[1] Kerimov I A, Daukaev A A and Borisenko Z G 2011 Historical aspects of the development of petroleum potential of the territories of Eastern Ciscaucasia *News of universities. North Caucasus region. Natural sciences* [in Russian – Izvestiya vuzov. Severo-Kavkazskiy Region. Estestvennye nauki] **4** 96-98

[2] Sukharev G M and Taranukha Yu K 1979 *Minerals of the Caucasus* (Moscow: Nedra) p 175

[3] Daukaev A A 2018 *History and Prospects for the Development of Exploration and Research Work on Oil and Gas in the North Caucasus (XIX-XXI Centuries)* (Moscow: Sputnik +)

[4] Kerimov I A, Daukaev A A, Gairabekov U T and Daukaev Asl A 2017 Natural-recreational potential of the mountain part of the Chechen Republic as a factor of sustainble development of the region *Sustainable development of the mountain territories* [in Russian – Ustojchivoe razvitie gornykh territorij] **9 3(33)** 211-215

[5] Daukaev A A 2015 The resource potentiai of the Upper Cretaceous complex deposits TCH *Scientific almanac. Earth sciences [in Russian – Nauchnyj almanah. Nauki o Zemle]* **9(11)** 1096-1099

[6] Litvinov A S, Kopyltsov A I and Murdalov L A 2012 Prospects for building the hydrocarbon potential of the Chechen Republic *Modern problems of geology, geophysics and geology of the North Caucasus. Materials of the II All-Russian scientific and technical conference, Grozny, Academy of Sciences of the Chechen Republic* pp 332-337

[7] Yandarbiev N Sh 2012 Prospects for oil and gas content in the territory of the Chechen Republic and adjacent areas *Oil and gas industry of the Chechen Republic: state, problems and priority directions of development. Materials of the All-Russian scientific and technical conference, Grozny: Academy of Sciences of the Chechen Republic* pp 156-170
[8] Kerimov I A and Daukaev A A 2017 Integrated use of minerals: historical and modern aspects *Modern issues of geology, geophysics and geoecology of the North Caucasus* **VI** 122-127
[9] 2007 The report "Comprehensive analysis of geological and geophysical materials with substantiation of the main directions for the development and increase of the hydrocarbon resource base of the Republic of North Ossetia-Alania" (Geological foundations of the RNO-Alania)

IOP Conf. Series: Earth and Environmental Science **350** (2019) 012009 doi:10.1088/1755-1315/350/1/012009

Formation and development of seismological research in Russia

E V Minina

S. I. Vavilov Institute of the History of Science and Technology, 14, Baltijskaya st., Moscow, 125315, Russia

E-mail: minina@ihst.ru

Abstract. The paper is devoted to the first stage in the development of seismic observations in Russia, covering the period from the late 19th century to the 1930s. The study aims to reconstruct the process of organisation of systematic instrumental seismic observations. The objectives of the study included identifying the main organisational forms of seismic observations, methods and equipment used in these observations, as well as analysing the main scientific results. It was established that the leading role in the emergence and development of seismic observations in Russia was played by a prominent physicist, academician B.B. Golitsyn. He had not only organised a network of I and II category seismic stations but also developed a seismograph with galvanometric recording of his own design for these stations. It is shown that a vast and full-rigged seismic network, the world's only, enabled tackling a number of important scientific problems. As a result of the organisational and scientific works conducted during this period, the basic principles for the well-rounded assessment of seismic hazard were developed that remain relevant to this day.

1. Introduction

The history of seismic observations in Russia is devoted to some publications, dedicated, as a rule, to anniversary dates. Ckhakya A.D. considered the development of seismic services in Georgia [1]. The articles of Sidorin A.YA. and Kozyreva L.I. presented the results of studying historical seismic events [2, 3]. Some authors analyzed the contribution of prominent scientists to the creation and development of seismological observations in Russia and other countries [4, 5].

Our country faced the problem of earthquakes and their catastrophic consequences after the seismically active regions of Transbaikalia, the Far East, the Caucasus and Central Asia incorporated into the Russian Empire. In 1840, a catastrophic earthquake occurred in Shamakha (Azerbaijan), in 1887 – in Verniy (today Almaty, Kazakhstan). The strongest earthquake happened in Krasnovodsk (today Turkmenbashi, Turkmenistan) in 1895, and in 1902 – in Andijan (Uzbekistan) and again in Shemakha. Frequent tremors in these areas led to the destruction and death of people.

2. The beginning of seismological research in Russia

In 1887 after the earthquake in Verniy, on the initiative of famous Russian geologist I.V. Mushketov, a temporary seismic commission was organized in the Russian Geographical Society. The purpose of this commission was to collect and systematize different information about seismic events on the territory of the Russian Empire and bordering countries. For this purpose the special questionnaires were developed. They were filled in according to the results of surveys of earthquake witnesses. Also to study the effects of devastating earthquakes some expeditions were organized. One of the important results of its activity was the preparation of the "Catalog of the Earthquakes of the Russian Empire" (by I.V. Mushketov and A.P. Orlov), which contained information about more than 2000 earthquakes that occurred in Russia and other countries during the period from the 6th century BC until 1880 [6].

The first regular observation of seismic phenomena with the help of a special pendulum was carried out by the German astronomer E. von Rebeur-Pachwitz (1861–1895) at Strasbourg university observatory in 1892. In Russia, observations of seismic phenomena using the horizontal Rebeur-Pashwitz pendulum were organized at the Nikolayev Marine Observatory (since 1892), at the Observatories of Kharkov (from 1892) and Yuryev (from 1897) universities and at the Tiflis Observatory (from 1899) [7]. Organization of seismic observations at observatories was due to the fact that the observatories were equipped with precision time instruments needed to accurately record the time of aftershocks.

On January 25, 1900, by the decree of Nicholas II, for the systematic study of earthquakes, the Permanent Central Seismic Commission (PCSC) was established under the Imperial Academy of Sciences. The director of the Pulkovo Observatory, O.A. Backlund, became its chairman. The purpose of the commission was a systematic study of earthquakes, collecting and publishing seismic data. To organize permanent seismic observations, it was supposed to create some seismic stations and provide them with the necessary equipment.

3. B.B. Golitsyn and creation of Russian seismic network

In seismology, as in no other science, international cooperation is of great importance. Only by analyzing data related to the same earthquake from different seismic stations, it was possible to determine the position of the earthquake epicenter, the depth of its focus and the intensity of shocks. During this period, horizontal pendulums of various designs with different accuracy of recording earthquakes were used for observations in different countries. The acquisition of the equipment abroad demanded big financial expenses, so the problem of the seismic instruments quality was extremely important. Boris Golitsyn, a famous physicist and head of the Physics Laboratory of the Imperial Academy of Sciences, played a major role in solving this problem.

B.B. Golitsyn was one of the first who suggested using physical and mathematical methods for studying the earthquake mechanism. He understood that strong earthquakes occurred rarely, but in order to catch the shocks from weak earthquakes that occur much more often, it is necessary to increase the sensitivity of seismic instruments and develop a reliable method to record seismic waves. Golitsyn was sure that the modernization of existing seismographs would not give the necessary accuracy of observations. He proposed changing the mechanical principle of seismic devices to electromagnetic one and developed seismograph of his own design. This approach not only increased the sensitivity of the instrument, but also allowed to use galvanometer and photographic paper fixed on a rotating drum for more accurate recording of seismic vibrations. In addition, by means of Golitsyn's seismographs another important task was solved – reducing the influence of the own vibrations of the device on the recording of aftershocks. Golitsyn proposed a new method of damping pendulums – electromagnetic method [8].

On November 26, 1906, the recording of remote earthquake signals using the Golitsyn seismograph began at an experimental seismic station organized at the Pulkovo Astronomical Observatory [9]. In 1909, B.B. Golitsyn and the head of the Pulkovo seismic station I.I. Wilip began researches deal with the constraction of a vertical seismograph (with a horizontal pendulum) with galvanometric registration [10]. He believed that equipping each seismic station with a set of two horizontal and one vertical seismograph would be optimal to provide the necessary accuracy of observations.

B. Golitsyn solved another important scientific problem – determining the azimuth of an earthquake epicenter according to data of one seismic station. He could solve it due to the fact that the vertical and horizontal electromagnetic seismographs had the same parameters.

Testing vertical and horizontal seismographs at the Pulkovo seismic station showed the reliability of these instruments, their higher sensitivity, as well as the clarity and legibility of the seismograms. After that, it was decided to equip all the existing and newly created seismic stations of I and II category with Golitsyn seismographs (Figure 1).

a b

Figure 1. Horizontal (a) and vertical (b) seismographs with galvanometric registration, Polytechnical museum.

The I category stations (including the Pulkovo Central Seismic Station) were located in Tiflis, Irkutsk, Tashkent, Yuryev, Yekaterinburg and Vladivostok. Their aim was to record and study remote earthquakes for the exploration of the overall seismicity and internal structure of the Earth. The location of these stations in the direction from west to east ensured that no earthquake would be missed. It was intended to equip stations of the I category with a set of two horizontal and one vertical Golitsyn's system seismographs.

Seventeen stations of the II category, designed to study and forecast strong earthquakes, were supposed to be located in seismically active areas. These stations were equipped additionally with Golitsyn's horizontal seismographs with mechanical recording to register strong shocks. All devices were made in the workshops of the Physics Laboratory of the Academy of Sciences.

Boris Golitsyn put a lot of effort to achieve financial support from the Russian state for the implementation of this project. On June 17, 1910, Emperor Nicholas II signed a law on the allocation of 45,440 rubles annually for the maintenance of seismic stations. For the management of all seismic stations, processing and publication of seismic data, as well as training special staff and formation of topics for further researches the Central Bureau of PCSC was elected. It consisted of B.B. Golitsyn, mining engineer A.P. Gerasimov and Lieutenant-General I.I. Pomerantsev [11, p. 196–197]. Financial support for the seismic service organization in Russia was also provided by private individuals. Thus, a seismic station in Makeyevka was organized with the funds allocated by the Congress of Miners of the South of Russia. Famous petroleum industrialist E. Nobel financed the creation of seismic stations in Baku and Balakhani [11, p. 200].

By 1915 Golitsyn seismographs were installed at seismic stations of the I category in Tiflis, Tashkent, Baku, Yekaterinburg, Irkutsk, Vladivostok, Baku, Makeyevka and seismic stations of the II category in Aleksandrovsky (Sakhalin), Petropavlovsk (Kamchatka), Kabansk (Transbaikalia), Barnaul, Verniy, Osh, Samarkand, Kandahar, Zurnabate, Shemakha, Borjomi, Pyatigorsk, Tomsk, Omsk and Nizhne-Olchedayev (Podolsk region). Such an extensive and well-equipped seismic network, the only one in the world, led Russia forward in the field of seismology. Bulletins with data from Russian seismic stations were sent not only to special seismic institutions, but also to leading universities and other scientific organizations in more than 30 countries. As a result, leading seismic stations in England, France, Germany, Italy, Sweden, Japan and other countries were equipped with sets of seismographs of the Golitsyn system. Equipping seismic stations with the same type of devises not only provided high quality analysis of seismic data obtained in different countries and regions, but also allowed to solve some important scientific problems.

A young graduate of St. Petersburg University P.M. Nikiforov (1884–1944) took an active part in organizing seismic stations and supplying them with the necessary equipment. Since 1908, he was invited by B.B. Golitsyn to participate in the scientific research at Physical Laboratory, in 1909 he was elected to the post of PCSC Secretary, and he held this post during the entire time of the commission existence.

In 1911, to provide seismic stations with qualified staff, B.B. Golitsyn gave a special course included 89 lectures. These lectures were of great importance for the development of the national seismological school. Preparing this course Golitsyn summarized and structured extensive theoretical material, which was then published in Russian and German. This book was the first textbook on seismology [12]. The training also included practical exercises at Pulkovo seismic station. In the future, such courses were held annually.

In 1911 B.B. Golitsyn was elected the Chairman of the International Seismological Association, which included 24 countries. It was the evidence of international acknowledgment of Russian seismology successful development. It was decided to hold the next congress of International Seismological Association in August of 1914 in St. Petersburg, but the beginning of the First World War thwarted these plans.

In 1913, B.B. Golitsyn summed up the results of using seismographs with galvanometric registration at Russian seismic stations. He noted that compared with 1911 (when the stations were equipped with another instruments), the number of recorded earthquakes increased from 383 to 576 (Pulkovo), from 219 to 456 (Tiflis), from 243 to 738 (Irkutsk). In the same period, one of the best seismic stations in Germany (in Göttingen) recorded only 134 earthquakes per year, and 139 ones in Strasbourg, and 64 – in Budapest [13, p. 360]. In addition, the number of cases of determining the coordinates or azimuth of earthquake epicenter according to one station data was increased and more than doubled.

In 1913–1915, studying some seismograms of distant earthquakes obtained at Pulkovo station, Golitsyn discovered the boundary of physical properties change in the Earth's interior – at a depth of 492 km. This boundary characterized by an intense increase of the seismic waves velocity. Later, the upper mantle layer, bounded by this depth, was called The Golitsyn layer.

May 16, 1916 B.B. Golitsyn died suddenly from pneumonia. P.M Nikiforov began leading Russian seismic service.

4. Development of seismological researches in USSR (1920–1930)

During the revolution and civil war, the activities at many of seismic stations were interrupted. Howeer, in the middle of 1920s, thanks to the efforts of P.M. Nikiforov, the work of the seismic service of our country has been restored.

The active construction of industrial enterprises and railways in earthquake-prone regions of Central Asia, the Caucasus and the Far East of Russia increased the importance of seismic observations. On the initiative of P.M. Nikiforov Seismological Institute of Academy of Science (SIAS) was established in March 1928. He became its first director. Under the guidance of P.M. Nikiforov the system of seismic observation was expanded by creating regional networks of seismic stations for the detailed study of earthquake-prone regions of the USSR. For such stations, P.M. Nikiforov invented a new type of seismograph for recording close earthquakes. In 1937 as a result of the regional seismic stations data analysis, a map of the seismic regionalization of the USSR was prepared. It had no analogues in the world [14].

5. Conclusion

The results of the organizational and scientific activity during this period became the base of the main principles of a comprehensive seismic hazard assessment development. They included the study of background seismicity at stationary seismic stations and the creation of temporary local seismic networks for more accurate seismic zoning. On a new instrumental basis, these principles continue to operate presently.

IOP Conf. Series: Earth and Environmental Science **350** (2019) 012009 doi:10.1088/1755-1315/350/1/012009

References

[1] Ckhakya A D 1950 *Essay on the development of seismology in Georgia (50 years of the Tbilisi Central Seismic Station)* (Tbilisi: Georgian Academy of Sciences)

[2] Kozyreva L I, Serova O A and Sidorin A Ya 2007 On the 120th anniversary of the Verniy earthquake in 1887 *Geophysical research* **7** 148–150

[3] Kozyreva L I and Sidorin A Ya 2007 On the 105th anniversary of the Shemakha and Andigan earthquakes in 1902 *Geophysical research* **7** 133–136

[4] Ponomarev A V and Sidorin A Ya 2012 Boris Golitsyn – the founder of modern seismology *Bulletin of the Earth Sciences Department RAS* **4** 60–61

[5] Branagan D 2010 Earth, sky and prayer in harmony: aspects of the interesting life of Father Edward Pigot (1858–1929) *Earth Sciences History* **29(2)** 232–263

[6] Mushketov I V and Orlov A P 1893 *Catalogue of earthquakes of the Russian Empire* (St. Petersburg: Imperial Academy of Sciences)

[7] Kozyreva L I, Mishina R L and Sidorin A Ya 2007 The organizers of the first instrumental seismic observations in Russia: I.E. Cortazzi (1837–1903) and G.V. Levitsky (1852–1917) *Geophysical research* **7** 146–147

[8] Sidorin A Ya 2012 Seismographs by B.B. Golitsyn for recording remote and local earthquakes *Seismic devices* **48(2)** 67–82

[9] Golitsyn B B 1907 The opening of a seismic station in Pulkovo *Izvestiya of Academy of Sciences* **1(1)** 25–27

[10] 1909 *Annual Report of Physico-Mathematical, Historical and Philological departments of the Imperial Academy of Sciences* (St. Petersburg: Imperial Academy of Sciences)

[11] 1910 *Annual Report of Physico-Mathematical, Historical and Philological departments of the Imperial Academy of Sciences* (St. Petersburg: Imperial Academy of Sciences)

[12] Golitsyn B B 1912 *Lectures on seismometry* (St. Petersburg: Imperial Academy of Sciences)

[13] 1914 *Annual Report of Physico-Mathematical, Historical and Philological departments of the Imperial Academy of Sciences* (St. Petersburg: Imperial Academy of Sciences)

[14] Sidorin A Ya 2007 The first map of seismic zoning: to the 70th anniversary of the seismic map of the USSR *Geophysical research* **8** 130–132

IOP Conf. Series: Earth and Environmental Science **350** (2019) 012010 doi:10.1088/1755-1315/350/1/012010

Landscape and historical conditions for the emergence and formation of the city of Smolensk

V A Nizovtsev[1] and N M Erman[2]

[1]Lomonosov Moscow State University, Geographical faculty, Department of Physical Geography and Landscape Science, 1, Leninskie gory, Moscow, 119991, Russia
[2]S.I. Vavilov Institute for the History of Science and Technology of the Russian Academy of Sciences, 14, Baltiyskaya str., Moscow, 125315, Russia

E-mail: nizov2118@mail.ru

Abstract. Complex landscape-historic studies were carried out at key sites of the historical waterway "from the Vikings to the Greeks" in the Smolensk region: in Smolensk and Gnezdovo. The purpose of this study is to identify the landscape and historical conditions of the development of the city of Smolensk. The main practical task was to identify the landscape structure of Smolensk and its surroundings and to conduct a comparative analysis of the natural properties of the landscapes of Smolensk and Gnezdovo as a factor in the development of these centers. More favorable landscape and ecological conditions together with a rich resource base and favorable defensive and strategic position of Smolensk at the present place were identified as compared to its original position (Gnezdovo). Perhaps, in the ancient Russian period this was the reason for accelerated development (or "transfer") of Smolensk on the left bank of the Dnieper river as compared to Gnezdovo.

1. Introduction

Smolensk is one of the most ancient cities of Russia, known by the first chronicles of mentions since 863. Already then Smolensk had defensive works, which was the most important sign of the status of the city in Ancient Russia. By the way, in the X century in his treatise "On State Management", the Byzantine emperor Konstantin Bagryanorodnyi [1] calls Smolensk "a fortress". The city was formed at the crossroads of the main trade routes of the time: "from the Varangians to the Greeks" and the "Great Volga Route" - from the Baltic Sea to the Mediterranean Sea and on the way from Europe to the Volga region and further to Asia.

Despite the extensive historiography devoted to the history of the development of Smolensk, there are questions that are practically in the initial stage of studies such as issues related to its geography and the influence of natural factors on it, first of all, the properties of landscapes "enclosing" and surrounding it, which were the resource base for its formation, existence and development. Therefore, the main objective of this study was to identify the landscape and historical conditions of the development of the city of Smolensk as a key center on the most important trade route of antiquity. The main practical tasks were: 1) locational and geographical analysis of the Smolensk-Gnezdovo hub on the way from the Varangians to the Greeks, 2) identification of the landscape structure of Smolensk and its surroundings, 3) comparative analysis of the natural properties of the landscapes of Smolensk and Gnezdovo as a factor in the development of these centers.

2. Materials and methods

The study is based on a paired analysis of the existing source base, reflecting complex landscape, component, paleogeographic and archaeological published and stock materials, as well as field studies in key areas. The boundaries, the main properties of the structure and features of the dynamics of landscape complexes of different rank for Smolensk and Gnezdovo have been revealed. Particular attention was paid to paleo-reconstructions and the assessment of the landscape conditions of the life of settlers in the Old Russian period, the retro-reconstruction of the existing landscape and economic systems. This has made it possible to identify the main patterns of environmental management, the formation of Gnezdovo and Smolensk and the formation of the settlement structure.

3. The history of the development of the city of Smolensk and the trade route "From the Varangians to the Greeks"

Even at the very beginning, Smolensk was the main city of the large tribal association of Krivichi and was a major trade and craft center. In 882, Prince Oleg conquered it and annexed to Kievan Rus. Later in the XII century, Smolensk became the capital of a large and influential independent Smolensk principality and rapidly developed over the next three centuries. And all this time, Smolensk was the most important reference point on the way "from the Varangians to the Greeks".

Already in the 9th century, Russia established full control over the Dnieper trade route, the way "from the Varangians to the Greeks" became not just the main transportation artery, but also the most important pivot of the Ancient Russian state, uniting Novgorod and Kievan Rus. This way connected the tribes of tivertsy, ulichi, polyane, northerners, drevlyane, dregovichy, radimichi, krivichy, and ilmenian slovens. Besides, by this way until the middle of the XIII century Kiev made connections with Byzantium, and, starting from the XIII century, it was the main trade route with the Ganza [2]. The most ancient Russian cities appeared on this way: Ladoga, Novgorod the Great, Staraya Russa, Smolensk, Lyubech, Vyshgorod, Kiev, Kanev, etc. It was not by chance that the Black Sea in the Eastern and Russian sources of that time was called the Russian Sea.

At that time, the way "from the Varangians to the Greeks" was both long and very difficult with many dangerous places. In "The Tale of Bygone Years" [3] it is stated that the way lay along the following route: the ancient shopping centers of Scandinavia and the southern coast of the Baltic - the Baltic Sea - the Gulf of Finland - the Neva River (which at tha time had rapids) - the stormy Ladoga Lake - the Volkhov River (with Volkhovsky, Gostinopol and Peche rapids) - Lake Ilmen (distinguished by its extremely restless and treacherous character for navigation). From the lake, the path went along the Lovat or along the Kunya or the Serezha, from which vessels were already dragged in the area of the present village of Volok to the River Toropa, the tributary of the Western Dvina.

Along the Dvina people descended to the Kasplya, to its spring in the lake Kasplya. And there was another portage in the river Katyn, which flew into the Dnieper. Further, the way went to the Black Sea, bypassing the Dnieper rapids (there were 7 rapids) and a narrow rocky place - the Krariansky ferry, where the Pechenegs often ambushed. In the lower reaches of the Dnieper, before going into the Black Sea, the ships were additionally equipped (sea sails were put on the boats) on the island of Khortytsya on the Dnieper, or on the island of Berezan near the mouth of the Dnieper. There was another stopping point on the Snake Island near the Danube Delta. By sea that was along the European coast (Rumelian coast) to Constantinople [4]. The total length of river routes and portages from the mouth of the Neva to the mouth of the Dnieper is more than 2,200 km. The way «from the Varangians to the Greeks», besides the Great Volga Route, was also connected to Pripyat-Buzhsky, which gas going to Western Europe. The main means of transportation were large (for 30-40 people) dugout single-tree rooks. The evidence of the Byzantine emperor Constantine VIII Bagryanorodhyi, krivichi and other ancient Russian tribes brought such boats to Smolensk, Chernihiv and other cities which were then floated down the Dnieper to Kiev, where they were additionally reequipped, loaded with goods and sent down the Dnieper [1].

]

As we see, one of the most important points of this route was located in the upper reaches of the Dnieper (Kasplya-Katyn in the area of modern Smolensk), from where the route passed to the Western Dvina and further directly to the Baltic Sea. Apparently, the city appeared here as the most important outpost on this road, performing not only managerial and economic, but security functions [5].

4. Landscape and historical development of Smolensk and Gnezdovo

Initially, Smolensk, as most modern researchers believe, originated on the right bank of the Dnieper River about 12 km away from the modern city (the current Gnezdovsky archaeological complex). This settlement occupied the maximum area in the X century and lost its importance as a handicraft and trade center in the first quarter of the XI century [6]. The settlement consisted of two hillforts, a village and a huge cemetery (the largest burial ground of the early Middle Ages in Europe (4.5 thousand barrows). Most of the central villages occupy the first terrace on both banks of the River Svinets, part of it is located on the Dnieper floodplain. The central hillfort occupies the cape of the first terrace on the left bank of the Svinets river with a height of 12-14 m above the river's gorge. Perhaps already then, at the foot of the settlement on the northeast shore of Bezdonka lake, there was an inner harbor with wooden flooring that could be used as a pier for small ships [7].

The first fortifications of Smolensk (wooden fortress) were built at the turn of the XI - XII centuries. In 1101, Prince Vladimir Monomakh laid on the top of the hill dominating over the valley of the Dnieper Cathedral Hill (Cathedral Hill), the first stone building of the city - the Cathedral of the Assumption of the Virgin Mary (Monomakh Cathedral), which was blown up during the siege of Smolensk by the troops of the Polish king Sigismund III in 1611.

Settlement of the trade and craft area of the city centre refers to the middle of the XI century and at this time begins the intensive development of the city and its becoming the most important supporting center on the way «from the Varangians to the Greeks». By the end of the XII century, in Smolensk, three main nuclei are formed, around which the ancient city grows [8]. At the same time, the center of the upper city is formed near the Assumption (Monomakh) Cathedral. The Church of St. Michael the Archangel was with the neighboring Boris-Gleb monastery a military defensive outpost of the city from the west. On the bank of the Dnieper in the 60s – 70s of the XII century there were built the princely temple of Ivan the Theologian (1160-1176) and the "German goddess" (1168-1190).

Apparently, here was the shopping center of Smolensk: a pier near the mouth of the river Churilovka, settlement of North European merchants and external bargaining, located outside the city walls. 2 km west of the city, the Borisoglebsky Monastery was situated at the mouth of the Smyadyn (the main temple was built in 1145), the Trinity Monastery (with the temple dated back to 1188-1209) was located 3.5 km away from the city. Lower down the Dnieper, the city was bordered by the Smyadyn monastery. Around the same time Spassky and other monasteries began to appear. They carried out the most important ideological and security functions and gave Smolensk the character of also the major political and economic center. By the XIII century there were 5 monasteries and more than 35 stone temples in the city.

In almost simultaneous attenuation of Gnezdovo (the place where Smolensk initially originated from) and the heyday of Smolensk, in our opinion, the natural resource factor played an important role [9]. Gnezdovo is located on the site of the Dnieper valley, represented by a segment of the floodplain and a system of broad low floodplain terraces and valley sandwiches with sandy infertile soils. This may explain the fact that part of the villages developed on the floodplain. It should be noted that at this time, due to the aridity of the climate and low water content, the floodplain was practically not flooded. That is, there were practically no relatively large plots for farming in the immediate vicinity. And this could be a significant limiting factor in the development of settlements in the nearest area and population growth of the hillforts.

5. Results

Smolensk developed in conditions with a completely different landscape structure and resource base than Gnezdovo. The city was founded on the high left bank of the Dnieper (up to 80-90 m of relative height), dissected by numerous ravines, valleys of stream-like beams and small rivers. Along this bank, only fragments of low floodplain terraces have been preserved, but in some places there is a well-marked low valley zander (III above-flood terrace according to D.I. Pogulyayev [10]) and, adjacent to the board of the valley, sloping surfaces of interstitial moraine, moraine-glacial and glacial plains with deep (3-7 m) groundwater bedding. Practically all these tracts from the surface are composed of a considerable thickness, with high trophism, loess-like loams with a thickness of up to 5-6 m [8]. On the left side, as in Gnezdovo, the system of low floodplain terraces and floodplain is common. Most of the soil-forming rocks are represented by loams with underlying sand of different genesis, less often moraine.

The relatively high trophism of the substrate in combination with good surface drainage (sloping surfaces, short lines of flow of surface water, no stagnation of moisture, etc.) and, consequently, a favorable water-air regime for agriculture created an almost unlimited base for agricultural development in the area . Interstudial and inter-river high ridges at the exit to the valley of the Dnieper form capes of various widths, called "mountains" by the local population: Cathedral, Pokrovskaya, Kazan, Voskresenskaya, Shklyanaya, etc. The first fortress of Smolensk was based on the most difficult mountain (Cathedral) with steep slopes.

6. Conclusion

The way "from the Varangians to the Greeks" had tremendous political and economic importance for the ancient Russian state. It was a means of management functions to be carried out, external and internal trade relations were realized. Naturally, along this way, large and small cities arose, being strongholds providing it with reliable protection and smooth functioning, since it was necessary to maintain dies, port facilities, etc. in proper condition.

We consider, it is very important, that for adequate functioning of these strongholds in the conditions of subsistence farming, the livelihoods of their settlers should have been based on the local resource base. For a long time, it was Smolensk that remained one of the most important and key strongholds.

Smolensk was one of such most important and key strongholds for a long time, which, at the first stage of its development, was probably located in the place of modern Gnezdovo. More favorable landscape-ecological conditions with a rich resource base and a favorable defensive and strategic position caused the accelerated development (or "transfer") of Smolensk on the left bank of the Dnieper, as compared to Gnezdovo in the ancient Russian period. Using the example of Smolensk, we can assume that one of the most important reasons for the "transfer of twin cities" was more favorable landscape and resource conditions of the new location.

7. Acknowledgments

The work is performed under the project of the Russian Foundation for Basic Research, № 19-05-00233.

References

[1] Bagryanorodnyi K 1991 *About empire management* eds G G Litavrin and A P Novoseltsev (Moscow: Nauka) p 493
[2] Bernstein-Kogan S V 1950 *Matters of Geography. The waterway from the Varangians to the Greeks* vol 20 pp 239-270 (Moscow: Geographgiz)
[3] Likhachev D S 1950 *Tale of Bygone Years* (Moscow-Leningrad) p 556
[4] Lebedev G S and Zhvitashvili U 2000 *Dragon "Nevo". On the Way from the Vikings to the Greeks: Archaeological and navigational studies of ancient water communications between the Baltic and the Mediterranean* (Saint Petersburg: Nordmedizdat) p 196
[5] Erman N and Nizovtsev V 2018 *Lecture Notes in Geoinformation and Cartography. The history of cartography of western border areas of Russia in eighteenth to early twentieth centuries (using the example of the Smolensk Province)* (Springer) pp 129-144
[6] Pushkina T A 2001 *Gnezdovo: results and objectives of the study,* in *Proceedings of the State Historical Museum Gnezdovo: 125 years of exploration of the place of interest* vol 124 (Moscow) pp 4-11

[7] Murasheva V, Bronnikova M, Panin A, Pushkina T, Adamiec G and Sheremetskaya E 2012 *Geoarchaeology of the upper Dnieper river valley at Gnezdovo: field excursion,* in *Geoarchaeological issues of the Upper Dnieper-Western Dvina river region (Western Russia): fieldtrip guide* (Moscow-Smolensk: Universum) pp 20-48
[8] Shkalikov V A and Bobrov E A 2009 *Socio-environmental problems of the city (on the example of Smolensk and other towns of Smolensk region)* (Smolensk: Smolensk University of Humanities) p 272
[9] Erman N M and Nizovtsev V A 2012 Landscape and historic prerequisites of the origin of the town of Smolensk *Izvestia: Herzen University Journal of Humanities & Science* (Saint Petersburg) **153(2)** 117-121
[10] Pogulyaev D I and Grozdov B V 1965 *The nature of Smolensk and its surroundings* (Smolensk) p 240

Landscape studies of ancient Russian cities on historic waterways

V A Nizovtsev[1] and N M Erman[2]

[1]Lomonosov Moscow State University, Geographical faculty, Department of Physical Geography and Landscape Science, 1, Leninskie gory, Moscow, 119991, Russia
[2]S.I. Vavilov Institute for the History of Science and Technology of the Russian Academy of Sciences, 14, Baltiyskaya str., Moscow, 125315, Russia

E-mail: nizov2118@mail.ru

Abstract. The aim of the study was to identify the role of the natural (landscape) factor in multifactor processes of structure formation, functioning and dynamics of urban landscapes. The practical tasks of this work were the reconstructions of the original landscape structure, identification of ecological properties and the resource base of host landscapes and cultural and historical landscapes of ancient Russian cities. Landscape studies of ancient Russian cities were carried out on the basis of the landscape and historical approach, where the entire territory of ancient Russian cities is viewed as a single, holistic entity, with inherent specific landscape organization which develops in space and time. A distinctive feature of the ancient Russian cities that formed on the historic waterways is that in the early stages of functioning of the water communications, cities were formed as the most important strongholds, whose vital activity was associated with subsistence farming and depended entirely on the local resource base and, accordingly, on the landscape structure of the territory. At different intervals, it differed sharply depending on the zonal-landscape conditions. The differences in the landscape features of the formation of ancient Russian cities on historic waterways in the IX-XI and XII-XIII centuries have been revealed.

1. Introduction

The formation and development of cultural and historical landscapes of ancient Russian cities is a natural result of the evolutionary processes of a particular territory. Studies of the basic patterns of formation and evolution of urban landscapes on the initial stages of their formation, as well as work on the study of the mutual influence of the city and landscape, are becoming increasingly important in historiography of both Russia and abroad.

The formation of urban landscapes is a spatial form of interaction between society and nature, man and landscape, and is a natural result of the evolution of the relationship between society and nature. The natural factor in the formation and development of cities is an integral part of the relationship between man and the surrounding landscape complexes. The natural component not only serves as a resource foundation for the formation of cities, but also creates its natural basis. The hierarchy of urban ancient Russian cultural landscapes is reflected in the fact that the city of the ancient Russian period exists partly independently in specific local geosystems, the city largely provides itself, within its natural-economic environment, it also provides its functioning through communication with subordinate settlements, - together they form a system of settlements, inscribed in one landscape or a system of landscapes.

2. Historiographic review

Scientific studies in the urban landscapes are the most complete and diverse, they are especially well studied from a historical point of view. There is also an abundant reservoir of sources studying the issues of urban planning and planning aspects of cities.

Not less sources include the bibliography of geographical aspects of the study of ancient Russian cities, including encyclopedic publications [1].

Relationship problems of the modern city and landscapes are in the center of attention of scientists of different areas (for example, the work of Vladimirov V.V., Mikulina E.M. and Yargina Z.N. [2] has become classic). It is worth mentioning the works touching upon the historical-landscape aspect of the development of ancient Russian cities by E.Y. Kolbovsky [3], E.A. Sukmanova [4].

A number of issues related to the understanding of the factors that led to the formation of the functioning of ancient Russian cities and the changes that took place in them are controversial, often insufficiently substantiated. This primarily relates to the geography of ancient Russian cities, the elucidation of the influence of natural factors and, above all, the properties of the "enclosing" and surrounding landscapes, which are the resource base for their formation, existence and development. One can just recall the ongoing controversy in historical science about the "transfer" of ancient Russian cities, in which there are practically no arguments concerning the properties of landscape structure as a factor in their development.

3. Tasks and methods

The most important aim of the research was the development of conceptual and methodological foundations for studying the role of the natural (landscape) factor in multifactor processes forming the structure, functioning and dynamics of urban landscapes. The main practical tasks were the reconstruction of the original landscape structure, the identification of ecological properties and the resource base enclosing the landscapes and the formation of cultural and historical landscapes of ancient Russian cities.

In recent decades, the Chair of Physical Geography and Landscape Studies of the Geographical Faculty of Moscow State University named after M.V. Lomonosov has started to develop a landscape-historical approach in the study of the landscape conditions of formation and development of ancient Russian cities, the development of clear criteria for the identification and diagnosis of urban anthropogenically modified landscapes, including the study of their intra-landscape structure [5, 6].

The main distinctive feature of the landscape-historical approach is that the entire territory of the ancient Russian cities is viewed as a single holistic entity with only its inherent specific landscape organization, landscape structure. The natural and anthropogenic components of the evolution of urban landscapes are considered in a single specific space and time, and all research is performed at the regional and local levels (for the historical centers of cities). That is, the landscape differentiation of the territory is necessarily taken into account and the study of the territory is carried out according to certain parts of the landscape (morphological units), which served as the territorial basis for the formation of ancient Russian cities and for the formation of the structure of urban landscapes. The identification of patterns of formation of ancient Russian cities and their dynamics and evolution is based on a "cross-cutting" landscape-historical study of the territory with the associated study of the landscape and time, the landscape and the economic and cultural activities of man in it.

An important method of landscape research of ancient Russian cities is a retrospective reconstruction of the original landscape structure and the formation on its basis the cultural and historical landscapes of ancient Russian cities [6].

The cultural-historical landscape is understood as an integral historical-cultural and natural formation, formed on a specific territory with certain homogeneous natural (landscape) properties as a result of the long-term interaction of man and landscape. It can be said that the cultural-historical landscape is to some extent optimized for a specific historical cross-section of time, the anthropogenic (cultural) landscape, which has a highly organized territory and optimal environmental management, reflects the specificity ("culture") of environmental management and spiritual life of an ethnos under specific landscape conditions. Along with the natural and anthropogenically transformed components and elements of the morphological structure, cultural and historical landscapes include heterogeneous elements inherited from various life cycles of the human being in the landscape: artifacts, social facts, mentifacts, the so-called "anthropogenic landscape layer". The characteristic features of cultural and historical landscapes are: originality, uniqueness (or typicallity) of the natural environment and historicity of the landscape (preservation of historical appearance), richness of monuments of historical, architectural and spiritual heritage, integrity and unity of the natural and anthropogenic component, as well as ecological compatibility of the landscape (organic combination natural and anthropogenic component, ecological and aesthetic properties) [7].

The reconstruction of the landscape structure of territories into concrete chrono sections is based on the use of a landscape edaphic approach. The result of such reconstructions was the mapping of conditionally restored (indigenous) landscape complexes. Special attention is paid to the analysis of anthropogenic relief changes (artificial terracing, plantation, “anthropogenic” erosion forms, etc.), the power of accumulation and distribution of the cultural layer and the change in the hydrological regime in a given territory. Such a map shows the ideal (theoretical) distribution of landscape complexes [7].

An important method of research is landscape historical mapping with making a series of multi-time maps of key areas reflecting the development of urban landscapes at a certain historical stage in specific environmental conditions. All maps for each key area are created in electronic vector form for visualization in Mapinfo system in a single coordinate system. All information in the maps is presented in the form of specific coherent map layers with numbered selections with attribute tables containing selection numbers and textual characteristics of legends to them. All the layers are integrated into the corresponding thematic map and a single historical and geographical GIS. In turn, thematic maps are also integrated into a single geographic information system. This allows you to quickly isolate the necessary information layer or, conversely, to integrate them for subsequent analysis or synthesis. In addition, the existence of a working set saves author's design of maps, during the creation of which the methods generally accepted in landscape science were used, therefore, in this form, they will be better and faster read by experts.

The conjugated analysis of landscape, natural component and historical-geographical maps of the Great Volga Route region, the way "from the Varangians to the Greeks" and the Severo-Dvinsky, Mariinsky, Tikhvinsky and Vyshnevolotsky historical waterways combined with research on key areas allowed us to establish some features of environmental management, the formation of ancient Russian cities and the formation of settlement structure, depending on the specifics of the functioning of waterways and the landscape structure of the area. For the analysis, we used a sample of ancient Russian cities from the works of A.V. Kuza [8], the map of physical-geographical zoning of the USSR edited by N.A. Gvozdetsky (1983) and the Landscape Map of the USSR edited by A.G. Isachenko (1988).

4. Results

Historical waterways are unique in their significance for the whole Russia as namely along them a vast territory was settled and developed economically. In the second half of the eighth century waterways began to form the main trade routes. The formation and functioning of waterways happened due to the formation of a network of support points on them, whose vital activity was associated with subsistence farming and depended entirely on the local resource base and, accordingly, on the landscape structure of the territory. It differed greatly on different segments depending on the zonal-landscape conditions [7].

Management and economic functions, external and internal trade relations were realized along the waterways. Naturally, along these routes there appeared large and small cities, open trade and craft settlements, guard fortresses - "gradki", rural settlements with waterway control function, ordinary rural settlements involved in the life support of communication, i.e. a lot of strong points that provide them with reliable protection and uninterrupted operation, because it was necessary to maintain dies and port facilities in proper condition. The strongholds along the river way were unevenly distributed, but in general they provided control over almost every day transition, such as along the Volkhov-Dnieper route (the famous way "from the Varangians to the Greeks") [9, 10]. The most ancient Russian cities arose on this route: Ladoga, Novgorod the Great, Velikiye Luki, Staraya Russa, Smolensk, Orsha, Lyubech, Vyshgorod, Kiev, Kanev, etc. In the future, these strongholds became the basis (skeleton) of the formation of the settlement structure that has existed until the present time.

The landscape analysis of the spatial distribution of historical cities in the Moscow glaciation area (Moscow, Vladimir, Murom, Starodub (Klyazminsky Gorodok), Yaropolets (Vyazniki), Gorokhovets Zvenigorod, Mozhaisk, Smolensk, Borovsk, etc.) showed that all these cities developed almost in the same way: fortresses were built on low or high valley sands with steep coastal slopes, on arrows with tributaries flowing into the main river or streams with girder valleys. That is, there were chosen landscapes with a spatial pattern, representing in defensive terms an almost ideal place in the whole area. All these cities occupy an ecotone position along the boundaries of two or three or more landscapes with a large set (over 40 species) of tracts with diverse natural properties. It allowed the local settlers to maintain flexible, integrated, diversified subsistence farming. In addition, these boundaries are the boundaries of physiographic regions and even provinces [6].

On the basis of a detailed landscape analysis with the compilation of the corresponding maps of the topography of Smolensk, Veliky Novgorod, Totma, Veliky Ustyug and Solvychegodsk, there was an attempt to explain the problem of transfer of a number of ancient Russian cities from landscape point of view (the problem of "paired cities" in historiography). One of the main reasons for the transfer of the city to a new place is related to the fact that initially, when choosing a place, first of all, defensive-strategic goals were taken into account: control over the surrounding area and waterways, close proximity to the river (sections of high floodplain, first floodplain terrace or on the floodplain with areas limited by steep slopes). However, the possibilities for the territorial growth of these proto-cities were extremely limited, as well as the limited resource base of the surrounding "feeding" landscape.

Studies of the landscape location of the ancient Russian cities of the forest areas of the European Plain, carried out at the level of typological landscape groups, showed that the overwhelming majority of the ancient Russian cities also have an ecotonic position along the boundaries of areas and even provinces. Evaluating the conditions of the location of cities in landscape complexes of different types, 4 groups can be distinguished: 84 cities located in river valleys, 76 - in erosion zones (mostly- in forest-steppe zone), 50- in moraine (about 60% of them are outside the Valdai glaciation zone) and 54 – in sandstone landscapes [6].

At the initial stage (IX-XI centuries) most of the cities (Rurikovo settlement - Veliky Novgorod, Gnezdovo-Smolensk, Beloozero, etc.) were located mainly on gently sloping surfaces of low floodplain or lake terraces, areas of low valley zandras, rarely on the near-valley slopes between the river plains, sometimes on high floodplains emerging from the flood regime.

From the 12th century, cities began to be formed mainly in the river valleys on steep banks on the relatively isolated sites of valley zandras (most often on low valley zandras, which corresponds to the third floodplain terrace), cut from the sides by valleys of small rivers or streams. Posads was located at lower levels: on the second and the first above floodplain terraces. Since almost all the cities were self-sufficient in terms of food or were formed as centers of the surrounding agricultural areas, when choosing the location for the city, preference was given to ecotone territories with a complex landscape structure and rich resource base that allowed the first settlers to conduct a flexible integrated economy. In general, the lands surrounding cities, as a rule, had optimal properties for the tillers of that time: leveled, well drained surfaces, loamy-sandy soils of relatively high trophicity with favorable water-air regime for farming, "warm" habitats with early field readiness for spring field work. The main limiting factor of nature management of that time is also revealed: poor drainage of land. It is also important that cities were developing according to a manor type: a house, outbuildings and an adjacent plot with gardens, gardens, etc. [6].

5. Conclusion

The landscape analysis of the spatial distribution of historical cities showed that, at an early stage in the Old Russian period, places for building cities (protocities) were usually chosen on the most important waterways taking into account their safety, with a relatively simple landscape structure, on low elements of the relief. Later, their expansion or even "transfer" to higher areas (high terraces, valley sandwiches, river moraine plains) with a more complex landscape structure and richer resource base occurs. Spatial differences in the location of cities are due to provincial-zonal conditions and their position in specific landscape conditions, which is often a decisive factor in the formation and development.

6. Acknowledgments

The work is performed under the project of the Russian Foundation for Basic Research, № 19-05-00233.

References

[1] Lappo G M 2012 *Cities of Russia: view of a geographer (*Moscow: New chronograph) p 514
[2] Vladimirov V V, Mikulina E M and Yargina Z N 1986 *City and landscape. Problems, constructive objectives and solutions* (Moscow: Mysl') p 238
[3] Kolbovskiy E Y 1993 *The history and ecology of landscapes of Yaroslavl Volga areas* (Yaroslavl) p 114
[4] Sukmanova E A 2007 Stages of development of urban space and urban landscapes of a large city on the example of Tver, in *Theory, methods and innovation in historical geography* (Saint Petersburg) pp 320-326
[5]. Nizovtsev V A 1996 Landscape prerequisites for the emergence of ancient Russian cities near Moscow, in *A man in the reflection of modern Geography* (Smolensk) pp 93-96
[6] Nizovtsev V A, Galkin Y S, Graves I V and Erman N M 2013 Old Russian city and landscapes *Actual aspects of the socio-economic development of Russia: experience, modern realities and prospects* vol 6 (Velikiy Novgorod) pp 159-169
[7] Nizovtsev V A 1999 Anthropogenic landscape genesis: the subject and goals of investigations *Bulletin of the Moscow University, Geography Series* [in Russian – Vestnik Moskovskogo Unviersiteta, Seriya Geografiya] (Moscow) **1** 26-30
[8] Kuza A V 1989 *Towns of the Ancient Russia (*Moscow: Nauka) p 168
[9] Lebedev G S and Zhvitashvili Y B 2000 *Dgaron Nevo: on the way "from the Varangians to the Greeks". Archaeological and navigational studies of ancient water communications between the Baltic and the Mediterranean* (Saint Petersburg: Nordmedizdat) p 196
[10] Eremeev I I and Dzyuba O F 2010 *Essays on the historical geography of the forest part of the way "from the Varangians to the Greeks". Archaeological and paleogeographic studies between the Western Dvina and Lake Ilmen* (Saint Petersburg: Nestor-History) p 670

IOP Conf. Series: Earth and Environmental Science **350** (2019) 012012 doi:10.1088/1755-1315/350/1/012012

The use of hydropower potential of the rivers of Moscow basin

N A Ozerova

S I Vavilov Institute for the history of science and technology Russian Academy of Sciences, 14, Baltiyskaya street, Moscow, 125315, Russia

E-mail: ozerova-nad@yandex.ru

Abstract. The purpose of this article is to study the changes in the ways of using the energy potential of the rivers of the Moscow River basin in the XVIII-XXI centuries. Analysis of the data of the General Survey suggests that in XVIII century the energy of flowing water was used for several kinds of mills: gristmills, saw mills and fulling mills. Energy of flowing water was applied for oil mills and blacksmith hammers. Analysis of literary data shows that the number of mills reduced at the end of XIX century. Only mills built on large rivers continued to work. At the turn of the 19th and 20th centuries small hydroelectric power stations were built instead of mill dams, and the modern hydropower facilities in Moscow basin are small stations of local meaning only. As a result, the maps showing the location of different kinds of mills in the Moscow river basin in XVIII century were composed for the first time. These maps are unique for this territory.

1. Introduction

Moscow basin occupies 17640 km^2 and unites 912 rivers. The largest left tributaries are rivers Inoch, Iskona, Ruza, Istra, Skhodnya, Pehorka, Gzhelka and Nerskaya; the largest right tributaries are rivers Setun, Pakhra, Ostra, Severka and Kolomenka. In ancient time hydrographic network of the Moscow River fulfilled many important functions. Due to the fact that timber harvesting, charcoal harvesting, billet resin were made in forests of Moscow Province, the Moscow River and its large tributaries above the town served forest rivers [1, 2] and below the town for shipping.

One of the most important resources of Moscow basin rivers is their hydropower potential. In spite the fact that Moscow river and rivers of Moscow basin are not major rivers and do not flow in the highlands, their waters were always used as an energy source.

2. Materials and methods

Different kinds of sources served the base of this work. In the XVIII century these sources were economic notes created during a general survey in the second part of the XVIII century. They are kept in the Russian State Military Historical Archive (Moscow). These notes concern 7 counties of Moscow province: Volokolamsky, Dmitrovsky, Klinsky, Kolomensky, Mozhaisky and Ruzsky [1] within the boundaries of the regional reform in 1775 [3, 4]. The materials of economic notes contain a lot of detailed information, for example, the amount of tax levied on the mill and other objects, which may serve as an indirect indicator for example of the power of the mill. Information obtained from these resources was supplemented with data from economic notes for Moscow province compiled as of 1800 [2].

The economic notes that were written in the second part of the XVIII century gave a fairly complete picture of all kinds of mills in the Moscow River basin. One of the attempts to summarize the data on the mills in the Moscow River's basin contained in the economic notes was made by surveyor F.A. Okhtensky who had a task to create the general plan of Moscow city. In May 1775 Okhtensky composed a description manuscript of Moscow city and Moscow province. According to Okhtensky, on the rivers of Moscow county there were 223 mills: 12 mills made wheat flour, one made pearl barley and 3 mills didn't work [5].

The research made by Okhtensky has become one of the many works that were written in the end of the XVIII century, based on Economy notes and contained data about sights and economy of Moscow province. But only in his research there was information about the number of water mills. Now these data allow us to know about the scale of economic use of the energy resource of the rivers in the Moscow county. Unfortunately, it is very difficult to determine the location and the exact number of mills situated in the Moscow river basin on the lists made by Okhtensky. The reason is that Moscow county includes the river basins of Klyasma and Lopasnya. It is possible that such description manuscripts and lists were written for other counties of Moscow province, but they have not been published yet. For example, the work of Okhtensky was prepared for publishing, but in the XVIII century it didn't come out. It was published only in 1997 in the collection prepared by S.S. Ilizarov [6].

So the work of Okhtensky is of great interest, but it doesn't give a complete picture of the mills in the basin of the Moscow River. That is why the economic notes are the main resource about these data for the XVIII century.

The information about the river use in the XIX-XX centuries and nowadays is taken from books, articles and some old maps. These data are not complete because in the XIX and XX centuries there were not such large-scale studies as in the XVIII century; so information is scattered. But it is not a secret that mills played a large role in economic life, not only in Russia, but also in European countries in the XIX and XX centuries [7]. The analysis of these data allows changes in economic use of the rivers of Moscow river basin.

3. The usage of hydropower potential of the Moscow river basin in the XVIII-XXI centuries

According to materials of the general survey [1, 2], about 472 gristmills were built on Moscow river and its tributaries – in the period of 1760-1800; however 5 of them were out of work. Gristmills were almost evenly distributed about Moscow river basin. The exclusion is the territory of Meshcherskaya lowland with rivers running there: Gzhelka, Nerskaya and other right tributaries of Moscow River because this territory differs by flat terrain inconvenient for dam construction.

The majority of flour mills (385) are noted as gristmills. 44 chaff mills produced rye flour and 41 grist (mills) - wheat flour. One grist mill on the Samynka river was reconstructed for the chaff mill and one more on the Ozerna river was used for threshing rye. It is remarkable that the majority of chaff mills and almost all grist mills were built in the nearest suburbs of Moscow (on the Khimka river, Khodynka river, Gorodenka river, in the basins of the Skhodnya, Setun, Pekhorka, Yauza rivers and on the Desna river (left tributary of the river Pakhra).

The majority of gristmills functioned all year long except the time of spring flood. The mills, which worked temporarily (during spring flood and other floods), were built in the upper reaches of the river and on the streams less than 10 kilometers long. There were 84 such mills. Gristmills in the basin of the Moscow river were "plantar" (water was running under mill's wheel moving it) or "bulk" (water was falling down at the mill's wheel turning it). The type of the mill's construction was mentioned very seldom in the general survey; however the period of active work was indicated always.

An average number of sets (grain processing and flour making machinery) were two on water mills of the Moscow river basin, the minimal number was one. Record breakers by the number of sets were Bykovka mill in Pekhorka basin, which had 18 sets, and the mill near Nikolskoe (Strelkovo) village with 12 sets on the Pakhra river. Not all sets were functioning at the same time on mills. Sometimes there were data that a mill, which had several sets (two or three, as a rule), was working with "change" using one set. So we can conclude that historical notes about the number of sets at mills may not reflect the efficiency of the use of hydropower potential of rivers.

Probably we could get a more exact picture of the efficiency of the use of hydropower potential of rivers in the XVIII century analyzing notes about tax amounts, paid to the treasury or to landowners. However if the mill was small and used for personal needs of the owner only (landowner or state peasant), they were exempted from tax. Just built mills were free of tax for five years. As a result, economic notes contained not all data about incomes from mills.

The data about tax amounts were represented in the earlier version of economic notes, which united seven counties of Moscow province only. Even based on these incomplete notes we can conclude that gristmills of Moscow province situated at Yauza, Pekhorka and Pakhra rivers were the most profitable. For example, the mill near Taynitskoe village brought 400 rubles per year to treasury. The landowner got 500 rubles per year from the mill Bykovka near Ostrovtsy village and tax amount from all mills in Pekhorka basin gave about 1800 rubles per year to landowners. So rented out gristmills generated great income for landowners and treasury. Judging by the sums of money, the mills near Moscow worked constantly and made the most of the entire river water unlike mills situated farther from the capital.

Besides gristmills there were 18 saw mills in the basin of the Moscow river. Almost all of them were situated in the upper reaches of the river, at the Ruza and Istra rivers – on the territories where there were a lot of forests. On average they had 2 saw frames and 4 saw frames maximum. Tax amounts from saw mills were paid rarely. The maximal sum of the money sent to treasury was about 20 rubles per year; landowners got about 40 rubles.

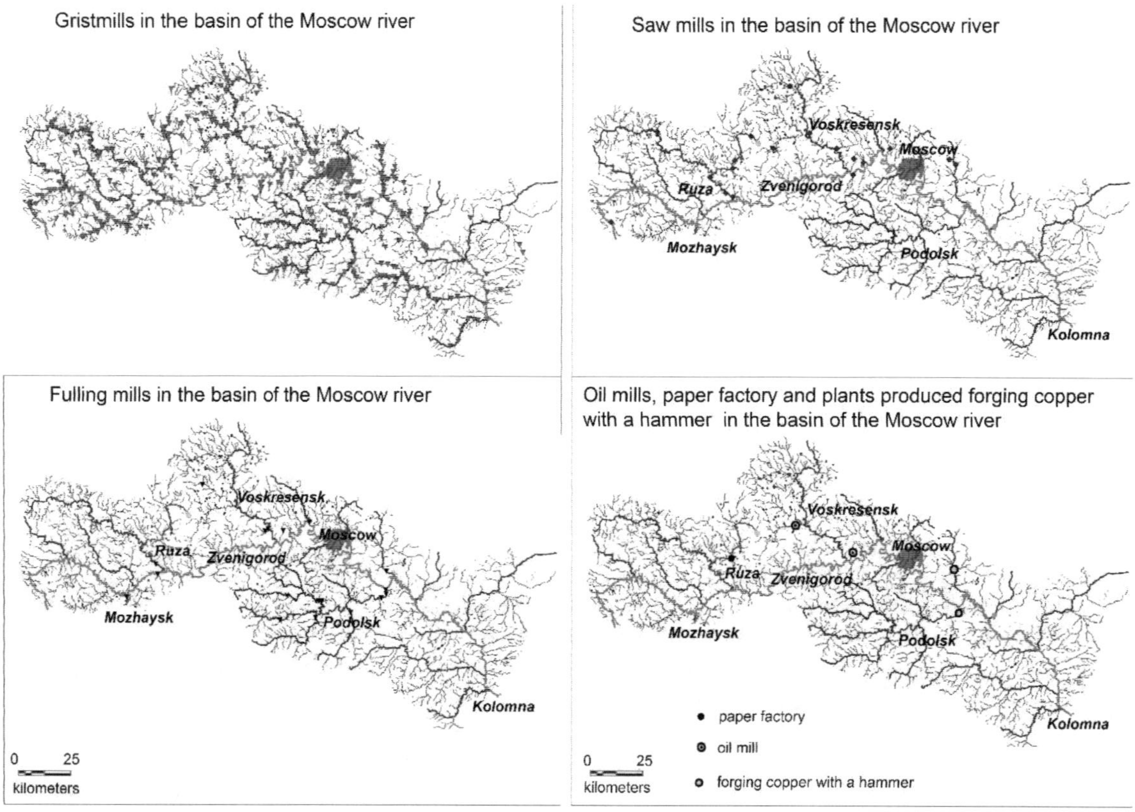

Figure 1. Mills in the basin of the Moscow river in the end of the 18th century

Nine of twenty fulling mills were built in the basin of the Pakhra river. Average 4 mortar stupa (shoe mill frame) were installed at each of them; the latter were designed for the production of woolen cloth.

The maximal known number of stupa was 19. Landowners got 400 rubles per year from rent of some fulling mills.

Three oil mills were situated at rivers of Moscow river basin, one of them did not function; one paper factory worked. One copper sheet plant in the village Green Sloboda operated at the lower mill at the mouth of the river Pakhra. The plant produced forging copper with a hammer, which moved with the help of running water. The same plant functioned in Bedrino village (the basin of the Pekhorka river) [2, pt.1, sh. 40-1].

Mills retained the main way of the use of hydropower potential of the rivers of Moscow basin till the end of the XIX century. The study of the map made by F.F. Shubert demonstrated that in comparison with the second half of the XVIII century the number of mills reduced in the middle of the XIX century; only the most cost-effective mills working most of the year remained till the end of the XIX century.

The era of the construction of small hydroelectric power plants has started since the end of the XIX century; the heyday of the era has come between 1920-1940 years. Many of hydroelectric power plants were created instead of old mill's dams at the same place. The first such station in the basin of the Moscow river started working in 1893. It was built at the Pakhra river in Leninsky Gorky suburbs in Novlenskoye village. In the 1920s it was reconstructed according to GOELRO plan. Producing electric power for V.I. Lenin residence and several neighboring villages, it became the symbol of GOELRO plan. "Light bulb of Ilyich" had been associated with this hydroelectric power plant for a long time. In 1959 this plant stopped working but the dam at the Pakhra river exists nowadays functioning as a regulating water system.

In the 1920s one more dam with "electric mill" was built at the Zhuravenka river – tributary of the Bitza river between villages Bolotnikovoe and Zagorje. In 1948, Yakshinskaya HPP and Krasnogorsk HPP were built on the Ruza river; each of them had a power of 75 kilowatts [8]. In January 1953 the Ruza river was blocked by a dam of Gorbovskaya hydroelectric station near village Gorbovo. All these hydropower facilities appeared instead of reconstucted old mill's dams.

During construction of Moscow Canal in 1932-1937 Skhodnaya waterworks were at the Skhodnya river. Skhodnya hydroelectric power station with adjacent straightening of the Skhodnya river became its main construction. The main feature of this hydroelectric power station was a pipeline constructed of two wooden pipes 180 meters length and with a diameter of 5.4 meters. They serve for the flow of water to the turbines of the station from the pressure pool.

In 1935-1989 several reservoirs were constructed at Moscow river and its tributaries Istra, Ruza and Ozerna. The hydroelectric power station of small power was constructed at Istrinskoye, Mozhayskoye, Ruzskoye and Verkhneruzskoye waterworks of these reservoirs. After starting of Vazuza hydropower system at full capacity in 1997, the waterwork of Ruza reservoir was strengthened with an additional hydroelectric unit. All the above-mentioned hydroelectric power stations generate little electricity and service hydraulic structures only [9].

4. Conclusion

Summarizing we can conclude that at the second half of the XVIII century hydropower potential of the Moscow river was used effectively. To the greatest extent, the energy of the running water was used for gristmills in Moscow suburbs. This can be explained by the fact that Moscow was the place where the grain was delivered from southern provinces. The location of saw mills is evidently was close to territories covered with forests and that of fulling mills was determined by availability of labor and raw material (wool), which was delivered from other regions of Russia. All other enterprises, which used the energy of flowing water for driving mechanisms, were in large and rich landowners' possessions. Generally, at the second half of the XVIII century the use of the energy of flowing water promoted the development of processing and food industry and served as an important source of income.

In the XIX century the number of mills reduced significantly; only the most cost-effective mills working constantly remained. The era of mass building of mills was over. The end of the XIX century and the beginning of the XX century are characterized by replacing of mills by small hydroelectric power plants.

However, the economic value of these plants fell as a result of the development of electrical networks and the possibility of electricity supply from other areas. These hydroelectric power plants were closed. The plant in Bitsa was destroyed completely. Ruins of the dams of Yakshinskaya HPP and Krasnogorsk HPP remained as a historical landscape monument [10, 11]. HPP Gorbovskaya stopped functioning, but the dam has preserved till nowadays. All hydropower facilities in the basin of the Moscow river are small facilities of only local meaning.

It seems very interesting that in pre-industrial time in European countries the number and the ratio of different types of mills were the same as in Russia. So Mr. Louis C. Hunter says that the "pre-industrial water-mills, supplemented in some regions by others propelled by wind, are numbered by the thousands and possibly by the tens of thousands. By far the greater number are gristmills, engaged at appropriate seasons in the reduction of the farmer's grain into meal, often ticking on unattended much of the time during the daytime hours. Much smaller numbers of water-mills are employed in the filling of hand-loomed cloth, the raising of water for irrigation, the crushing of olives, and, much less frequently, the carding of wool and sawing of lumber" [7].

References

[1] Russian State Military Historical Archive **846** 16 18859 Parts 1-10
[2] Russian State Military Historical Archive **846** 16 18861 Parts 1-8
[3] Ilushina T V 2007 *S I Vavilov Institute for the history of science and technology. Annual scientific conference* Moscow 422-425
[4] Kusov V S 2004 *The lands of Moscow region in 18th century* vol **1** (Moscow: Moskovia)
[5] Okhtensky F A 1997 *Moscow in the descriptions of XVIII century* (Moscow) pp 121-154
[6] Ilizarov S S 1997 *Moscow in the descriptions of XVIII century* (Moscow) pp 110-120
[7] Hunter L C 1967 *Technology and Culture* **8(4)** 446 DOI: 10.2307/3102113
[8] Yaytsova T A Shakhovskoy region, Available at: http://www.vrigash.ru/content/region/shakhovskaya.html
[9] Khramenkov S V, Volkov V Z, Gorban O V, Kalashnikova E G and Fomushkin V P 1999 *From source to Moscow* (Moscow)
[10] Saethornorsdottir A D and Hall C M 2018 *Sustainability* **10** 7 DOI: 10.3390/su10072315
[11] Vasquez V, Valdebenito G and Ogalde M 2018 *Revista de Urbanismo* **38** DOI: 10.5354/0717-5051.2018.48831

Introduction of sosnowsky's hogweed as a cause of landscape transformation

M G Krivosheina[1] and N A Ozerova[2,3]

[1]A N Severtsev Institute of Ecology and Evolution Russian Academy of Sciences, 33, Leninsky prospect, Moscow, 119071, Russia
[2]S I Vavilov Institute for the history of science and technology, Russian Academy of Sciences, 14, Baltiyskaya street, Moscow, 125315, Russia
[3]State University of Land Use Planning, 15, Kazakova street, Moscow, 105064, Russia

E-mail: kriv226@rambler.ru; ozerova-nad@yandex.ru

Abstract. Archival documents on the introduction of sosnowskyi's hogweed in the USSR in the Achieve of the Russian Academy of Sciences and the data from published resources (books and articles) were investigated. The role of scientists and different scientific as well as political organizations in this event was demonstrated. Four stages of introduction were highlighted: local introduction from the end of the 1940s till the end of the 1950s; mass introduction from the end of the 1950s till the middle of the 1970s. The completion of introduction was from the middle of the 1970s till the end of the 1980s; uncontrolled invasion since the 1990s. The maps showing stages of introduction were composed. These maps reflect the territories with transformed landscapes as a result of *H. sosnowskyi's* introduction and invasion.

1. Introduction
In the 1940-1980s the science in the USSR was an integral part of policies of the government. The tasks of *production* to increase the planted area and yields, as well as the introduction of new crops, were addressed to scientists [1]. On the 5th of October, 1931, according to the Decree of the Presidium of the Academy of Sciences of the USSR (AS USSR), the Polar-Alpine Botanical garden was founded in Kirovsk to fulfill these tasks. Later in 1938 a special "fodder" group was created on Kola base of AS USSR, and it was planned to investigate Fabaceae and other crops [2]. The plant which was determined as *Heracleum pubescens* appeared in the garden at the beginning of 1939 [3] and became one of the objects for study. *H. sosnowskyi* was described by I.P. Mandenova in 1944 and later the scientists realized that they dealt with *H. sosnowskyi*, not with *H. pubescens*, all that time.

2. Materials and methods
Data on the distribution of *H. sosnowskyi* in Russia were obtained by analyzing publications on the distribution of the plant and various aspects of its economic use. The main sources of information were publications on fodder and silage plants, including *H. sosnowskyi* [4].

Although these data cannot be considered as complete, it was possible to establish the main centers of cultivation of *H. sosnowskyi* on the territory of the USSR. Data from published resources (books and articles) were added from archival documents of the Archive of the Russian Academy of Sciences. These documents contain a lot of unique facts about the history of introduction of sosnowskyi's hogweed in the USSR.

3. Stages of introduction

Acute feed shortage was traced in the beginning of the Second World War in the USSR. In 1942 silage produced from *H. "Pubescence"* was offered to cows for the first time. The results seemed to be good. However, this result was not included in the official annual plan of Polar-Alpine Botanical garden and as a result nothing was written about it in reports of 1942-1945 [2].

Silage harvesting in collective and state farms of Murmansk region was not enough in post-war years. In 1946-1947 the scientists from the garden started the work on the choice and selection of perennial silage crops for Murmansk region; the first experiment was set up by A.A. Marchenko and L.Ya. Avrorina. This activity was supported and approved by the Secretary of Murmansk regional Committee of the CPSU (b), A.M. Kutyrev [5, p. 24]. As a result, twelve species of perennial crops from the introductory nursery of the garden were tested and three of them only were chosen for silage, *H. "pubescens"*, and *H. dissectum* was among them.

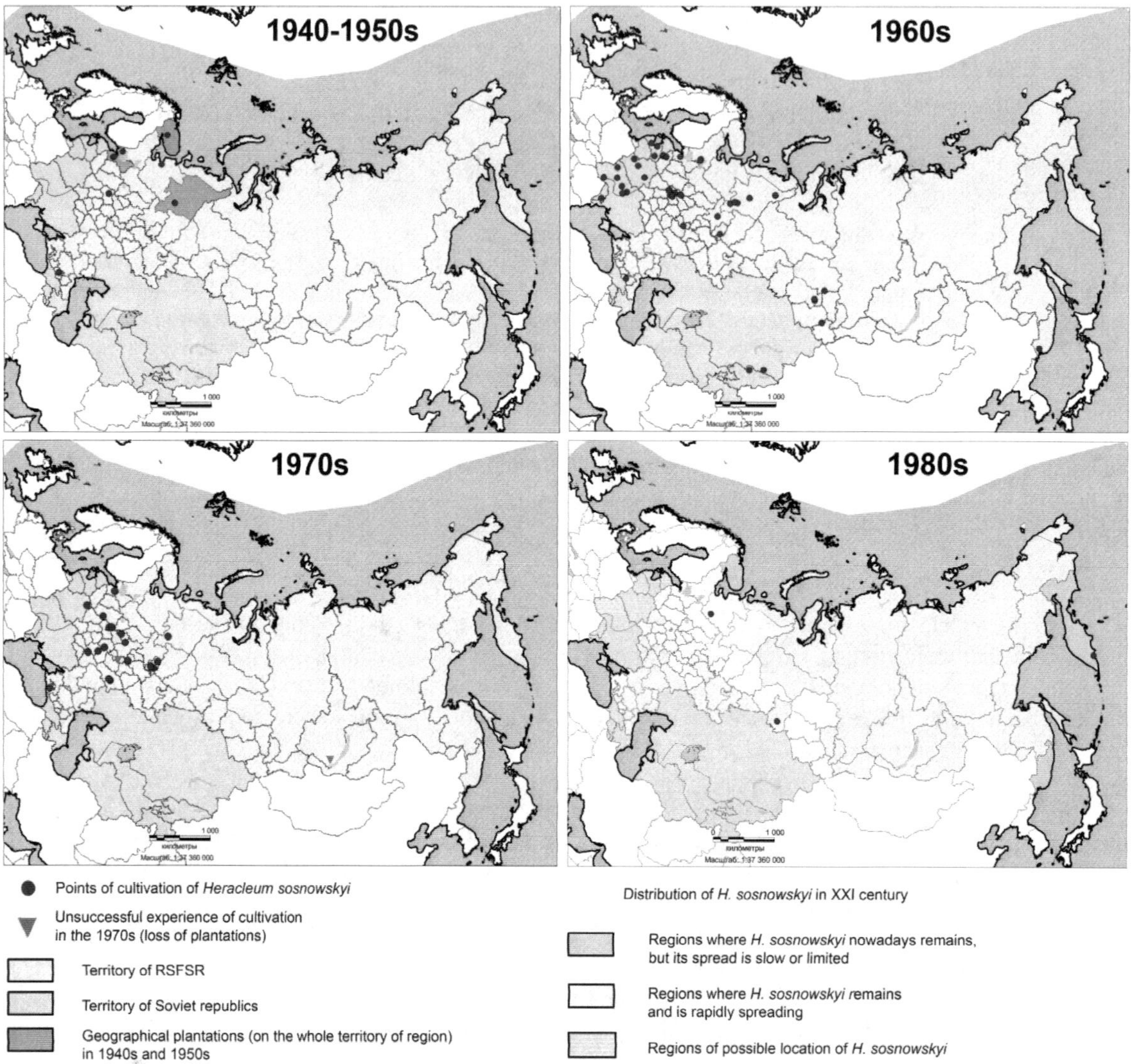

Figure 1. Stages of introduction of sosnowskyi's hogweed in the former USSR

In 1948 A. A. Marchenko continued the study on biological and economic qualities of these two *Heracleum* species. It was written in the annual report that "hogweeds do not contain harmful to livestock alkaloid substances and green storage and silage produced from hogweeds are eaten by animals with pleasure and increase milk yield being surely harmless to animals" [6, p. 534]. In his further reports A.A. Marchenko noted high frost resistance of the plants, high yields (to 80-100 tons/hectare) and good eatability by livestock [5]. According to the summary of the extended academic council of Kola base of AS USSR, the decision to carry out experienced crops of *Heracleum* species in collective and state farms of Murmansk region in 1949 was taken.

In order to guarantee the fulfillment of this decision, A.A. Marchenko went to expedition to the Kabardino-Balkar Republic in August, 1949, where he made natural observations on *H. "pubescens"*. The explorer picked up and stored 150 kg seeds during his expedition in Nalchik suburbs. He notes that the plant may cause burns, but he makes a conclusion that possibly this feature may change and depend on various climatic conditions, and "possibly *Heracleum*, after introduction... from south to north..., reduces an ability to accumulate essential oils" [5, p. 33].

In 1951 *H. "pubescens"* was included in the master plan for pilot works on fodder investigations in Murmansk region for 1952-1955 years. The Botanical Institute of the Academy of Sciences of the USSR (BIN AS USSR) became an appoint curator on the problem of foraging. In 1953 the Department of General Biology of AS USSR (DGB) assigned BIN AS to be a curator of introduction and acclimatization of plants; the Head of the Botanical garden in BIN AS, S. Ya. Sokolov, was to take personal responsibility for the results of the project. The Head of Department of Introduction of nursery-garden BIN AS, V. S. Sokolov, became the curator of the study of silage cultures in BIN AS in 1953.

At the turn of 1952-1953 the mistake of wrong definition of *H. pubescens* became clear. It is still unknown who was the scientist and from what institute was the first to establish correct name for *Heracleum* species, but in 1953 exactly seeds of *H. sosnowskyi* were sent from Kirovsk to Leningrad and Syktyvkar.

BIN AS got seeds of *H. sosnowskyi* for investigation in 1953. The work under supervision of V.S. Sokolov was finished in 1955. In 1954 the plant was included in the annual plan of BIN AS for a pilot industrial test and introduction. In 1955 *H. sosnowskyi* was introduced into broad industrial crops in 17 collective and state farms of Leningrad region. During cultivating of the plant in this region, the feature of causing burns was registered, but the scientists hoped that it would be eliminated during selective work [4].

In 1953 seeds of *H. sosnowskyi* were sent from Kirovsk to Syktyvkar to Institute of Biology of Komi (IB Komi AS USSR) in which complex investigations on biology, agricultural technology and nutritional value of silage from *H. sosnowskyi* started in the laboratory of plant introduction under the supervision of K.A. Moiseev [1] and continued till 1957. As a result experienced geographic crops through all territory of Komi Republic were made in 1957. Since 1959 the process of obligatory introduction of *H. sosnowskyi* to the culture started in collective and state farms of the republic and the crop was distributed from Inta to the southern border of Komi [4].

There were more centers dealing with *H. sosnowskyi* in the USSR. One of the most significant was the Local Botanical Garden in Nalchik in which "large cowberry bushes" were studied under supervision of Yu.I. Kos. Besides crops of *H.sosnowskyi* were planted on the territories of the B.R. Williams Institute of Feed in Lobnya of Moscow region [4].

In 1956 the All-Union meeting on the introduction of new useful plants to culture was held, and Brochure containing a list of more than 500 useful plants was published; *H. sosnowskyi* was included in this list [7].

The results of investigation of new silage plants were discussed on meetings of the Bureau of DGB AS USSR in 1962, 1965 and 1968 as well as on the First - Fourth All-Union Symposia on new silage plants, in the Ministry of Agriculture of RSFSR and in leading organizations of the Komi Republic and got positive rating everywhere.

Many Party and Government regulations were adopted to promote these decisions. On the 19th of June, 1962, the First secretary of the CPSU Central Committee, N.S. Khrushchev, sent a letter to the Institute of Biology of Komi in which he asked to send him reference about their work on growing silage crops. This letter which was caused by the need to strengthen the country's food security during the cold war became new impetus in introduction of *Heracleum*.

The first mass mailing of *Heracleum* seeds by the Institute of Biology of Komi happened at the beginning of the 1960s. The plant was cultivated from Pskov region to Sakhalin; farmers continued to multiply this plant everywhere. For example more than 10 tons of seeds were produced in the state farm "Lidino" of Ruza District of Moscow region in 1965-1969; the seeds were distributed among 55 farms of Moscow region and 67 farms of other regions of RSFSR [4]. Since 1961, the crop was cultivated in Belorussia and Ukraine, at the end of the 1960s - in Moldavia and other republics including Chui Valley of Kyrgyzia.

In 1970 P.P. Vavilov, the former director of IB Komi AS USSR, moved to work in Timiryazev Agricultural Academy (Moscow) and the center on investigation and introduction of *Heracleum* moved to Moscow too. Collective and state farms, experimental fields of educational and production facilities became testing grounds for a new crop all over the USSR. One of experimental stations in Udmurtia produced and distributed 1694 kg of *Heracleum* seeds among farms of the republic in 1973-1976. The harvest in only one state farm "Pravda" reached 4 centner/hectare, seeds production from 1 hectare was enough to plant 25 hectares [4].

BIN AS continued the study of *H. sosnowskyi* at this time too. One of the researchers, I.F. Satsyperova, discovered that it was impossible to get the form of *Heracleum* devoid of furocoumarins and not causing burns. Moreover, she noted that furocoumarins retained in silage and were found in milk and meat and reduced their quality [8]. Other scientists pointed that another insufficient character of the plant - *H. sosnowskyi* was monocarpic, giving seeds once during the life cycle. I.F. Satsyperova tried to offer another species for silage – *Heracleum ponticum*.

In fairness it should be noted that not only *H. sosnowskyi* but other *Heracleum* species were studied also in many institutes including IB Komi beginning from the 1970s. On the rise of these investigations one of the researchers – E.M. Shumova – defended thesis. She managed to get not a burning form of another giant hogweed, *Heracleum manteggazzianum*. However, this result was ignored on some reason and remained without attention [4].

The Board of the Ministry of Agriculture devoted to introduction of new feed crops in collective and state farms production was established in December 1975. The resolution was to organize wide production tests of *H. sosnowskyi* in different geographical zones of the USSR [1].

The second half of the 1980s is characterized by the reduction of the investigations on *H. sosnowskyi* as agricultural crop. However, the Secretary of the Department of General Biology of the Academy of Sciences of the USSR, M.S. Gilarov, mentioned *H. sosnowskyi* in his 1980-1984 report as the most significant result in biological sciences and noted that this plant gave production to 1000 centner/hectare and was cultivated to the Arctic zone.

4. Conclusion

So, we can distinguish the following periods and stages of the introduction of *H. sosnowskyi*:

1) Local introduction from the end of the 1940s till the end of the 1950s. Growing and study of the plant was accumulated in 5 scientific organizations: Polar-Alpine Botanical garden in Kirovsk, Botanical Institute of the Academy of Sciences of the USSR in Leningrad (St.-Petersburg), Institute of Biology of Komi in Syktyvkar, Local Botanical Garden in Nalchik and scientific institutes of Moscow region.

2) Mass introduction from the end of the 1950s till the middle of the 1970s. The great significance of the distribution of the plant was acquired by experimental agricultural stations, collective and state farms, educational institutions.

3) Completion of introduction from the middle of the 1970s till the end of the 1980s. At this time the danger of the plant, its negative impact on agricultural products became evident, but the plant was still cultivating in some farms.

4) In the 1990s, a new stage began in the distribution and transformation of hogweed into an aggressive invasive species [9–11]. Because of the decrease in agricultural production, this species

quickly began self-distribution, first near the centers of cultivation and then at a considerable distance from them. The plant is more confined to synanthropic habitats and forms monospecies thickets over large areas, has become widespread on the territory of the former USSR, destroying the habitual appearance of landscapes and transforming natural ecosystems.

5. Acknowledgments
This study was supported by the Russian Science Foundation (project no. 16-14-10323).

References

[1] Mironova N P and Koroleva V I 2017 *Historical Journal: scientific studies* **6** 178–191
[2] Archive RAS **188** 3 274
[3] Avrorin N A 1956 *Introduction of plants to Polar North* (Moscow-Leningrad: Publishing house of Academy of Sciences of USSR (Nauka)) pp 1–286
[4] Ozerova N A and Krivosheina M G 2018 *Russian Journal of Biological Invasions* **9(2)** 155–162 DOI: 10.1134 / S2075111718020091
[5] Archive RAS **188** 1(1949) 19
[6] Archive RAS **188** 2(1941-1950) 274
[7] Sokolov V S (ed) *New useful plants* 1956 (Moscow-Leningrad Publishing house of Academy of Sciences of USSR (Nauka)) pp 1–70
[8] Satsyperova I F 1984 *Hogweeds of the USSR are new fodder plants* (Lenindrad: Nauka) pp 1-224
[9] Silantyeva M M, Shmakov A I and Smirnov S V 2005 *Turczaninowia* **8(4)** 36–40
[10] Panasenko N N 2017 *Russian journal of biological invasions* **10(2)** 95-106 DOI: 10.1134 / S2075111717030110
[11] Ozerova N A, Shirokova V A, Krivosheina M G and Petrosyan V G 2017 *Russian Journal of Biological Invasions* **8(4)** 327-346 DOI: 10.1134 / S2075111717040075

Historical cartography of Muscovy: symbols and emblems

E V Pchelov

Institute for the History of Science and Technology n. a. S.I. Vavilov, RAS, 14, Baltiyskaya str., Moscow, 125315, Russia

E-mail: evg-pchelov@yandex.ru

Abstract. The article is devoted to the historical cartography of Moscow kingdom of the 16- 17th centuries and symbolic emblematic designation of Muscovy and adjacent territories. Four types of heraldic emblems were revealed. Some of them have various own versions. The rider was used by cartographers only in the 16th century. In the earliest version, the rider had a bow. The double-headed eagle was used since 1610. The original emblem of the beginning of the 16th century was an image of a lion and a dragon. The open gate was used on some maps since the second half of the 16th century until the end of the 17th century. The emblem is connected with ancient ideas of the northern peoples who were locked up by Alexander the Great with the help of the gates of the northern peoples identified with the peoples of Gog and Magog in the Christian tradition.

1. Introduction

Historical and cartographic research is one of the important directions in world and national historiography. Historical maps can provide a lot of important information on the history of geographical representations and geographical development, serve as a source for symbolic topography, semantics of conventional graphic or color symbols, images ps, including emblems.

The visual symbolization of territories on the maps could be carried out in several ways: images (real or conventional) of rulers, local people and / or animals, sketches illustrating rituals, emblems. In the latter case, historical cartography was closely connected with imaginary heraldry, the symbolic designation of those phenomena and persons who did not have these symbols, but should have, according to compilers and "users" of the maps [1]. In addition to emblems, the visual symbolization could be carried out with the help of vexillological images. This tradition is characteristic of portulans of the XIV – XVI centuries. [2].

The symbolic-emblematic representation of space on geographical maps with reference to the cartographic "Rossica" is understudied [3]. The article aims to identify, systematize and define the semantics of the emblems which were used on European maps of the XVI-XVII centuries to designate the Russian state, Muscovy. Special attention is paid to the emblems that reflect ethno-geographical representations of medieval and Renaissance geographic literature.

2. Materials and methods

The materials used for the study are maps and plans published in Western Europe from 1516 (Martin Waldsemüller map) to the end of the XVII century. Since the beginning of the XVII century, many maps were compiled by Dutch travelers and cartographers. Not all the maps featured heraldic emblems of Russia.

Reports and catalogs of maps of the XVI – XVII centuries published in pre-revolutionary and Soviet historiography [4, 5] are still improtant. Historical and geographical science contains a general source study review of European cartography in Muscovy [6]. The following methods were used: analysis of graphic and cartographic sources; comparison of emblems on maps and in armorial places, on material monuments, in geographical descriptions.

3. The study of the symbols and emblems on the maps of Muscovy

Analysis of the heraldic emblems of geographical maps of the Russian state of the XVI-XVII centuries identified four types of heraldic images designating Muscovy.

The first type is an armed rider known as the emblem of the great princes of Moscow since the beginning of the XV century. The rider kills a dragon with a spear was perceived as a symbol of the great sovereign, the king who defeats enemies. The figure of the rider corresponded to the title "sovereign" [7]. The rider was present on the seals of the great princes since the end of the XV century, and on the coins of the Moscow kingdom since the 1530s. There are two types of images of the rider - with a spear and with a sword (saber). This emblem was well known to foreigners who visited Muscovy. It was perceived as a symbol of Moscovy. However, on geographic maps the rider is known only in three cases. These are maps created by O. Magnus, SS. von Herberstein who personally visited Moscow, and brothers van Deutekom. They all relate to the time between 1539s and 1560s.

Ohe map created by O. Magnus, the rider is depicted with an arrow and a bow which does not find analogues in the Russian visual material. In our opinion, the comparison of Russia with Scythia and Sarmatia (typical of the geographical literature of the Middle Ages and the Renaissance) could have affected this image. This view could also affect the image of rider's armor and weapons on the seals of the Moscow Great Princes beginning with Ivan the Third. On the maps attached to various editions of Herbershtein's "Notes", a shield with a horseman accompanies the image of the Great Prince sitting on the throne. Thus, the affiliation of this emblem to the sovereign rathher than the state is reflected in cartographic sources.

Unlike the maps by O. Magnus and S. Herberstein, on the map by the van Deutekom brothers, the image of the dragonslayer-horseman in the coat of arms is presented without any connection with the figure of the sovereign. The van Deutekom's map is completely devoid of images of rulers, although it has images of local residents and animals that sometimes make up whole scenes. Thus, it demonstrates the next stage in the cartographic understanding of the emblem of the Moscow horseman. Apparently, by the middle of the XVI century, the emblem with the dragon-fighter was perceived not so much as the personal emblem of the Moscow sovereign, but as the emblem of the country subject to him – the Moscow Grand Princedom. The prevalence of the horseman on European maps in the first half of the XVI century, before the double-headed eagle became the emblem of Russia on them, can be explained by the fact that it was the image of the horseman that was placed on the front of the seals of the Russian monarchs, and only in 1561 did the horseman and the eagle combine into a single heraldic emblem. Apparently, the decisive role in the formation of the tradition of the image of the dragon fighter on the maps was played precisely by the work of S. Herberstein, which became very popular in Europe and was repeatedly reprinted during the 16th century.

The second type of Russian emblems is the image of a two-headed eagle which symbolized the Russian state (rather than the sovereign). For the first time, this image was used on the Sigismund Plan of Moscow in 1610. Due to the popularity of the Dutch map by H. Gerritsz in 1613, the double-headed eagle with a rider on his chest became a heraldic designation of the Moscow kingdom in the XVII century. His images are present in Dutch cartography, including maps and atlases by I. Massa and the Blaeu dynasty. There are 8 types of images of this emblem. They correspond to the images of a two-headed eagle used in the official Russian heraldry throughout the XVII century. [8]. Other symbols are also used (e.g., sables that served as one of the main Russian export items).

It is important to note the fact that some elements in the image of the Russian double-headed eagle by European cartographers carried a certain modernization. For instance, on the map of Russia by G. Gerrits (1613) and on the plan of Moscow by M. Merian (1638), the wings of the double-headed eagle are raised, while on the official sphragistic images, the wings of the Russian eagle were still down (they rose only from 1654-1667). Such modernization can be explained by the influence of the European heraldic tradition (primarily, the state heraldry of the Holy Roman Empire), and it finds analogues in some works of decorative and applied art of European masters who worked in Russia. For example, the saadak of Tsar Mikhail Fedorovich, made in 1627, is also decorated with enamel images of double-headed eagles, but with wings raised according to the European model.

Another detail is the crown topping the cartouches with the coats of arms. On the maps by G. Gerrits and I. Massa, this crown has the appearance of an imperial one with two hemispheres, again similar to the crown of the emperor of the Holy Roman Empire. Meanwhile, in reality, the crowns of the Russian tsars were wreaths, and only in the era of Peter I did a European-type crown appear in Russia. Giving the Russian emblem imperial elements on European maps testifies to the perception of Russia – or rather, the Moscow kingdom – as an empire state, that is, equal in status to the Habsburg empire.

At the same time, maps and plans by those people who personally visited Muscovy demonstrate greater accuracy in the image of emblems. Such, for example, is the coat of arms of Russia on the plan of Moscow by A. Oleariy, who visited Moscow twice in the mid-1630s and made an extremely detailed and accurate description of it.

The third type of heraldic emblems is a very original image. It was found only on one map of the beginning of the XVI century. It was built by M. Waldseemüller in 1516 [9]. This emblem is placed next to the conventional image of the “Great pronce and emperor of Russia and Muscovy” who seats on a throne and represents a lion and a dragon opposing each other. Some German emblems of the first half of the XVI century can be considered as analogues.

The same heraldic image in the shield flanks a travel arch of the Borovitskaya Tower of Moscow Kremlin. It was built in 1490 [10]. The semantics of this image is unclear. It goes back to one of the types of emblems on the personal seals of Ivan the Third depicting a lion tearing a snake. This composition became known in Europe thanks to foreigners visiting Mowcow Kremlin at the turn of the XV and XVI centuries. It was not included in the most sublime official emblems of Moscow kingdom, but it was reflected in Western European heraldic and cartographic materials.

The fourth type of emblems was found on Dutch maps. First of all, this is a unique map built by brothers van Deutekom in 1562, i.e. after the publication of the map by Anthony Jenkinson. In the modern edition of the map compiled by van Deutekom, this emblem was incorrectly attributed to Veliky Novgorod [11, p. 28].

Secondly, these are maps compiled by the successors of C. Vischer published since the beginning of the 1680s. The emblem is an image of an open silver or gold gate on a black field. It discovers numerous analogues in the German emblems of the XVI-XVII centuries. This heraldic tradition goes back to the "Chronik des Konstanzer Koncils" by Ulrich von Richental (the first half of the XV century). In subsequent emblems, it was attributed as an emblem of Muscovy (or Russia).

As the emblem of Russia, it was also described in heraldic treatises, including Polish ones. This information, thanks to translators of the Ambassadorial order, at the end of the XVII century came to Russia and was reflected in one of the heraldic writings.

In our opinion, the semantics of this fictional coat of arms is associated with ancient ideas about northern peoples Gog and Magog which were locked by the gates by Alexander the Great (or locked by a wall). These peoples correlated with many real peoples of antiquity and the Middle Ages, including the Scythians, Sarmatians and Musagets [12]. The contingency of these peoples with the conditional prince (tsar) of Rosh from the Bible made it possible to compare them with Russian people. The open gates meant that these peoples came out of their forced “imprisonment”.

On the map compiled by Van Deutekom, the coat of arms with an open gate is accompanied by the inscription “Russia”, territorially correlating with the northern lands of the Russian state. The heraldic image of a rider who kills a dragon with a spear is also present on this map, but it refers to Moscow. In other words, the first emblem is a visual symbol of the entire country, and the second one is only of Moscow land.

The fact that the map compiled by Van Deutekom was based on the emblematic tradition dating back to the "Chronicle" by W. von Richental is confirmed by the fact that the southern Dnieper region or the right-bank Ukraine is marked with a coat of arms consisting of with three crosses in the upper field and a two-headed eagle in the bottom field. Starting with the work by W. von Richental, this emblem was symbolized "Chervonnaya Rus" (Roth Reussen) and used in northern European. Its origin is not clear. It is as fantastical as the "Russian" coat of arms with open gates. The Golden horde is marked with an emblem containing the tutor of Juchids.

One more emblem indirectly connected with Muscovy should be attributed to fantasy heraldry on geographical maps. This is the emblem of Tartaria, which is an image of a black owl in a golden field. One of its earliest images was recorded in the "World Geography" by P. Duval, which first edition was published in 1658. The presence of an owl in this coat of arms is explained by the legends of the Tatars, but does not find confirmation in the historical and ethnographic material. Nevertheless, the coat of arms of Tartaria with an owl gained great popularity as a sign of the representation of Tartaria in European culture of the late XVII - early XVIII century. It is present as an image on one of the flags of Tartaria in numerous naval reference books of the turn of the XVII - XVIII centuries. It is found on geographical maps of the beginning of the XVIII century. This is the emblem of Tartaria on the maps by I.B. Homann, created no later than 1723, and N. de Fer (1737). In the former, it is compositionally correlated with the emblem of Russia. In the latter, it is among other emblems, including the Moscow horseman, surrounding the cartouche with an explanatory inscription.

Perhaps this coat of arms dates back to the 16th century, and to cartographic monuments. On a map of the eastern regions of Russia, Siberia and Tartaria by Anthony Wied, dating back to around 1537, a cartouche with a legend is decorated with images of several birds, the central place among which belongs to a sitting long eared owl. This is not yet a coat of arms, but a decoration of the map, but it is extremely significant. It seems that the correlation of Tartaria with the emblem of an owl has been forming in the Western European tradition of cartography of Russia and neighboring eastern territories, at least since the 16th century.

4. Conclusion

There are four types of heraldic emblems symbolizing Russia on European maps of the XVI and XVII centuries. They differ in chronological localization and semantic interpretation.

The first two types are variants of state heraldic emblems. In one of the options (the image of a rider with a bow and arrow), ethno-geographical representations of the Middle Ages and Renaissance relating Russia to Scythia and Sarmatia are reflected.

The rider is present on the maps compiled during the XVI century. This image is associated with Muscovy, the Moscow sovereign rather than with Russia as a whole.

The emblems with a double-headed eagle were used on the maps of the XVII century to designate Russia. Their variability with a large number of examples is insignificant.

The other two types are almost unique. The first one is the earliest among the examples of heraldic symbolization of Muscovy. It corresponds to the early version of the state emblems of the end of the XV century which was reduced in official Russian symbolism. However, this option became known to foreigners and used on the maps in the beginning of the XVI century.

The fourth type of emblems has analogs in the broad tradition of European imaginary heraldry of the XV-XVII centuries. It was used on the geographical maps only in the second half of the XVI century. Sporadically, its application can be traced to the end of the XVII century. This emblem is peripheral compared to the double-headed eagle, but stable. It also correlates with certain ethno-geographical ideas about the Gog and Magog peoples which have biblical and ancient origins. The rich tradition of localizing these peoples in Christian geographical literature made it possible to compare them with Russia and attribute the emblem with an open gate to Russia as a whole.

Thus, ethno-geographical representations have found a visual embodiment in the heraldic and cartographic traditions of the XVI-XVII centuries.

References

[1] Pastoureau M 2009 *L'Art de l'héraldique au Moyen Âge* (Paris)

[2] Shchetnikov A I 2015 *Portulans are- nautical charts of the XIV–XVI centuries ΣΧΟΛΗ* **9(1)** 24–34

[3] Bulatov A M 2005 Cartographic Rossica in terms of heraldry *Workshop on heraldry and auxiliary historical disciplines* (Moscow) 105–11

[4] Kordt V A 1910 *Materials on the history of Russian cartography* 1899, 1906 **2(1)** (Kiev) Klepikov S A 1956 *Bibliography of Moscow printed plans for the 16th–19th centuries* (Moscow)

[5] Bagrow L 1962 At the sources of the cartography of Russia *Imago Mundi* **16(1)** 33–48

[6] Avdeev A G 2018 Sovereign or ruler? On one element of the titles of the rulers of Moscow Russia *Russian history* [in Russian – Russkaya istoriya] **5** 9–16

[7] Pchelov E V 2017 The State Seal of Alexei Mikhailovich of 1667 *Archivist's Bull.* [in Russian – Vestnik Arkhivista] **4** 147–58

[8] Van Duzer C A 2010 Northern Refuge of the Monstrous Races: Asia on Waldseemüller's 1516 *Carta Marina Imago Mundi* **62(2)** 221–31

[9] Petrov D A and Yakovlev D E 2014 White-stone emblems on the Borovitskaya Tower of Moscow Kremlin The results of the preliminary survey *Russian archeology* [in Russian – Rossiyskaya arkheologiya] **3** 147–55

[10] Fomenko I K 2010 *The final map of Muscovy of the XVI century from the collection of the State Historical Museum* (Moscow)

[11] Westrem Scott D 1998 *Against Gog and Magog Text and Territory: Geographical Imagination in the European Middle Ages* (Philadelphia) pp 54–77

IOP Conf. Series: Earth and Environmental Science **350** (2019) 012015 doi:10.1088/1755-1315/350/1/012015

Expedition of Ivan Isleniev to Yakutsk and a handwritten map of the Lena River compiled in the mid-XVIII century

O S Romanova[1] **and O A Lazebnik**[2]

[1]S.I. Vavilov Institute for the History of Sciences and Technology Russian Academy of Sciences, Moscow, 14, Baltiyskaya St., Moscow, 125315, Russia
[2]Saint Petersburg State University, 7/9, Universitetskaya emb., St. Petersburg, 199034, Russia

E-mail: olgroma09@gmail.com; o.lazebnik@spbu.ru

Abstract. The article discusses the course and results of the astronomical expedition headed by I.I. Islenyev. Using archival materials, an attempt to localize the Astronomical Observatory built in Yakutsk in 1768 was made. New information on the handwritten map of the Lena River compiled in the mid-XVIII century was provided. Analysis of the toponymic map load was carried out and assumptions about its application for compiling generalized maps of Yakutia at the end of the XVIII century were made.

1. Introduction

Astronomical expeditions of the second half of the XVIII century were a basis for a broader study of the northeast of Russia, including the territory of Yakutsk region. On May 23, 1769, a rare astronomical event was expected - the passage of Venus through the solar disk. To observe this phenomenon, it was decided to send astronomical expeditions to different cities of the Russian Empire – Peterburg, Kola, Ponoy, Umba, Orenburg, Orsk, Guriev, Yakutsk. S. Ya. Rumovskiy, A. Malle and L. Pikte went for observations to Kola, Ponoy and Umba, respectively. V.L. Kraft went to Orenburg, G.M. Lovits went to Guriev, and H. Eiler – to Orsk. Yakutsk was the most distant place of observation. The leader of this (the most difficult one) expedition was Ivan Ivanovich Islenyev. Fedor Cherny was his assistant podsturman. Later, they both became associates of the Academy of Sciences. The expedition had to conduct astronomical and meteorological observations in the city and collect materials on "natural history". The course and results of the expedition, and activities of Ivan Islenyev were documented in letters, journals, and reports deposited in the Archives of the Academy of Sciences. The astronomical materials of the expedition were partly published in Russian and in Latin, German and French for informing the European academic community [1]. For historical studies, Grekov, Shibanov, Shirina, Postnikov and others [2–5] addressed the analysis of the expedition. However, not all details of the expedition are known. The article aims to identify where and under what conditions astronomic-geodesic observations were made, what cartographic works were performed and how their results were used in the subsequent works of the Academy of Sciences.

2. Research methods

The comparative historical method was used. We compared a handwritten map of the Lena River compiled by I. Islenyev with later maps of the XVIII century which depict the territory of Yakutia. The maps were compiled in 1776 and 1792. In addition, the documents in the Archive of the Russian Academy of Sciences allow for localization of the Yakutsk observatory built for astronomical observations.

It is known that the expedition left St. Petersburg on February 11, 1768. Rumovsky provided them with instructions on astronomy, methods for collecting and preserving rare Siberian animals", tools, instruments and books. The route of the expedition passed through Moscow, Kazan, Tobolsk, Krasnoyarsk, Irkutsk, Kachug, Verkholenskiy stockaded town, Ust-Kut, Kirenskiy stockaded town, Vitimskiy stockaded town, Olekminskiy stockaded town. They arrived in Yakutsk on June 8, 1768.

In accordance with the "Mandate ...", an observatory was built in the place where full-scale instrumental astronomic-geodesic observations were carried out.

The documents say about the location of the Observatory: "The Observatory is located in Yakutsk near the Bogorodskaya Cathedral Church at a distance of 380 fathoms" [6].

The Bogorodskaya Cathedral Church was located at the entrance to Yakutsk along Irkutsk highway. The wooden Church was built in 1760. The location of the Observatory can be calculated from this Church. The church existed until 1922 (it was dismantled during the siege of the city). In 1773 (after the observations were completed), at the expense of the merchant D. Barabanov, to the south of the wooden church, the Yakutsk Bogorodskaya stone church was built. Both churches had rich church plates [7].

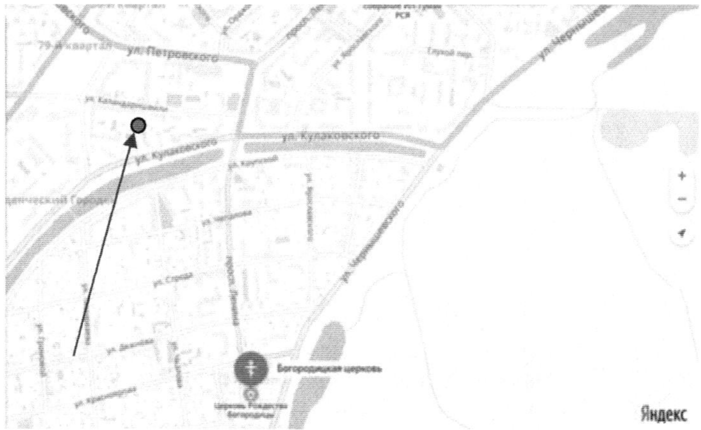

Figure 1. Estimated location of the Observatory on a modern map of Yakutsk

This place gained a historical and spiritual significance - the stone church was restored, the worship was resumed; in 2010, the Yakut Theological Seminary and the Spiritual and Educational Center n.a. St. Innocent (Veniaminov) were created there.

This is the Pledged part of the city located on the shore of Teploe Lake. In this part, there are manor buildings built in the XIX - beginning of the XX centuries.

In the Observatory, Islenyev was conducting observations from August 12, 1768 to June 11, 1769. The "Journal of astronomical observations contains information about instruments, methods, scope and conditions of observations. In August, the observations were carried out for 11 days (on 12, 13, 14, 15, 16, 17, 25, 26, 27, 28, 29 August; 20 two-hour observations); in the remaining days, observations were not conducted due to cloudiness and rain [8].

An important result was the high-precision identification of the latitude and longitude of Yakutsk. For the first time, the latitude and longitude were determined - 62 degrees. 01 min. 50 sec. nl; .8 hours 29 min 35 sec on eclipses of the satellites of Jupiter. In March 1989, the Yakut airborne geodetic plant installed a memorial plaque to memorize this event on the initiative of Gennady Vasilyevich Bocharov, director of the Yakutsk aero-geodesic enterprise from 1980 to 2005.

The measurement data obtained by I. Islenyev were included in the catalog of coordinates compiled by Rumovsky. It became a basis for changing maps complied by the Geographical Department of the Academy of Sciences.

In addition, the expedition collected information on the flora and fauna of the region, fossil animals in the Lena basin. They collected ethnographic folklore materials about the local population. A description of Yakutsk was compiled, and meteorological observations were made [3, 4].

Reports of the astronomical expedition were partially published, but much of them are stored as manuscripts which were studied in the XX century.

Figure 2. A plaque in memory of the first accurate evaluations of the coordinates of the city of Yakutsk. Photo by O.A. Lazebnik, 2018.

3. Handwritten Map of the Lena River

Information about the Lena River obtained during the expedition is of particular interest. In the XVIII century, the mapping of rivers and lakes was intensively developing in Russia. In 1736, a special instruction was created at the Petersburg Academy of Sciences that regulated the range of issues related to the characterization of the studied river and its mapping [9].

The Archives of the Russian Academy of Sciences (St. Petersburg) stores documents on the map compiled by Islenyev "Explanation about the Lena River map. Irkutsk, Yeniseisk, Ilimsk, Kirensk, Olekminsk, Yakutsk were used as a basis. The Lena River flows from Kachug to the mouth of Dubrovka. Using a compass and measures according to the rules of navigational maps, lands are marked on both banks of the river" (From portfolio XXXIX No 28, number 32; From portfolio XLIX No 48; From Portfolio XXXXIV No. 63 of May 23 of 1782, From Ivan Islenyev" [10]. Thus, the surviving documents indicate that Isleniev compiled a map of the Lena River from the Kachug village to the mouth of the Dubrovka River using old cartographic materials and indicated which maps in particular he used, and from the mouth of the Dubrovka River to the city of Yakutsk, Isleniev made a map using only his own observations.

The map was not attached to the Explanation; its location was not established. Our studies of handwritten cartographic materials carried out in the manuscripts department of the Library of the Russian Academy of Sciences allow us to state that the "The Lena River flow map from Dubrovka to Verkholensk" and the Lena River flow map from Vitim prison to Yakutsk" were compiled by I. I. Islenyev.

In terms of geography, the maps are detailed; they show tributaries, islands, residential places, relief elements, types of banks, river channel nature.

The Lena River map from Vitim to Yakutsk shows 36 tributaries, ostrogs, winter huts, stations, 23 villages, islands, relief elements, "high peaked cliffs (famous Lena Pillars which are the main attraction of Yakutia (Fig. 1)). I. Islenev used the materials collected by S.U. Remezov.

In the XVIII century, the Lena River, the largest water artery of Eastern Siberia, was surveyed. Islenyev surveyed the river from the mouth of the Dubrovka river to Yakutsk; the survey of the river from Yakutsk to the estuary was made in 1739-1742 during the Second Kamchatka Expedition.

Soon after all the necessary observations were made in Yakutsk, Isleniev received an order to conduct geographical observations in Western Siberia. In 1769-1771, he was already shooting the Irtysh River.

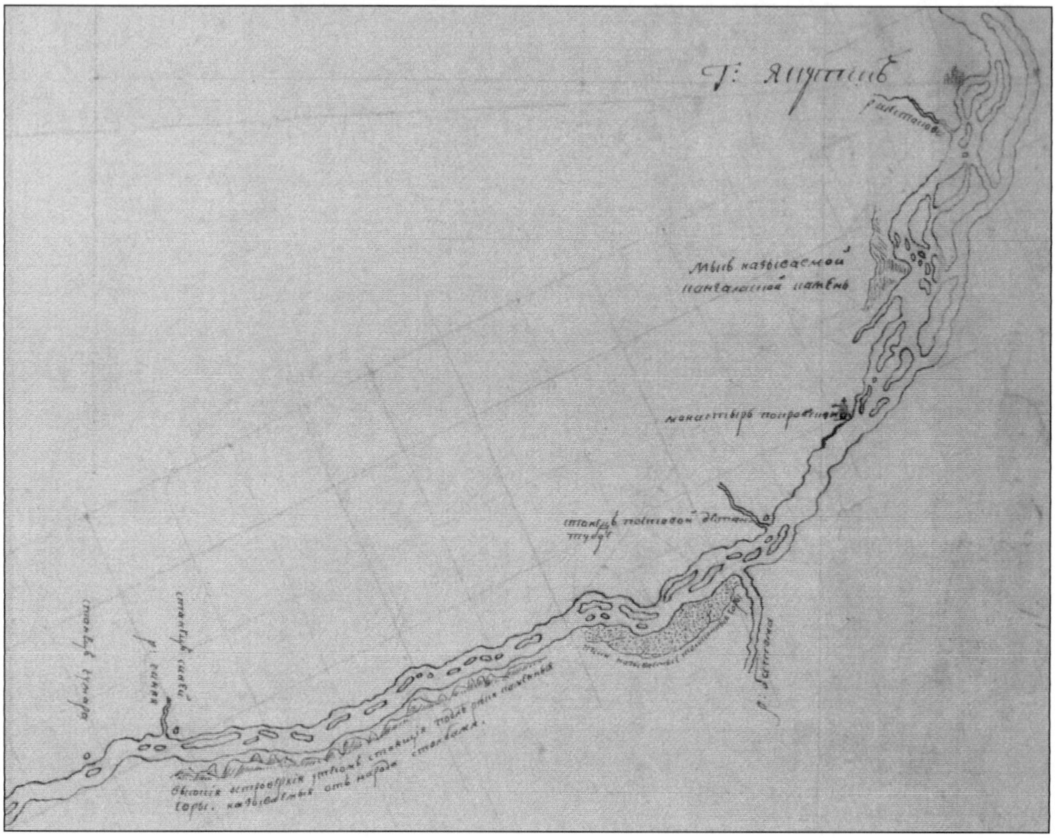

Figure 3. Map of the flow of the Lena River from Vitim prison to Yakutsk". Fragment. [11]

4. Using data from the map of the Lena River collected by I. Islenyev

The study of map toponyms makes it possible to use the map for compiling maps of Irkutsk province. On the "General map of Irkutsk province of 1776 which shows Irkutsk, Yakutsk, Uda provinces" (the scale is about 1: 8,400,000), there are the same names of tributaries and settlements. However, their number is 15 which indicates the generalization of the map. The Goose Mountains were marked. In the area of the Lena Pillars, small unsteady mounds were shown [12].

On the Russian Atlas of 1792 ("The Map of the Western Part of Irkutsk Province"), there are few names from the Islenyev's map - only four signatures of the Lena tributaries - Olekma, Talb, Botoma, Blue, and "Vitim", "Olekminsk", and "Yakutsk" [13].

Thus, the handwritten map of the Lena River compiled by Islenyev was not used by the Academy of Sciences to create maps of northeastern Russia in 1770-1800. Analysis of the toponymic dynamics of the average flow of the Lena River showed the relative static nature of hydronyms.

5. Conclusion

An assumed location of the Astronomical Observatory built in Yakutsk in 1768 was established. Assumptions about handwritten maps stored in the BRAN manuscript fund were made. The maps show the Lena River from the headwater to Yakutsk. The analysis of the toponymic dynamics of the average flow of the Lena River showing the static nature of hydronyms was carried out. The ignorance of data obtained by I. Islenyev when creating maps of northeastern Russia by the Academy of Sciences of 1770–1800 was revealed.

Maps of the Lena River will become part of the electronic collection of cartographic images of Yakutia in the XVII - early XX centuries.

6. Acknowledgements

The study was carried out within the state assignment. It was financially supported by the Russian Foundation for Basic Research as part of the research project No. 17-03-00840-OGN.

References

[1] Gnucheva V F 1940 Materials for the history of the expeditions of the Academy of Sciences in the XVIII and XIX centuries *Proc. of the archive* **4** 110–2 (Moscow; Leningrad)
[2] Grekov V I 1971 From the history of the organization of academic expeditions 1768–1774 *News of the Academy of Sci. of the USSR ser. of geography* **4** 134–43
[3] Shibanov F A 1975 I I Islenev – astronomer – geographer and cartographer of the XVIII century *Bull. of the Leningrad University* **18(3)** 133–9
[4] Shirina D A 1983 *Chronicle of the expeditions of the Academy of Sciences in the north-east Asia in the pre-revolutionary period* (Novosibirsk) pp 31–6
[5] Postnikov A V 1989 *The development of large-scale cartography in Russia* (Moscow)
[6] *Saint-Petersburg branch archive Russian Academy of Sciences* F 3 Op 28 Case 7
[7] *Treasures of culture of Yakutia* 2011 pp 588–99
[8] *Saint-Petersburg branch archive Russian Academy of Sciences* F 3 Op 28 Case 9
[9] Postnikov A V and Ozerova N A 2018 On the history of the cartographic study of the rivers and lakes of Asiatic Russia in the 18th and early 19th centuries *Bull. of the Academy of Sci. of the Chechen Republic* **3(40)** 90–5
[10] *Saint-Petersburg branch archive Russian Academy of Sciences* F 3 Op 10 Case 183 sheet 29
[11] Library of the Russian Academy of Sciences, manuscripts research department *Collection of hand-written maps, basic inventory* 197, 198
[12] Russian State Library (RSL) *Department of cartographic publications* Co 111/II-77
[13] *The Russian atlas, consisting of forty-four cards consisting of an empire that divides into forty and two provinces. Writing, engraving and printing at the Mining School* 1792 (St. Petersburg)

Ice phenomena on the rivers of Russia: the role of academic expeditions

V M Savenkova

S.I. Vavilov Institute of the history of science and technology, 14, Baltiyskaya ave., Moscow, 125315, Russia

E-mail: Savenkovavm@mail.ru

Abstract. The systematic study of the ice regime of the Russian rivers began with the organization of gauges at the end of the XIX century. It was preceded by a long preparatory period of expeditionary works performed from the second half of the XVIII century. The article discusses the contribution of members of academic expeditions of the second half of the XVIII century to the study of ice phenomena on the rivers of Russia. Special attention is paid to the dangerous phenomena that often accompany opening and freezing of rivers, ice jams, floods, etc. The author summarized and systematized data and observation materials obtained during the journeys. They can be used for studying ice phenomena on the rivers.

1. Introduction

In the second half of the XVIII century, large-scale expeditions were conducted. They were based on the projects developed by M.V. Lomonosov. These are famous "physical expeditions" of Petersburg Academy of Sciences of 1768–1774. Expeditionary academic studies provided materials on geography, geology, flora and fauna of Russia; customs, languages and business traditions of the population; data on ice phenomena on the rivers of Russia (the opening and freezing times, the nature of floods, etc.).

2. Sources and methods

The study identifies, systematizes and analyzes a huge number of sources.

3. Academic expeditions of the second half of the XVIII century

The most important result of academic expeditions was publication of such fundamental works as "Journey to different provinces of the Russian Empire" by P.S. Pallas; "Daily notes of travel to different provinces of the Russian state" by I.I. Lepekhin; "Journey to Russia to explore three kingdoms of nature" by S.G. Gmelin; "Travel notes" by I.P. Falk; "Notes about the Travel from St. Petersburg to Kherson in 1781 and 1782" by V.F. Zuev and "Journey through Ladoga and Onega Lakes" by N.Ya. Ozeretskovsky. They contained descriptions of nature and observations of "the nature of the earth and waters", features of the ice regime of rivers associated with seasonal phenomena, a set of regularly processes of the development and destruction of ice formations on water bodies. The travelers paid attention to hazardous river phenomena creating obstacles for economic activities: spring flooding, very early or late ice drift and freezing, ignition and congestion.

The expedition of 1768–1774 headed by P.S. Pallas examined European Russia, the Lower and Middle Volga region, Orenburg Territory, the Urals, Western and Eastern Siberia, and Transbaikalia. During the expedition, Pallas noted the time of freezing and opening of rivers, explained the reasons determining these terms, gave the first empirical "forecasts" of flooding (floods) as a result of rising water levels due to rapid snow melting and traced spatial and temporal patterns of changes in ice phenomena in various natural areas. Moreover, Pallas tried to describe and distinguish phases of the ice regime of the majority of rivers. He described autumn ice phenomena (freezing) and spring ice phenomena (opening).

On October 20, 1768, the scientist recorded autumnal ice drift and freeze-up on the Volga River near Simbirsk. He paid attention to the direct dependence of the fixed ice cover (freezing) on the wind and lowering air temperature. On December 14, 1768, in Simbirsk and Tetyushi, Pallas paid attention to the dependence of the ice thickness on morphometric characteristics of the river. He wrote that near Simbirsk, it was still impossible to cross the Volga River, while in Tetyushi located where the river was much narrower, people crossed the frozen river [1].

During the freeze-up period, there are ice-free areas that have a double origin: dynamic and thermal ones. Thermal areas arise under the influence of warm groundwater described by Pallas described. He noted that the keys flowing from the coastal slopes into the river are the causes of the lack of strong ice [1].

Spring ice phenomena were observed on the Samara, Belaya, Miass, Angara, Selenga, Kame Rivers. In 1769, in Samara, due to warm weather, the snow melted for a very short time, and on April 9, the spring drift began, the level of water increased. On April 11, the ice was broken on the Volga River. By April 15, due to the strong north wind, the river ice melted. When describing the opening of the rivers and the rise of the water level, the scientist paid attention to the height of the snow cover affecting the level of the spring flood [2]. On the Oka River near Murom, Pallas identified horizontal channel deformations that cause soil erosion. To fix the soil and prevent landslide processes, he proposed to plant willow on the river banks. In 1774, on the Volga River near Tsaritsyn, Pallas observed ice clogging phenomena — blockage of the river by floating ice floes.

The 1768–1773 expedition headed by I.I. Lepyohin studied the Middle and Lower Volga, the Lower and Middle Urals and part of Arkhangelsk province. In the "Daily Notes", Lepyokhin specified the dates for the opening and freezing of the rivers of Kanin and Timan lands. The Usa River opened in mid-May and froze at the end of September. For Pechora, observation data were in several sections which made it possible to trace spatio-temporal changes in ice phenomena. On the basis of his own observations and data obtained from the local people, Lepyokhin wrote that Pechora opened near Usttselma (Ust-Tsilma) at the end of May, and in Pustozersk, it opened a week later; near Ustzelma, it froze at the beginning or in the middle of October, and in Pustozersk - a week earlier; near Izhma Slobodka (Izhma), it opened in the middle or at the end of May, and froze at the beginning or in the middle of October [3].

Lepyokhin paid great attention to dangerous ice phenomena accompanying the opening of rivers. On May 7, 1773, he observed a congestion on the Northern Dvina during the spring ice drift which increased the water level, and caused flooding in Arkhangelsk [4]. The traveler paid attention to the fact that the annually recurring phenomenon prompted local residents to predict the height of the flood, linking it with the behavior of seabirds: "the higher the seagulls are flying, the higher the water level will be" [3].

Changes in the morphological structure of the river bed due to the action of flowing water (channel processes) related to instability and deformation were observed on the Kama, Vyatka, Yarenga and Sysola rivers. He wrote about the latter that its banks are sloping, sandy, swampy, subject to the detrimental effect of spring waters. He paid attention to the transport of sediment and displacement of channel forms on theVychegda river near Yarensk which made local residents to move to less dangerous places. The Kynyrma and Lipka are like creeks in other seasons.

The 1768–1774 expedition headed by S.G. Gmelin (junior) in explored parts of European Russia and the Lower Volga region, the North Caucasus and Transcaucasia. Gmelin paid attention to the nature of spring floods affecting the life and working conditions of the coastal population. After snow melting, the Voronezh river became extremely deep, spreaded several miles and flooded the floodplain, the first and second terraces. He wrote about flooding on the Don River near Esaulovskaya that spring and autumn floods caused a lot of harm to coastal villages. On the Kuban River, spring floods were very large, the water stood until the beginning, and sometimes the end of July. Houses along gentle banks were built on stilts, and the local population kept boats for spring communication. The same phenomena were observed on the Cuba and Kure Rivers [5].

The 1768–1773 expedition headed by I.P. Falk surveyed Orenburg Territory, the Middle and Lower Volga regions, the foothills of the North Caucasus, the Southern Urals, and Western Siberia to Tomsk and Kuznetsk. The Falk's Notes provide data on observations of ice phenomena on the rivers. Based on his own observations, Falk recorded the dates of opening and freezing and identified phases of the water regime of the rivers. The Belaya River freezes in early November and opens in mid-April; The Tom River freezes at the end of October, and opens at the end of March. For the Irtysh and Volga Rivers, the opening date depends on the location: in 1772, near Tara, the Irtysh opened on April 17, near Barabinskaya steppe, - on April 10-20. For the Volga River, the scientist tried to trace spatio-temporal changes in the onset of opening in various natural zones: in 1770, in Tsaritsyn, the Volga river opened on March 3, and int Kazan - in mid-April. For the Moscow River, the data on the duration of the freeze-up were provided [6].

Falk provided information about the width of the rivers during the ice cover: above Kuznetsk, the Tom River was 30 sazhens, below Kuznetsk - from 50 to 60 sazhens, above Tomsk - 30 sazhens, and below - 60 sazhens. Observing the spring ice phenomena, in the Terek River, Falk recorded the rising of the water level by 30-40 feet; in the Saratovka River, the left tributary of the Volga, - by 40-50 feet; in the Volga river near Tsaritsyn – by 20-30 feet. He compared flood levels in different places on the Volga river: in the spring, the water in Astrakhan rises by 8-10 feet, in Kazan - by 25-35 feet, in Saratov - by 40-50 feet.

Falk paid attention to the changes in the direction of the flow of the river due to the spring flood. At the confluence of two rivers during the peak flood, huge masses of water can change the direction of flow of one of the rivers, directing the water upstream, i.e. to the source. The scientist observed this phenomenon on the Kazanka and Terek rivers. He wrote about the first one: "during the initial spreading, the Volga rover leaves its banks so quickly that the water enters Kazanka which flows backward for several days" [6].

Falk paid attention to the conditions of navigation which was often possible during the spring floods. According to his data, there are loaded barks from Penza to the Volga river during the spring flood; in the summer, the shipping stops; The Lesnoy Voronezh river is navigable only in spring; In spring, the Iset River is navigable for large vessels, and in summer - only for small ones; The Tura River is navigable to Turinsk, and in summer, it is navigable only to Tyumen. The Moscow River is navigable only when the level rises by one sazhen during the spring flood.

In 1781–1782, Zuev was travelling in the European part of Russia, left-bank Ukraine (to Kherson) and across the Crimea. In his travel notes, Zuev carefully described rivers, lakes and springs. When describing water bodies, he paid attention to the use of rivers for the water supply and the time of their opening and freezing. The Seim and Tuskar rivers freeze in late November or early December, and open in March; the Oka river and its tributary Orlik freeze in November or December, and open in March, sometimes in April. The traveler describes the floods in the Oka basin which result from spring ice phenomena. Travelling through the Nara river, he noted that in June 1781 it was quite wide, and during the spring flood, it is three times wider, as evidenced by sands forming the border of the "hollow water". At the end of June, the Protva river was no more than 12 sazhens wide, and during the spring flooding, its width was more than half a verst [7]. When describing Oryol province, Zuev noted that on the Oka, Desna and Sosna rivers, navigation was possible only during the spring flood.

The 1785 expedition headed by N.Ya. Ozeretskovsky travelled through the territories of the basins of Ladoga and Onega Lakes. The scientist paid attention to the difference in time of the opening of the Shaly and Vodla rivers at the point of their confluence. He noted the destructive power of the spring flood on the Lyaskile River and formation of a "waterfall" (threshold) at a distance of five kilometers from the mouth upstream. The height of the threshold reached 17 m, and the fall of water, especially strong in the springtime, threw huge boulders. He encountered a similar phenomenon in the spring on the Suna River near the Kivach threshold.

The information on the influence of snow cover on the height of the spring flood on the Vuoksa river obtained by Ozeretskovsky is valuable. He paid attention to the fact that local residents were observing the water regime of rivers flowing into Ladoga Lake. They noted that the amount of snowfall in winter affects the height of the spring flood. The water level in the lake increased and decreeased. In 1785, the water level was above normal. In order to explain this phenomenon and establish the probable cause of changes in the water regime, Ozeretskovsky pointed out the need to monitor the water level at several points, and the circumstances that may affect the changes in the water level (rainfall, snow, fogs, the number of sunny days). This would help establish causes of the change in the water regime [8].

4. Conclusion

The data obtained during academic expeditions of the second half of the XVIII century are still relevant. These observations are a basis for the retrospective analysis of the rivers in the XVIII century. They can be sued for assessing climate changes. P.S. Pallas, I.I. Lepyokhin and I.P. Falk paid attention to the spatial and temporal patterns of changes in the periods of opening and freezing of rivers in different natural zones, described phases of the ice regime in certain sections of rivers in different regions.

We see their desire to explain the causes of spring floods depending on changes in climatic and ice conditions of the water system. They tried to establish the dependence of the ice cover on morphometric characteristics of the river and hydrometeorological elements.

The works are valuable because they can be a basis for further studies on the ice regime of Russian rivers.

References

[1] Pallas P S 1773 *Journey to different provinces of the Russian Empire* Part 1 (St. Petersburg) pp 180–217

[2] Ozerova N A, Snytko V A and Shirokova V A 2015 Expedition to study the sources of the main rivers of European Russia (1894–1902) *News of the Rus. Academy of Sci. Geographic Ser.* **4** 113–27

[3] Lepekhin I I 1805 *Day Notes of the Journey of the Doctor and the Academy of Sciences Adjunct Ivan Lepyohin to various provinces of the Russian state* Part 4 (St. Petersburg) pp 2–247

[4] Tersky P N and Frolova N L 2011 Floods on the rivers of the north of the European territory of Russia (on the example of the Northern Dvina river basin) *News of the Rus. Academy of Sci. Geographic ser.* **3** 94–105

[5] Gmelin S G 1771 *Journey through Russia, to explore the three kingdoms of nature* Part 1 (St. Petersburg) pp 160–271

[6] Falk I P 1824 *The complete collection of travel scientists in Russia, published by the Imperial Academy of Sciences, at the suggestion of its president* vol 6 (St. Petersburg) pp 132–422

[7] Zuev V F 1787 *Travel notes of Vasily Zuev from St. Petersburg to Kherson in 1781 and 1782* (St. Petersburg) pp 2–116

[8] Ozeretskovsky N Ya 1792 *Journey through Ladoga and Onega lakes* (St. Petersburg) pp 59–60

The surrounding environment monitoring in the Soviet Union: a review of establishing new "ecological" science

A V Sobisevich[1], V A Snytko[1,2] and A V Postnikov[1]

[1]S. I. Vavilov Institute for the History of Science and Technology RAS, 14, Baltijskaia street, Moscow, 125315, Russia
[2]V.B. Sochava Institute of Geography SB RAS, 1, Ulan-Batorskaia street, Irkutsk, 664033, Russia

E-mail: alex.v.sobis@gmail.com

Abstract. The programs of surrounding environment monitoring were presented in 1972 at UN Conference on the Human Environment in Stockholm. When the Soviet Union had adopted a resolution of the conference, it became obligatory to establish biosphere reserves and provide environmental monitoring. The ideas of such monitoring in the Soviet Union were promoted by two scientists - Innokentiy Gerasimov and Yuri Israel.

1. Introduction

According to Innokentiy Gerasimov (1905 – 1985), who was the head of the Institute of Geography at the Academy of Sciences of the USSR from 1951 to 1985, ecology did not enter the scientific vocabulary until the second half of the XIX century and was defined as the interaction of vegetation and animals with their environment. Then the biologist used that term extensively as Darwin's theory of evolution, so, for a long time "ecology" was considered to be a part of Biology. From the second half of the XX century, the term "ecology" was understood in a much wider sense. The term can be applied to nature and various forms of human activity so that preserving the existing environment is a result of both natural processes and man's conservation efforts [1].

From the 10th to 20th of March, 1970, Soviet delegation headed by soil scientist Viktor Kovda (1904 – 1991) participated in the compilation of the program for the upcoming conference at the United Nations headquarters in New York City. Soviet delegation with the support of the delegations from France and Czechoslovakia promoted a conception of alarming destruction of the environment, which was connected with many social and economic reasons when capitalist countries and big corporations vastly exploited the nature of developing countries. At the top of the agenda of the forthcoming conference, it was proposed to put the report of the UN Secretary-General U Thant (1909 – 1974) about an analysis of the causes of the biosphere destruction and measures for its preservation [2].

In the same year, 1970, biogeographer Viktor Sochava (1905 – 1978) turned his attention to ecological issues in the Soviet Union. At the V-th Congress of the Geographical Society of the USSR he made a presentation "Geography and Ecology", which showed some further problems of surrounding environment monitoring. The term "monitoring" was not known at that time in the Soviet Union, so Viktor Sochava said about ecological approaches to geosystems modeling. According to his views the most advanced geosystems modeling was made abroad by George M. Van Dyne (1933 – 1981) at desert-steep territories of Colorado [3]. Viktor Sochava understood human ecology as the most sophisticated geographical direction because geographers were making research of territories and nature in terms of its relation to the human population. The advanced studies of different aspects of environment which influence human health were made in Cambridge University, where the department of human ecology was established. The importance of ecological research according to Viktor Sochava was connected with many problems of humans' adaptation to the fast environmental changes, so geographers had to prevent the undesirable transformation of nature [4].

In 1971, at the International Council of Scientific Unions members of a Special Commission of Scientific Committee on Problems of the Environment (SCOPE) announced the main thesis of environmental monitoring which was published at a brochure "Global environmental monitoring". Environmental monitoring was considered as systematic identifying of any environmental changes, which were caused by human activity. In 1972, at the United Nations Conference on the Human Environment in Stockholm, a professor of Chicago University Gilbert F. White (1911 – 2006) and a meteorologist Thomas Malone (1917 – 2013) proposed the program of environmental monitoring. Soviet delegation together with the delegations of other socialistic countries boycotted that conference because the Organizing Committee refused the participation of the German Democratic Republic as a non-member of the United Nations. However, despite the boycott the Soviet Union adopted the resolution of the conference and took the responsibility of establishing biosphere reserves and conducting ecological monitoring. The environmental monitoring processes were initiated by Innokentiy Gerasimov and Yuri Israel (1930 – 2014), who was at that time the head of the Hydrometeorological Service of the Soviet Union.

2. Methods and materials

The process of establishing the ecological monitoring system in the Soviet Union is the object of our research. We are using the methods of environmental history. One of them is a content analysis of scientific reports and other archival sources. According to our concept, the international science networks where Soviet scientists and officials collaborated with their Western colleagues during the time of the Cold War and Perestroika were crucially important for environmental policy. World famous scientists and politicians who understood that time was a really vital factor of environmental monitoring were important actors in the process of environmental monitoring establishing and creating biosphere reserves.

We paid attention to the problem of different dimension of scaling in environmental monitoring which was discussed at the workshop in the University of Tübingen on April 11th and 12th, 2018. At that workshop Alexey Sobisevich presented his thesis "Scaling the Black Earth. Nature Preservation in the Soviet Union". After that presentation, it was very important to answer the following questions: *How do the actors themselves perceive scales? Which viable alternatives exist to the power of the state as a focal point in the historian's narrative?* [5].

3. Results and discussion

The term "monitoring" was used in relation to the natural environment in the early 1970s. In the framework of this concept natural environment was considered to be a multicomponent aggregate of natural phenomena subject to natural dynamic changes and various effects under the influence of human activity. In 1970 the Swedish Natural Science Research Council and Special Problem Group from the USA on the creation of a global monitoring network made significant contribution to the development of environmental monitoring problems.

In 1971, the ideas of environmental monitoring were announced by Gilbert White and Tomas Malone in the brochure «Environmental Monitoring: Physical and Chemical Measurements», where environmental monitoring was understood as a systematic observation of the environment and identification of possible changes, especially in connection with human activities [6]. At the meeting participants proposed to build a system of controlling anthropogenic changes of the environment which would allow planning environmental management measures.

In 1972, at the UN Conference on the Human Environment the program of environmental monitoring and establishing biosphere reserves was accepted. Biosphere reserves were understood not only as a very essential part of conservation policy but also as a very efficient instrument of environmental monitoring. In September 1972, after UN Conference resolution ratification, Supreme Soviet of the Soviet Union issued the resolution on the measures for protection of nature and rational use of natural resources [7].

In 1973, the issue of environmental monitoring was also discussed by a group of experts from UNESCO's Man and the Biosphere Program, who admitted the necessity of preserving biodiversity and protect the most typical territories in each bioclimatic zone of the world. In the same year a Scientific Council on the biosphere problems, headed by academician Alexander Sidorenko (1917 – 1982), was organized in the Academy of Sciences of the USSR [8].

In 1974, an intergovernmental meeting convened by the United Nations Environment Programme (UNEP) of the World considered the environmental monitoring system. In the same year, Yury Izrael became the new head of the USSR Hydrometeorological Service and after his appointment published the article «Global Observing System. Forecast and assessment of changes in the natural environment. Basics of monitoring». In this work, for the first time in the USSR, the definition of environmental monitoring was given as a system of observations, allowing identifying changes in the state of the biosphere under the influence of humans [9].

Vladimir Sokolov (1928 – 1998) headed biological section in the Scientific Council on the biosphere problems. It coordinated research on the use of biological resources, the protection of natural areas and the conservation of rare species of animals and plants. Innokenty Gerasimov headed the section of global and regional forecasts which gave forecasts on changes in environmental conditions both in the entire world and in certain areas. Yuri Izrael headed the biosphere monitoring section which coordinated problems related to the organization of biosphere reserves and assessment of the environmental state [8]. During the first meeting of the section, held on July 15, 1975, Innokentiy Gerasimov and Yuri Izrael made a joint report on the basic principles and modern ideas about the state of the biosphere.

The results of Innokenti Gerasimov's work as a part of the scientific section on biosphere monitoring were shown in the article "The Scientific Basis of Modern Environmental Monitoring" published in 1975. The article analysed international programs for creating a global environmental monitoring system but Innokenty Gerasimov believed that such observations should be carried out by the national environmental control services. Innokenty Gerasimov also advocated the creation of a separate service that would deal with environmental monitoring in the Soviet Union because the existing hydrometeorological or earthquake control services had their own tasks [11].

Innokenty Gerasimov made a proposal on the steps (blocks) of modern environmental monitoring: the first block - a bioecological, the second - a geoecological, and the third - a biospheric. The role of the bioecological block is mainly to permit observations and control over the state of the environment regarding its influence on the health of the population (SO_2, CO, NO, NO_2 and others components). The functions of the geoecological (natural-economic) block concern observation and control of changes in those geosystems (including both natural and antropogenic ecosystems) that constitute the environment. It is particularly important to stress the interrelations and mutual interactions among these blocks of the monitoring system. It is only in the second block (the geoecological one) that the indicators of the first (bioecological block) can acquire the needed level of scientific validity and a scientific basis for spatial extrapolations. The third block of biospheric monitoring serves for observations and control of global background changes in the environment as well as an ecological evaluation of these changes. Its basic indicators must be therefore in the atmosphere, the hydrosphere, and the lithosphere [1, 12].

In 1978, Yuri Izrael proposed dividing environmental monitoring into two parts, "abiotic" and "biotic". According to him, "abiotic monitoring" meant systematic long-term observations of the pollution level in the environment, as well as concomitant geophysical observations. Yuri Israel considered it advisable to include atmospheric air and precipitation, surface water, soil, and bottom sediments and biota, when Hg, Pb, Cd, As, N, P, DDT, benzopyrene and others pollutants were monitored. Biotic monitoring assumed the studies of biota taking into account the possibility of the spread of pollutants through food chains and concentrated toxic effects of pollutants on ecological systems. Yuri Israel argued the importance of the experimental program, which had the goal to determine the sensitivity of individuals of various species in the observed ecosystem to the effects of low, close to background concentrations of pollutants [9].

During these years, Soviet scientists take part in a number of environmental conferences. In September 1975, the Soviet-German environment forum took place in Münich. In November 1975, Soviet and British scientists took part in the meeting of the Committee for cooperation in the field of environmental protection. That cooperation had its extension when in July, 1976 members of the scientific section concerning monitoring biosphere shared their materials about ecological monitoring in the Soviet Union with British scientists. Those materials were necessary for future First Soviet-English symposium on biosphere monitoring [10].

The Presidium of the Academy of Sciences of the USSR defined the main dimensions of research, which were necessary for environmental monitoring, for the period from 1976 to 1980. The development of environmental monitoring strategies led the Academy of Sciences of the Soviet Union together with the USSR Hydrometeorological Service to the research concerning creation of a global monitoring network. The study of international environmental laws and the analysis of the advantages and disadvantages of the existing system of management in the field of environmental protection were a basis for further improvement of environmental protection in the Soviet Union [8].

4. Conclusion

According to Elena Aronova, environmental monitoring was boosted by the environmental movement of the late 1960s when people learnt about negative effects of such carcinogens as DDT and Agent Orange [13]. Gilbert White supported that environmental movement in the idea that, «*We can work together and with our common natural environment for the greater benefit of ourselves and future generations. There can be water for all; the world can be a safer place with less loss of life and property from natural hazards*» [14]. That was the main reason when he promoted the program of environmental monitoring at UN Conference on the Human Environment in 1972 and shortly after that he and Thomas Malone brought together SCOPE under the aegis of the International Council of Scientific Unions' intention to create an independent scientific organization, which could offer assessments of global environmental problems [15].

In 1970, Viktor Sochava was the first in the Soviet Union who promoted environmental research in geography and had the ideas very close to «environmental monitoring». Innokentij Gerasimov emphasized that though ecology had been developed as a special discipline within the biological science then it obtained the scientific approaches in such disciplines as geography, geology, soil science and others [12]. After UN Conference on the Human Environment Yuri Israel promoted the ideas of environmental monitoring in the Soviet Union and soon Innokentiy Gerasimov joined him in that activity.

Yuri Israel was supported with the resources of Hydrometeorological Service of the Soviet Union, so he had an initiative to establish the system of «antropogenic monitoring», where the effective control over the flow of pollutants in air and surface waters would be conducted. According to his ideas, all territory of the Soviet Union should be covered with observation stations, which could provide environmental forecasts based on long-term observations. Innokentij Gerasimov proposed the method of «geosystem monitoring», which aimed primarily at the rational use of natural resources and the protection of ecosystems.

Denis Shaw, Jonathan Oldfield and Julia Lajus noted that Soviet scientists participated in many international initiatives aimed at understanding natural resources, and despite many political limits they could play a significant role in environmental initiatives [16]. Those political limits could be connected with the time of Cold War when physical data about the environment was in the sphere of military agencies' interests. Elena Aronova noted that information on oceanography, radiology, meteorology, and data in some other fields of research had great importance and could not be shared, but biological sciences were less prohibited area for international cooperation [13]. Laura A. Henry brings attention to environmental activism in the late Soviet period and according to her, such state-sponsored scientific organization as the All-Russian Society for Nature Protection (VOOP) and the Moscow Society of Naturalists (MOIP) were mostly the exception from the state industrial initiatives criticism [17].

The Academy of Sciences of the USSR was a state-sponsored organization ranked much higher than VOOP and MOIP. From that point academic's institutions could be much more powerful and oppositional players in environmental initiatives. Scientists could be supported in their environmental initiatives for the purpose of abroad propaganda that the Soviet state is an environmental-friendly country. Those facts show that further studies in the history of surrounding environments monitoring should focus on the study of such research questions: *what were the most important factors in the existence of that international science networks? Had it been the state's interest, or scientists' intention to cooperate with their international colleagues? When such personal matters as friendship or political views were of crucial importance?*

References

[1] Gerasimov I P 1983 *Geography and ecology. A Collection of Articles* (1971-1981) (Moscow: Progress)
[2] Archive of Russian Academy of Sciences (ARAN) 2081-1-169.
[3] Van Dyne G M 1969 Grassland Management, research and training viewed in a systems context *Range Sci., Dep. Sci.* **3** 50
[4] Sochava V B 1970 Geography and Ecology *Papers of V Congress of the Geographical Society of the USSR* 25
[5] Schönfelder T and Sobisevich A V 2019 Playing with Scales in Environmental History: International Workshop Held at the University of Tübingen on 11-12 April 2018 *History and Biology Research* [in Russian – Istoriko-biologicheskiye issledovaniya] **1** 114–118
[6] Malon T 1971 Environmental Monitoring: Physical and Chemical Measurements *Proceedings of the Scientific Program, XVII General Assembly of the International Union of Biological Sciences* pp 62-65
[7] Sobisevich A, Snytko V and Savenkova V 2018 The role of biosphere reserves in the environmental protection at the Soviet Union *GeoConference SGEM* **18(5.1)** 963–969
[8] ARAN 2092-1-5
[9] Israel Y A 1974 Global system of observations. Forecasting and estimating of environment changes. Basics of monitoring *Meteorology and hydrology* [in Russian – Meteorologiya i gidrologiya] **7** 3-8
[10] ARAN 2081-1-167
[11] Gerasimov I P 1975 Scientific basics of modern environmental monitoring *Izvestiya of Academy of Sciences of USSR. Geography Series* **3** 13-25
[12] Gerasimov I, Annenkov V, Davitaya F, Konstabtinov O, Masbits Y, Pokshishevsjy V, Preobrazhansky V and Ryabchikov A 1986 Scientific and Technological Revolution and Soviet Geography *Geographical Prognostication. Problems and Prospects* (Moscow: Progress) pp 31-48
[13] Aronova E 2015 Environmental monitoring in the making: from surveying nature's resources to monitoring nature's change *Historical Social Research* **40 (2)** 222-245
[14] Kates R W and Burton I 2008 Gilbert F. White, 1911–2006, Local Legacies, National Achievements, and Global Visions *Annals of the Association of American Geographers* **98(2)** 1-8
[15] Kates R W 2011 Gilbert F. White, 1911-2006, A Biographical Memoir (Washington: National Academy of Sciences)
[16] Oldfield J, Lajus J and Shaw D 2015 Conceptualizing and Utilizing the Natural Environment: Critical Reflections from Imperial and Soviet Russia *Slavonic and East European Review* **93(1)** 1-5

[17] Henry L A 2018 Russia's Environment and Environmental Movement *Understanding contemporary Russia* ed Bressler M L (Boulder, Colorado: Lynne Rienner Publishers) chapter 9 pp 275-302

Dr. V. A. Varsanofieva writings on the protection of the environment

O A Valkova

S I Vavilov Institute for the History of Science and Technology, Russian Academy of Science, 14, Baltiiskaia str., Moscow, 125315, Russia

E-mail: o-val2@yandex.ru

Abstract. The paper presents data on the ideas and efforts of well known soviet geologist and mineralogist, historian and popularizer of science Dr. Vera Varsanofieva (1889–1976) in such area as environment protection. Despite the small number of published works about this topic she was a prominent figure in nature preservation struggle in the Soviet Union. Her efforts were mostly focused on the protection of nature preserves from destructions caused by desire of unrestricted economic use. She was convinced that integrated natural reserves should be created to protect both living and inorganic nature because they were closely intertwined and that such reserves were the best places for the nature exploration.

1. Introduction

Vera Varsanofieva (1889–1976) was the first female in the USSR granted doctoral degree in geology and mineralogy in 1935. She was a professional geologist and spend a large portion of every year in the "field" starting from her student's years in 1907–1915 and till her retirement in 1970 when she was already 81 years old. She regularly published results of her field research and was well known and well respected in the professional community. Simultaneously she was teaching different geological courses in different universities beginning from 1916, organizing later, in 1925, the department of geology in the II Moscow State University (*II Moskovskiĭ gosudarstvennyĭ universitet*) and leading it for more than three decades.

She was also a popularizer of science, written several popular-science books and many articles that were published and republished. Interested in humanities, specifically in history from her childhood she in time became a historian of science and published several books and articles in the history of science in the USSR, history of natural science education in the Russian Empire and women education.

Again, from the very first days of her professional career in geology she obtained memberships in several Russian natural science societies and enthusiastically participated in their activities. In time she became a part of the leading bodies of a few of them and in 1943 was elected a vice-president of the eldest and most renowned of all Russian societies – Moscow Society for the naturalists (*Moskovskoe obshchestvo ispytateleĭ prirody*).

As many women of her generation who archived significant career success in previously strictly men's sphere she was invited to participate in political activities and was elected people's deputy first in one of the Moscow local districts and then a member of Moscow Soviet of Deputes. From her biographical data it is known that she really cared about people she represented and tried to aid them as much as was possible as well as she cared about her students and her research [1].

But her real passion was wild nature. Through all her life she was spending an extraordinary amount of time and effort working on protection of living and inanimate nature. Despite that she published only a few items on the subject and her special book about Pechora-Ilych nature reserve (Pechero-Ylychskiĭ zapovednik) was also left unpublished. As of today, I don't know the reason but I strongly suspect that it was dew to the governmental policy of the time. In the article I will analyze her forgotten or unknown writings and ideas on the protection of the environment.

2. Materials and methods

The main sources for my research were original papers published by Dr. V. A. Varsanofieva and also her unpublished writings such as speeches given on different occurrences, books and papers left unpublished by different circumstances, her private and official correspondence. All this data was accumulated in the Russian State Archive of Economics (*Rossiĭskiĭ gosudarstvennyĭ arkhiv ėkonomiki*) in Moscow [2].

3. Dr. V. A. Varsanofieva writings on the protection of the environment

From its very early days Vera Varsanofieva was a member and an active participant in the activities of the All-Russian society for nature protection (*Vserossiĭskoe obshchestvo okhrany prirody*) founded in 1924. For her this effort was not a social or professional or even political obligation but a heart desire. In her private correspondence with friends and colleagues she frequently mentioned that the time she was spending in communication with nature was the best and the happiest in her life. She even wrote that she became a geologist in order to have a legal opportunity to regularly spend time in wild nature environment. Among her unpublished documents one can find texts of several speeches from different years dedicated to the subject of nature's protection. They were given during sessions of the All-Russian society for nature protection, special conferences and official meetings in state departments, for example, such as the State Planning Committee of the USSR Council of Ministers (*Gosudarstvennyĭ planovyĭ komitet Soveta Ministrov SSSR*) [2].

a

b

Figure 1. Portraits of Vera Varsanofieva: a – in her youth; b – in her middle age.

There were several subjects she mostly concentrated on. Among them the need to protect objects of inanimate nature as much as living nature; the necessity of protecting nature in complex including inanimate nature, living nature, and humanity as an integral part of that last; the terrible need in the reasonable and accurate use of natural resources; the need to teach children and teenagers to love and protect nature and last but not the least the necessity to create and protect natural reserves.

One of her ideas which she actively promoted was the need to protect not only animate nature but inorganic nature as well. Namely mountains as well as interesting and rare outcroppings of rocks, etc. In 1959 she published an article entitled "Monuments of inanimate nature" (*"Pamiatniki nezhivoĭ prirody"*) where she outlined her thoughts on the subject [3]. First, she mentioned that layers of hard and loose rocks which make up the earth's crust are in fact beautiful chronicles of Earth. She argued that they were in fact pages of an interesting book where Earth itself wrote its own history. She outlined natural factors destroying even mountains and named several very interesting for geological research rocks outcroppings situated in the USSR.

She then mentioned that sedimentary rock which contained fossils were unique monuments of nature as they presented information about development of life on earth, allowed to reconstruct the history of modern plants and animals and tells us a story about changes which occurred on the surface of the globe. She argued that rock outcroppings rich with fossils should be preserved from destruction as the monuments of geological history of great value. She named then several very interesting for researches and rare places on the territory of Pechora-Ilych nature reserve. She also argued that it was important to preserve natural outcrops on the sites where extractions of minerals were performed as they allowed to trace outcrops relationships with the surrounding rocks and other peculiar properties.

Her next point was that the inorganic nature was closely connected with plant and animal life as well as the artifacts of prehistoric man. She concluded that sites of wild nature where significant geological and mineralogical monuments were mixing with archaeological artifacts and interesting specimens of vegetation and fauna should be declared complex nature reserves and be protected by law. She argued that there it would be possible to study a diverse and complex existence of the inanimate and live nature simultaneously as they are closely connected between themselves. That is why according to Varsonofieva it was so important to organize and protect nature reserves in different regions of the country.

In 1961 Varsonofieva participated in the special conference dedicated to the nature protection in Komi Autonomous Soviet Socialist Republic (*Komi Avtonomnai͡a Sovetskai͡a Sot͡sialisticheskai͡a Respublika*) where she was one of the organizers and main speakers. The purpose of the conference was an attempt made by scientific community to protect the Pechora-Ilych nature reserve from elimination or severe reducing of its territory. Varsonofieva began her speech with the statement that it was well known that everything was interrelated and interdependent in nature and to learn the laws of nature evolution was possible only by taking into account all complex and diverse connections between different elements of the landscape [4]. And to learn those laws is essential as for the development of natural science as well as for sensible and clever use of natural resources. She stressed that rational and profitable use of natural resources is possible only with parallel care about their recovery and increase. And in turn that is possible if nature is exploited and then restored on the base of complex and comprehensive research. Numerous failures in rearrangement of nature undertaken by soviet government without understanding of many and various connections between different parts of relief and consequences of destroying of any link in this chain were the sad result of such ignorance. As examples of such failures with dire consequences for environment she mentioned the deliberate lowering of the water level of lake Sevan in Armenia, land reclamation of the floodplain of the Oka river. Then she argued that optimum conditions for a comprehensive study of nature existed exclusively in nature preserves. She finished her speech with the statement about importance of the environment protection for people of different countries. The humanity, she argued, couldn't afford not to care about depletion of natural resources, threatened extinction of rare animal species, about the need to preserve areas of wild nature as it is our duty to future generations. And next she logically suggested that the USSR with its large and diverse territory should take the leading role in the environment protection. And that was a quintessence of Varsonofieva's ideas about the environmental protection.

The conference was a success. The opinion of scientific community was communicated to the government and taken into account. As a result, not only the Pechora-Ilych nature reserve was left alone, its territory was somewhat increased. And it is staying that way till nowadays.

The idea about rational nature management was among the most important in Varsonofieva's opinion. Conference held in 1961 was not the first where she spoke about the matter. She was not afraid to name sad or even tragic consequences of some of the governmental projects, as I already mentioned, and also common consequences of greed, neglect, uncaring, and pure stupidity of the business managers, company directors, and poachers. She was really angry about destruction of forests in Carpathians, central Ural, North Ural etc. She frequently spoke about the consequences of such actions for rivers of these regions, for its flora and fauna. She despised poachers who killed animals for example moose with cruel methods and then even did not use the whole animal leaving large parts of it to rot [2, p. 52]. She argued that it was impossible to condone such behavior. Varsanofieva was convinced that there were ways to minimalize damage inflicted on nature by economic development firstly by way of consulting with scientists before the start of new gigantic and potentially dangerous nature-relevant projects and secondly by teaching children to love and respect nature.

The idea about dare need to raise children loving wild nature was also among dearest to Dr. Varsonofieva. She argued that handmade parks and gardens would never replace wild nature in upbringing of the well-balanced and decent human beings. As an official of the All-Russian society for nature protection Varsonofieva supported nature protection education in secondary school's curriculum. In her opinion a child raised with the love of nature would never commit acts of atrocities against her as an adult even when he will be in charge of large and profitable economy projects. He or she would care and that was the most important for Varsanofieva, the result she wanted to achieve.

It is not strange at all that authorities didn't support her ideas much, that her papers on the subject often were left unpublished in strictly regulated soviet periodicals and that her activities with time vanished from memory. Nevertheless, largely forgotten in this day and time, efforts of Dr. Varsonofieva in the field of environment protection were significant and had been continued by those who shared her views even if they never heard her name [5–7].

4. Conclusion

After analysing the papers and unpublished writings of Dr. Vera Varsanofieva it is possible to conclude that she was a conscious supporter of ideas of environmental protection. And first of all, ideas of protecting remaining on the planet wild nature by creating nature preserves. She spent a lot of time and effort fighting for her ideas, created a chain of logical arguments based on natural laws and although her publishing's on the subject are rare they were significant as was an example of her active position and some of her winnings. She was a prominent figure in soviet scientific community, a teacher in her own right and was able to influence not only her colleagues but the young generation as well.

References

[1] Valkova O A 2012 Interview with the historian of science: questions that have not been asked. Materials to the biography of Vera Alexandrovna Varsanofyeva (1889–1976) *Issues of history of natural sciences and engineering* [in Russian – Voprosy istorii estestvoznaniya i tekhniki] **1** 100–127
[2] *Russian state archive of Economics* Fund **3** List 1 File 24
[3] Varsonofieva V A 1959 Monuments of inorganic nature, in *Cultural monuments of Komi ASSR* pp 92–109
[4] Varsonofieva V A 1961 Comprehensive study and protection of nature and their importance in the development of the natural sciences and the rational use of natural resources, in *Environmental protection in Komi ASSR* pp 8–13
[5] Crespin S J and Simonetti J A 2019 *Ambio* **48(2)** 131–138
[6] Stokes D L 2018 *Biodoversity and Conservation* **27(6)** 1539–1544
[7] Watson J E M, Venter O, Lee J, Jones K R, Robinson J G, Possingham H P and Allan J R 2018 *Nature* **563** 27–30

Cadastral land surveying as a tool to study and develop Russian lands of the 16th – 17th centuries

S N Volkov, I I Shirokorad, E S Kievskaya and E V Ryzhkov

State University of Land Management, 15, Kazakova str., Moscow, 105064, Russia

E-mail: Shirokorad_irina@mail.ru

Abstract. The article reveals the role of cadastres surveyings, as the most important land arrangement and planning activities in the framework of the study and economic development of the Russian territory during the final stage of collecting lands around Moscow and the formation of a estate-representative monarchy in Russia.

1. Introduction
Land planning (land surveying) arose long before the formation of the Old Russian State. Its appearance was associated with the development of social approach to production in general, the emergence of statehood and economic laws in the field of land planning, which began to manifest themselves in the early stages of social organization [1]. Since the Russian Empire had been manifested, the land surveying activities were carried out to study and develop the lands, establish the order in lands' using, set the boundaries and areas of land tenure, the rational use of lands. Among these activities can be identified cadastral surveying.

2. Methods
The work was prepared on the basis of the narrative, historical-genetic, comparative methods of scientific research, including the principles of historicism and objectivity.

3. Study of cadastral surveying of the 16th – 17th centuries
The period of the 16th – 17th centuries is characterized by the rapid growth of the territory in the Russian State. The total area of the country in the middle of the 16th century reached 2.8 million km, and by the end of the century – 5.4 million km [2]. By conquest, on the basis of vassal relations, Bryansk lands, Ugra land, Pskov Republic, Volotsk, Ryazan and Smolensk Knyadzestvo (principalities), lands of Kazan, Astrakhan Khanates, Great Nogai Horde, Siberian Khanate, Kodsk Knyadzestvo, Apostolic lands, Izhora and Pegaya Hordes were annexed voluntarily to Moscow. [3].

In the next century, the expansion of the Russian State borders continued. It included the Lands of the Don Cossacks, Left-Bank Ukraine, Zaporozhye, Southern Urals, Ishim, Yakutia, the Okhotsk Sea, Kolyma, Anadyr, Pribaikalye and Trans-Baikalia, Priamurye, Ingermanlandia, Livonia and Estland and a number of other territories. Thus, by the beginning of the 18th century (1708), the Russian State occupied the terrotory of 15.02 million square km. [4]. This means that during this period, the territory of Russia increased more than five times.

The annexation of new lands was accompanied by the large-scale reforms in military affairs, the fiscal system, state administration and legislation, which aims at building a State on the principles of estate representation.

An important tool in studying and economic development of new lands became cadastral land surveyings, which were massive documentary descriptions of lands and their possibilities in terms of organization of a local management system, taxation system, and economic development of individual territories.

The first Tsar of All Russia, Ivan IV, with the adoption of the Code about the service including the period of 1555-1556, marked the beginning of the mass documentary descriptions of lands. At the same time, the Book of Sokha Letter was published, in which the calculations towards lands planning with applications of the geometric drawings were indicated (instructions on how to organize land planning works).

Tsar Ivan the Terrible, in his message to the Stoglav Cathedral, "ordered to send seven 'notariats' (scribes) to describe all their land ...". And later, by the Decree dated on September 20, 1556, he ordered: "who among the nobles and the boyars' children took possession of many lands, and are scanty in state servicing ... on estates with surveying, comprehend and establish what they deserve, and divide the non-deserved lands among the poor" [5].

Land surveying works were carried out in accordance with some special documents describing a procedure, which were called 'scribes' (prescriptions). In the course of these works, the notariats made records about local and patrimonial lands, some lands belonging to villages and small villages in conformaty with their landholding, described the population and its occupations, checked the availability of lands according to previously issued documents of ownership (state certificates, records from land surveying books, bill of sale and etc.) with the actual use of lands based on the descriptions and measurements data, the testimony of witnesses from local people; established the legality of land tenure and seized hidden or illegally used lands in favor of the Sovereign with subsequent distribution; described lands according to landholding types, qualitative and cultural-technological state, including the types of agricultural use; solved land conflicts.

One of the main tasks of cadastrel surveying was to establish the boundaries between landowners. In one of the mandates to the notariats, referring to the middle of the 17th century, there was said: *"woodland, and villages, and floodland, and lands inhabited and used as means to living which have become the landlords lands with plowed lands and mowings and with any other land, although they are not the subject of a conflict.... those woodlands are needed to divide (make boundaries between them) and in the book these boundaries are to be fixed in order to avoid further land-related conflicts"* (art. 29 Cadastres order 1622) [6].

During the entire reign of Ivan IV, the lands of more than 22 cities in the central regions of the country were described and cadastred - the Dvina region (lands lying on the western coast of the White Sea), Novgorod, Pskov, Polotsk, and Livonia.

In the Order of 1622, a number of articles were addressed the border relations and land surveying, and it was emphasized that cadastral descriptions – the important governmental task, which the State takes as its own responsibility, was fixed in the Local Order by the notariats of the department.

Cadastral land surveyings were supplemented with some graphic images of a locality, made extremely schematically, due to the lack of the geodetic instruments in that time. An example of such drawings is shown in Figure 1.

In the first half of the 16th century, the first cartographic works were carried out in Russia. The main instrument for measuring land plots was a measuring rope with the length of 80 sazhens, and the width – 30 sazhens. I. Ye. Herman gives the citation from the document: *"It is fitted that the leader in the Moscow State is given to notariats the measuring rope with the length of 80 sazhens, and the width - 30 sazhens. And in those two measuring ropes, there are 2400 quadrangular sazhens"*[1] .

[1] We are talking about the State tithes in 2400 square sazhens, which served as the main measure of land in Russia until 1917.

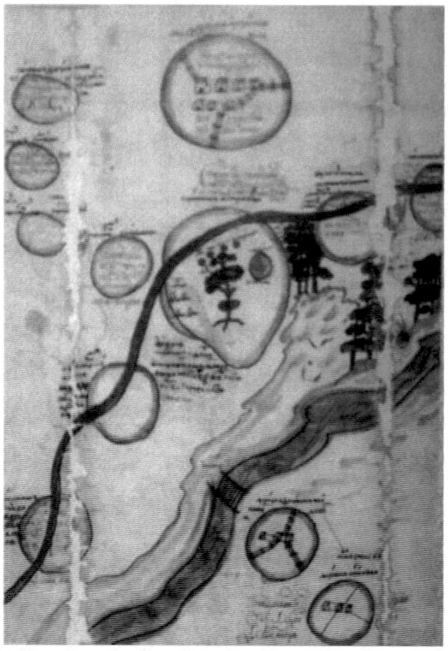

Figure 1. Drawing of lands in the Bezhetsk district, in Gorodetsky and Meshchersky camps, taken from a conflict case between I. I. Buturlin and I. I. Monomakhov. The conflict theme was about the search for the floodlands of Yastrebuhi and Strebukha in Meschere and in camps

According to the results of the first cartographic works, a map depicting *Muscovy* (Moscow Land), as well as a map of the Moscow State with a part of Siberia and Tataria was compiled [7]. The map is presented in Figure 2.

Figure 2. Map of Muscovy by G. Bessel (Dutch geographer and cartographer)

In 1697, the book "Geometry, or land surveying by radix and compass, deeply wise, giving an easy way to measure the most inaccessible places, lands and forests," compiled by hand, was published. In the book, the considerable attention was paid to the measurement of land and issues which the notariats and city builders were facing.

At the beginning of the 17th century, during the Time of Troubles, there were large changes in land use related to the illegal occupation and land embezzling. During the numerous fires of the Moscow Kremlin, the documentation of the National Order also suffered significantly. Addressing issues related to the ordering in land laws, became one of the priorities of the Government of Mikhail Fedorovich Romanov (22.07.1596 – 23.07.1645).

After Mikhail Fedorovich ascended to the throne, the sentinel guards were sent to the western and northern regions of Russia to describe the country's regions devastated by the Polish. However, these works were practically not started. Then in 1619, the Order was issued, according to which *"it was again determined to send the notarints to cities, which were not raveged or under ravege from the part of the Lithuanians and Cossacks; to send good patrolmen, providing them both with full instructions, so that they describe and dismantle all cities by the truth, without bribes"* [8] .

The Decree dated on October 16, 1629 "On land surveying in the Moscow district", and then in Art. 50-53 ch. 17 of the Cathedral Code of 1649 there were legally enshrined rules and procedures for land surveying.

First of all, the order of carrying out surveying works on private requests of land owners was determined, the procedure for handling conflicts about land division was regulated, the procedure for resuming overgrown lands was established. Also, the Legislation defined severe punishment for the law violation:

a) *"who breaks the land surveying, no matter state or private land, particulaly who: throws out border-pillars, level pits, or plow up the land, and be found guilty, he will be punished at a place mercilessly by whipping and thrown into a prison for a week; in favor of a claimant, it is charged 5 rubles for each land line, and boundaries, lines and pits are renewed as before. Whenever the peasants in the same village did this, they are punished with a whip, and the boundaries and land lines are renewed as before";*

b) *"who forcibly takes away the land or redoes the boundaries and lines and will be found or caught in doing it, he will be taken away this land and this land will be given to someone who has been taken it away; additionally, the land property of a person, who is guilty, is reviewed in line with the Decree of the Sovereign, and for violence that he has done, he is given the punishment, what the Sovereign will assign" [9].*

In the reign of Alexei Mikhailovich Romanov (March 19, 1629 - January 29, 1676) the land surveying continued in various regions of the country in accordance with the above legislation. In 1673-1674, in the Zaoksky region, the notariants carried out extensive work related to the decision of the State to colonize these lands.

Forty years later, after the adoption of the Cathedral Code, the serious difficulties again arose in the land system of Russia. To a large extent, this was related to the legislation of that time about so-called "overgrown lands". According to the Cathedral Code, adopted in 1649, it was allowed to request ownership of floodlands and "overgrown lands" that were within the estate or beyond their borders in order to expand the tribute collection, but the lack of descriptions of this category of lands led to misunderstandings [10]. There were precedents when several owners claimed the same piece of land. These pieces of land that already had their actual owners, whose rights were not documented, fell into the category of floodlands. The same lands had different names in the cadastral books and was recorded as lands of different owners, etc. The archive of the National Order has preserved many documantes relating to some conflict cases, concerning the consolidation of tenure rights in respect of vacant land, suitable for agriculture. The sample material of such cases is presented in Figure 3 [11]. Due to the situation relating increasing confusion in ownership rights to the "overgrown lands" by the Decree of 1679, the transfer of floodlands into ownership was suspended until the end of their surveying. However, having begun land surveying in relation to "overgrown lands", the State was confronted with the necessity of gross land surveying of local and patrimonial lands [1].

In 1680, the gross notariants began to conduct the descriptions of land, whose works continued until 1686 and were discontinued due to the outbreak of the Russian-Turkish war. During this period, Voevods were engaged in surveying (fixing boundaries) of the local lands.

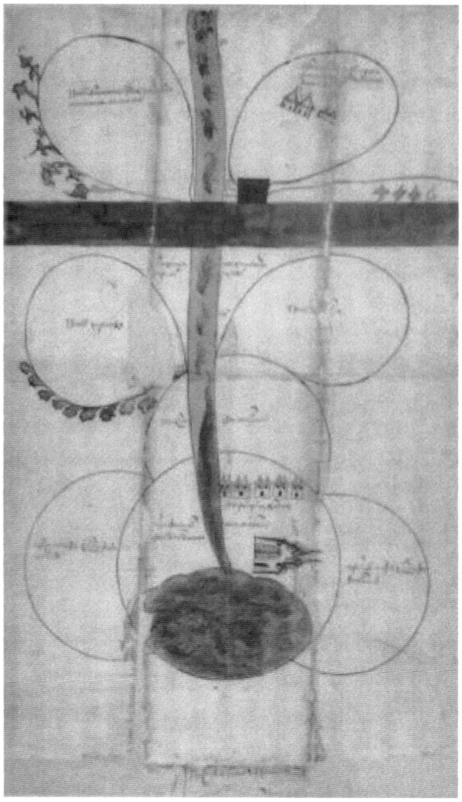

Figure 3. A drawing of the lands of Filippovskiy village in Pereyaslavsko-Zalessky district from the conflict case between G. A. Bashmakov and the head of the Moscow Strelci (soldiers) R. S. Efimiev about the floodlands of Kuznetsovka and Kharyovka in Serrebozhsky camp.

Land surveys in the course of the gross description of lands were regulated by legislative acts, instructions and pattern-based documents, including the notariants Orders issued in August 1681 and April 1684, as well as with the additional articles that were published in March 1685 and in March 1686, given to the notariants for land surveying towards Mordovian, Cheremis and Chuvash lands, as well as the cities subordinated to the Order of The Treasury [6].

The notariants (or land gatherers) or patrolmen were engaged in checking, correcting and updating the descriptions made earlier. In cases where detailed descriptions and measurements of a land were made, these people were called 'Big'. They were notable, endowed with substantial powers.

The main purpose of these written descriptions was documentary fixation, identification of owners of lands and other real estate, the layout of taxes to the Treasury.

The notariants determined the boundaries of land plots, the categories of land use (the number of arable land, fallow lands, meadows, forests), the quantity and quality of arable land (good, average, thin), calculated taxes for land use, resolved land conflicts and issued property titles [12].

Also, they resolved arguments about the possible leasing of state-owned industrial and other institutions, among which salt manifacturie facilities and mills, etc. can be noted.

For the designation of the lines in the land surveying documents, there were used different symbols. Among them: natural - rivers, streams, lakes, swamps, ravines, trees, roads, mounds, fences, hedges, neighboring lands, yards, etc., as well as man-made marks cut into trees and pillars, holes filled with stones, coal, ash of industrial type.

In the forest area, the marks were carved on separate trees. Other trees that separated the survey, were hewn. On arable and hay lands, the survey was a line of land that was not cultivated [1].

Borders of districts, volosts, camps, and pyatin were also separated by land surveying marks.

The acts that recorded the information about land holdings were very diverse. All data on land ownership were recorded in books including the cases of refuses, separation, and gifting.

Executive decisions of the Sovereign or the Local Order to voivods, town majors, other governers about rejection towards property rights on land ownership were entered/recorded in special books (refusal one).

The decisions relating to transferring the property rights to another person within the same district were recorded in the other special books (separations and gifting).

In books containing the data about summer houses the information indicated: to whom, of what kind, in what quantity, in what locality the land was allotted for the patrimonial or local ownership by Order of the Government.

A general and complete picture of the organization of surveying works can be imagined on the basis of one of the notariants' order dated by the end of the 17th century, for example, the Order of 1684.

In accordance with the Order of 1684, the land surveying was carried out by a Local Order.

Surveying took place with the obligatory participation of owners, the old-timers, local and neighbour-living people. By the beginning of the land survey procedure, land owners had to provide property titles – “fortresses”.

The scribes/notariants had with them the following set of documents: the scribal order as a legislative document, the former scribal, surveying books, a three-plane sazhen, a rope for measuring a land.

A scribe/notariant who arrived to do the land surveying read out the Order to the patrimonial landowners, explaining a land surveying procedure.

In case of any misunderstandings, a scribe could have filed a complaint within the Order under which he was working. To clarify the circumstances, the high-level officials were sent to places and they were called "detectives". Their duties included conducting an investigation regading ths or that case.

In 1686, additional articles for the Land Surveying (in 1684) were legislatively approved – recommendations for controversial cases.

Summing up, it is necessary to dwell on the general features of land surveying of the end of the 17th century. The survey in that time was already separated from the description of the land, the layouts of the population according to the sokhas and other functions which were characteristic of the surveying of the beginning of the century. From now on, the scribes/notariants only checked and supplemented the information obtained earlier.

4. Conclusion

Written land surveys carried out during the 16th – 17th centuries allowed to study the territories newly attached to the Russian State from the point of their economic opportunities dealing with spreading the local, fiscal and military systems [13].

The results of written land surveys of the 16th – 17th centuries formed the basis of the General Land Survey in the Russian Empire, which began in the reign of Catherine II and lasted more than a hundred years. Notariants' surveying, sentinel books compiled during the surveying work of this period are the most important mass historical source allowing to study the various aspects of state and public life during the time under consideration [14].

5. Acknowledgments

The authors express the special gratitude to all those who provided the information and technical support.

References

[1] Volkov S N and Shirokorad I I 2011 *History of land planning in Russia: the experience of the millennium. Educational and scientific publication* (Moscow: State Healthcare Institution) p 13

[2] Zimin A and Khoroshkevich A 1982 *The Russian Kingdom Russia of the Time of Ivan the Terrible* (Moscow: Science)

[3] Hanway J 1754 *Herrn Jonas Hanway zuderlassige Beshreibung seiner Reisen* (Hamburg; Leipzig) iss. 1
[4] *Encyclopedic dictionary of Brockhaus and Efron* in 86 volumes vol 54 1899 (St. Petersburg) pp 211–3
[5] Ivanov P I 1846 *The experience of historical research on land surveying in Russia* (Moscow: Publishing House of S Selivanovskiy) p 19
[6] *History of Russian land surveying in materials and documents* 1930 ed I D Shuleikin (Moscow; Leningrad: Selkhozgiz) p 49
[7] Shaikhislamov R B and Aznabayev B A 2018 Landlord colonization of Bashkortostan *Bulletin of St. Petersburg University. The history* **63(4)** 987–1000
[8] Zabelin I E 2009 *Russian history Russian antiquities or archeology* Lecture course (Moscow: Gralia) p 19
[9] *Cathedral Code of 1649 Text Comments* 1987 (Leningrad: Science) p 87
[10] The 17th century and the European context 2018 *Ural Historical Vestnik* **61(4)** 6–13
[11] *RGADA* F 1209 Op 77 trans. 92070 L 231
[12] Ilyushina T V and Nazarov L S 2017 The state significance of land surveying and old linear-land measuring in the history of Russia development *News of higher educational institutions Geodesy and aerophotography* [in Russia – Izvestiya vuzov. Geodeziya i aerofotosjemka] **3** 6–13
[13] Ragulina M V, Rogovskaya N V, Grigorieva M A and Ippolitova N A 2018 Geographical factors of the 17th and 19th centuries *Past Years* [in Russian – Bylye Gody] **48(2)** 496–504
[14] Ilyushina T V 2016 Surveying and mapping of the first province of Russia *News of higher educational institutions Geodesy and aerophotography* [in Russia – Izvestiya vuzov. Geodeziya i aerofotosjemka] **4** 71–7

The Information System *"History of Geology and Mining"* as a scientific tool for historians of Earth sciences

I P Vtorov[1]

[1]Institute of Geology, Russian Academy of Sciences, 7, Pyzhevsky lane, Moscow, 119017, Russia

E-mail: vip@ginras.ru

Abstract. The Information System "History of Geology and Mining" is described in this paper. It briefly presents the theory of scientific information in modern IT and the study of history of geology in Russia. The collected scientific information about geoscientists are developed and published at first in online history of the Science information system. The Internet system is the most convenient and useful way to use and develop scientific biographies, bibliographies, photos and other information for research, education and international cooperation in history of sciences.

1. Introduction

General properties of the scientific information are novelty, originality, completeness and objectivity. This information is mainly used for search, collecting, comparing, analysis and distribution. Sources of the information are necessary for the historians of sciences. Modern information data resources are expanding and getting more common in any science. Information technology (IT) helps to fill and develop this content. Work with such databases may become a standard research processes and computer systems allow scientists to avoid the "information barrier" of storage and move to paperless processes and IT [1].

From the information point of view, general science describes it as a complex dynamic, relatively separate system made for the collection, analysis, and production of scientific data [2]. Methods of storing and processing information may also be the subjects and sources of research [3]. Modern scientific information systems with Internet access are a structured set of interrelated components containing accumulated data with the collection of several valuable features, such as relevance, usability, expert assessment and reassessment, as well as the openness in combination with information law. The system combines, organizes, and provides access with search in computer data sources and multimedia files. These information theory statements provide a great opportunity for scientific information systems used as a modern online tool for historians of sciences.

2. Study of the history of geology in Russia

Professor Vladimir Vernadsky (1863-1945, academician since 1912) was the first historian of geoscience in Russia. He studied many problems in the history of scientific knowledge and published papers about scientific heritage and scientific biographies. In 1902, he started a lecture course about the history of natural science at the Imperial Moscow University. He planned to start description of the world history of natural sciences. He was a first chairman of the "Commission for the study of the history of science, philosophy and technology", then organized the "Commission in the history of scientific knowledge" in the Russian Academy of Sciences (1921), which was later transformed into the Institute of the history of natural science and technology [4].

Vladimir Obruchev (1863-1956, academician since 1929) published series of issues "History of the geological study of Siberia" (1931-1949) [5]. They were the first collected works about the history of geology in Russia. Obruchev also proposed to set up the Commission of the Geological Study of the USSR (established in 1955). The commission members from more than 80 geological institutions of the USSR Academy of Sciences, the Ministry of Geology, the Ministry of Higher and Secondary Specialised Education, and others took part in this work. They compiled over a million sources and published series called "Geological exploration of the USSR" (1961-1991). It was in 50 volumes (over thousand books) of the annotated bibliographies and reports covered all regions of the USSR [6]. The Commission was based on the Department for the History of Geology, Geological Institute, USSR/Russian Academy of Sciences. The commission was led (since 1958) by a historian of geosciences, Vladimir Tikhomirov (1915-1994), who was the head of the Department and the chief editor of the series.

In 1956, USSR joined the International Union for the History and Philosophy of Science (IUHPS) and created the Soviet National Union of Historians of Science and Technology (now the Russian National Committee for the History and Philosophy of Science and Technology). In 1967, the International Committee for the History of Geological Sciences (since 1980 – the International Commission on the History of Geological Sciences; INHIGEO) was initiated by Tikhomirov (first president, 1967-1976). His idea was to publish 30 volumes of the series "World History of Geological Sciences". The merits of the contributions of Tikhomirov to the history of geosciences and to the foundation of INHIGEO were recognized by the International Union of Geological Sciences – the IUGS Science Excellence Awards – "V.V. Tikhomirov History of Geology Award" (since 2012) [7].

3. Materials and methods

The above-mentioned Department for the History of Geology was collecting information since 1949, it was the first specialised division in the history of geology as a scientific theme [8, p. 77]. In 1965-1978, eight volumes of series "Materials about the history of geology in the USSR" were published. It was the beginning of production of dictionaries (included only 3 first Russian letters) with geoscientists' short biographies and bibliographies. The next step was the book series "Essays about the history of geological knowledge" published in 32 volumes (1953-2006).

Information accumulated over the years in papers (forms, reports, letters, autobiographies, event programs, list of papers, correspondence and other publications), photos and other illustrations. Over 2,500 personal folders of scientists are accumulated in our department, also several cabinets with bibliographic cards and photos are collected. All this information was required for processing and digitizing documents for preservation and presentation.

Our goal is to save and publish this information as a source for historians of sciences. The best way to solve this problem was the creation of the online Information System "History of Geology and Mining" [9]. The System is located on URL: *http://higeo.ginras.ru* (since May, 2019).

The System is based on scientific books' catalog software platform "SciRus" which was developed by the Library for Natural Sciences, Russian Academy of Sciences [10]. The Department for the History of Geology made a structure of the system and its content, the Library provides IT support.

4. The Information System

Information loaded to the database is connected to the dynamically generated web pages. It combines personal data about scientists and includes biographic notes, links to the organisations where a researcher worked, documents, photos, membership in the Academy of Sciences and other scientific societies. It also includes geography of research, areas of work (according to the "Russian state rubricates of scientific and technical information"), some scientific biography or autobiography, lists of publications, obituaries, papers about scientists and other information.

More than 900 scientists were added to the system in 2019, and the number is still increasing. It is

17th– 21st century's naturalists, geologists, mining and soil scientists, some geographers, and other geoscientists. First of all, we included all academicians, corresponding members and foreign members of the Russian Academy of Sciences, geological and mining departments.

The main advantage of the online information system is integrated search tools, which allow you to find any information. Table 1 shows search options in: scientist's names, bibliography, documents, personnel, organisations, sources of information, and photos. If you leave search window blank and press "search" ("искать" in Russian) button, then you can see total list of links to persons, documents or photos. The system is also open to main international Internet search engines. Scientists' names and system interface is added in both Russian and English. Bibliography titles are presented in original European languages (oriental languages are translated to Russian). We use some standard abbreviations in bibliography serial titles and organisations, transcript available.

Table 1. Information System "History of Geology and Mining" online search options

Documents	Scientists	Organizations	Sources	Photos
Title (in Russian)	Full name (in Russian)	Original title	Title	Title
Original title	1st name (in Russian)	Title (in Russian)	Type	Description
Document type	2nd name (in Russian)	Brief info	Description	
Document language	Surname (in Russian)	Activities		
Document year	1st name	History		
References	Surname			
Comments	Centuries (XVII-XXI)			
Full text	Date of birth			
Comments	Place of birth			
File (PDF)	Date of death			
	Place of death			
	Scientific degree			
	Fields of interest			
	Exploration regions			
	Scientific institutions			
	Biography			
	Bibliography			

The System is also developed to collaborate with colleagues from the INHIGEO in research and bibliographic projects. 27 Russian INHIGEO members are presented in the System.

Table 2 shows the number of organizations and institutions included with lists of associated personnel. Each scientific bibliography and biography has its author-compiler; publications are checked by our editors and updated as content is expanded.

The long-term work of professional bibliographers of the Department for the History of Geology has resulted in hundreds of thousands of bibliographic cards (Russian and foreign), which we process into standard lists of bibliography and place them in the Information System. 50 thousand important cards were scanned for further development.

Table 2. Information System "History of Geology and Mining" content fills for mid 2019 (with future plans)

Documents	Scientists	Organizations	Sources	Photos
710	870	460	105	1,270
(> 10,000)	(> 2,500)			(>20,000)

Documents include original scans (mostly in PDF files) of forms, autobiographies, obituaries, some old newspaper publications, correspondence, manuscripts, programs of the anniversary events, reports, awards, and more historical materials. These online documents may be the first step to review from our information sources, many other documents and papers we cannot publish because of copyright regulation. Then welcome to contact us for further unpublished information about our collection of publications, papers and photos. Our offline history of geoscience information sources includes personal scientist's files and folders, literature, translations, illustrations, autographs, photographs, and other documents. All of this additional information are available for all forms of cooperation with historians of sciences, researchers and students.

5. Connected information sources

Bibliographies, obituaries, and other titles in our System often contain Internet references (web URL links) to the sites with official publications. In particular, the Information System is closely integrated with the digital library "Scientific Heritage of Russia" (URL: *http://e-heritage.ru*), created under the program of the Presidium of the Russian Academy of Sciences. It contains over 2,100 publications on geosciences.

We have already scanned about 9,000 photos of geoscientists and mining engineers, over 20,000 more left in paper and film format. Online hidden copy of scanned photos are easy to arrange, classify, tag and search for its description. Some of them as thumbnail pictures gallery we add to the Information System for preview. Digital images may use in exhibitions and publications.

6. Conclusion

Any historians of geosciences can get online Internet access to the Information System "History of Geology and Mining". They can use it to search and study information about scientists, publications, organisations, sources, etc. It can also serve as a reference database for bibliographies, biographies, publications, images, and documents. Analysing primary sources of information, searching for links, reconstructing the chronology of events are very useful to historians of geosciences. Dates and places of conferences and congresses in publications are includes, reviews, editors, discussions, reprints and translations of works added in bibliographies.

Sorting and searching are especially convenient for searching anniversary dates. Since 2017, students of the Geological Faculty of the Moscow State University use the System in the lecture course "History of Geology". The System already cited in source studies by scientists, museums and in cultural projects. We can see web traffic statistics from last year with 2-6 thousands monthly visitors, mostly from former USSR (in descending order) Russia, Ukraine, Belarus, Kazakhstan, Armenia, and Azerbaijan, but also from Germany, Norway, USA, France, UK, Israel and other countries.

Further development of the "History of Geology and Mining" Information System may be in database expanding with personalities of scientists from around the world, and the inclusion of related geosciences. While this is the first and only system in the history of science, its future lies in interdisciplinary and integration with future systems and in the expansion of functionality. This will allow turning it into an "expert system" that could analyse and solve with logical forms, global questions of estimation and modelling.

With the future contacts, communications, joint projects, and open sources information exchange, the idea of Vladimir Vernadsky about the "World History of Natural Science", as well as Vladimir Tikhomirov's plan to cover the "World History of Geological Sciences", may come true in open digital format and brake languages, educational, ideological, financial and bureaucratic barriers.

References

[1] Glushkov V M 1987 *Basics of paperless computer science* (Moscow: Nauka) p 557
[2] Dobrov G M 1970 Information concept of the scientific process *Science of the Science* (Kyev: Naukova Dumka) pp 36-71
[3] Ufimtsev G F 2006 Science content *Seven words about the theory of geology* (Moscow: Scientific World) pp 18-27
[4] Yanshin A L and Yanshina F T 1997 Preface to Vernadsky V I *Scientific thought as a planetary phenomenon* (Moscow: Vernadsky foundation) pp 6-20
[5] Obruchev V A 1931-1949 *History of geological research of Siberia* (Moscow, Leningrad: USSR Academy of Sciences Press) [5 periods, 12 books]
[6] 1961-1991 *Geological state of exploration of the USSR* [50 volumes in 1050 books]
[7] Malakhova I G 2017 The founding of INHIGEO: documents and letters, with comments *Celebrating 50 Years of INHIGEO* (London: GSL Special Publication 442) 9-19
[8] Tikhomirov V V, Soloviev Yu Ya, Panutina L B, Gordina I A, Malakhova I G and Bugelskaya L V 1980 *History of the Geological Institute of the USSR Academy of Sciences: Development of the Institute, its scientific schools and bibliography of the published works* (Moscow: Nauka) p 223
[9] Kalenov N E and Malakhova I G 2017 Integrated public information resource "History of Geology and Mining" *Information resources of Russia* [in Russian – Informatsionnye resursy Rossii] **1** 19-23
[10] Yakshin M M 2015 Development of the platform SciRus *Information support of science: new technologies* (Moscow: BEN RAS) pp 203-207

First plan of the area near the source of the river Don (1701)

I N Yurkin

S I Vavilov Institute for the History of Science and Technology, RAS, 14, Baltiyskaya street, Moscow, 125315, Russia

E-mail: ig-yurkin@yandex.ru

Abstract. We have found records on the drawing of the map of *Ivan-ozero* (Ivan lake) and the upper course of the River Don dated 1701 in the archive of *Razriadnyi prikaz*, RGADA. These records are the earliest evidence of mapping of the region of Don in its upper course at the moment. We outline the circumstances of drawing of a plan. We reconstruct geographic information of the plan according the revealed description. We compare data of the description to data from later cartographic sources. We conclude that sizes and distances reported in the description are reliable data. We surmise about the authorship of the plan.

1. Introduction

Geographers of the ancient world paid attention to the River Don from the very beginning. Cartographers of Western Europe depicted it on the better part of maps of Eastern Europe. Although the ideas of what was the River Don in its upper course differed from one another, many of cartographers, according the tradition that went back to Herodotus, depicted a lake as a source of the Don.

Mapping was common in Russia in the 17th century. But the best parts of geographic drawings were compiled to solve specific problems, most of them to resolve land disputes. The region of Don in its upper course was a sparsely populated area in the 17th century, especially in the first half of the century. The intensity of economic life was low. Location plans of the area of that time were not found.

At the beginning of the 18th century the construction of *Ivanovskii kanal*, the waterway to connect the River Don to the tributaries of the River Oka, was launched in the region. The canal route passed along the rivers Upa, Shat and Don. In some areas water was passed through artificial channels where locks were built. Ivan-ozero was a small lake located in the area between the rivers Shat and Don. The lake was cleaned from mud and deepened.

Prince Matvei Petrovich Gagarin was appointed for a duty as the head of works and he went to Ivan-ozero from Moscow in July 1701. He traveled along the future canal route and examined it [1]. Prince Gagarin supervised the construction of Ivanovskii kanal from headquarters on Ivan-ozero. Thousands of people worked on the construction of Ivanovskii kanal for 10 years. Technical maintenance of the project was done by Dutch craftsmen and engineers.

We have got brief and inconsistent accounts of the Ivanovskii kanal from the literature of the 18th and the beginning of the 19th century. Ivan Shtukenberg, German by origin, was the first scientist in hydrography and statistics, who published reliable information about the history of it [2]. The most complete study on the subject in historiography of the time before the revolution of 1917 was the article of A.I. Milovodov [3]. Later the history of the project was not the subject of studies, but we can see more precise definitions and additions to the subject in some articles [1, 4–6]. Some of the authors

IOP Conf. Series: Earth and Environmental Science **350** (2019) 012021 doi:10.1088/1755-1315/350/1/012021

analyzed maps and plans of the area of Ivanovskii kanal route [4, 5]. They described early experiences of the mapping of the region of the Don in its upper course.

Map of hydrographic system of the region of Ivan-ozero and projected canal route were drawn likely in preparation for construction. Perhaps this kind of work was done also during the process of construction. When and how was this initial mapping of hydrosystem of Don in its upper course done? What were the results? It is important to give an answer because information about the geography of the region was not correct, especially in Europe. S. von Herberstein in Rerum Moscoviticarum Commentarii (Commentaries on the Moscow principality) stated that Ivan-ozero, the source of the rivers Shat and Don, was 1500 *versta* (measure of length) long and 1500 versta wide. Herberstein wrote down information gained during the visit to Moscow at the time of Vassilii the III, Grand Prince of Moscow. Description of Herberstein was one of the main sources of knowledge about Russia for the Europeans for a long time. Commentaries were cited as an authoritative source. The same mistake in measuring of Ivan-ozero (in fact the lake was small) was repeated many times. Even the Dutch cartographer N. Witsen described Ivan-ozero as a big one [7].

Meanwhile Russian maps and records, in which the real size of the lake was described, were already existed at the beginning of the 18th century, and perhaps even earlier. They were done during the construction of Ivanovskii kanal. We will not venture to affirm that persons who assisted in the work (those who were on the site or those who arrived for a short period of time) could make high-quality geographical drawings of the area. Having regard to the lack of maps of the region in Western Europe and even in Russia, find of a drawing is an advance in the research of the geography of the region.

2. Materials and methods

To answer the question, we have done research in the archives. We looked for the maps and descriptions. We focused on documents reporting on the initial stage of construction of Ivanovskii kanal.

Prince Gagarin worked closely with central and local bodies of government. Correspondence was maintained in the executive office on Ivan-ozero (*Ivanoozerskaia kantseliariia*). Executive office had its own archive, with incoming documents, drafts of outgoing documents, as well as, possibly, technical documentation. This archive was not preserved. All information about the construction is extracted from documents remained in the archives of other institutions. Most of them reported on the material maintenance of the construction, equipment, tools, workers and specialists.

We used records of Razriadnyi prikaz *fond* (collection) of Russian State Archive of Ancient Acts (RGADA). Razriadnyi prikaz expedited orders from Prince Gagarin in most cases. Razriadnyi prikaz decided the contractor and sent the appropriate order. Archival files that contain correspondence of Razriadnyi prikaz and Prince Gagarin of three primal years of construction are of special interest. They reveal data on cartographic works and persons who could make a job. To reconstruct the lost documents we used their descriptions. Data from descriptions was compared to data on water objects (Ivan-ozero lake and the adjacent area of Don watercourse) from records of the 2nd part of the 18th century and the early 19th century.

3. Results and Discussion

We have not found drawings of the design works or plans in the records of Razriadnyi prikaz. It seems that the said institution was involved in the issues of construction only from the start of works. Prince Gagarin did no report on the implementation of works till the autumn of 1701. We guess he considered himself accountable only to Tsar as he received the assignment personally from the Tsar. Razriadnyi prikaz provided Prince Gagarin with funds and material resources, and was worried about the productiveness of resources, but did not take liberties to require the report from commissioner of Tsar.

On September 30th, two months from the departure of Prince Gagarin from Moscow, *boiarin* (noble of high rank) Streshnev, head of Razriadnyi prikaz, ordered to send *podiachii* (minor official) Polikostritskii to Ivan-ozero. Polikostritskii was instructed to ask Prince Gagarin for report to Tsar on *Horomnoe* (palace, residential building). Prince Gagarin had to depict Horomnoe and to make a drawing of it with instructions keyed to accompanying drawings, to report with a letter to Tsar on estimation of the end of works. The same should have been reported on *Perekopnoe delo* (digging and construction of the canal). The drawing, the letter and the description had to be given to Polikostritskii, who would immediately go to Moscow [8].

The drawing (the plan) mentioned in this message to Polikostritskii is probably the earliest graphic document, the first mapping of the upper course of the Don. It is the subject of our report.

Polikostritskii was sent by Razriadnyi prikaz, but we have not found the said plan in the Archive of it. Perhaps the plan and the related papers were delivered directly to Tsar and were not send to Razriadnyi prikaz.

We have not found letters from Prince Gagarin from earlier date with data on works he supervised as an appointed executive. We think these letters do not exist. Gagarin reported to Streshnev in detail only in October. He mentioned the drawing (the plan) he sent to Razriadnyi prikaz in the report. He underlined he had no person to draw it better because "the place was a desert" [8]. As we can see, the history of this plan was not limited to the fact the plan was ordered to draw, the plan undoubtedly existed.

We do not know who drew it. Probably, officials of Ivanoozerskaia kantseliariia worked on it. It is also possible that Polikostritskii had a hand in it, officials of Razriadnyi prikaz did this kind of work. Gagarin probably participated as well as the head of work, consultant and supervisor. We have not found evidence of the presence of foreign experts on the lake in October 1701 in the records. So we doubt they took part in the drawing.

Although the geographical drawing of the area around Ivan-ozero of 1701 is not available for research now, it is possible to get an idea of its content. We can do it with the help of an excerpt which was preserved in the records of Razriadnyi prikaz Archive in two editions. In one of them, it is reported that Gagarin sent a *Chertezh perekopnoi rabote i zhilomu stroeniiu* (Plan of the canal construction with a drawing of a building). The excerpt gives the detailed description of the plan. It shows the proportions of Ivan-lake, it reports which rivers connected with the lake were cleared, it gives data on the constructed canals, it describes the adjacent land routes, describes the palace built for Tsar, measurements of the chambers and their specifications.

Officers of Razriadnyi prikaz took a close look at the report and put Polikostritskii a question. How far was the Tsar palace from the lake Ivan-ozero and the canal? Was there a bridge in the place where the road intersects the canal? What was the place where the river Urvanka flows into the lake? Policostritskii could not answer the question on water route of Urvanka before the beginning of earth-moving works [8].

No matter how imperfect was the plan of 1701 (the survey was done without the instrumental work), we regret for its loss. First of all, the document fixed data on the Don River's headwaters and the terrain. In other circumstances it would enrich the geographical knowledge of that time. In the second place, it contained information about the initial stage of construction of Ivanovskii kanal, a waterway project which could initiate developmental growth of the region.

A small insert map that shows headwaters of Don, Voronezh and Oka rivers is close in time to the plan of 1701. We can find this map as map-inset in an atlas *Novaia chertezhnaia kniga, soderzhashchaia velikuiu reku Don ili Tanais... (1703-1704)* (New drawing book depicting great river Don or Tanais... (1703-1704)). Drawer of the atlas had no idea what was the source of the Don River at a time. There was no Ivan-ozero on the map [4]. It was strange, as it went against the antique tradition of Don description, and also considering the fact that the mapper probably had sharp interest to this region at the moment of construction of man-made waterway. The plan of 1701 could have been used by the compiler of the atlas, yet it was not used.

The lake was present on the drawing created under supervision of Prince Gagarin. The description reveals the measurements of Ivan-ozero, it has 96 *sazhen* (measure of length) in its length, 64 sazhen in its width, 3 sazhen in its depth [8] (for sazhen of the 17th century it has 207.4, 138.2 and 6.5 meters respectively). Apparently these were first accurate measurements of the lake, which was the source of the River Don. They are valuable for the reason that since the time of Herodotus the lake, the source of Tanais, was considered very large.

Were the measurements of the plan of 1701 fixed accurately?

a) Are they a reliable source?

b) Whether they depict Ivan-ozero before the start of construction of Ivanovskii kanal or not?

We can compare the description of the plan of 1701 with the maps of the reign of Catherine the Great. In the period between the reigns of Catherine the Great and Peter the Great work on water project on the lake was suspended. So maps of the second half of the 18th century most likely fixed the layout as it was at the end of the reign of Peter the Great. Of course we should keep in mind possible changes in the location of rivers and in landscape happened for natural reasons, above all seasonal changes in natural watercourse.

We can see the lake of irregular form surrounded by filling dams on a chart *Plan kanalov, kotorye idut ot Ivan-ozera po reke Donu i Shatu* (Plan of canals that go from Ivan-ozero to the rivers Don and Shat). The drawing was made in between 1740 and 1775 (perhaps in 1763). A length of the lake is 100 sazhen and a width is 70 sazhen (216 and 151 meters respectively). On the other chart from *Atlas Tulskogo namestnichestva* (Atlas of Tula region) of 1787 the area in between the embankments with the lake inside it measures 100 by 60 sazhen (we see the same measurements in the explanatory text of the atlas) [5]. So, the data received from these charts and the description of the plan of 1701 is close enough. These maps were drawn 50 years apart so the exchange of information between the drawers was highly improbable event. Preservation of proportions of the lake for almost ninety years gives us an opportunity to say that data cited by Prince Gagarin is a reliable source. It applies to lateral dimensions of the lake. A depth of the lake was not indicated in Plan kanalov, kotorye idut ot Ivan-ozera po reke Donu i Shatu. In the explanatory text of Atlas Tulskogo namestnichestva a depth of the lake is 6.5 *arshin* (measure of length) (i.e. 4.7 meters). If these figures were exact we could say that the lake was grown shallow. We cannot state this definitely as we do not know the time of measuring.

To answer the question whether the plan of 1701 depicts Ivan-ozero before the start of the construction of Ivanovskii kanal or not, we refer to the records of Razriadnyi prikaz. Records contain information about vehicles (boats) and tools (wooden pails) for working on water. Tools are mentioned only in connection with the construction of the canal [8]. Records of the years 1701 and 1702 contain no information about works on the lake, modification of its shape or depth. It means that during these years such work was not carried out on Ivan-ozero. However, we cannot be sure of it.

With the description of the plan we can form clear picture of Ivan-ozero and earth-moving works at the moment of drawing of the plan. In the middle of October of the year 1701 two artificial channels (canals) were existed in the area of the lake, the construction of another one started. The canals were 3.5 and 0.5 km long, water depth was from 2.2 to 3.6 meters. At a distance of 650 meters from the natural channel of the Don River the construction of another one artificial canal was started. The width of it was four times as much as the width of the previous canal. The artificial canal was 17.3 meters wide against the 4.3 meters wide of an old one. The text of the document is not clear enough, so we cannot say where the earth-moving works were done. In addition to the construction of new artificial channels old river-bodies were improved. The natural channels of the Shat River and the Urvanka River were cleared for the space of 30 meters and for the space of more than 20 meters for the Don River [8]. Report on the works on the Urvanka River is unique, we have not found information about the works on the Urvanka River in other historical sources.

4. Conclusion

Russian geographical plans (drawings) are a valuable historical source. They contain information on the history of geographical knowledge, but also on politics, economics and technology [9; 10].

Prince Matvei Petrovich Gagarin, officer of mapping of Ivan-ozero in 1701 (Ivan-ozero lake was the source of the River Don at the time), fixed accurately the natural dimensions of the basin. Most likely the fixation matched well actual dimensions of the lake before the start of water project works. Drawing of the plan initiated further studies of the terrain along the body of the River Don, one of the largest rivers of Eastern Europe.

References

[1] Yurkin I N 2009 Prince Matvei Petrovich Gagarin on Ivan-ozero *Transactions of the State Hermitage Museum* **47** 243–244

[2] Stuckenberg J Ch 1841 *Beschreibung aller, im Russischen Reiche gegrabenen oder projectirten, schiff- und flossbaren Canaele…* (St. Petersburg: Druckerey des Militair-lehr-anstalten)

[3] Milovidov A I 1892 Ivanovskii canal which was constructed by the order of Peter the Great to connect the river Volga to the river Don (Essay in history and geography) *Reading at the Imperial Society of History and Russian Antiquities* **1** 1–30

[4] Shafranovskii K I 1951 Projects of Volga-Don canal on Russian maps printed in the 1st half of the 18th century *Proceedings of the Russian Geographical Society* **6** 631

[5] Shafranovskii K I 1957 Ivanovskii canal and two plans of Ivan-ozero, 18th century *Proceedings of the Russian Geographical Society* **5** 477

[6] Ozerova N A, Snytko V A and Shirokova V A 2015 Expedition to Explore the Sources of the Main Rivers of European Russia (1894–1902) *News of the Russian Academy of Science. Geography* **4** 113–127

[7] Shtukenberg I 1891 The river Don and the tributaries of the river Don (Voronezh: Tipo-litografiya gubernskogo pravleniya)

[8] Russian State Archive of Ancient Acts **210** Moskovskij Stol Stolbcy 790 155

[9] Yurkin I N and Kalita S P 2018 Ironworks images on the 17th and early 18th century Russian geographic maps as a source for the history of technology *Studies in the History of Science and Technology* **4** 660–680

[10] Medved A N 2019 Russian cartographic materials from the 17th and the 1st quarter of the 18th century as sources for the history of technology *Studies in the History of Science and Technology* **1** 129–142

Information in the document of the 1730s about iron ores that were mined near the city of Tula

I N Yurkin

S I Vavilov Institute for the History of Science and Technology, RAS, 14, Baltiyskaya street, Moscow, 125315, Russia

E-mail: ig-yurkin@yandex.ru

Abstract. The paper presents data on the ore and the location known to the ore-finders working in the vicinity of Tula in the 1st half of the 18th century. The archival document of the year 1733 (the list compiled by Demidov), the list of types of iron ores of the area, was found and used as a material for the investigation. We specify that the majority of titles of iron ores were associated with the sites of ore occurrence. The inference can be made that the geological surveying and mining were carried out intensively. The paper presents an investigation of the question whether miners and iron makers had linked together the qualities of the discovered ore and the quality of iron or not. The classification of the list (of types of iron ore of the area) was compared to the classifications of Russian and Norwegian authors of the 18th century. We cite data from other records. We underline that the work of ore-finders was successful, average lifetime of some mines was about 100 years. The results of the study can be used in interpretation of the results of archeological studies.

1. Introduction

Iron making is one of the oldest crafts of the rural population of Tula region. First records about it went back only to 1580s. There is no doubt, however, that by this time iron making had already had a definite history in the region. Ore was mined by mine-workers. Their activities were as ancient as the activities of iron-makers.

Sustained supply of ore was not possible without studies of the natural resources of the region. This paper presents an investigation of the search and the study of ore occurrence in the region.

Historical sources of the 18th century on the search of ore deposits in Russia are few and are of poor contents. There is only a detailed description of the practice of iron ores search of the middle of the 18th century in the manual of I.A. Shlatter, the Head of the Mining board (*Berg-kollegiia*), *Obstoiatelnoe nastavlenie rudnomu delu* (Full manual on ore mining) [1]. In the manual Shlatter described places of ore occurrence, ore layers and seams, ore search method (chapter "On signs of ore seam presence and the search of seams"), operating of new mines, mines machinery, preparation for the processing of ores, ores picking, breaking, washing. The manual had engravings that could have been used as a referral source on the subject. Shlatter described in detail the technical aspects of the mining but he had shown no interest in the history of the search and the find of iron ores. The same method holds true for the scientific papers of M.V. Lomonosov "Pervye osnovaniia metallurgii ili gornyh del" (The fundamentals of metallurgy or mining engineering).

Scientists of the 19th and 20th centuries gave attention to the history of mineral exploration and minerals studies in definite geographical regions of Russia. A.A. Ievlev examined history of finding of ores on the river Tsilma that started at the end of the 15th century. He thought it possible to describe some types of relief as traces of old mining works [2]. E.A. Kurlaev and I.L. Mankova studied the subject of the search and the development of iron ores deposits in Ural and Siberia region. They found how the discovery of mines correlated to the direction of colonization flows, what was the process of mineral exploration during the expansion of new lands, what triggered the exploration of ores, what share had private and state initiative in the process of search of mineral deposit, what was the process of mineral exploration on the new lands [3].

The European part of Russia or the vicinities of Tula were not examined in the larger half of studies of the history of mineral exploration. The nearest place to Tula mentioned in these studies was a place near Dedilov town on the river Olen. There were mines referred to in the records of the 17th century. But these records reported only of the small territory and of the ore production. For this reason the history of the mineral exploration in this region and the increase of knowledge of the natural resources of the region were not covered in the literature. The survey of K.N. Serbina on the peasant metallurgic industry can be taken as an exception. The author touched the subject of the provision of ore. K.N. Serbina did the survey of large territory of three neighboring counties, *Aleksinskii uezd*, *Dedilovskii uezd*, *Tulskii uezd*, one chapter for each county. These chapters contain references to the places of ore production [4]. However, the survey of K.N. Serbina did not examine the history of mineral exploration and the increasing knowledge of the natural resources of the region.

One branch of this human knowledge was the knowledge of the types of ores and of the link between the type of iron ore and the quality of iron. In Russia this experiential expert knowledge was summarized in a multivolume study of I. Shlatter *"Obstoiatelnoe opisanie rudnago plavilnago dela"* (Full description of ore smelting…) (1763-1784). Shlatter described the identification of the type of ore by its appearance in Chapter 1, Volume II of the book. The author gave us the description of *obyknovennoi kamennoi zheleznoi rudy* (iron ore), reported distinguishing features of the valuable ore according the nature of fracture. He described other rocks and ores suitable for the production of iron: ore of *oshpatovatogo vida*, *krovavik* (similar to *glaskopf*), *nazhdak*, *braunstein*, *eizenman* (the same as *eizenglimer*), *volfram*, three types of *kolchedan – sernyi, zheltyi,belyi*, *zheleznaia vohra*, *chernyi pesok*, *magnit* [5]. It is possible that the description was based partly on the experience of ore-finders in Russia, but the classification as we see from the names of types of ore represented knowledge of German geoscientists and specialists in mineralogy.

We cannot exclude the possibility that expert knowledge of Russian ore-finders, *rudoznattsy* (experts in ores and the search of them) was recorded. But we have not found the historical records of the 17th and 18th centuries on the subject or the mentioning of them. We can expect the find of such historical records for Tula region and the adjacent areas.

The main part of Tula region is Tula ore region with its characteristic outcrops of iron ores (mainly brown iron ores). The existence of *domnitsa* (hearths where the direct reduction of iron from ore take place) is mentioned in the records of 16th, 17th and 18th centuries in *Tulskii, Dedilovskii, Krapivenskii, Aleksinskii uezd* (Tula, Dedilov, Krapivna, Aleksin counties). Three of them were within the boundaries of the main ore field of Tula iron-ore area. As we can see the location of rural iron making centers are in line with the place of ore occurrence. Minor enterprises of local peasants were main consumers of local ores. Extraction of ore was made mainly by peasants. They had no resources to seek for advice in ore finding from outside specialists. But they had no need to do this because the outcropping of ore was a common place there. Peasant miners were satisfied by the results of their own search work. Surely locals had material knowledge representing their skills and experience in ore finding.

We assign the objective to find out what kind of data on Tula ores and the locations of the ore were known to iron-ore finders of the region in the 1st half of the 18th century.

2. Methods and materials

Relation of the location of centers of rural iron making to the location of ore mining is an established fact. We hope we can extract information about places of ore-mining from the sources of history of local metallurgy. This could add to the history of the research and the development of natural resources of the region.

We studied documents of two Archives. We looked for the documents of 1720 and of later time in *Gosudarstvennyi arhiv Tulskoi oblasti, Fond 55 (Tulskaia provintsialnaia kantseliariia)* (State archive of Tula region, Collection of papers 55, Tula Province Chancellery). We studied documents of Russian State Archive of Ancient Acts (*RGADA*). Records of the *Fond 271 (Berg-kollegiia)* (Collection of papers 271 (Mining board), RGADA) are of great interest. We have found data on the locations of ore mining in the records of the beginning of the 1730s regarding the collection of tax for iron makers. Data was compared to data on iron ores of the studies of norwegian and russian scientists of the 18th century. Data on the mining locations was compared to cartographyc documents.

3. Results and discussion

The study revealed a document containing the list of ores mined around Tula in the 1730s. Here are the circumstances of its compilation.

In the late 1720s a controversy occurred between the gunsmiths of Tula and Nikita Nikitich Demidov. Demidov was a furnaces tax money official at a time. The conflict escalated into personal controversy between Nikita Demidov and Makar Polovinkin, head of the *Tulskaia oruzheinaia kontora* (Tula armory office). Polovinkin proposed to close all ironworks in Tula region since they lessened the supply of timber and ore in the region, which could make difficult the manufacturing of weapons in the future. Demidov had not agreed with it. In his letter to *Kommerz-kollegiia* (Commerce-board) of July 14, 1733, he stated that Tula iron should not be used for the making of arms. Siberian iron was best suited for this. Coal and ore of better quality were used in the production of Siberian iron.

Demidov reported details on Tula ores. He wrote there was a lot of ore in the region of Tula. The location of the ore was in the lands of landlords and in *zaseka* (Zaseka are old thick woods, felled-tree barriers used to protect Tula from the raids of southern neighbors). Demidov cited the names of the types of iron ore mined in the vicinity of Tula. The List included the following *zvaniia* (titles) of iron ores: 1 *medved,* 2 *ovsianik*, 3 *smolian,* 4 *strekalov*, 5 *ivashkin*, 6 *sherap*, 7 *kostomar, 8 aleshenskoi*, 9 *kostomar akuninskoi*, 9 *retin*, 10 *pirogov*, 11 *leskoi*, 12 *frolovka*, 13 *degatna*, 14 *baturka*, 15 *liubogoshch*, 16 *oslon*, 17 *kochan*, 18 *baturinskoi*, 19 *alen*, 20 *nikolskoi*, 21 *salomas*, 22 *kolpna*, 23 *shchekin*. Demidov noted that iron made from twenty of the cited ores was not good for the manufacture of weapons. The ore of Tula region, which was good enough for use in weapons, was of lower quality than the ore of Ural region [6].

Data on the region reported in this document is unique for today. Analysis of the text allows us to make several conclusions.

First of all we deal with the unique written records of empirical data of local ore-finders and iron-makers of the region of Tula. List of Demidov was in fact a primitive classification of ores found and mined in the region. Later, specialists in the field of metallurgy and mineralogy developed much more advanced classifications. Such was the classification of Norwegian scientist O. Ewenstad based on the study of local ores that was published half a century later [7]. Information about the kinds of iron ores was given in surveys of M. Lomonosov and I. Shlatter. The information on the ores of Russia had to be summarized in *Rossiiskaia mineralogiia* (Russian mineralogy), an unfinished project of M. Lomonosov. Samples of minerals and rocks were sent to M. Lomonosov from every place where the iron-making factories were existed [8]. The list of ores reported by Demidov was simple, but it dated from the earlier time and described the ores of the Tula region that was not present in the surveys of Lomonosov and Shlatter.

In the second place, iron ore seams and lenses of ore in the vicinity of Tula had unstable chemical compositions. Ore characteristics in the volume of ore body were inconstant. All these characteristics determined the quality of iron. Did the iron-finders link chemical compositions of ore with the qualities of iron made of it?

We reply in the positive to the question. Demidov reported that ores were different in quality, some of them could be used in the production of weapons. This can be considered as the first proof. If ores differed from each other only in the place of location, it would be logical to give the list of mining locations not the list of the names of ores. This can be considered as the second proof.

In the third place, ore-finders participated in search of the ore occurrence but also in mining. Mining was done in some of the locations found by ore-finders. If we assume that one kind of ore was mined in one exact mine, we could say that at least 23 of mines operated simultaneously. This indicated intensive mining in the vicinity of Tula in the 1730s. It was possible because of the presence of ores in Tula region and the exploration of ores.

This conclusion agrees with the known active mineral exploration of the region which was not limited to the exploration of iron ores [9].

In the fourth place, ores were mined, according to Demidov, in landlords land and in *zaseka*. Names of *zaseka,* for example, *Malinovaia*, are often found in the records of the 1700 - 1730s on the ore mining. As an example we could bring the records on appropriation of land in Malinovaia zaseka for mining in 1727. Owners of the metallurgical furnaces of armory district of Tula presented a petition. They reported they made iron and steel from the ores they bought from peasants who mined ores in Malinovaia zaseka. According the report of the owners of furnaces, recently the mining of these ores were forbidden for everyone except the family of Demidov. Metallurgists pleaded to permit peasants to mine the ores and to supply ores to Tula for "use in weapons". They were afraid, the quantities of ores were not sufficient to cover the needs of the armourers of Tula and proposed to forbid mining for those, who sold ore to another places. [10].

The list of Demidov contained information about the location of sites of iron ore mining. The names of ores were given often according the places where they were collected, namely linking the name of ore mine to the name of human settlements. Names of the kinds of ores, such as *medved, ovsianik, strekalov, pirogov, salomas, shchekin* and others, matched the names of settlements that were situated nearby zaseka. We could compare them to names of settlements fixed in *Ekonomicheskie primechaniia k Generalnomu mezhevaniiu* (Economic annotations to General land survey). It could help to localize mines and places of ore processing, such as places of roasting and the reduction of ores.

The same names of settlements we could see in the records concerning conflicts of the ore-mine workers. We could give an event of 1730 as an example to illustrate the point. Servants of Akinfii Demidov caught peasants of the village *Ovsiannikova* of landlord S. Izvolskii for the reason that they mined ores on the land granted to Demidov in Malinovaia zaseka.

4. Conclusion

There is a growing interest in archeological studies of the lands of zaseka and zasechnaia cherta (felled-tree barriers and fortifications of the 16th-17th centuries) in the recent years. Not only remnants of fortifications but the artifacts of different provenance and age were found on the lands of zaseka and the adjacent areas. Some of them are the archaeological evidence on the iron ore mining (also known by written evidence) and rural iron production in the area (in particular in Malinovaia zaseka). We cannot implement the task of finding the sites of mining and processing of iron ores entirely with the help of written documents. The list of Demidov provides us with the reliable data on the places of ore mining in the 1730s.

Information of the written evidence like the list of Demidov could be used in the interpretation of archaeological finds. Blast-furnace factories of Tula region together with the furnaces of rural iron-makers consumed less ore than it was able to mine in the region. Demand satisfied supply. Some ore mines remained working mines for a long time. The mine near the village Solomasova, mentioned as *salomas* ore in the list of Demidov, was a producing mine up to the middle of the 1820s [4]. This settlement existed on the plans of General land survey and other cartographical documents of the end of the 18th century and the 19th century. The long term of functioning makes it difficult to date archaeological finds of the mine. But the time data binding (dates were given by the list) seems to be useful since it gives us the reliable chronological data on the mines.

References

[1] Shlatter I A 1760 *Full manual on ore mining in four parts, where we describe places of ore occurrence, veins and the ways of ore-finding...* (Saint-Petersburg)

[2] Ievlev A A 2014 Investigations of silver mines in the era of Ivan III *Studies in the History of Science and Technology* **1** 50-68

[3] Kurlayev Y A and Mankova I L 2005 *Development of the Ural-Siberian ore fields in the 17th century by the Origins of the Russian Industrial Policy* (Moscow)

[4] Serbina K N 1978 *The peasant metallurgic industry in central Russia, from the sixteenth to the first half of the ninetenth century* (Leningrad) p 64

[5] Shlatter I A 1765 *Full description of ore smelting, how to smelt metals in large amounts from ores and enclosing rocks...* vol 2 (Saint-Petersburg) pp 3–8

[6] Russian State Archive of Ancient Acts **271** Description 1 Book 1400 260-260 rev

[7] Espelund A 2010 The early stage of ferrous metallurgy in Norway *Russian Archaeology* **3** 44-50

[8] Yurkin I N 2011 Factories are the "collegues or coworkers" of Lomonosov *Studies in the History of Science and Technology* **1** 14–45

[9] Yurkin I N 2008 "Better than the silver of another countries" (Silver-making factory of the 17th century on Tulitsa river: the unknown pages of the history of search and exploration of ores in Russia) *Studies in the History of Science and Technology* **3** 59–80

[10] State archive of Tula region **55** Description 1 1554

Changes in floodplain soils of the Nizhny Don under the influence of flow regulation and anthropogenic load

A F Gurov [1], V A Shirokova[1,2], A O Khutorova[1] and Yu D Yurova[1]

[1]State University of Land Use Planning, Kazakova str., 15, Moscow, 105064, Russia
[2]S.I. Vavilov Institute for the History of Science and Technology of the Russian Academy of Sciences, Baltiyskaya str., 14, Moscow, 125315, Russia

E-mail: Yuliya.yurova.1996@mail.ru

Abstract. Formation of hydromorphic floodplain soils of the valley of the Nizhny Don occurs mainly under the influence of anthropogenic erosion and due to lowering the water content of the floodplains below the dams as the result of the construction of the Tsimlyansk reservoir. Road transport and overgrazing contribute to the degradation of natural meadow vegetation until its complete destruction. As a result, the properties of soils, as well as the content and composition of humus, significantly change, the water-physical properties of soils deteriorate and the processes of secondary salinization develop with the nearby proximity of mineralized groundwater.

1. Introduction

Analysis of historical data shows that for a long time this territory was mainly a nomadic animal husbandry area. It was only from the 18th — 19th centuries that the development of agriculture began. At the same time, up to the beginning of the 20th century, the fallow farming system was dominant. The construction of the Tsimlyansk reservoir had a significant impact on the functioning of floodplain landscapes. The volume of spring runoff decreased to 30-40%, and in some cases – even to 20%. But significantly increased runoff in other seasons of the year suggests a further decrease in the value of spring runoff.

The waterfall of the floodplain below the reservoir dam decreased, which led to profound changes in floodplain landscapes. Currently, there is a further decrease in the flow of the Don under the influence of anthropogenic (due to the increase of irrecoverable water losses in agriculture, industry, municipal water supply) and climatic factors. As a result, we can expect a decrease in the water content of floodplain landscapes and their further transformation under the influence of anthropogenic factors [1].

The decrease in the spring part of the runoff in the present and future leads to a decrease in the role of flood elements in the formation of floodplain landscapes: surface flooding and sediments of the solid phase by water spills.

In the lower reaches, the valley of the Don is represented by five terraced levels: modern Holocene and four floodplain terraces composed of quaternary loess sediments. Floodplain terrace within the study area also has a step structure. There are four floodplain levels: three, four, five and six-meter marks at the absolute height. Floodplain vegetation is represented at low levels by cereal and forb-grass associations with dominance of bluegrass (Poa) and awnless campfire (Brōmus inērmis).

At the highest flood level there is a wormwood association with a mixture of steppe species, which indicate the process of vegetation steppe formation.

Currently, after rare floods, the vegetation increases area hygrophilous associations. Bonfire and bluegrass are supplanted by wheat grass. In relief depressions, in former channels and oxbows, wetland plants (reeds and cane) are developing. As a result of the regulation, the productivity of the meadow hayfields has decreased greatly (from 15-20 to 5-8 kg/ha).

2. Methods and materials

The objects of study are hydromorphic soils and wetlands of the Nizhny Don valley. The main methods of research are comparative-geographical, comparative-analytical, soil-archaeological, as well as mapping wetlands with the involvement of remote sensing. To characterize the composition and properties of soils, a complex of conventional agrophysical, chemical, physico-chemical, granulometric, structural, mineralogical analyses was used.

3. Results and discussion

The study of the soils of the floodplain of the Nizhny Don was carried out by B.P. Polynov, V.N. Budko, G.V. Dobrovolsky, et al. Soil formation in the studied area of the floodplain develops in two main directions, determining the formation of soils of meadow and meadow-steppe series. The meadow series include alluvial meadow-dark-colored and wet-meadow dark-colored soils. Also silt is widespread in marsh soils.

These soils develop at low floodplain levels, on clay sediments under cover of cereal and grass-cereal vegetation with a ground water level of no more than 3.0-3.5 m [2, 3].

These types of soils are characterized by low or medium humus content, high absorption capacity, with a wide variety of soil absorbing complex. The most common saline to the difference in area is divided into the fall. Salinization of soils makes them sulphatic with the content of salts to 1.0-2.058.

The soils of the meadow-steppe series are represented by meadow-chernozem and meadow-chernozem solonetzic soils in a complex with solonetzic-saline soils, formed in microdepressions 5-15 cm deep. This type of soil develops based on sediments of the lightweight granulometric composition with the groundwater level above 3.0-3.5 m.

Studies have shown that meadow-chernozem solonetzes in their properties are largely similar to saline solonetzes. The main differences between them are the absence of the upper part of the humus horizon and their higher salinity. Soils are characterized by a low content of organic matter, low absorption capacity. The composition of the soil absorbing complex of meadow-shernozem solonetzes and salt marshes is often by more than 50% saturated with sodium (Na). These soils, especially saline soils, are characterized by a high salt content, reaching 1.0-1.5% [2, 3].

Excessive grazing pressure on some parts of the floodplain has led to the destruction of vegetation and soil erosion. The area of eroded areas, such as cattle tracks on some saline and waterlogged soils, reaches 10-16 %.

In eroded soils there is a decrease in the humus content, accumulation in some cases of easily soluble salts and the entry of sodium into the soil absorbing complex.

Similar soil changes occur in the course of haphazard passage of vehicles. As a result of the development of secondary processes, meadow-black soil solonetzes under the trails and roads transform into salt marshes. As studies have shown, natural salt marshes are similar in their properties to salt marshes-salt marshes under paths and roads.

Based on studies of the biological cycle of salinization of meadow soils, a number of assumptions can be made. Formation of natural salt marshes-salt marshes of the studied territory proceeded in a way similar to salt marshes-salt marshes located under paths and roads. The process of formation of salt marshes is associated with changes in their surface, especially with the destruction of vegetation of floodplain soils [4].

At the same time, on some other soils, such as meadow and chernozem, etc., violation of the surface, in particular laying of tracks and roads, does not lead to the significant changes in properties of soils.

To study the effect of erosion on the properties and regimes of floodplain soils in the field, an experiment was conducted. During the experiment on experimental plots, herbicides were introduced to the undisturbed soil surface according to the main soil differences of the floodplain. The result was the destruction of natural vegetation [5]

As shown by the results of observations, the lack of vegetation shading the surface of the soil increases their warming on summer days by 10-20°C. In the meadow, the dark-colored saline soils and meadow-chernozem solodized solonetzes wave capture additional heat in the capillary fringe. There is an increase in evaporation of groundwater, resulting in an increase in the amount of water-soluble salts in these soils. As a result of secondary salinization, strongly saline meadow-chernozem solonetzes soils turned into saline solonetzes soils, and meadow-dark-colored soils; medium-saline turned to strongly saline [5, 6].

In meadow-chernozem and meadow-dark-colored fused soils, additional heat does not reach the capillary rim, so secondary salinization did not occur.

In the first years of the experiment, the number of salts in the secondary salted solonetzic soils without vegetation had considerable hesitations, and in the damp years they approached sizes which are characteristic of control specimens of soils. Further flushing of salts has become more difficult, which was determined by the development of other secondary soil processes.

Due to the mineralization of organic matter and the impact of deflation in dry years, the destruction of soil aggregates of the upper horizons occurred. The surface of the plots without vegetation deepened by 11 cm. The products of destruction were partially carried by wind processes outside the experimental sites. A part of the silt fraction with amorphous iron oxides and humus substances in wet years was washed out together with salts down the profile, and they deposited in the illuvial horizon, filling the pores. The predominance of soda in the composition of secondary accumulating salts, as well as the increase in the content of mobile organic matter, contributed to the increase in the mobility of the silt fraction.

Thus, the strengthening of the saline process occurs in wet periods. The illuvial horizon is enriched with silt particles. As the amount of illuminated matter in the horizon "B" increases, voids, especially pores, become more and more clogged, which worsens the possibility of soil washing from water-soluble salts.

In the horizon "B" the pores of the natural solonetz-saline are almost completely filled with illuminated clay, which determines its negative water-physical properties. However, strengthening of solonetzic process happens against the background of a rather stable high level of ground waters. This factor defines the evaporative water mode and accumulation of salts in a soil profile. And deterioration in conditions of the descending substance migration happens in stabilization of secondary salinization. Thus, the solonetzic process leads to strengthening of the saline process and to formation of difficult reclaimed soil differences in a flood plain of the Nizhny Don.

Destruction of vegetation on meadow-dark-colored merged and meadow-black-earth soils did not cause secondary salinization. On the contrary, salt reserves in meadow-chernozem soils without vegetation decreased slightly. There were no significant changes in the structure of soils and their properties. On lightweight meadow-chernozem soils, the particle size and the absence of sodium in the soil absorption complex consisting of soluble salts have not led to the displacement of the clay fraction and the formation of the relevant secondary properties of soils. On meadow-dark-colored fused soils, the clay granulometric composition and deep occurrence of the groundwater level proved to be an obstacle to changing the properties [5, 6].

The absence of significant adverse properties of vegetation-deprived meadow-chernozem and meadow-dark-colored merged soils affected the possibility of rapid vegetation renewal on them (after 4-5 years). The emergence of higher vegetation led to a rapid recovery of humus reserves and humus composition. On the contrary, on meadow-chernozem solonetzes and meadow-dark-colored saline soils, due to a number of unfavorable initial properties and the acquisition of new ones, the overgrowth of plots did not occur, and the soils remained not overgrown for 8 years.

The results of the study of soil properties under the trails and dirt roads, as well as field experiment data, allow us to identify two main groups of floodplain soils. The impact of erosion on the first group of soils, which includes meadow-chernozem solonetzes and meadow-dark-colored saline soils, leads to the development of irreversible changes in their properties, so that they move to other classification units.

In the second group of soils, which include meadow-dark-colored fused and meadow-black soils, anthropogenic erosion causes reversible changes, which quickly, within 2-3 years, leveled to the original state after the resumption of vegetation, although the composition of the emerging phytocenosis is actually different from the indigenous one.

4. Conclusion

Allocation of groups of floodplain soils by their resistance to anthropogenic erosion allows developing measures for their rational use. In the area of distribution of the group of soils resistant to anthropogenic erosion, which includes meadow-dark-colored fused and other soils, it is possible to carry out grazing, mowing and construction of irrigation systems. In order to increase the productivity of floodplain lands for animal husbandry, the establishment of cultural pastures should be recommended. The construction of rice irrigation systems requires a well-developed drainage network, which will eliminate the rise of groundwater levels above the critical one.

For a group of soils unstable to anthropogenic erosion and sensitive to vegetation destruction, grazing should be strictly regulated. Given that there is a high probability of adverse changes in the soil, it is advisable to mow and organize cultural pastures, where by means of irrigation, fertilization and using other measures one can achieve dense vegetation and strong turf. When creating cultural pastures on the soils of this group, it is necessary to carry out a set of measures to reduce the level of groundwater, as well as its chemical melioration. For this purpose, a number of meliorants can be used: sulfuric acid, sulfuric acid waste, phosphogypsum, gypsum, organic waste, etc.

References

[1] Georgiadi A G, Koronkevich N I, Milyukova I P, Kashutina E A and Barabanova E A 2014 *Modern and scenic changes of river flow in the basins of the largest rivers of Russia Volga and Don river basins* (Moscow: MAX Press) **2** 214
[2] Polynov B B 1920 *Soil Aksai floodplains* (Novocherkassk: Don pedagogical Institute)
[3] Budko E N 1956 *Soil Don loans* (Novocherkassk: Southern research Institute of hydraulic engineering and reclamation)
[4] Ivanova E N 1939 Genesis and evolution of saline soils due to geographical environment *The soils of the USSR* (Moscow, Leningrad: USSR Academy of Sciences) **1** 349-402
[5] Gurov A F 1985 Influence of anthropogenic impact on the soil of the valley of the Nizhny Don, *Abstract of candidate thesis* (Moscow)
[6] Gurov A F, Shirokova V A, Khutorova A O and Yurova Y D 2019 Features of remote determination of humus content in arable soils *Advances in engineering research* **182** 138-141

The Central Asian orography studies by G.E. Grumm-Grzhimailo's expedition, 1889–1890

T I Yusupova

Institute for the History of Science and Technology, St. Petersburg Branch, Russian Academy of Sciences, 5, Universitetskaya emb., St. Petersburg, 199034, Russia

E-mail: ti-yusupova@mail.ru

Abstract. The paper deals with the history of the Grigorii Grumm-Grzhimailo's Central Asian expedition, 1889–1890, and its main results in orography. The tasks of the expedition included an exploration of West and Central part of Inner (Central) Asia and the gathering of the naturalistic collections. The expedition was organized by the Russian Geographical Society. Grigorii Grumm-Grzhimailo used Przhevalskii's reconnaissance methods to explore the uncharted lands. At the same time, he conducted geographical investigations, including orographic ones. This expedition covered a vast territory and made two major geographical discoveries of the Turfan depression and the Beyshan Mountains, the latter providing an orographic link between the Eastern Tian Shan and the middle Kuen Lun.

1. Introduction

Grigorii Efimovich Grumm-Grzhimailo (1860–1936) was one of the outstanding Russian explorers of Central Asia, the followers of N.M. Przhevalskii (1839–1888) [1]. Unlike Przhevalskii and other Russian travelers, the majority of whom were army officers, G. Grumm-Grzhimailo was a civilian with a university degree in natural sciences.

The exploratory activities of G. Grumm-Grzhimailo included: 1) his field lepidopterological work in the European part of Russia in 1881–1883; 2) the four expeditions into the Pamir country in 1884 and 1887; 3) the Central Asian Expedition in 1889–1904; and 4) several trips he made between 1903 and 1914 to Mongolia, Uriankhaiskii Krai and the Far East. He started his career as an entomologist and a traveler to the Pamir, Bokhara and Western Tian Shan. From there he brought the large lepidoptera collections, part of which he had presented to his sponsor Prince Nikolai Mikhailovich Romanov. Other parts of the lepidopterological collection he partly presented to the Zoological Museum of the St. Petersburg Academy of Sciences, and partly sold to European museums.

His most important expedition was the Central Asia expedition, 1889–1890, that was organized by the Russian Geographical Society (henceforth RGS). In this expedition Grigorii Grumm-Grzhimailo was assisted by his brother, an army officer, Mikhail (1861–1921). Therefore, it is often called the 'Expedition of Brothers Grumm-Grzhimailo'. Grigorii Grumm-Grzhimailo regarded himself as a follower of Przhevalskii and his scientific reconnaissance methods of field work. In his report of this expedition he wrote that "we looked at our expedition through the eyes of Przhevalskii" [2, p. IX].

It should be noted that in 1889–1890 the RGS supported the three major expeditions to Central Asia: led by brothers Grumm-Grzhimailo, Grombchevskii, and Pevtsov. While the focus of Grumm-Grzhimailo was the eastern part of the Tian Shan, Pevtsov explored the northern outskirts of western Tibet, Kuenlun, and Grombchevskii investigated the Pamir.

This paper briefly presents the history and the results of the Central Asia expedition by G. Grumm-Grzhimailo, 1889–1890, the most prominent among his travels.

2. Materials and methods

The activities of Grigorii Grumm-Grzhimailo has not been properly researched thus far. The first full biography of the explorer was written in 1947 by his son, Aleksei Grumm-Grzhimailo [3]. He used a wide range of sources – primarily the materials of the Russian Geographical Society, as well as the documents he found in other archives in Leningrad.

New publications about G. Grumm-Grzhimailo have appeared only recently. One of them is a preface to the reprinted version of the report of the brothers Grumm-Grzhimailo [4], where the A. Grumm-Grzhimailo's book was widely used. Others reflect G. Grumm-Grzhimailo's scholarly activities as a geographer, economist, historian and entomologist [5–7].

This essay follows Aleksei Grumm-Grzhimailo's narrative, which in its turn was based on his father's detailed autobiography written in early 1924. We also used some new archival sources (of the Archive of the RGS), as well as the latest publications on G. Grumm-Grzhimailo.

3. Organizing the Expedition

In January 1889 Grigorii Grumm-Grzhimailo was approached by the president of the RGS, the Grand Duke Nikolai Mikhailovich Romanov (1859–1919), who offered him to continue his previous research of the Pamir and suggested that he make a journey to the southern slopes of the Pamir, into the Cis-Hindukush region, namely to Shugnan, Vakhan, Chitral and Kunzhut.

This proposal concurred with G. Grumm-Grzhimailo's own plans, so he agreed. Grumm-Grzhimailo wanted to complete his zoogeographic exploration of the Pamir and the Cis-Pamir countries. However, the Council of the RGS recommended him to change his route "for political and other reasons" and choose the Eastern Tian Shan and the Nan Shan as an objective of the new investigations [2, p. VII].

The "other reasons" were the Society's desire to fill in the remaining 'blank spots' on the map of Central Asia. The RGS also wanted to link up the routes of Przhevalskii and G.N. Potanin (1835–1920) in the Nan Shan, which had been insufficiently explored by Russian travelers (Przhevalskii traversed this region during his first and third Central Asian expeditions in 1870–1873 and in 1879–1881, and Potanin passed across its north-eastern outskirts on the way back from his Tangut-Tibetan journey of 1884–1886).

The route of his expedition was discussed together with the vice-president of the RGS Petr Petrovich Semenov (1827–1914) and the well-known cartographer from the Military-Topographic department of the Russian Main Staff, Colonel A.A. Bolshev (1828–1904). Grumm-Grzhimailo's expedition was to pass through the Eastern Tian Shan and the Bogdo-Olo range, thence across the Dzungar Gobi up to Lopnor and, having passed over the Nan Shan in its central-western part, to reach Lake Kokonor and the Valley of Huang He. Apart from Grigorii, the expedition staff included his brother Mikhail and an escort of Cossacks.

The RGS funded the expedition, but that was not enough. The Zoological Museum of the Academy provided Grumm-Grzhimailo with the accessories that he needed to collect samples of fauna: gun-powder, small shot, spirits, instruments for preparing birds etc., plus a small sum of money [8, p. 60].
In the middle of March 1889, brothers Grumm-Grzhimailo left St. Petersburg for Vernyi (now Alma-Ata), where they began to organize and equip the expedition caravan. Here they were joined by other members of the staff – their escorts and translators. They were eleven people altogether. In April G. Grumm-Grzhimailo received the Chinese travel passports from the Ministry for Foreign Affairs. However, the documents did not guarantee the travelers a safe passage along their chosen route. According to the Russian consul in Kashgar Petrovsky, "the Chinese permitted the expedition to work in Kashgar Province rather reluctantly" (see: Archive of the RGS, F. 1–1889, Op. 1, D. 13, l. 22–22 ob.).

4. The Expedition activities

On 27 May 1899 the expedition left the frontier Kazakh town of Dzharkent and crossed into the Chinese territory. Grigorii and Mikhail Grumm-Grzhimailo spent the summer exploring the Eastern Tian Shan and Southern Dzungaria. They collected a large amount of geographical data within the Eastern Tian Shan, where they had discovered an uncharted mountain, the Dos-Megen-Ola, 6100 meters high. Then the Grumm-Grzhimailo's expedition explored another mountain range, the Bogdo-Ola, to the east of Urumchi. They were the first Europeans to do so. The Bogdo-Ola range divided the Turfan depression from the Dzungar Valley, into which they subsequently descended to hunt for the wild horse. The animal had been originally discovered by Przhevalski, in 1881, and was given the name of *Equus Przewalskii* by zoologists. The brothers turned out to be the first observers of the horse in its natural habitat and their hunt was quite successful. They later delivered to the Zoological Museum four skins, three sculls and parts of the skeleton of several wild horses, both male and female, of different age, which were of much value to zoologists.

From the Dzungar Valley the expedition moved on to the oasis of Turfan. Some 250 km south-east of Urumchi brothers Grumm-Grzhimailos explored the Lukchun (Turfan) depression, having ascertained hypsometrically that it was located 130 meters below sea level. Nearly at the same time, the expedition of Pevtsov, working in the opposite (western) part of the depression, in the oasis of Toksun, attained the same result. The discovery of the Lukchun depression became one of the Grumm-Grzhimailos' most important geographical discoveries and, quite naturally, it attracted the attention of geographers world-wide. The investigations of the depression continued in the subsequent years by V.A. Obruchev (in his expedition in 1892–1894) and V.I. Roborovsklii together with Kozlov (the expedition in 1893–1895).

From Turfan the expedition made its way via the oasis of Kham directly to Nan Shan. Their route ran diagonally (from south-west to south-east), almost in parallel to the routes of the expeditions of Potanin and Przhevalskii. This stretch of their journey was also crowned with great success, this time in the field of orography. The expedition discovered the Beyshan upland, 2700–3000 m high, in a place where the maps showed the Khami desert. As G. Grumm-Grzhimailo was able to ascertain, Beyshan provided a connection between the Tian Shan and the Nan Shan. Three years later the geologist V.A. Obruchev would conduct a detailed geological investigation of the Beyshan Mountains. In the opinion of geologist K.I. Bogdanovich (1864–1947), the discoveries of the Lukchun (Turfan) depression and of the Beyhan Mountains by brothers Grumm-Grzhimailo were very important for the theoretical understanding of the geological evolution of Central Asia.

In the spring and summer of 1890 the expedition explored the southern slopes of the Nan Shan and thence descended into the valley of the Hwang He river. The expedition proceeded further along the valley of the Heihe river (the upper and middle courses of Edzin Gol) towards Suchow, and hence, again across the Nan Shan and the northern foothills of the Tian Shan following their old route, to Urumchi. On the 8th of November the travelers arrived in Kulja, and a few days later (on 13th of November) they were finally back to Russia.

The results of the Grumm-Grzhimailos' expedition were quite outstanding. The travelers covered 7250 km, 6000 of which along the hitherto uncharted route. They made 140 hypsometric and aneroid measurements and determined the geographical coordinates of 30 points. Apart from these, the brothers conducted systematic meteorological observations. Their discovery of the Lukchun depression and of the Beyshan Mountains was especially valuable to the geographical science.

The expedition also collected some ethnographic material and took many photographs of local people and of the lands they had visited. The ethnographic photographs were displayed, in 1891, at an exhibition set up by the Russian Technical Society and were awarded the bronze medal. Also, Grigorii Grumm-Grzhimailo made a collection of seeds, at the request of the Ministry of the State Property, which planned to use these for agricultural purposes in Russia.

The report of the expedition's results was introduced at the meeting of the RGS, which took place on March 13, 1891 in the building of the State Duma in St. Petersburg. G. Grumm-Grzhimailo's report aroused much interest among those present, numbering about 600 people. The scientific societies and institutions also highly appreciated the work done by the expedition. P.P. Semyonov, for example, was of the opinion that "the expedition of brothers Grumm-Grzhimailo is one of the most interesting and successful expeditions to Central Asia in the past decade" [9, p. 14].

The full account of the expedition led by G. Grumm-Grzhimailo made three large volumes, including numerous illustrations, photographs, maps and other appendices, published in 1896, 1899, and 1907. This was his major work in the field of geography and his most prominent contribution to the study of Central Asia.

5. Conclusion

The peak of Grigorii Grumm-Grzhimailo's activities as an explorer was his Central Asian expedition made together with his brother Mikhail. This journey stands out prominently as his greatest scientific achievement. Using Przhevalskii's method of reconnaissance of uncharted lands, the brothers covered a vast territory and made two major geographical discoveries – of the Turfan depression and the Beyshan Mountains, the latter providing an orographic link between the Eastern Tian Shan and the middle Kuen Lun. These discoveries, according to geologist K.I. Bogdanovich, had a profound theoretical importance for understanding the evolution of Asia, as mentioned above.

The geological interpretation of the territories explored by G. Grumm-Grzhimailo, especially his view of the continental – and not marine – character of the so-called Khanhai tertiary deposits, were supported by the prominent Austrian geologist Eduard Suess (1831–1914).

In addition, the expedition led by G. Grumm-Grzhimailo made an important contribution to the geographical study of the Nan Shan, determined the southern borders of Central Gobi, the Takla Makan desert, the shape of the source of Edzin Gol – the river Kheihe, as well as of the Tarim and Lobnor.

Grigorii Grumm-Grzhimailo's explorations did not go unnoticed by Western scholarly community. The Parisian Academy of Science (Académie des sciences) on 4 December 1893 awarded him, the first among Russian scholar-travelers, the Chikhachev Prize ('le prix Tchihatchef') "for his research of Western China, the Pamir and the Tian Shan" [10, p. 289]. And the Royal Dutch Geographical Society (Koninklijk Nederlands Aardrijkskundig Genootshap, or KNAG), in 1895, elected him as the Society's honorary member.

The RGS highly appreciated G. Grumm-Grzhimailo's explorations in Central Asia by awarding him the Konstantin gold medal in 1907 for the sum total of his works on geography and particularly for the final volume of his "Account of the Journey to China". And on April 6, 1910 the Society elected him as an honorary member, together with N.G. Potanin and P.K. Kozlov. In the opinion of the RGS Council, these three explorers completed the cycle of Przhevalskii's expeditions.

The bibliography of works written by G. Grumm-Grzhimailo is extensive and thematically diverse; it includes his books and articles in geography, entomology, economy, history, and a wide range of bibliographic, political, ethnographic, and other notes. His accounts of his travels were not simply descriptions of his routes. His works were based on the analyses of the vast literature of his predecessors, which allowed him to compare his own work with that of other explorers. Grigorii Grumm-Grzhimailo was the first among Russian geographers who used the Chinese sources.

In recognition of G. Grumm-Grzhimailo's explorations in Central Asia, his name was given to three geographical objects – the largest glacier discovered by the Grumm-Grzhimailo brothers in the upper Tanymas river in the Pamir, one of the passes in the Sikhote Alin Mountains, and one of the glaciers of the Bogdo-Ola range in Eastern Tian Shan.

References

[1] Andreev A, Baskhanov M and Yusupova T 2018 *The Quest for Forbidden Lands: Nikolai Przhevalskii and his Followers on Inner Asian Tracks* (Leiden; Boston: Brill)
[2] Grumm-Grzhimailo G E 1896 *Description of the Journey to East China* **I** (St. Petersburg: Tipo-litografia S M Nikolaeva)
[3] Grumm-Grzhimailo A G 1947 *Deeds and days of G.E. Grumm-Grzhimailo, the traveller and geographer, 1860–1935* (Moscow: Moscow society of nature researchers)

[4] Khoroshevskii A 2014 Introduction in: Grumm-Grzhimailo G E, *Via a ladder of "God's throne"* (Moscow: EKSMO) pp 5–27
[5] Latushko Yu V and Briantseva S O 2010 "Deeds and days" by Grumm-Grzhimailo: To the 150th anniversary of the researcher *Bulletin of the Far-East Department of Russian Academy of Sciences* **2** 106–14
[6] Koroliov V A and Murzin V S 2005 *History of lepidopterological research in Russia* (Moscow)
[7] Yusupova T I 2016 On the unrealized project on developing the textbook on Mongol history by G.E. Grumm-Grzhimailo *Mongolica* **16** 79–90
[8] Slepkova N V 2014 "To thank in the name of Academy": Central Asia collections of the Zoological Museum in the chronicle of sessions of the physics and mathematics department of Emperor's Academy of Sciences, 1874–1894 *History and biology research* [in Russian – Istoriko-biologicheskie issledovaniia] **6(4)** 80–104
[9] Semenov P P 1892 Review on the works of G E and M E Grumm-Grzhimailo *Report of Emperor's Russian Geographical Society for the year 1891* (St Petersburg) pp 14–17
[10] Gauja P 1917 *Les Fondation de l' Académie des sciences (1881–1915)* (Hendaye: Basses-Pyrenees)

Forecasting the use of enhanced oil recovery methods in oilfields of Bashkortostan

V E Andreev[1,2], A P Chizhov[1,2], A V Chibisov[1,2], V Sh Mukhametshin[3]

[1]SASI "Institute of strategic research of the Republic of Bashkortostan", 129/3, October avenue, Ufa, Republic of Bashkortostan, 450075, Russia
[2]Ufa State Petroleum Technological University, 1, Kosmonavtov St., Ufa, Republic of Bashkortostan, 450062, Russia
[3]Ufa State Petroleum Technological University, Branch of the University in the City of Oktyabrsky, 54a, Devonskaya St., Oktyabrsky, Republic of Bashkortostan, 452607, Russia

E-mail: vsh@of.ugntu.ru

Abstract. Today oil and gas complex in Bashkortostan is characterized by a deterioration in the quality of raw material base, which is associated with a significant degree of highly productive fields. It is a long-term operation, resource increment due to the discovery of fields unconventional oil, which are confined to low-permeable and weakly thick reservoirs oil deposits with extensive underwater gas and waterfowl zones. At the current stage of unconventional oil development of Bashkortostan oilfields, the urgent tasks are to study the experience of their development; identify peculiarities of the influence of geological and technological factors on the development efficiency of productive layers reserves; geological and commercial substantiation of various methods for enhanced oil recovery. The classification of development objects is based on the methods of mathematical statistics. Recommendations to improve the impact effectiveness on residual and unconventional oil reserves are based on the criteria-based analysis of enhanced oil recovery. The paper describes characteristics of reservoir systems and results of object grouping for oilfields development in Bashkortostan. Taking into account the established criteria for effective use of enhanced oil recovery and geological and field conditions of selected groups of objects with similar characteristics, there are some recommendations to increase the efficiency of oil recovery with targeted formation effects technologies.

1. Introduction

The analysis of more than 100 oilfields in Bashkortostan (686 oil deposits) showed that there is a variety of deposits types, reservoir characteristics, and physical and chemical properties of oil, gas and water. Oil deposits are located in a wide range - from the Middle Devonian to the Lower Carboniferous, for which a significant amount of information was obtained on the geological and technological conditions of their operation.

The current state of oilfield development requires a detailed geological field analysis of dynamics for main technological indicators of exploitation objects under conditions of significant variation for geological and physical characteristics of reservoir systems and technological parameters of impact systems. Such analysis purposes the selection of development objects with similar geological and physical parameters of reservoir systems [1-4].

In this situation, when a natural decline in oil production is not compensated by a corresponding increase in geological reserves due to the discovery of new fields, an alternative way to maintain oil production at the level achieved is the large-scale introduction of advanced methods of enhanced oil recovery into the practice of oil production [5-10].

2. Research Methods

Any method to improve the efficiency of oil recovery process is not universal for "all cases". The method effectiveness largely depends on geological, physical and oilfield conditions in which it is used [11–14].

In the conditions of objects consideration, the operation to identify (group) acquires special significance, that is, the selection of relatively homogeneous groups of objects.

Grouping helps to solve some of the most important tasks of oil and gas geology and development. It gives the way to assess the similarities and differences of productive layers in identifying development objects, justify development systems and methods for enhanced oil recovery, establish measures for control and regulation, etc. [1, 3].

When selecting homogeneous groups in conditions of a significant number of objects of study and parameters characterizing them, the process of effective and reliable grouping becomes possible only with the use of a method based on logical and mathematical analysis.

Currently, there are widely used methods from the theory of pattern recognition — factor analysis, principal component method, discriminant analysis, cluster analysis, etc. The choice of one method or another is determined by problem statement and method's advantages.

When solving a similar series of problems in geology and development of oil and gas fields, the method of principal components and discriminant analysis finds the greatest application.

The choice of principal component method to solve the problem of grouping objects is due to the following reasons:

- grouping of research objects is carried out according to generalized, independent indicators (main components) and is more objective than grouping according to separate initial parameters;
- grouping is made depending not only on general peculiarities of a deposit geological structure, but also on combinations of parameters, which to a greater extent affect the efficiency of development process;
- research objects are described by a number of factors (main components), significantly smaller than the number of initially taken parameters, that is, multidimensional space is compressed, within which grouping is extremely difficult;
- main components adequately reflect the original parameter and, at the same time, contain more information than individual parameters;
- studying the factors structure (main components) helps you to test existing and put forward new hypotheses about the causes of the relation between the parameters, give a causal interpretation of results, and develop and adopt a scientifically based control effect that contributes to the efficiency of oil development processes;
- main components are not mutually correlated, which greatly simplifies the task to construct various models of development process according to the obtained main components.

The paper does not describe in detail the essence of principal component method, but many researchers' works describe this method [1-4].

3. Results and Discussion

The results analysis of principal component method solution showed that out of 15 main components, the first four account for 76.3% of the total dispersion of parameters, that is, when selecting relatively homogeneous groups of objects, it is enough to consider them in the space of only these components. Each of four main components is informative, amenable to semantic interpretation, reflecting a particular property characterizing occurrence conditions, reservoir-filtration, physicochemical properties of layers and fluids saturating them, current state of development of these objects.

The first component reflects the properties of reservoirs and fluids, since sandiness ratio (15.9%), oil recovery ratio (14.7%), effective oil-saturated thickness (13.9%) and oil density (13.3%) add up to 57.6% of total dispersion of parameters in this component.

The second component is characterized by 56.9% technological parameters and conditions of occurrence of operational facilities: average annual water content of extracted products is 17.2%; inventory utilization rate - 16.5%; reservoir temperature - 11.7%; porosity - 11.5%.

The third component at 55.7% reflects physicochemical properties of oil - viscosity (19.6%), sulfur content in reservoir oil (18.2%), and type of reservoir (17.9%).

In the fourth main component, the greatest contribution is made by paraffin content in oil reservoir (18.1%) and reservoir temperature (14.8%).

As you can see, each component reflects geological and technological peculiarities of object development at one or another hierarchical level.

The geometric representation of research objects in coordinate axes of main components $Z_1 - Z_2$, $Z_1 - Z_3$, $Z_1 - Z_4$ helped to distinguish six groups of objects. When selecting groups and drawing boundaries, the condition was fulfilled under which each group of objects must occupy a certain and limited zone in the space of the main components.

Qualitative characteristics and peculiarities of selected groups of objects according to the initial parameters require the calculation of their values for “average” hypothetical deposits. Each of the selected groups of objects has its own peculiarities.

The first group included 130 objects. The group of objects is represented by terrigenous reservoirs of lower Carboniferous deposits and is characterized by the highest values - effective oil saturated thickness of layers, coefficients of porosity, oil saturation, permeability and oil recovery, tar content in reservoir oil and asphaltene; high rates - watering and use of reserves; the smallest indicator of the density of oil.

The second group consisted of 99 objects. The group of objects is represented by terrigenous reservoirs of Devonian deposits with small thickness of productive formations, the highest values of oil density, paraffin content in it and water production of produced products, low values of oil viscosity. Here you can also distinguish two subgroups, in which, by geological field data, different oil recovery factors are identified: 0.48 for large fields and 0.28 for medium and small fields. In the first subgroup, the maximum oil recovery factor was achieved (on average 0.5). For smaller fields, oil recovery factor reached more than 0.37. In group II, a high level of watering was achieved - about 80%.

One hundred and twenty-two objects of the third group are represented by carbonate reservoirs and characterized by the highest values of total thickness of productive sediments, viscosity of reservoir oil and sulfur content, utilization rate of reserves and water content of products produced; the lowest values of porosity coefficient, oil saturation permeability, content of paraffins and asphaltene-resinous compounds, reservoir temperature.

In the fourth group entered 131 objects. The group is also represented by carbonate reservoirs, where the smallest values of oil recovery and use of reserves.

In general, group III-VI include deposits with carbonate reservoirs. Many deposits are developed in natural modes. A low oil recovery factor is planned for all fields of these groups - an average of 0.23, actual ratio is much lower, no more than 0.07–0.08.

Currently, oil deposits in Bashkortostan are being developed quite rationally. However, the main disadvantage of the implementation of field development systems is the use of only one method of oil extraction - flooding.

A differentiated approach to the choice of enhanced oil recovery methods for each group of fields can increase the final oil recovery factor, and the effect is most productive with combined methods.

Based on the criteria-based analysis of the effectiveness of oil-jet device and comparing them with geological and physical characteristics and development characteristics of considered groups of fields, we can draw preliminary conclusions about the use of enhanced oil recovery methods in oil fields in Bashkortostan.

Group I (productive strata are terrigenous reservoirs of Devonian age, low-viscosity oil, fairly homogeneous reservoir). It is advisable to evaluate the effectiveness of the oil recovery devices to increase the rate of displacement - water-gas effect, injection of alternating CO2 fringes and water, oil displacement by highly concentrated fringes of surfactants. Predicted increase in enhanced oil recovery 4–5 points;

Group II (terrigenous reservoirs of coal age with a high degree of macro-inhomogeneity; moderately viscous oil) is promising for enhanced oil recovery, aimed at increasing the rate of sweep, using silicate-alkaline solutions, polymer systems, and vibrational exposure. In addition, a highly developed system to maintain reservoir pressure at facilities of groups I and II makes the use of biometric methods on them relevant. Forecast of growth of oil recovery factor in the group of 3-4 points;

Group III (carbonate, mainly fractured collector of Famennian age) - objects to apply hydrodynamic, wave and regulating technologies. Predicted growth of oil recovery factor for a group of 5-7 points;

Group IV (pore-type carbonate reservoirs of Kashiro-Podilsky age, medium viscosity oil) - conditions of objects are favorable for the application of water-gas action and forcing out oil with highly concentrated surfactant rims. Predicted growth of oil recovery factor for a group of 4-5 points;

Group V (carbonate pore-cavernous reservoirs of Bashkir and Tournaisian age) - use of water-gas effect is recommended at the sites. Predicted growth of oil recovery factor for a group of 5-7 points;

Group VI (reef-bearing carbonate reservoir of Sakmaro-Artinsky age with a high oil-bearing floor and a high degree of anisotropy in reservoir properties) - objects are promising for the application of gas technologies (hydrocarbon gases and alternation of hydrocarbon gas lines and BFLH). Forecast growth in oil recovery factor for a group of 7-8 points.

Thus, a wide application of the proposed methods to influence the objects of the selected groups in relation to geological peculiarities of Bashkortostan deposits, and the choice of optimal water flooding system will further extract 5.3-8.1% of initial geological reserves of Bashkortostan's oil.

4. Conclusion

Based on a detailed geological field analysis of basic indicators dynamics for exploitation in Bashkortostan fields in conditions of significant variation of geological and physical characteristics of reservoir systems and technological parameters of systems affecting development objects using the method of principal components, identification (grouping) of development objects has been carried out on complex of geological field data, which helped to identify six homogeneous object groups with similar characteristics of reservoir systems of objects group and applied development systems.

Typical objects in the selected groups were identified, a criterial analysis of effective use of enhanced oil recovery was performed, and a set of recommendations was proposed to improve the efficiency of producing unconventional oil and residual oil reserves and an additional increase in oil recovery factor was predicted during the implementation of the proposed measures. The increase in oil recovery factor from the implementation of recommendations at operational facilities in the Republic of Belarus will be on average 5-8 points.

References

[1] Chizhov A P, Sultanov Sh H, Vafin R I and Nugaibekov A G 2007 Classification of oilfield productive formations in Birsk saddle *Problems of Gathering, Treatment and Transportation of Oil and Oil Products* **2(68)** pp 16–21

[2] Zeigman Yu V, Mukhametshin V Sh, Khafizov A R and Kharina S B 2016 Prospects of Application of Multi-Functional Well Killing Fluids in Carbonate Reservoirs *SOCAR Proc.* **3** pp 33–39 DOI: 10.5510/OGP20160300286

[3] Mukhametshin V V 2018 Rationale for trends in increasing oil reserves depletion in Western Siberia cretaceous deposits based on targets identification *Bull. of the Tomsk Polytechnic University. Geo Assets Engineering* **329(5)** 117–24

[4] Chudinova D Yu, Dulkarnaev M R, Kotenev Yu A and Sultanov Sh Kh 2017 Diffentiation of wells in zones with residual reserves of oil, using neural network modeling *Exposition Oil and Gas* **4(57)** 10–4
[5] Vafin R V 2007 Fractured-Pore Reservoirs by Water-Gas Impact on. the Reservoir. St. Petersburg: Nedra Publ., 217 p
[6] Nguyen T-V and Júniora S de O 2018 System evaluation of offshore platforms with gas liquefaction processes *Energy* **144** 594–606
[7] Salavatova T Sh, Strekovb A S, Karazhanovac M K and Koilybayevd B N 2017 Decision making during treatment of bottomhole zone by polymeric systems on the basis of indefinite cluster analysis *Procedia Computer Sci.* **119** 478–80
[8] Deyaba S M, Taleb-berrouaneab M, Khana F and Yangac M 2018 Failure analysis of the offshore process component considering causation dependence *Process Safety and Environmental Protection* **57** 220–32
[9] Berge L I, Stonsen J A, Crapez B and Qaule E A 2002 SAWG Infectivity Behavior Based on Siri Field Data *SPE/DOE IOR Symposium*
[10] Christensen J R, Stenly E H and Skauge A 2001 Review of WAG Field Experience *SPE REE* **4(2)**
[11] Mukhametshin V V and Andreev V E 2018 Increasing the efficiency of assessing the performance of techniques aimed at expanding the use of resource potential of oilfields with hard-to-recover reserves *Bull. of the Tomsk Polytechnic University. Geo Assets Engineering* **329(8)** 30–6
[12] Tyncherov K T, Mukhametshin V Sh, Paderin M G, Selivanova M V, Shokurov I V and Almukhametova E M 2018 Thermoacoustic inductor for heavy oil extraction *IOP Conf. Ser.: Mater. Scie. and Engineer. Proc. Equipment, Mechanical Engineering Processes and Metals Treatment (11th International Conference on Mechanical Engineering, Automation and Control Systems* vol. 327(4)) pp 1–8 DOI:10.1088/1757-899X/327/4/042111
[13] Mukhametshin V V 2017 Eliminating uncertainties in solving bottom hole zone stimulation tasks *Bull.of the Tomsk Polytechnic University. Geo Assets Engineering* **328(7)** 40–50
[14] Khayredinov N Sh, Popov A M and Mukhametshin V Sh 1992 Increasing the flooding efficiency of poor-producing oil deposits in carbonate collectors *Oil industry* **9** pp 18–20

IOP Conf. Series: Earth and Environmental Science **350** (2019) 012026 doi:10.1088/1755-1315/350/1/012026

Upper Jurassic subsalt complex of the Terek-Caspian trough due to oil and gas content

T Kh Bachaeva[1,2], A A Daukaev[1,3], I A Kerimov[2,3] and A A Shaipov[2]

[1]Integrated Research Institute of RAS named after H I Ibragimov, 21A, Staropromyslovskoe Highway, Grozny, 364906, Russia
[2]Grozny Oil Technical University named after academician M D Millionshchikov, 100, H Isaev ave., Grozny, 364061, Russia
[3]Academy of Sciences of the Chechen Republic, 13, M Esambaev ave., Grozny, 364001, Russia

E-mail: bacheva@bk.ru

Abstract. The article considers a complex of factors determining the oil and gas content of the Upper Jurassic subsalt deposits of the Terek-Caspian trough. The potential resources of hydrocarbons are associated primarily with the zones of the Nazran and Harbizhsk saddles, the western end of the Terek and Sunzhensk and Soviet-Right-Bank anticlinal zones. Synclinal and near-side zones are assigned to zones with unexplained oil and gas potential, where it is recommended to conduct regional geophysical studies in conjunction with parametric drilling.

1. Introduction

Subsalt deposits in relation to their oil and gas potential were studied by many scientists. The most detailed description of subsalt deposits in connection with the prospects for their oil and gas content is given in the works of N Sh Yandarbiyev in which quantitative parameters of capacities are given in different parts of the territory under consideration, depths, resource potential, etc., as well as data on global practices of exploration for subsalt deposits [1, 2]. The tectonic features and oil and gas potential of deep-seated sediments, including subsalt, were considered in the works of A I Letavin, P A Petrenko [3], B G Voblikova, G I Lebedko and geochemical aspects are presented in the works of A A Yaroshenko [4], A N Stepanova, V V Dotsenko.

The purposeful study of Upper Jurassic deposits within the western part of the TCT has been conducted in the territory under consideration since the 60s with the laying of well No. 100 within the Karabulak area. Later, the following areas were introduced into drilling: Malgobek, Zamankul, Khayan-Kort, Dattykh and others. In the southeast, the subsalt sediments were studied by deep drilling in the Sureta, Bass, Benoy and r. Argun basins.

East of the Sharo-Argun River along the rivers of Elistanzhi, Kharachoi, Benoy-Yassi, Andy Koysu, Avarskoe-Koisu, a reduction in the subsalt content and a change in the lithological composition in the direction of a decrease in the amount of terrigenous material in the roof is noted. In the junction zone of the Dagestan wedge and the Montenegrin monocline, subsalt deposits were discovered by wells in areas – Benoi, Shamkhal-Bulak, Miatly, Zauzanbash, Talgi, etc. parts of dark gray layered limestone (12-15 m). The thickness of the subsalt Jurassic is reduced to 41 m in well No. 2 of Talgi and 38 m in well No. 2 of Zauzanbash.

According to the ammonite finds in the sections of the North Caucasus, the age of the subsalt layer is determined in the range of the Upper Callovian-Lower Kimeridzh [5]. A large middle part of the cut is of Oxford age.

In the well sections, the upper limit of the pre-salt Jurassic region is bent on a sharp change of relict-organogenic dolomites by chemogenic rocks of Kimeridi-Titon, the signs of which are reservoirs and the inclusion of the mineral polyhalite. In the Suret area, in well No. 1, the upper Jurassic section was completely passed, including the subsalt salt thickness of 738 m, it opened the section with rocks of tires (anhydrite) and reservoir rocks (dolomite and fractured limestone).

The results of drilling at Benoy Square (well No. 47) showed the absence of a salt Jurassic stratum in the area. The well revealed a thick mass of clay-argillite deposits of presumably Triassic age below the Jurassic rocks.

On the Dattykh Square in 1987 when testing the interval of 3112-3076 m of Upper Jurassic intersalt deposits in well No. 15, an industrial flow of oil with a flow rate of 55 m3 / day and gas with a flow rate of 5180 m^3/day was obtained. In well No. 12, an industrial gas inflow from subsalt deposits was obtained on the same area.

Drilling operations, started in the 70s on these deposits, were suspended due to the difficult conditions of well placement and the lack of drilling equipment in anticorrosion performance. Seismic surveys carried out at the end of the 80s in the western part of the TCT traced horizons, stratigraphically confined to the top of the Oxford and more ancient sediments, were traced. In the subsalt sediments, the Nazran structure, presumably reef-like, with an amplitude of up to 500 m was established, within which the drilling of a parametric well was recommended.

Terek-Caspian trough (TCT), characterized by a complex tectonic structure and a wide stratigraphic range of oil and gas oil and gas content. The stratigraphic section is represented by a thick stratum (up to 12 km) of Permo-Triassic, Jurassic, Cretaceous, Paleogene-Neogene sedimentary rocks, in which regionally oil and gas bearing carbonate complexes (Upper Jurassic, Valanginian, Upper Cretaceous) and terrigenous (Lower Cretaceous, Karagan-Chokrak) are developed) reservoirs and rock-bearing rocks (titon evaporites, maikop and chokrak clay strata). The development of the region began at the end of the 19th century by well-oil production and the discovery of oil deposits in the Karagan-Chokrak sandy strata of Starogroznenskaya area.

At present, industrial oil and gas content has been established here from Miocene to Upper Jurassic subsalt deposits in the depth range from 200 to 5900 m, more than 150 deposits have been discovered, including more than 25 in the Mesozoic complex of rocks. Many oil deposits are confined to the Karagan-Chokrak sand beds and some of the Cretaceous deposits are practically developed.

In general, the Mesozoic sediment complex within the TCT is characterized by a high degree of exploration both by deep drilling (190 m / km^2 or 0.05 well / km^2) and by seismic exploration (1.1 running km / km^2), which predetermines the orientation of the search for new hydrocarbon deposits at great depths (5500-6100 m) in difficult temperature and pressure conditions.

One of the most promising areas with which the long-term prospects for the development of the oil and gas industry in the Chechen Republic (CR) are connected is the subsalt salt complex of the Upper Jurassic [6]. According to many researchers, the high prospects for the pre-salt Upper Jurassic complex are due to a number of factors:

- the presence of large subsalt elevations and disjunctive disorders;
- large thick carbonate rocks with favorable reservoir properties;
- availability of reliable tires;
- industrial petroleum potential subsalt deposits and high productivity of wells in several regions of the world.

2. Materials and methods

Research in this region was carried out in the following direction: collection and analysis of the actual and stock material on the geological structure of the area. Particular attention is paid to poorly studied monoclinal and near-board zones of the TCT.

IOP Conf. Series: Earth and Environmental Science **350** (2019) 012026 doi:10.1088/1755-1315/350/1/012026

3. Results and discussions

As a result of deep drilling and interpretation of seismic survey materials, it was found out that the geological structure of the Upper Jurassic sediments turned out to be more complex than expected (Figure 1).

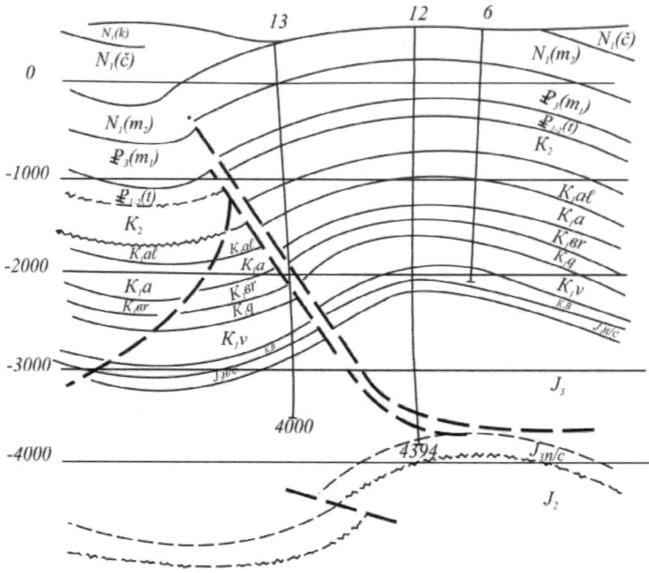

Figure 1. Geological profile of wells No. 6, 12, 13 of the Datty deposit

The thrust structure of the Cretaceous and Upper Jurassic supersalt strata and their displacement along the gaps in the north direction relative to the subsalt salt strata has been established. In the Upper Jurassic subsalt deposits, large sub-thrust structures were noted — Bassovskaya, Korinskaya, Bamutskaya, etc. Obviously, the arches of these structures will be located to the south relative to the thrust structures. In the western part of the TCT, the following structures were fixed in the subsalt strata: Zamankulskaya, Karabulak-Achalukskaya, Severo-Zamankulskaya, Kharbizhinskaya, Akhlovskaya, Malgobek-Voznesensky-Aliyurtovskaya. The axes of these folds are displaced to the south relative to the Cretaceous by 0.5-3.5 km (Table 1).

Table 1. Morphological parameters subsalt structures of the western part of TCT

Name	Closed isohypsum	Linear dimensions, km		Amplitude, m	Other
		length	width		
Zamankul	-5600	17.6	2.5	550	asymmetric
Karabulak-Achalukskaya -	-5600	17.3	2-6	900-950	asymmetric
Severo-Zamankulskaya (Psedakhskaya)	-5800	18.5	3.3	500-550	-
Kharbizhinskaya	-5600	4.6	1.9	200-250	-
Akhlovskaya	-5800	14-16	2.3	-	-
Malgobek-Voznesensko-Aliyurtovskaya	-5800	39-40	3	550	asymmetric

In some areas, according to the results of the method of reflected waves, the geological bodies were recorded according to the method of a common depth point and drilling, identifying developing organogenic structures in shallow areas along faults (Zamankulsky and Kharbizhinsky uplifts, etc.). Favourable geotectonic conditions of reef formation were predicted within the Sjuretsko-Akhlovskaya and Benoisko-Eldarovskaya zones of the TCT [3].

On seismic profiles, carbonate and clay-carbonate deposits are usually represented by a series of continuous and intermittent parallel reflective sites. Organogenic constructions (reefs) on time sections have a number of characteristic features: the presence of an encrustation structure in the ridge sediments with a decrease in its amplitude up the section; loss of correlation (or its deterioration) of reflecting horizons in the reef zone; the presence of diffracted waves directly under the organogenic structure.

Slope reef deposits are expressed in seismic profiles less clearly, but clearly enough, to distinguish them. Usually, at the periphery of the organogenic structure, a series of short clinoform-lying reflected sites, often crossing each other, can be traced. The slope angles of the sites gradually flatten out towards the center of the basin and acquire a subparallel pattern of wave recording.

On the territory of the TCT due to large depths of over 5500–6000 m, powerful evaporitic strata that are difficult to pass through wells, reef and bioherm structures have not been opened, with the exception of the Datykh area, where gas is obtained from a depth of 4300 m.

The Oxford Age was characterized exclusively by carbonate sedimentation. In the south of the region, sponge-algal dome-shaped bioherms of up to 60 m thick are observed at the base of the longline. The formation of such organogenic structures usually took place near the shore. On the middle and late Oxford, at the intersections of deep faults of latitudinal and meridional orientation, coral reefs were born and developed. The connection of reef formation processes with deep faults is proved by the results of the studies of V E Khaina, M S Burshtar and others [7, 8].

In the territory under consideration, similar structures were fixed in Pervomaisk, Datykh, Argun, Suret, Elistanzhinsky and Benoy districts. Everywhere reef and bioherm constructions are mainly represented by dolomites and dolomitic limestones of different-grained relict porous and fractured. In well No. 12 Datykhskaya in the range of 4300-4265 m, industrial gas flow was obtained with a flow rate of 250 thousand tons/day. Borehole No. 1 opened reef facies with a thickness of more than 700 m.

The zones of the most powerful oil and gas accumulation consider the strip of the barrier reef system surrounding the Late Jurassic basin, the predicted linear zones of biogenic sedimentation at the base of the Terek-Sunzhensk folded zone. The synclinal zones (Predtersky trough, Alkhanchurskaya syncline, Petropavlovsk and Chechen depressions) are referred to territories with unclear prospects. The areas lacking or significantly reducing the thickness of the subsalt Jurassic in the northern and south-eastern part of the TCT to 40–60 to 150 m are considered unpromising.

The thickness of the Upper Jurassic subsalt complex is comparable to the productivity of the Upper Cretaceous complex, characterized by regional oil and gas content.

Filtration-capacitive properties of subsalt carbonate rocks [9] are due to secondary processes and are improved due to partial dissolution of carbonates under the influence of high temperatures and aggressive waters that affect the formation of secondary porosity. The improvement of reservoir properties of rocks is also promoted by secondary dolomitization, which is usually characteristic of reef formations. In the reefs, a rigid skeleton (frame) preserves the free volume formed, in the process of dolomitization [10].

Some of them are presented on the upper Jurassic section and the reservoir properties of rocks give the study of natural outcrops on Mountain Rivers in particular along the river Chanty-Argun. As a result of studies of rock samples in the thickness of cavernous dolomitic limestone in the bottoms of the Oxfordian stage, high values of reservoir properties associated with secondary processes — recrystallization and elevation [11]. This made it possible to predict at great depths the favorable capacitive properties of these rocks in the study area. This was confirmed by the results of drilling well No. 12 in Dattytskaya area.

In Dattyshskaya area, the average value of saturation porosity is 0.1–0.3%, gas permeability is 0.00204x10-3 μm^2, and the type of reservoir is fractured-porous-cavern. In the Sureta and Bass areas, the saturation porosity varies from 0.4 to 1.8% the gas permeability varies from 0.0306x10-3 μm^2 to 1.53x10-3 μm^2.

The subsalt sequence was traversed in a number of deep wells within the southern and western sides of the TCT and its nearest frame. This is a single carbonate, mainly dolomitic stratum, which is an analogue of the Kion and Iron suites. Sections of the subsalt Jurassic in the wells of the southern side of the TCT are represented by dolomites with layers of dark gray clay or algal limestone. The maximum thickness is noted in the well number 1 Syuretskaya - 651 m.

Oxford sediments (subsalt Jurassic), in which reef massifs are predicted, are characterized by significant potential hydrocarbon resources. Speaking generally about global oil reserves, the researchers note a significant proportion of reserves in them associated with reef tracts. To one degree or another, such well-known, largest deposits of the Middle East as Kirkuk (2.015 billion tons), Agadzhari (1.3 billion tons), Ahvaz (1.2 billion tons) and Posa are associated with the reef structures. Rica (0.387 billion tons) in Mexico.

As for potential hydrocarbon resources, there are different figures for the pre-salt Jure TCT, according to Samoylovich and others. Initial potential resources are: oil - geological 597993 thousand tons, recoverable - 302829 thousand tons; dissolved gas - geological 368436 million m^3, recoverable - 185999 million m^3; free gas - 701765 million m^3.

According to N Sh Yandarbiyev [1] the total initial hydrocarbon resources generated by Jurassic oil source sediments are estimated at approximately 500 million tons of oil and 1.5 trillion m3 of gas. According to a set of criteria (paleotectonic, geophysical, geochemical, technical and technological), the western parts of the Montenegrin monocline and the Sunzha anticlinal zone (Argudanskaya, Dattykhskaya, Nazranovskaya, Zamankulskaya areas) seem most promising.

In general, the upper Jurassic hydrocarbon resources (recoverable) category $D_1 + D_2$ oil are 27.8 million tons, dissolved gas is 14.3 billion m^3 and free gas is 55 billion m^3.

4. Conclusion

In order to make a comparative assessment of the degree of prospect of the Upper Jurassic subsalt complex of deposits in certain areas of the study area and the selection of priority objects for geologic exploration for oil and gas, the territory was zoned based on the use of signs favorable for the formation of local hydrocarbon accumulations in the Upper Jurassic deposits.

According to the complex of features (structural, lithofacial and depths), the prospects for the development of potential hydrocarbon resources are linked by the Nazran and Harbizh saddles zone, the western end of the Terek and Sunzhenskaya and Soviet-Right-Bank anticlinal zones.

Synclinal and near-side zones are assigned to zones with unexplained oil and gas potential, where it is recommended to conduct regional geophysical studies in conjunction with parametric drilling. The areas of high-priority drilling include: Ali-Yurt, Starogroznenskaya and Ischerskaya.

It is proposed to continue the search seismic survey by the method of reflected waves using the method of a common depth point together with high-precision gravimetric magnetometric studies that allow tracing active fault zones in the areas of Kharbizhin, Arak-Dalatarek, Akhlovo, Karabulak-Achaluki. This, in particular, will allow a detailed study of the velocity characteristics of the section of the Upper Jurassic sediments.

The search for oil and gas deposits in the “pre-salt Jurassic” is possible with a clear knowledge of the section of these deposits, for which purpose it is advisable to drill a deep parametric well in the western part of the Terek anticlinal zone.

References

[1] Yandarbiyev N Sh 1999 *Estimation of the prospects for oil and gas content of subsalt Jurassic deposits of the Terek-Caspian trough* (Moscow) p 69

[2] Yandarbiyev N Sh 2012 Prospects for oil and gas potential of the territory of the Chechen Republic and adjacent areas *Conf. Oil Industry* (Grozny)

[3] Petrenko P A and Letavin A I 1990 *Geotectonic features of the development of marginal zones of the Late Jurassic salt pool of the Terek-Caspian edge trough due to oil and gas content* Collection of works of the Institute of Geology and the development of combustible minerals: oil and gas content forecast and geological exploration directions based on geological and geophysical methods (Moscow) pp 53–9

[4] Yaroshenko A A, Fernaya A A, Savina Yu A and Skrypnyuk O V 2018 Assessment of the main oil and gas-material properties of the jural deposits of the Terek-Caspian advanced deflection *Sci. Innovation Technol.* **2** 197–210
[5] *The mineral resource base of the Chechen Republic* 2016 ed I A Kerimov, E M Aksenov (Grozny: Academy of Sciences of the Chechen Republic)
[6] Li Yu and Burlin Yu K 2005 Oil and gas geological characteristics and conditions of formation of large oil and gas fields in the South of the Caspian Basin *Geology of Oil and Gas* [in Russian – Geologia nefti i gaza] **6** 840–6
[7] Burshtar M S, Arbatov A A and Chernobrov B S 1967 Connection of the Upper Jurassic reef massifs of the Caucasus with fault zones *Geotectonics* [in Russian – Geotektonika] **3** 49–54
[8] Milanovsky E E and Khain V E 1963 *Geological structure of the Caucasus. Essays on the regional geology of the USSR* iss 8 (Moscow: Moscow State University)
[9] Sakharov A S 1992 *Litofacii and the Oxford collector of the Terek-Caspian regional trough. Geology and oil and gas content of Eastern Ciscaucasia* iss 57 (Grozny: North Caucasus Scientific Research and Production Institute of Petroleum) pp 14–21
[10] Kuznetsov V G 1978 *Geology of reefs and their petroleum potential* (Moscow: Nedra)
[11] Vezirova A D 1975 Types of carbonate rocks of the Upper Jurassic deposits of the Terek-Sunzhensk region in connection with the prospects of their petroleum potential *Collection of IGIRGI articles: Lithological studies of reservoir rocks in connection with the exploration and development of oil fields* ed A D Vezirov et al (Moscow) pp 36–48

IOP Conf. Series: Earth and Environmental Science **350** (2019) 012027 doi:10.1088/1755-1315/350/1/012027

Influence of direct seeding technology on typical chernozem structure

V P Belobrov[1], S A Yudin[1], N R Ermolaev[1], V K Dridiger[2], R S Stukalov[2], V A Kholodov[1], N V Yaroslavtseva[1] and A Yu Aidiev[3]

[1]Soil Institute named after V.V. Dokuchaev, Bld. 2, 7, Pyzhevsky lane, Moscow, 119017, Russia
[2]North Caucasus Federal Taxation Center, 49, Nikonov Street, Mikhailovsk, Stavropol Territory, 356241, Russia
[3]Kursky Research Institute of AIP, Cheremushki, Kursk District, Kursk Region, 305526, Russia

E-mail: Belobrovvp@mail.ru

Abstract. The paper discusses the impact of direct seeding technology on volumetric weight of arable layer, content (%) of air-dry and water-resistant aggregates (mm), as well as agronomically valuable aggregates during dry sifting The fact of using the direct seeding technology for four years in grain crop rotation did not lead to an increase in volumetric weight of an arable horizon. This increases the content of dry agronomically valuable aggregates and the coefficient of structure increases. The direct sowing positively affects the water-resistant structure of typical chernozem, increasing the content of water-resistant aggregates in dry fraction > 10 mm from 68.7% (plowing option) to 80.7%.

1. Introduction

Traditional "arable" farming is usually based on mechanical tillage. The retrospective of agricultural farming in the world and Russia shows that the mechanical tillage with a plow and other equipment (flat cutter, chisel, cultivator, etc.) is accompanied by significant degradation of black soil [1]. In the initial stage of application of "arable" farming, as a result of the process of mineralization of the organic matter of soil, there is an increase in the nutritional elements providing an increase in potential fertility. However, over time the intensive farming techniques lead to a decrease in the organic matter in soil, as a result of which they degrade. The soil organic matter (OM) not only provides nutrients for cultivated plants, but also is a key element in stabilizing soil structure. The destruction of soil aggregates as a result of mineralization of organic matter leads to the formation of upper compacted horizons, reduction, deterioration of gas exchange, and finally the loss of soil resistance to adverse environmental factors. The most important integral characteristics of physical fertility of chernozem are the indicators of their structural state - the distribution of structural individual sizes, their mechanical strength and water resistance, which ultimately affect the chemical, physical and biological processes in the soil [2–4]. These indicators are informative not only in relation to the complex physical parameters of fertility, but they are interesting in solving the issues of soil treatment technology.

The degradation of black soil with the current intensive agricultural use makes it necessary to look for technologies and methods of cultivating crops that would restore the original fertility of soil. The technology change for biologization of farming in the form of refusing to plow with a reservoir turnover, the use of surface treatment (disc harrows) or deep loosening (chisel) helps reduce the degradation of black soil as shown by numerous experiments and practical examples [3, 5, 6]. For several decades in the countries of South and North America, Australia, Europe and New Zealand they have been using environmentally-friendly technology of direct seeding (DS) or no-till, which is associated with high yields and minimal damage to the environment. The use of DS technology, which preserves plant residues on the soil and reduces the mechanical load, entails an increase in the share of macroaggregates and generally improves the soil structure [5, 6]. Due to the fact that this new approach to agriculture is only gaining the experience in Russia there was set a goal to assess the impact of DS on the structural state of typical chernozem.

2. Materials and methods

The object of the study was a typical heavy loamy chernozem [7, 8] located on the divide of the Kursk Region near the village of Cheremushki (51°37'46" N; 36°15'40" E). Since 2013 four fields were founded on a plot of 9.2 ha each of which is of 2.4 ha. The fields are separated from each other by a 3 meter protective strip. Each field is divided into 4 plots of 0.6 ha (100x60 m). The effect of 4 treatment options (plowing, combined treatment, minimum - surface treatment, no treatment - DS) on the properties and productivity of typical chernozem in a four-field grain crop rotation is studied. To assess the heterogeneity of soil morphological, agrochemical, and physical properties, a statistical analysis was carried out, which showed that all the studied parameters refer to the same general population with a probability of 95%.

Upon the completion of rotation, the second soil survey was carried out at the control points. The samples were taken from the depth of 2–7 and 10–15 cm to determine the equilibrium density of chernozem along two different options, i.e. plowing and DS [9, 10]. For comparison and as a control measure on density, the samples were taken from the same depths in a forest belt located 200 m from the site. The forest belt was founded in 1964 and is characterized by high fertility and structural characteristics of chernozem [1].

In addition, to analyze the capacity of macroaggregates on plowing and DS options, samples of chernozem were taken from the depth of 0–15 cm, approximately 15×15×25 cm in size and weighing about 5 kg each from five points within the area of 5 m in radius using an envelope method. Such an approach made it possible to avoid losses and artificial redistribution of size fractions of aggregates [11]. The samples were dried to an air dry state.

The obtained five air-dry samples were analyzed separately for both options and then the weighted average value for the desired index was calculated. Dry sieving was carried out according to the method of Savvinov [11, 12]. A set of 0.25, 0.5, 1.0, 2.0, 3.0, 4.0, 5.0, 7.0 and 10.0 mm sieves was used, through which each individual sample was passed. To estimate the water resistance, the Hahn approach was used [12] according to which the aggregates of the target size fraction (usually 1–2 or 1–3 mm) were sieved in water.

3. Results and discussion

In 2013, the equilibrium density of the arable horizon of chernozem was characterized by optimal values and was at a depth of 2-7 cm - 1.08±0.10 g/cm^3; 10-15 cm - 1.18±0.07 g/cm^3. Upon the completion of the first rotation of grain crop rotation, the state of the arable horizon in the plowing option almost remained the same. 4-year-application of DS technology did not have a significant impact on density of the arable horizon. The mean values over the depths of 2–7 cm and 10–15 cm, in the options of plowing and DS are close to each other and stay within the confidence limit (Table 1).

The agricultural use of typical chernozem affects the distribution of dry aggregates into fractions. In the forest belt option the content of the blocky fraction (> 10 mm) is minimal and amounts to 19.1%, while in the plowing option the maximum is 30.9%. The direct seeding option occupies an intermediate value, i.e. 26% (Figure 1).

Table 1. Influence of direct seeding technology on the volume weight of arable horizon of typical chernozem

Options	Depth, cm	Volume weight, g/cm^3	Confidence limit 0.05
Plowing, 2013	2-7 cm	1.08	0.10
	10-15 cm	1.18	0.07
Direct seeding, 2017	2-7 cm	1.11	0.06
	10-15 cm	1.17	0.09
Plowing, 2017	2-7 cm	1.11	0.06
	10-15 cm	1.09	0.09

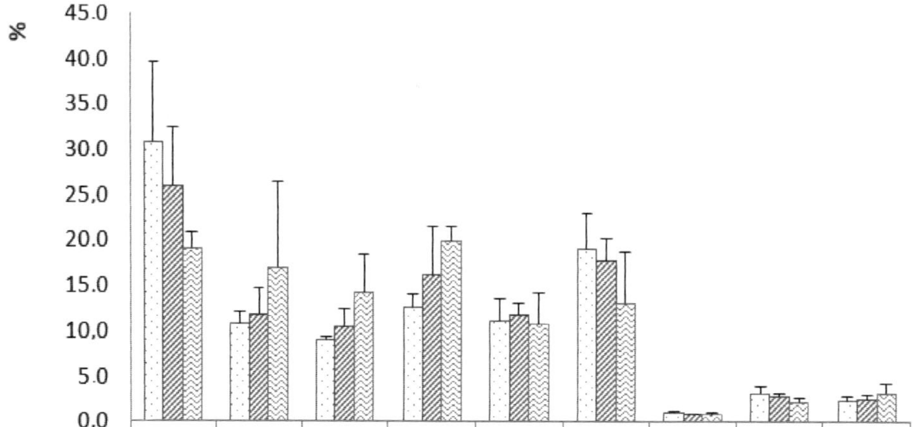

	>10	10-7	7-5	5-3	3-2	2-1	1-0.5	0.5-25	<0.25
Plowing	30.9	10.8	9.0	12.6	11.2	19.1	1.0	3.2	2.3
Direct seeding	26.0	11,7	10.4	16,2	11.7	17.8	0.8	2,8	2.5
Forest belt	19.1	16.9	14.3	19.9	10.8	13.0	0.8	2.1	3.1

Figure 1. Influence of tillage on the content (%) of air-dry aggregates, mm

Considering the fractions of aggregates with dry sifting, it can be noted that the refusal from plowing leads to a gradual restoration of the soil structure, as it can be seen from the content of any of the fractions in the DS option, tending to the options of the forest belt option (Figure 1). Similar situation is observed with the coefficient of the structure and the content of agronomically valuable aggregates in terms of the sum of fractions in the forest belt option and direct seeding, which can be attributed to the optimal indicators according to the standards (Table 2).

Table 2. Influence of direct seeding technology on the content of agronomically valuable aggregates during dry sifting

Option	0.25 -10 mm	<0.25 ->10 mm	Structural coefficient
Plowing	66.9	33.1	2.0
Direct seeding	71.5	28.5	2.5
Forest belt	77.8	22.2	3.5

Long-term intensive plowing of chernozem led to a decrease in the structural coefficient (down to 2) and a decrease in agronomically valuable aggregates (down to 66.9%), which corresponds to an acceptable level of aggregate structure according to standards [10].

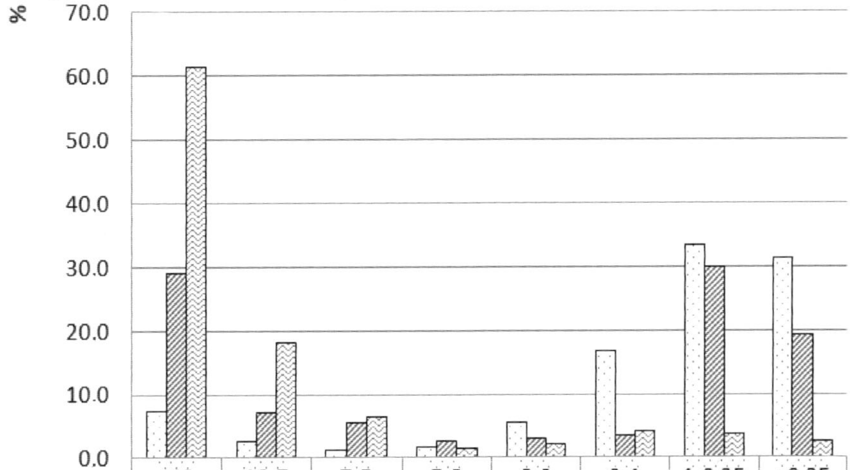

	>10	10-7	7-5	5-3	3-2	2-1	1-0.25	<0.25
Plowing	7.3	2.6	1.3	1.7	5.6	16.8	33.3	31.3
Direct seeding	28.9	7.1	5.5	2.7	3.0	3.5	30.0	19.3
Forest belt	61.3	18.1	6.4	1.4	2.2	4.3	3.8	2.6

Figure 2. Influence of tillage on the content (%) of water-resistant aggregates, mm.

The tillage has had a significant impact on the maintenance of water-intensive aggregates. there was a significant transformation of waterproof units in the barable soil. First of all, the attention is drawn by sharp decrease in the number of water-intensive aggregates of large fractions and an increase in fine fractions (Figure 2). In the forest belt option, the content of fractions > 10, 10-7 and 7-5 is several times higher than the indicator compared to the plowing option. An inverse pattern is observed in the following fractions: 3–2, 2–1, 1–0.25, and <0.25. The agricultural use of chernozem typical in the technology (DS) has a beneficial effect on the restoration of the water-resistant structure of the arable horizon. The content of the fraction> 10 mm increased up to 28.9%, and the fractions 10-7 and 7-5 increased up to 7.1% and 5.5%, respectively with DS compared to the plowing option.

Table 3. Content of water-resistant units, fractions > 0.25 mm depending on the tillage

Option	Fraction			
	>10 mm	7-10 mm	5-7 mm	3-5 mm
Plowing	68.7	69.0	60.5	67.3
Direct seeding	80.7	68.3	58.5	56.9
Forest belt	97.4	97.6	96.2	95.1

As it has been already noted, the bulk of the water-resistant aggregates of typical chernozem is concentrated in fractions of more than 3 mm, moreover, this figure exceeds 90% in the forest belt option. 4-year-application of DS technology in grain crop rotation leads to an increase in the water-resistant structure of the fraction > 10 mm relative to plowing up to 80.7%.

4. Conclusion

4-year-application of DS technology in grain crop rotation did not lead to an increase in the volumetric weight of the arable layer. Its values at the depth of 2-7 cm and 10-15 cm are characterized as optimal ones and are 1.1±0.06 g/cm^3; 1.17±0.09 g/cm^3, respectively.

Under the influence of direct seeding technology, the content of dry agronomically valuable aggregates increases. In this case, the coefficient of structure increases up to 2.5.

The technology of direct seeding has a positive effect on the water-resistant structure of typical chernozem. The content of water-resistant aggregates in the dry fraction > 10 mm increases from 68.7% (plowing option) to 80.7%.

5. Acknowledgments

This work was supported by the Russian Science Foundation, Project No. 19-16-00053

References

[1] Kholodov V A, Yaroslavtseva N V, Farkhodov Yu R, Belobrov V P, Yudin S A, Aydiev A Ya, Lazarev V I and Frid A S 2019 Changes in the ratio of aggregate fractions in humus horizons of chernozems in response to the type of their use *Eurasian Soil Science* **52(2)** 162–170

[2] Munkholm L J, Schjønning P, Rasmussen K J and Tanderup K 2003 Spatial and temporal effects of direct drilling on soil structure in the seedling environment *Soil Tillage Res.* **71** 163–173

[3] Bronick C J and Lal R 2005 Soil structure and management: a review *Geoderma* **124** 3–22

[4] Piron D, Boizard H, Heddadj D, Pérès G, Hallaire V and Cluzeau D 2017 Indicators of earthworm bioturbation to improve visual assessment of soil structure *Soil Tillage Res.* **173** 53–63

[5] Dridiger V K, Kulintsev V V, Stukalov R S and Gadzhiumarov R G 2018 The dynamics of changes in the agrophysical properties of soil during cultivation of field crops using No-till technology *News of the Orenburg State Agrarian University* [in Russian – Izvestiya Orenburgskogo agrarnogo universiteta] **5 (73)** 35-38

[6] Six J, Paustian K, Elliott E T and Combrink C 2000 Soil structure and organic matter: I. Distribution of aggregate-size classes and aggregate-associated carbon *Soil Sci. Soc. Am. J.* **64** 681–689

[7] Egorov V V, Fridland V M, Ivanova E N, Rosov N N et al 1977 *Classification and Diagnostics of Soils of the USSR* (Moscow: Kolos) p 223

[8] IUSS Working Group WRB World Reference Base for Soil Resources 2014 Update 2015 International Soil Classification System for Naming Soils and Creating Legends for Soil Maps *World Soil Resources Reports* Rome **106**

[9] Haidapova D D, Chestnova V V, Shein E V and Milanovski E Yu 2016 Rheological properties of typical chernozem (Kursk Region) with different land use *Soil Science* **8** 955-963

[10] Frid A S, Kuznetsova I V, Koroleva I E, Bondarev A G, Kogut B M, Utkaeva V F and Azovtseva N A 2010 Zone and province standards for changes in agrochemical, physical and chemical and physical indicators of the main arable soils of the European territory of Russia under the anthropogenic impact *Agrochemistry* [in Russian – Agrokhimia] (Moscow: Soil Institute named after V.V. Dokuchaev) 176

[11] Vadyunina A F and Korchagina Z A 1986 *Methods for Studying Physical Properties of Soils* (Moscow: Agropromizdat) p 416

[12] Khan D V 1969 Organic-mineral compounds and soil structure *Science* 140

IOP Conf. Series: Earth and Environmental Science **350** (2019) 012028 doi:10.1088/1755-1315/350/1/012028

Wastewaters from ethylenediamine production and assessment of their toxicity by bioindicator testing

N A Bykovsky, E A Kantor, L N Puchkova, P A Rahman and N N Fanakova

Ufa State Petroleum Technological University, 1, Kosmonavtov Street, Ufa, 450062, Bashkortostan, Russia

E-mail: nbikovsky@list.ru

Abstract. This article discusses the occurrence of toxic wastewater from the production of ethylenediamine by aminating 1,2-dichloroethane with ammonia. It is shown that wastewater from ethylenediamine production is formed at the stages of evaporation of ethylenediamine dihydrochloride and rectification of a mixture of amines obtained as a result of evaporation. In the first case, the wastewater contains a saturated NaCl solution with a content of 1÷2% of polyethylene polyamines, and in the second case, the wastewater is a solution containing about 1% of ethylenediamine and about 2% of ammonia. To study the toxicity of these wastewaters, watercress (*Lepidium sativium*) varieties Zabava and Krupnolistovoy were used. The assessment of toxic properties of wastewater was carried out according to such indicators as seed germination, the average length of seedlings and the dry weight of seedlings. It is revealed that the studied wastewaters have an acute toxic effect on the watercress of both varieties. It is shown that the regression equations obtained for the dry weight of seedlings and seed germination, in contrast to the length of the seedlings, do not adequately describe the experimental results and cannot be used to determine the safe dilution ratio. The safe dilution ratio, calculated using the average length of seedlings, ranges from 489.1 to 892.9 for various runoffs and watercress varieties. It is shown that the most toxic is the runoff containing a saturated NaCl solution with a content of 1÷2% polyethylene polyamines. Zabava is the most sensitive to the degree of effluent toxicity.

1. Introduction

Due to its unique properties, ethylenediamine (EDA) has found its use in various fields. EDA using is required to obtain some antioxidant additives to motor oils, latex stabilizers, fungicides, plasticizers, etc. [1–3]. EDA is synthesized by the reaction of 1,2-dichloroethane and ammonia [4-7]. As a result of the synthesis, ethylene diamine dihydrochloride is formed:

$$C_2H_4Cl_2 + 2NH_3 \rightarrow C_2H_4(NH_2)_2 \cdot 2HCl.$$

As it is shown in [6, 7], the reaction of dichloroethane and ammonia proceeds in 20÷70% aqueous solution. The following parameters must be observed: the process temperature is about 1800 °C, the pressure varies from 0.8 to 7 MPa, the molar ratio of dichloroethane to ammonia is 1:2÷6.4.

To isolate ethylenediamine, the dihydrochloride is destroyed with sodium hydroxide:

$$C_2H_4(NH_2)_2 \cdot 2HCl + 2NaOH \rightarrow C_2H_4(NH_2)_2 + 2NaCl + 2H_2O.$$

The result is an aqueous solution, which consists of ethylene diamine, polyethylene polyamines and sodium chloride. Further, this solution is fed to the evaporation and rectification [7]. The evaporation leads to the formation of gaseous ammonia, amines solution and saturated solution of sodium chloride with a content of 1÷2% polyethylene polyamines. Gaseous ammonia is used to obtain an aqueous solution used in the amination process. The amines solution for its separation is rectified. During the rectification an aqueous fraction containing up to 1% ethylene diamine and up to 2% ammonia forms in the upper part of the column.

In the first case, an effluent is formed, which contains a saturated solution of sodium chloride mixed with polyethylene polyamines. In the second case, it is an effluent containing ethylenediamine and ammonia. Production of 1 ton of ethylenediamine is accompanied by the formation of up to 3 m^3 of effluent from the evaporation stage and up to 14 m^3 of effluent from the rectification stage. The concentration of amines, ammonia and salt in these runoffs is much higher than their maximum allowable concentrations. The purpose of the research is to assess the waste toxicity and the degree of its dilution before discharge.

Toxicity of wastewater was determined by bioindicator testing, using 2 varieties of watercress as a phyto-indicator [8-16]. Experiments were carried out according to the method [17].

The results of studies of the wastewater toxicity, formed in the upper part of the rectification column and containing ethylenediamine and ammonia, as well as wastewater formed during the evaporation stage and containing saturated NaCl solution and polyethylenepolyamines, are shown.

2. Material and research methods

The toxicity analysis of wastewater from ethylene diamine production was carried out with two effluents. The first effluent contained 1% of ethylenediamine and 2% of ammonia. The second effluent contained a saturated solution of NaCl with 2% polyethylene polyamines. The watercress varieties Krupnolistovoy and Zabava were used as the phyto-indicators. Experiments carried out as follows: 30 seeds of the plant were placed on the filter paper, placed in Petri dishes and moistened with 5 ml of the test solution.

The analyzed solution was the undiluted wastewater and the same solution diluted by a factor of 2, 4, 8, 16, 32, 64, 128, 256, 512 and 1024. In the control sample, the filter paper was moistened with distilled water.

Each experiment was conducted in 3-fold repeatability. After eight days, seeding was evaluated, dry weight and length of watercress sprouts were measured [17]. The relative measurement error was determined from the results of each experiment. In this case, the confidence probability was 95%. According to the method [17, 18], a linear regression equation was used to obtain the dependence of the parameters under study on the dilution degree.

To determine the effect of ethylene diamine and polyethylene polyamines on the toxicity of wastewater, experiments were performed with a 2% ammonia solution and with a saturated NaCl solution. The cress Zabava was used as a phyto-indicator.

3. Experimental results and discussion

The effect of the wastewater dilution factor on seed germ is shown in Tables 1 (var. Zabava) and 2 (var. Krupnolistovoy).

Seed germination begins with a 16-fold dilution of wastewater. At the same time, germination ranges from 29.7% to 40.6% at 16-fold dilution of effluent. Further dilution of wastewater leads to an increase in seed germination to a value of 98.9%. However, starting from 32-fold dilution, the dependence of seed germination on the effluent dilution ratio is not observed. This is fair for both the Zabava and the Krupnolistovoy varieties.

The magnitude of the error in determining the germination of seeds of cress of both varieties vary widely from 3.4% to 65.5%. It should be noted that the seed germination values in the case of the control sample both for wastewater containing NH3 + EDA and for the wastewater containing NaCl + PEP differ from each other by no more than 1.2%. This indicates a fairly good reproduction of the results obtained in identical conditions.

Table 1. Effect of the wastewater dilution factor on the seeds germ of cress var. Zabava.

Dilution	Effluent, containing NaCl + PEPA		Effluent, containing NH_3 + EDA		Saturated NaCl solution		2% Ammonia solution	
	Germinating, %	Error, %	Germinating, %	Error, %	Germinating, %	Error, %	Germinating, %	Error, %
16	32.3	19.5	42.7	23.6	16.71	12.5	8.76	20.5
32	84.5	11.2	95.4	5.1	32.69	11.3	13.44	18.4
64	88.8	3.5	98.8	4.8	83.32	14.2	47.56	15.6
128	97.9	7.4	96.7	14.9	73.34	8.9	86.72	19.1
256	97.6	7.5	94.3	10.3	73.31	10.5	76.74	17.5
512	92.4	15.3	95.6	5.2	76.75	15.8	73.32	10.6
1024	98.8	3.4	–	–	96.73	16.4	–	–
Control	93.5	22.2	94.5	13.5	73.36	18.3	83.33	17.4

Table 2. Effect of the wastewater dilution factor on the seeds germination of cress var. Krupnolistovoy.

Dilution	Effluent, containing NaCl + PEPA		Effluent, containing NH_3 + EDA	
	Germinating, %	Error, %	Germinating, %	Error, %
16	29.7	18.7	30.0	65.5
32	89.9	17.9	92.2	32.4
64	90.0	6.8	95.5	13.2
128	94.5	7.6	87.8	19.6
256	93.3	11.4	84.4	49.4
512	92.2	7.7	88.9	14.2
1024	94.2	25.2	–	–
Control	93.6	11.4	93.3	15.4

To assess the possibility of using the experimental results of seed germination in determining the safe dilution ratio of runoff, the regression equations for the seed germination dependence on the dilution ratio were calculated. In accordance with the methodology [17, 18] the linear regression equation was used to process the experimental results. The correlation coefficients between the experimental results and the equations obtained for its description were calculated. The regression equations and correlation coefficients are shown in table 3.

Table 3. Regression equations and correlation coefficients of seed germination depending on the dilution ratio.

Effluent	Zabava		Krupnolistovoy	
	Regression equation	Correlation	Regression equation	Correlation
NH_3 + EDA	$Y = 96.63 - 0.003 \cdot X$	0.212	$Y = 72.15 + 0.049 \cdot X$	0.347
NaCl + PEPA	$Y = 90.40 + 0.008 \cdot X$	0.465	$Y = 92.71 - 0.005 \cdot X$	0.288
Saturated NaCl solution	$Y = 49.70 + 0.052 \cdot X$	0.431	–	–
2% ammonia solution	$Y = 32.06 + 0.113 \cdot X$	0.409	–	–

Here, Y – seed germination, %; X is the multiplicity of dilution.

It is known [18] that the critical values of the sample correlation coefficient at 95% confidence probability for 18 experiments (NH3 + EDA sink) are 0.468, and for 21 experiments (NaCl + PEPA sink) are 0.433. Analysis of the results shows that the correlation coefficients are in the range from 0.212 to 0.465. This indicates that the regression equations of seed germination dependencies for both cress

IOP Conf. Series: Earth and Environmental Science **350** (2019) 012028 doi:10.1088/1755-1315/350/1/012028

varieties on the dilution ratio do not describe the experimental data reliably. Thus, the use of the dependence of the germination of watercress seeds on the dilution ratio cannot not reliably determine the safe dilution ratio for both effluents. This is true for a saturated solution of NaCl and for a 2% solution of NH_3.

Tables 4 and 5 show the average length of seedlings of watercress seeds at various wastewater dilution ratios for each parallel experiment.

The minimum length of watercress seedlings is 6.12 mm and is observed when the multiplicity of dilution of the effluent is 16. The maximum length of the seedlings of watercress is 115.61 mm and is observed when the multiplicity of dilution is 512.

There is a clear the dependence of the length of seedlings, both on the composition of the wastewater, and on the cress plants. So for the effluent containing a saturated solution of NaCl with 2% PEPA, the average seedling length is less than for a runoff containing 1% ED and 2% NH_3. The difference in the length of the seedlings is 58.7% for Zabava variety and 41.4% for Krupnolistovoy variety.

The error in determining the average length of seedlings for all experiments has approximately the same values and varies from 6.33% to 15.62%. This fact indicates a fairly high reproducibility of this parameter. At the same time, there are no visible deviations in the seedlings length for a saturated solution of NaCl and 2% NH_3 solution.

Table 4. Dependence of the average length of seedlings of cress seeds of Zabava variety on the dilution ratio.

Dilution	Effluent, containing NH_3 + EDA		Effluent, containing NaCl + PEPA		2% Ammonia solution		Saturated NaCl solution	
	Length, mm	Error, %	Length, mm	Error, %	Length, mm	Error, %	Length, mm	Error, %
16	29.14	10.36	11.56	7.86	8.34	10.3	10.25	5.2
	30.52	12.54	13.42	12.41				
	25.68	11.48	10.25	11.25				
32	38.21	9.43	19.16	13.78	25.37	11.2	18.41	4.9
	34.86	9.78	21.26	14.56				
	41.07	13.04	24.00	21.26				
64	77.10	13.37	56.92	8.87	81.24	9.4	83.25	9.5
	69.86	9.88	59.93	10.62				
	80.62	11.55	55.22	9.66				
128	73.11	10.14	90.21	9.75	85.31	7.1	107.47	8.3
	80.54	7.67	76.53	12.11				
	85.27	8.35	83.33	11.04				
256	84.82	10.03	84.70	11.17	90.27	5.3	115.26	7.8
	86.12	12.48	96.32	7.38				
	89.52	10.93	99.17	8.65				
512	85.58	10.59	97.38	11.47	100.32	9.2	146.39	10.6
	84.74	7.73	95.84	12.39				
	91.58	8.30	95.38	9.17				
1024	–	–	98.80	8.08	–	–	136.83	11.8
			99.13	10.56				
			91.79	7.73				
Control	99.13	11.27	99.32	11.34	97.46	8.6	121.27	9.2
	97.46	10.02	102.83	10.53				
	107.56	13.21	113.93	11.17				

Table 5. Dependence of the average length of cress seedlings of the Krupnolistovoy variety on the dilution ratio.

Dilution	Effluent, containing NH_3 + EDA		Effluent, containing NaCl + PEPA	
	Length, mm	Error, %	Length, mm	Error, %
16	11.12	14.91	6.12	10.25
	14.40	12.59	8.31	11.67
	14.73	15.62	9.15	9.84
32	41.38	11.80	24.17	14.35
	43.63	10.52	22.54	12.56
	45.89	13.05	23.33	14.88
64	61.74	13.05	62.82	10.42
	60.21	6.33	73.89	13.18
	62.94	9.31	70.96	16.15
128	65.13	15.01	80.03	7.67
	82.19	12.71	98.07	7.49
	79.14	8.40	87.76	14.83
256	94.79	15.24	84.30	14.84
	89.47	16.58	96.90	11.04
	95.03	13.47	83.93	10.06
512	106.59	10.39	96.11	13.23
	115.61	11.46	92.89	13.73
	108.64	15.09	97.48	11.59
1024			103.62	9.87
	–	–	107.08	9.13
			91.18	9.04
Control	102.41	9.74	103.52	11.66
	122.53	11.01	114.96	7.62
	124.33	9.81	106.46	8.21

Table 6 represents the regression equations and the correlation coefficients of the dependence of the average length of seedlings on the dilution ratio for both effluents.

Table 6. Regression equations and correlation coefficients of the dependence of the average length of seedlings on the dilution ratio.

Effluent	Zabava		Krupnolistovoy	
	Regression equation	Correlation	Regression equation	Correlation
NH_3 + EDA	$Y = 49.78 + 0.093 \cdot X$	0.686	$Y = 43.06 + 0.150 \cdot X$	0.876
NaCl + PEPA	$Y = 46.43 + 0.066 \cdot X$	0.668	$Y = 52.67 + 0.102 \cdot X$	0.669
Saturated NaCl solution	$Y = 57.72 + 0.105 \cdot X$	0.737	–	–
2% ammonia solution	$Y = 41.01 + 0.144 \cdot X$	0.711	–	–

Here, Y is the average length of the seedling, mm; X is the multiplicity of dilution.

The values of the correlation coefficients between the experimental data and the regression equations are much larger than their critical values [18]. This indicates a fairly good agreement between the experimental results and the regression equations obtained for their description. Therefore, the regression equations shown in Table 6 can be used to determine the safe dilution ratio. Thus, the safe dilution ratio obtained using Zabava variety was 554.8 for wastewater containing NH_3 + EDA and 892.9 for effluent containing NaCl + PEPA. The safe dilution rate for a 2% NH_3 solution makes 392.9, and this one for a saturated NaCl solution makes 625.1. The safe dilution obtained using Krupnolistovoy variety was 489.1 for wastewater containing NH_3 + EDA and 545.8 for effluent containing NaCl + PEPA.

Table 7 shows the seedlings dry weight depending on the dilution ratio of wastewater containing NaCl + PEPA for each parallel experiment.

The dry weight of seedlings varies from 1.30 mg to 2.15 mg. The error determined for the averaged values of the dry weight of the shoots in each parallel experiment varies from 7.57% to 15.68%, which indicates a good reproducibility of this parameter. But the dependence of the dry weight of the seedlings on the effluent dilution ratio is not observed.

The regression equations and the correlation coefficients of the dependence of the dry weight of seedlings on the dilution ratio are shown in Table 8.

Table 7. Dependence of the seedlings dry weight on the of in dilution ratio of wastewater containing NaCl + PEPA.

Dilution	Zabava		Krupnolistovoy	
	Weight, mg	Error, %	Weight, mg	Error, %
16	1.98	9.25	2.15	10.03
	1.86	10.13	1.96	8.26
	2.07	8.17	2.07	9.48
32	1.96	8.87	1.88	10.07
	1.76	8.28	1.94	9.01
	1.85	9.51	1.95	9.57
64	1.88	7.30	1.85	9.68
	1.78	8.83	1.66	7.57
	1.73	7.76	1.74	7.98
128	1.48	7.53	1.52	10.43
	1.59	9.79	1.60	9.48
	1.57	8.35	1.53	9.81
256	1.30	13.65	1.53	10.07
	1.38	9.79	1.64	8.66
	1.48	11.06	1.50	11.60
512	1.49	6.90	1.44	9.35
	1.45	8.26	1.54	11.68
	1.41	11.77	1.49	10.23
1024	1.51	9.78	1.64	7.39
	1.50	9.78	1.51	10.16
	1.58	6.75	1.45	14.32
Control	1.04	15.68	1.41	9.09
	1.25	12.53	1.34	9.17
	1.13	15.38	1.37	10.36

Table 8. Regression equations and correlation coefficients of the dependence of the dry weight of seedlings on the degree of dilution.

Effluent	Zabava		Krupnolistovoy	
	Regression equation	Correlation coefficient	Regression equation	Correlation coefficient
NaCl + PEPA	$Y = 1.677 + 0.00001 \cdot X$	0.467	$Y = 1.733 + 0.0001 \cdot X$	0.362

Here, Y is the dry weight of the seedling, mg; X is the multiplicity of dilution.

From table 8 it follows that the multipliers of the dilution ratio are quite low values. As a result, even with a 1000-fold dilution, the dry weight of seedlings varies from 0.6% to 5.7%. This indicates that the dry weight of seedlings practically does not depend on the multiplicity of dilution of the flow for both watercress varieties. In this case, the correlation coefficients have values below their critical values [18]. All this suggests that these dependencies cannot be used to determine the safe multiplicity of wastewater dilution.

4. Conclusion

Studies show the acute toxicity of wastewater from ethylene diamine production. The germination of seeds of watercress for both varieties is observed, starting with a 16-fold dilution both for effluent containing NH_3 + EDA, and for effluent containing NaCl + PEPA. The obtained results state that to determine the safe dilution ratio of both effluents, only regression equations obtained from the dependence of the average seedling length on the dilution factor can be reliably used.

It should be noted that the effluent containing NaCl + PEPA is more toxic than the effluent containing NH_3 + EDA. At the same time, the Zabava variety of watercress turns out to be more sensitive to the waste water toxicity than the Krupnolistovoy variety. Thus, the safe degree of dilution of runoff containing NH_3 + EDA, determined by the length of seedlings of the Zabava variety, is almost 12% greater than the dilution ratio, found from the length of seedlings of the Krupnolistovoy variety of watercress. In the case of wastewater containing NaCl + PEPA, the same result makes almost 39%.

In conclusion, it can be noted that the Zabava and Krupnolistovoy varieties of watercress can be used to study the toxicity of wastewater containing various substances. The studies of Zabava varieties conducted with a saturated solution of NaCl and 2% NH_3 solution showed that the presence of impurities of polyethylene polyamines and ethylene diamine in these solutions increases its toxicity. However, to compare the toxicity of wastewater containing different components, it is necessary to use the same species and variety of plant as a phyto-indicator.

Acknowledgements

This scientific paper was prepared on basis of research work within the government work No. 5.12863.2018 / 8.9 of the Russian Federation.

References

[1] Kuliev A M 1985 *Chemistry and technology of additives to oils and fuels* (Leningrad: Khimiya)
[2] Varnavskaya O A, Khvatov L K, Lebedev N A, Khlebnikov V N, Fakhrutdinov B R, Safin D H, Chebareva A I and V A Shepelin RU Patent 2,174,997 (20 October 2001)
[3] Levashov V I and Mudrik T P 2008 *Bashkir Chemical Journal* **15(4)** 102-104
[4] Testova N V, Sukhova O B and Jonah K G RU Patent 2,114,849 (10 July 1998)
[5] Bykovsky N A, Daminev R R, Fatkullin R N and Puchkova L N RU Patent 2,226,189 (10 January 2015)
[6] Zagidullin R N RU Patent 2,537,564 (27 March 2004)
[7] Lebedev N N 1988 *Chemistry and technology of basic organic and petrochemical synthesis* (Moscow: Khimiya)
[8] Bykovsky N A, Puchkova L N and Fanakova N N 2015 *Ecology and Industry of Russia* **19(10)** 48-51
[9] Daminev R R, Islamutdinova A A and Shayakhmetov A I 2012 *Ecology of urbanized territories* **2** 80-84
[10] Melekhova O P and Yegorova E I 2010 *Biological control of the environment: bioindication and biotesting* (Moscow: Academia)
[11] Dmitriev A I 1996 *Bioindication* (Nizhny Novgorod: Volgo-Vyatsk Academy of Government Service)
[12] Evgenyev M I 1999 *Soros Educational Journal* [in Russian – Sorosovskiy obrazovatelnyj zhurnal] **11** 29–34
[13] Olkova A S 2014 *Successes of modern biology* [in Russian – Uspekhi sovremennoy biologii] **134(6)** 614–622

[14] Persoone G 1992 Ecotoxicology and water quality standards *River water quality ecological assessment and control* (Brussels: Commission of European Communities) pp 461-482
[15] Braginsky L P 1977 *Problems of analytical chemistry* [in Russian – Problemy analiticheskoy khimii] **5** 27-35
[16] Braginsky L P 1981 Water quality assessment of natural water bodies by toxicological indicators *Proc. Conf on Scientific bases of quality control of surface waters by hydrobiological indicators* (Moscow: Sovetskaya nauka) pp 201-206
[17] PND F T 14.1:2:4.19 2013 *Biological methods of control. Methods for determining the toxicity of drinking, groundwater, surface water and wastewater, chemical solutions by measuring the germination, average length and average dry weight of seedlings of watercress* (Moscow: Federal service for supervision of natural resources)
[18] Urbah V Yu 1964 *Biometric methods* (Moscow: Nauka)

IOP Conf. Series: Earth and Environmental Science **350** (2019) 012029 doi:10.1088/1755-1315/350/1/012029

Electrochemical treatment of waste water from nickel in galvanic production

N A Bykovsky, E A Kantor, P A Rahman, L N Puchkova and N N Fanakova

Ufa State Petroleum Technological University, 1, Kosmonavtov Street, Ufa, 450062, Bashkortostan, Russia

E-mail: nbikovsky@list.ru

Abstract. In the production of metal coating by electroplating, wastewater containing the metal ions is used. The paper presents the study results of cleaning waste water of nickel after electroplating. Wastewater treatment was implemented in an electrodialysis plant. The initial wastewater contained from 20 to 80 mg/l of nickel II. Experimental results obtained by studying the electrodialysis extraction of nickel from wastewater of electroplating plants, indicate that the cleaning process is influenced by such factors as the effluent treatment time, the current load on the electrodialyzer, the initial metal concentration in the effluent. The increase in voltage in the electrodialysis plant leads to an increase in the treatment efficiency. So after six hours of processing, the processing rate during nickel extraction increases from 39 to 66.5% with an increase in voltage from 40 to 100V. Along with cleaning wastewater of nickel electrolyte, the electrodialysis method allows obtaining concentrates of this electrolyte containing from 260 to 1230 mg/dm^3 of nickel and suitable for use in the electroplating process.

1. Introduction

Electrochemical deposition of metal coatings on various products are widely used in various areas of production: mechanical engineering, automotive, electronics, precision instruments, and so on. It is enough to note that about 70% of machine building enterprises have galvanizing plant or bays in its structure.

The application of electroplated coatings in mechanical engineering is one of the most common methods to protect the products against corrosion. In addition, the electroplating significantly increases the wear resistance of parts, improves its electrical conductivity and other important properties. Most often for these purposes the material surfaces are coated by the chromium, nickel, copper, cadmium and other metals [1].

As a result of the electroplating plants operation, some wastewaters are produced containing non-ferrous metals and other chemicals, the release of which into the natural environment can lead to significant damage. In addition, with the wastewater of electroplating, plant facilities lose quite expensive non-ferrous metals.

In this regard, the treatment of electroplating plants wastewater and extraction of non-ferrous metals from it are of great environmental and economic importance.

The wastewater containing nickel (II) was monitored generated at the Sterlitamak Lenin Machine-Tool Factory, where electroplating is carried out in a hardening plant for hardening parts, and in the consumer goods plant (TKBN) to impart decorative properties to products. The nickel content in the wastewater of the galvanic section of the thermal workshop and the TKBN workshop on average exceeds the maximum permissible concentration from 1 to 30 times (MPC of nickel (II) 0.1 mg/dm^3 [2]). There are one-time jumps in nickel concentration exceeding the MPC from 100 to 450 times. Obviously, this occurs at the time of discharge of the electrolyte from the bath.

Thus, the problem of wastewater treatment from nickel (II) in machine-tool plants is acute. There are a number of methods to treat such wastewater. The most common of them is the method of nickel (II) deposition [3, 4].

Electrodialysis treatment allows not only purifying waste water, but also obtaining a concentrate suitable for the preparation of electrolyte solutions. As a result, expensive metals are returned to the production cycle [5-10].

2. Material and research methods

To study the process of electrodialysis extraction of nickel from the galvanic section wastewater, a plant was used, schematically shown in Figure 1. The electrodialysis plant is an assembly of cells separated by alternating cationic and anion-exchange membranes. The entire assembly is in a DC electric field created between the electrodes.

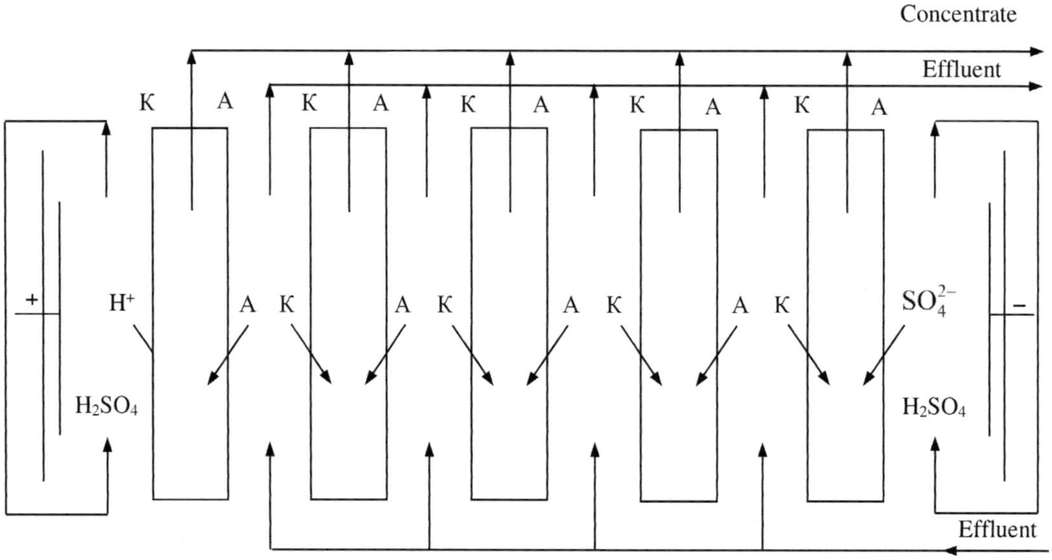

Figure 1. Scheme of electrodialysis plant

Under electric field, directed motion of ions occurs. In this case, the cations move in the direction of the lines of the electric field strength, and the anions in the opposite direction. If pumping the runoff through the plant is organized as shown in Figure 1, the cations and anions migrate through the cation-exchange and anion-exchange membranes, respectively, and from the desalting cells will fall into the concentration cells, where ion-exchange membranes with opposite exchange properties will impede their further promotion. Thus, in the concentration cells of the electrodialysis plant the solutions of the same composition will be generated as in the desalting cells, but with a higher concentration.

The hull frames made of PVC sheets, 2 mm thick, form cells for pumping effluent and collecting the resulting concentrate. For the sell separation it is applied the cation-exchange membrane of the mark MK-40 and anion-exchange membrane of the mark MA-40.

To turbulize the solutions and to prevent the contact of the membranes, a mesh of stretched calendered vinyl plastic was placed in each cell. The plexiglass plates were used to tie the whole structure. Electrode cells were also made in plates. Titanium plates coated with ruthenium oxide were used as electrodes. The distribution and collection of solutions flowing through the plant was carried out using supply manifold and gathering main made in the dialyser core. The experimental electrodialyzer contained 17 cells, in eight of which the effluent was desalted, and in nine the extractable salt was concentrated. The working surface of the membranes was 115 cm^2.

The waste water was pumped through the desalting cells of the electrodialysis plant. The concentrate as it accumulated in the corresponding cells flowed out of the electrodialysis unit by gravity. A 0.1 N solution of sulfuric acid was circulated in the electrode cells. The power of the electrodialyzer was carried out from a controlled rectifier.

3. Experimental results and discussion

Laboratory studies were performed with model solutions of effluent containing from 20 to 80 mg/dm^3 of chromium per metal. The volume of wastewater exposed to desalting in all experiments remained constant and amounted to 61 liters.

During the operation of the electrodialysis plant, samples of the treated effluent and concentrates generated in the electrodialysis concentration cells were periodically collected and analyzed for nickel content. The concentration of nickel was determined by the colorimetric method with dimethylglyoxime.

The dependence of nickel concentration in the effluent on the time of its electrodialysis treatment is shown in Figure 2.

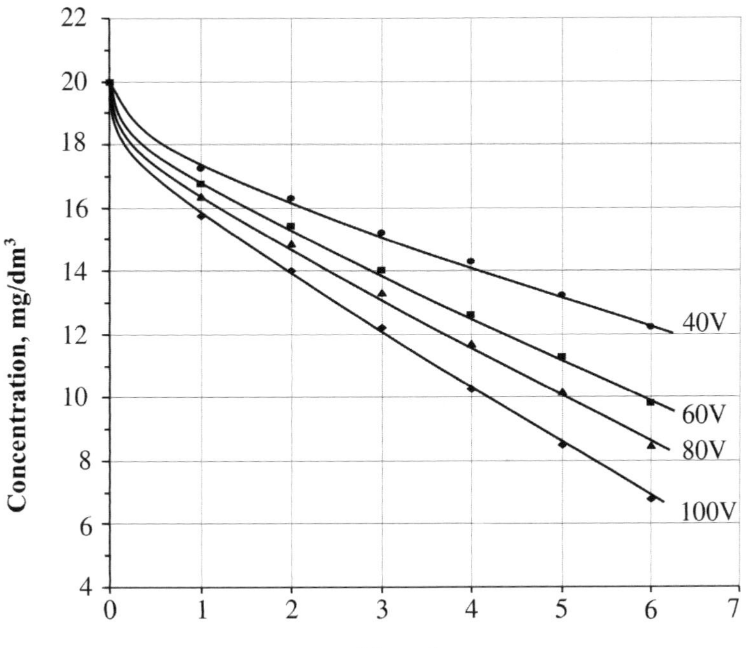

Figure 2. Dependence of nickel (II) concentration in wastewater on the treatment time in the plant

It may be seen that after the first hour of operation of the experimental installation, the dependence of nickel concentration in the effluent on the processing time is linear. With increasing voltage on the device, the degree of nickel extraction from the effluent increases. So after 6 hours of operation, the degree of extraction of nickel from the effluent varies from 39.0% at a voltage of 40V to 66.2% at a voltage of 100V.

Table 1 presents the values of the rate of nickel extraction from the effluent at different voltages on the electrodialyzer.

The data presented in the table show that with increasing voltage on the plant, the rate of nickel extraction increases also.

The dependence of nickel concentration in the electrodialysis plant concentration cells on the run-off treatment time with its initial concentration of 20 mg/dm^3 is presented in Figure 3.

Table 1. Nickel extraction rate from effluent

Voltage on plant, V	Nickel extraction rate, mg/dm^3·h
40	1.01
60	1.39
80	1.59
100	1.80

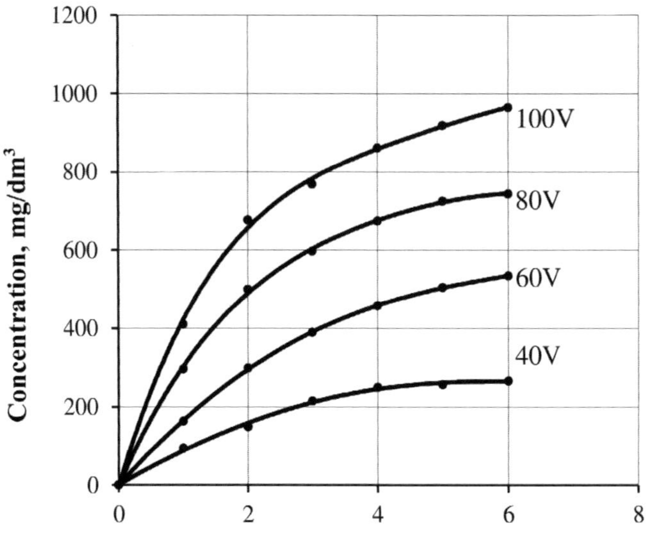

Figure 3. Dependence of nickel concentration in wastewater on the treatment time

Analysis of the presented results shows that the maximum rate of increase in the concentration of nickel in the concentration cells of the electrodialysis plant is observed in the first hour of its operation.

Thus, the rate of change of nickel concentration in the concentration cells during the first hour of operation of the plant is 91 mg/dm^3·h for a voltage of 40V and 413 mg/dm^3·h for a voltage of 100V. And during the sixth hour of work at a voltage on the plant equal to 40 V the rate of change of nickel concentration in the concentration cells is 5 mg/dm^3·h and 47 mg/dm^3·h at a voltage of 100V.

It should be noted that with voltage increasing on the electrodialyzer, the concentration of nickel in a concentrate increases as well. So after six hours of the process, the concentration of nickel in the concentration cells reaches 260 mg/dm^3 at a voltage of 40V and 961 mg/dm^3 at a voltage of 100V.

Figure 4 shows the dependence of nickel concentration in the concentrate on its initial concentration in an effluent after 6 hours of the plant operation.

It is seen that with an increase in the initial concentration of nickel in effluent, its concentration in the concentrate also increases. The greatest increase in concentration is observed at low initial nickel concentrations in the effluent. Indeed, when the concentration of nickel in the effluent changes from 0 to 20 mg/dm^3, its content in the concentrate changes from 510 to 840 mg/dm^3, depending on the voltage on the plant. An increase in the initial nickel concentration in the effluent to 80 mg/dm^3 leads to an increase in its concentration in concentrate from 755 to 1230 mg/dm^3.

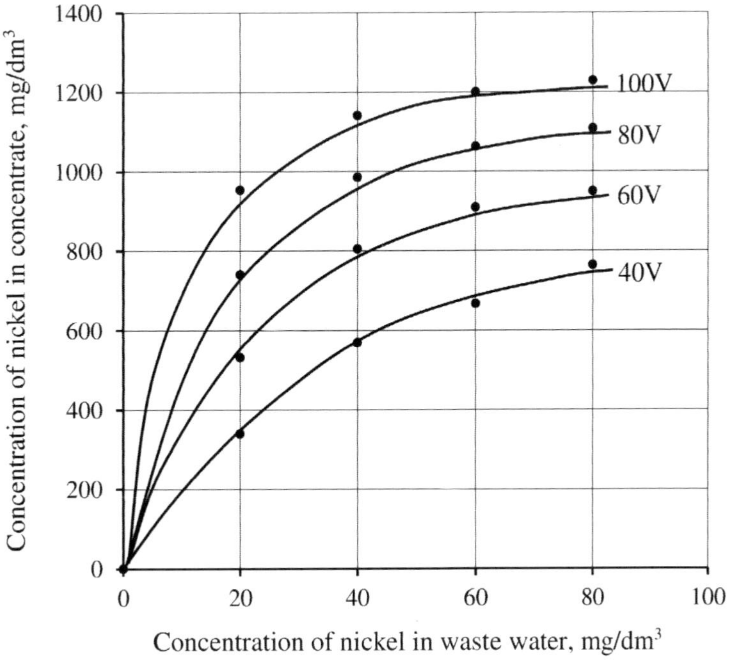

Figure 4. Dependence of nickel concentration in concentrate on its concentration in effluent

4. Conclusion

Analysis of the above results shows that to treat the wastewater generated during the electrochemical deposition of nickel coatings, the electrodialysis method is the most suited. It allows creating a closed cycle in water and producing concentrates suitable to use in the process.

Experimental results obtained at study of processes of nickel electrodialysis extraction from wastewater of electroplating plants, indicate that the treatment process is influenced by such factors as the treatment time of the effluent, the current load on the electrodialyzer, the initial concentration of the metal in the effluent.

The voltage increasing on the electrodialysis plant leads to an increase in the treatment efficiency. So after six hours of operation, the purification efficiency during nickel extraction increases from 39 to 66.5% with a voltage increase from 40 to 100V.

Along with wastewater treatment from nickel electrolyte, the electrodialysis method allows obtaining concentrates of this electrolyte containing from 260 to 1230 mg/dm^3 of nickel and suitable for using in the electroplating process.

5. Acknowledgements

This scientific paper was prepared on the basis of the research work within government work No. 5.12863.2018 / 8.9 of the Russian Federation.

References

[1] Fedotiev N P, Alabyshev A F, Rotinyan A P, Vyacheslavov P M, Zhivotinskiy P B and Galnbek A A 1967 *Applied Electrochemistry* (Leningrad: Khimiya)

[2] Bespamyatnov G P, Bogushevskaya K K, Bespamyatnova A V, Krotov Yu A, Zelenskaya L A, Plehotkin V F and Smirnov G G 1975 *Maximum permissible concentrations of hazardous substances in the air and in water* (Leningrad: Khimiya)

[3] Guillard D and Lewis A E 2001 *Ind. and Eng. Chem. Res.* **23** 5564–9

[4] Gohlke U and Stegmann S 2000 *Way of purification of water solutions of ions of heavy metals* DE Patent No. 19829827 (5 January 2000)
[5] Bykovsky N A, Nizov V A and Fanakova N N 2016 *Way of regeneration of the fulfilled sour etching solutions which are formed when processing products of the titan* RU Patent No. 2596564
[6] Tongwen X and Weihua Y 2003 *J. of Membrane Sci.* **220** 89–95
[7] Formanoir C, Suard M, Dendievel R, Martin G and Goden S 2016 *Additive Manufacturing* **11** 71–6
[8] Liu Z, Tsai I L, Thompson G E, Liu H and Donatus U 2015 *Materials Chemistry and Physics* **160** 329–36
[9] Marin E et al 2016 *Materials & Design* **108** 77–85
[10] Mullabaev I M, Bykovsky N A and Kantor E A 2010 *Bashkir Chemical J.* **17(5)** 153–4

Removal of sulfides from tannery-sourced wastewaters

N A Bykovsky, E A Kantor, P A Rahman, L N Puchkova and N N Fanakova

Ufa State Petroleum Technological University, 1, Kosmonavtov Street, Ufa, 450062, Bashkortostan, Russia

E-mail: nbikovsky@list.ru

Abstract. The paper presents the study results for the treatment of sulfide wastewater generated by tanneries. Wastewater treatment was carried out by electrolysis with a sacrificial ferrous anode. It is shown that during the treatment process, the ferrous anode dissolves and deposition of iron sulfide is formed. The process of effluent treatment is determined by the anode current density. The most efficient effluent treatment is carried out at a current density of 133 A/m^2. In this case, the purification efficiency reaches 99.2%, and the cost per unit of electricity makes 169.2 kW·h/kg.

1. Introduction

Sulfide wastewater treatment is relevant, especially for the tanning industry. The process solution used for leather liming includes sodium sulfide. After the operation of liming, the concentration of sodium sulfide reaches 8-10 g/l and, as a rule, the spent solution is discharged into the effluent. Lowering the concentration of sulfides in the effluent occurs mainly due to dilution when combining wastewater from various technological shop floors. In this case, quite often the concentration of sulfides in wastewater exceeds its maximum permissible value, which is equal to 0.003 mg/l [1].

From the literature, it is known that there are a number of methods to purify the wastewater from sulfide-containing compounds. The method of oxidation of sulfide ions by various oxidizing agents is the most widely applied, where ozone, chlorine, oxygen, hydrogen peroxide, and ammonium persulfate are used [2-6]. All of these methods involve the use of chemical reagents.

The oxidation of sulfides to sulfates by electrochemistry with anodes from PbO_x, TiO_2, IrO_x, Pt is possible [7, 8].

As is known, sulfides of many metals are sparingly soluble compounds [9]. It is attractive for its use in the sulfide-containing effluents treatment. In particular, iron can be used for this purpose by setting iron ions in solution by anodic dissolution [10]. The iron sulfide deposition formed during the process can be recycled at metallurgical plants.

The purpose of this work was to study the regularities of sulfides extraction from a solution by treating in an electrolyzer with a ferrous anode.

2. Materials and research methods

The experiments were carried out both with model solutions of sulfide effluent, containing from 30 mg/l to 200 mg/l of sodium sulfide, and with the wastewater of the Sterlitamak leather and footwear work. The electrolyzer was a reservoir with iron electrodes.

Power was supplied from a controlled rectifier, which allows varying the current load. The volume of the treated solution in all experiments remained constant and was of 300 ml.

3. Experimental results and discussion

The operation of the electrolysis plant was investigated at electrode current densities of 67, 133, 200 and 267 A/m^2. The dependence of sulfide concentration in the effluent on the time of its treatment in the electrolyzer is shown in Figure 1. It can be seen that when the effluent treatment time in the electrolyzer increases, the concentration of sulfides in it decreases. Moreover, the rate of extraction of sulfides from wastewater in the form of iron sulfide deposits is different at different electrode current densities. An increase in current load on the apparatus from 67 A/m^2 to 133 A/m^2 leads to an increase in the extraction rate. A further increase in the current load leads to a decrease in the rate of extraction of sulfides from the effluent.

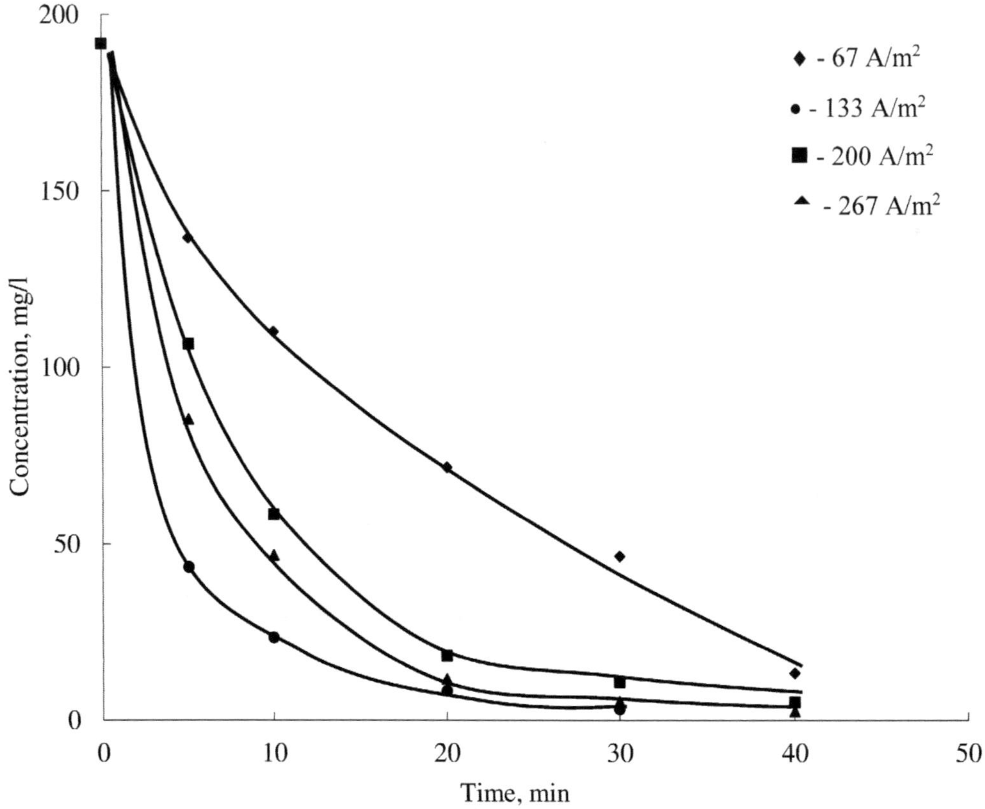

Figure 1. The dependence of the sulfide concentration in an effluent on the processing time

Figure 2 shows the dependence of sulfide concentration in the effluent on the quantity of electricity passed through the solution. It should be noted that the experiment results were divided into two areas. So, to achieve the same concentration of sulfides in the waste water with a current density of 133 A/m^2, is needed less electricity than with other current densities.

The dependence of the extraction ratio of sulfides from wastewater on the electrode current density is shown in Figure 3. The effluent treatment time in the electrolyzer in these experiments was the same and made 40 minutes. It is evident that a change in the electrode current density leads to a change in the extraction ratio of sulfides from the effluent. The maximum value of the extraction ratio of 99.2% is achieved at a current density of 133 A/m^2.

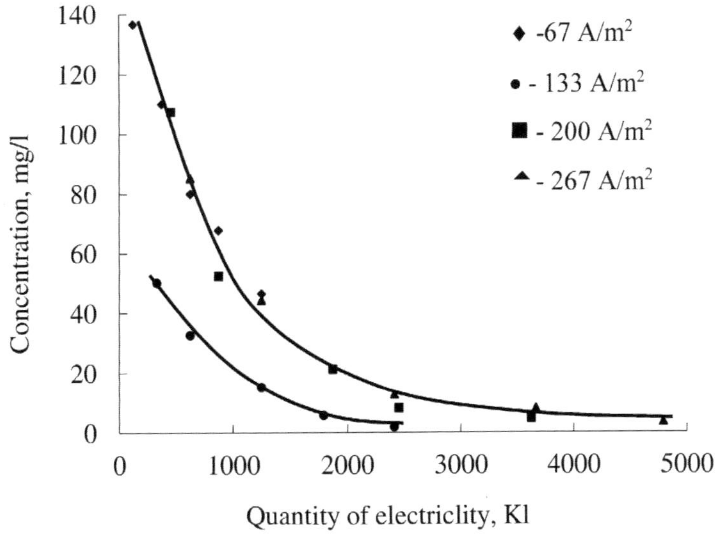

Figure 2. Dependence of sulfide concentration in the effluent on the quantity of electricity

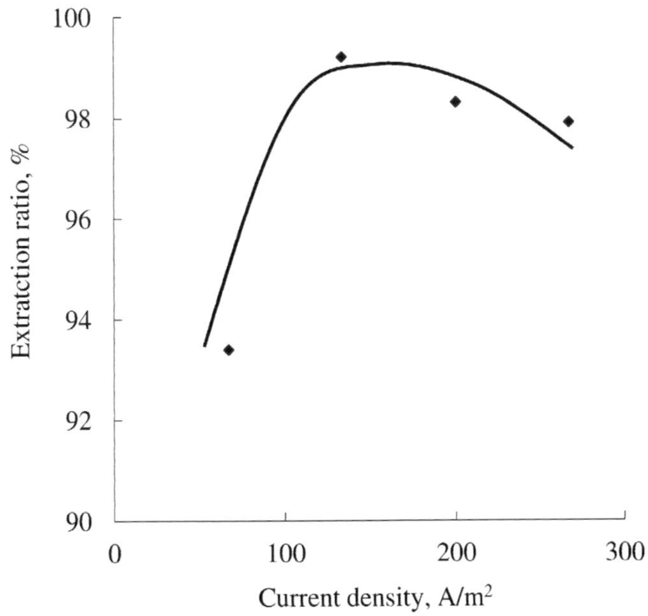

Figure 3. The dependence of the sulfides extraction ratio on the current density

Obviously, such behavior of the sulfide recovery process is associated with the following. The bivalent iron ions entering the solution due to an anodic dissolution react with sulfide ions, forming a deposition of iron sulfide, the main mass of which precipitates on the anode surface. Consequently, two processes will determine the rate of iron sulfide formation: the process of iron anodic dissolution, and the transport of iron ions and sulfide into the zone of iron sulfide formation. The rate of mass transfer of ions in the electrolyte volume, determined by convective currents, is rather high. Therefore, the anodic dissolution of iron and the diffusion of iron ions and sulfide through the layer of iron sulfide deposit covering the anode will be limiting. The course of the curves shown in Figures 1-3 indicates that at a current density of 67 A/m^2, the iron dissolution from the anode will be the limit, and at current densities of 200 A/m^2 and higher, the diffusion of ions through the deposit layer of iron sulfide. This is also indicated by the formation of an iron hydroxide deposition, observed as yellow-brown inclusions in the mass of black iron sulfide deposition. Obviously, at a current density of 133 A/m^2, the rate of iron dissolution from the anode and the rate of diffusion of ions through the layer of iron sulfide deposition will be fairly close.

Figure 4 shows the dependence of the concentration of sulfides in the treated effluent on its concentration in the waste water before treatment. The electrode current density in these experiments was 133 A/m^2, and the treatment time in the electrolyzer makes 40 minutes. It can be seen that this dependence is linear. Moreover, a change in the sulfides concentration in the effluent from 30 to 200 mg/l corresponds to a change in the sulfides concentration in the purified water from 0.08 to 1.97 mg/l.

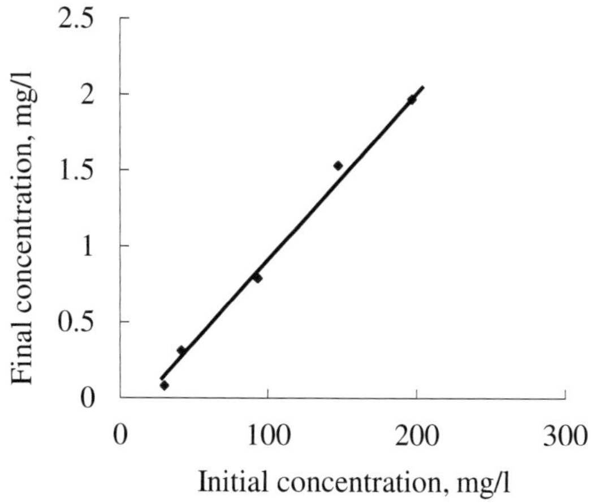

Figure 4. Dependence of sulfide concentration after purification on initial concentration

Since the economic indicators play an important role in the implementation of any process, it was investigated the dependence of the electricity expenditures in the process of extraction of sulfides from wastewater on the technological parameters. The data are shown in the table 1.

Table 1. Dependence of energy consumption at removal of 1 kg of sulfides on current density

Current density, A/m^2	Energy expenditures, kW·h/kg
67	75.8
133	169.2
200	379.2
267	653.3

As it should be expected, when in the current load on the electrolyzer is increased, the power consumption for the sulfide extraction also increases. Thus, with a volume of wastewater containing sulfides at the Sterlitamak leather and footwear works of 45 m^3/day and containing up to 30 mg/l of sulfides, the power consumption during the treatment process in an electrolyzer with an electrode current density of 133 A/m^2 will be 228 kW·h/day.

4. Conclusion

From the results presented in the paper, it follows that in the electrolyzer with a sacrificial iron anode, the sulfides are removed from the solution as an iron hydroxide deposition. The treatment process is determined by the anode current density. The optimum current density makes 133 A/m^2. Obviously, in this case, the process of iron dissolution of and the process of sulfides diffusion through the layer of FeS sediment covering the surface of the anode are in equilibrium. The purification efficiency under these conditions reaches 99.2%.

5. Acknowledgements

This scientific paper was prepared on basis of research work within the government work No. 5.12863.2018 / 8.9 of the Russian Federation.

References

[1] GN 2.1.5.1315-03 2003 *Maximum permissible concentrations of chemicals in water for household, drinking, cultural and community water use* (Moscow: Ministry of health of the Russian Federation)
[2] Kochinashvili M V, Anfinogenova T S, Menshikova N N and Karasev V N 1997 *Way of purification of sulphurous and alkaline sewage of sulfides* RF Patent No. 2078053 (20 January 1997)
[3] Gavrikov M A 1976 *Chemical industry* **2** 29–30
[4] Plechev A V, Safin R R, Voltsev A A, Konoshenko E V and Ismagilov F R 2000 *Ecology and Industry of Russia* [in Russian – Ecologija i promyshlennost Rossii] **6** 18–20
[5] Vitkovskaya R F, Panov V P, Petrov S V, Tereshchenko P Ya and Ukhanova E I 1997 *Way of sewage treatment from sulfides* RF Patent No. 2099292 (12 October 1997)
[6] Laskov Yu M, Fedorovskaya T G and Zhmakov G N 1984 *Wastewater treatment of leather industry enterprises* (Moscow: Legkaya i pischevaya promyshlennost)
[7] Caliari P C, Pacheco M J, Ciriaco L and Lopes A 2019 *Environmental Technology and Innovation* **14** 100345
[8] Ntagia E, Fiset E, Lima L S, Pikaar I and Rabaey K 2019 *Water research* **149** 111–9
[9] Glinka N L 2013 *General chemistry: a textbook for bachelors* (Moscow: Yurayt)
[10] Bykovsky N A, Bykovskaya L N, Shulaev N S, Abramov V F and Ryskulov S G 2001 *Way of sewage treatment from sulfides* RF Patent No. 2165892 (27 April 2001)

IOP Conf. Series: Earth and Environmental Science **350** (2019) 012031 doi:10.1088/1755-1315/350/1/012031

Development of a new deformation forecast method

A A Dordzhiev, G M Skibin , A G Dordzhiev, M M Sangadzhiev and A V Arashaev

Kalmyk State University n.a. B.B. Gorodovikov, 11, Pushkin st., Elista, Russia

E-mail: smm54724@yandex.ru

Abstract. Deformations of foundations of buildings and structures are calculated by the second limit state, i.e. by limiting deformations. These calculations are relevant for specific soils, in particular loess soils. The paper suggests a method for determining the structural strength of the soil in compression tests and presents a method for predicting base foundation deformations when soaking and flooding built-up loess territories using approved equipment and software.

1. Introduction

When calculating uneven deformations of foundations composed of loess with subsidence soils having the structural strength, the following factors should be taken into account:

- for soils with the structural strength, the SP method [1] provides overestimated values of settlement and underestimated settlement unevenness;
- subsidence properties of soils manifest themselves when the required humidity is achieved (initial subsidence (critical) moisture). When predicting uneven deformations, subsidence soil properties should be taken into account at S_r<0.8 [1–6].

2. Materials and methods

Methodology for prediction of foundation deformations:

1. Let us analyze adjacent foundations, under one of which the soil is in its natural state (S_r <0.5), i.e. the soil does not exhibit subsiding properties. The second foundation is subject to 1) $0.5 \leq S_r < 0.8$ and 2) $S_r \geq 0.8$.

Relatively non-uniform deformation is determined in two cases: 1 - local soaking; 2 - local full soaking. When designing new objects, to predict uneven deformations, it is recommended to consider case 1 (local partial soaking), since full water saturation of the whole subsidence is unlikely which is confirmed by the data of field studies [2, 5, 7, 8].

2. Deformations of the foundation with a partially soaked base are defined as the sum of deformations caused by the settlement and subsidence

$$S = S_i + S_{Sl}$$

where S_i is determined by the SP method, the soil does not have P_b (example 1). To determine S_{Sl}, the SP method is used (example 2). The deformation of the completely soaked (water-saturated) foundation is determined according to the standard SP method (Example 3). Let us calculate σ_{zg} and $0.5\sigma_{zg}$;$\sigma_{zi} = \sigma_{zgi-1} + \gamma_{IIi} \cdot h_i$;

where h – layer thickness i; γ – specific weight i слоя; point 0 – earth surface; $\sigma_{zg0} = 0; 0.5 \cdot \sigma_{zg0} = 0$; T.1 – at the level of the foundation base; $\sigma_{zg0} = \gamma_i \cdot h_{\frac{1}{1}} = 29.25$ kPa; $0.5 \cdot \sigma_{zg0} = 14.63$ kPa; T.2 – at the boundary of layers 1 and 2;$\sigma_{zg1} = \sigma_{zg0} + \gamma_1 \cdot h_{\frac{1}{2}} = 29.25 + 19.5 \cdot 0.5 = 39$ kPa; $0.5 \cdot \sigma_{zg1} =$ 19.5 kPa; T.3 – at the boundary of layers 2 and 3; $\sigma_{zg2} = \sigma_{zg1} + \gamma_2 \cdot h_2 = 39 + 19.5 \cdot 1.5 = 68.3$ kPa; $0.5 \cdot \sigma_{zg2} = 34.1$ kPa; T.4 – at the boundary of layers 3 and 4;$\sigma_{zg3} = \sigma_{zg2} + \gamma_3 \cdot h_3 = 68.3 + 18.2 \cdot 2 = 104.7$ kPa; $0.5 \cdot \sigma_{zg2} = 52.4$ kPa; T.5 – at the boundary of layers 5 and 6; $\sigma_{zg4} = \sigma_{zg3} + \gamma_4 \cdot h_4 = 104.7 + 17.7 \cdot 2 = 10.1$ kPa; $0.5 \cdot \sigma_{zg2} = 70.1$ kPa ;T.6 – at the boundary of layers 6 and 7; $\sigma_{zg5} = \sigma_{zg4} + \gamma_5 \cdot h_5 = 140.1 + 17.4 \cdot 2 = 174.9$ kPa; $0.5 \cdot \sigma_{zg2} = 87.5$ kPa; T.7 – at the boundary of layers 7 and 8;$\sigma_{zg6} = \sigma_{zg5} + \gamma_6 \cdot h_6 = 174.9 + 17.2 \cdot 2 = 209.3$ kPa; $0.5 \cdot \sigma_{zg2} = 104.7$ kPa

1. Based on the obtained values of ordinates on a geological section, we constructed the plot of natural pressure σ_{zgi} and auxiliary plot $0.5\sigma_{zgi}$.
2. We determined the vertical pressure on the base along the foundation base: $\sigma_{zp} = p = 150 \kappa Pa$, where p – average pressure under the foundation base.
3. We divided the soil thickness under the base into elementary sublayers with a thickness Δ_i=(0.2-0.4)b_f, where b_f – base width. Let us take Δ_i=0.2·2=0.4 m.
4. Let us determine vertical normal stresses σ_{zp} by depth z_i from the foundation base: $\sigma_{zpi} = \alpha_i \cdot \sigma_{zp}$, where α – stress dispersion coefficient for the corresponding soil layer depending on the shape of the base and the ratio $\zeta=2_{zi}/b_f$ и $\eta=l_f/b_f$, where z_i – depth of the i-th elementary layer from the base; $z_i=\Sigma\Delta_i$ is determined by Table 4.5$\xi = 2 \cdot z_i/2 = z_i$; $\eta = 2/2 = 1$

Table 1. Calculation of probable settlement

No	Name of the soil and its condition	Layer thickness h_i, m	Δ_i, m	z_i, m	ζ_i, m	α_i, m	σ_{zpi}, kPa	σ_{zpi}^{cp}, kPa	E_{zpi}, kPa
GEE-1	Refractory loam	0.4	0.0	0.0	0.0	1.0	150	147	5000
			0.4	0.4	0.4	0.96	144	132	
		1.6	0.4	0.8	0.8	0.8	120	105.5	13500
			0.4	1.2	1.2	0.606	90.9	79.2	
			0.4	1.6	1.6	0.449	67.4	58.9	
			0.4	2.0	2.0	0.336	50.4	44.5	
		4.0	0.4	2.4	2.4	0.257	38.6	34.4	26000
			0.4	2.8	2.8	0.201	30.2	27.1	
			0.4	3.2	3.2	0.16	24	21.9	
			0.4	3.6	3.6	0.131	19.7	18	
			0.4	4.0	4.0	0.108	16.2	15	
			0.4	4.4	4.4	0.091	13.7	12.7	
			0.4	4.8	4.8	0.077	11.6	10.9	
			0.4	5.2	5.2	0.067	10.1	9.7	
			0.4	5.6	5.6	0.058	8.7	8.2	
			0.4	6.0	6.0	0.051	7.7		

5. According to the data obtained, we constructed a plot of the additional vertical stress σ_{zp} from the base.
6. Let us determine the height of the compressed base thickness H_c, whose lower boundary BC is $z=H_c$, where σ_{zp}=0.5· σ_{zg}.
7. Let us determine the value of total settlement by formula

$S = \beta \sum_{i-1}^{n} \frac{(\sigma_{zpi}-\sigma_{zji})\cdot h_i}{E}$, where β – dimensionless coefficient β=0.8; σ_{zpi} – average value of vertical normal stress from the base in the i-th layer; Δ_i – thickness of the i-th layer; E – modulus of deformation of the i-th layer; n – number of layers.

8. Let us calculate settlements in a tabular form (Table 1)

$$S_1 = 0.8/5000 \cdot (147 \cdot 0{,}4) = 0.94cm$$

$$S_2 = 0{,}8/13500 \cdot 0.4(132 + 105.5 + 79.2 + 58.9) = 0.89cm$$

$$S_3 = 0.8/26000 \cdot 0.4(44.5 + 34.4 + 27.1 + 21.9 + 18 + 15 + 12 + 10.9 + 9.7 + 8.2) = 0.24cm$$

$$S_t = S_1 + S_2 + S_3 = 0.94 + 0.89 + 0.24 = 2.07cm$$

9. Let us compare settlement value S with the value of ultimate deformations of the base Su, depending on the structural system of the building according to [6, 8].

S=2.07 cm < S_u=12 cm, the condition is met.

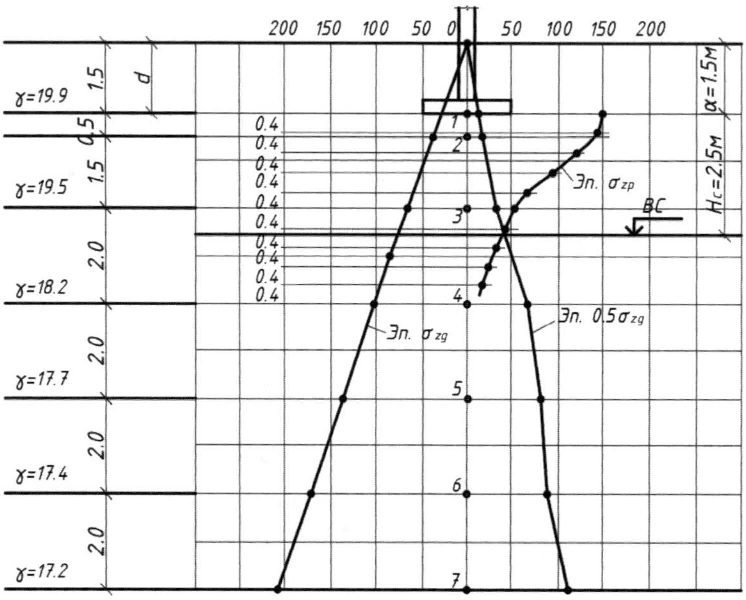

Figure 1. Calculation of settlement of the foundation: BC – lower bound of compressible stratum; d – foundation depth; H_c – compressible thickness; эп. σ_{zg} and $0.5\sigma_{zg}$ – main and auxiliary diagrams of vertical stresses of the own weight of the soil; σ_{zp} – vertical stress plot from the bottom of the foundation

Example 2. It is required to calculate deformations of the base with dimensions 2.0x2.0 m, depth d = 1.5 m with an average pressure along the sole pII = 150 kPa. The admissible settlement for this building is Su = 12 cm. The construction site is composed of loess loams 8.0 m in thickness underlain by non-subsiding soils. The main characteristics of soils are given in Table 1. The specific gravity of the soils is adopted for the soaked saturated water.

3. Solution

For a given base size, we determined the distribution of vertical stresses σ_{zgi} and σ_{zpi} over the depth of their base. The calculation results are summarized in Table 3. At the same time $\sigma_{zg0} = \gamma \cdot d_f = 19.9 \cdot 1.5 = 29.85$kPa.

The thickness of the elementary soil layer is assigned from the condition $h_i < 0.4 \cdot b_f = 0.4 \cdot 2 = 0.8m$, h_i=0.5 m.

Using Table 2, we determined the relative subsidence of the soil in the middle of each layer with an average actual stress $\sigma_{zi.m}$. Results are presented in Table 2. From the comparison of stress values $\sigma_{zi.m}$

at the base with the value of initial subsidence P_{sl}, we established that total stresses over the entire thickness are greater than pressure P_{sl}.

Let us determine the amount of base subsidence by SP formula CR using the data from Table 1 and 3.

$$S_{Sl} = \sum \varepsilon_{Sl} \cdot h_i \cdot K_{Sl.i} = 0{,}5 + 1{,}5(p - p_{Sli})/p_0 = 0.5 + 1.5(150 - 125)/100 = 0.88$$

Values of coefficient $K_{Sl.i}$ Are Calculated by SP formula for each layer. Results are presented in Table 3. The value of soil subsidence is

$$S_{Sl} = 50 \cdot 0.88(0{,}068 + 0.052 + 0.047 + 0.033 + 0.028 + 0.024 + 0.021 + 0.024 + +0.015 + 0.014 + 0.017) = 50 \cdot 0.88 \cdot 0.343 = 15.11cm$$

Let us calculate the settlement value for the foundation by the method of layer-by-layer summation according to CR 22.13330.2011

Table 2. Soil characteristics for the construction site

№	Soil	Depth h, m	Ground weight γ, kN /m³	Share of dry soil γ_d, kN /m³	ε_{Sl} at P, kPa				P_{Sl}, kPa	E, kPa
					50	100	200	300		
1	Loam	1	19.9	15.5	0.02	0.046	0.09	0.117	105	4
		2	19.9	15.8	0.02	0.046	0.09	0.117	105	
2	Loam	3	19.5	15.6	0.011	0.021	0.036	0.056	125	5
		4	19.5	15.9	0.011	0.021	0.036	0.056	125	7
3	Loam	5	18.2	15.2	0.007	0.014	0.024	0.043	130	12
		6	18.2	15.2	0.007	0.014	0.024	0.043	130	
4	Loam	7	17.7	15.0	0.007	0.014	0.024	0.043	135	15
		8	17.7	15.2	0.007	0.014	0.024	0.043	135	
5	Loam	9	17.4	15.1	0.004	0.009	0.017	0.032	140	22
		10	17.4	15.5	0.004	0.009	0.017	0.032	140	
6	Loam	11	17.2	15.8	0.003	0.008	0.009	0.009	150	30
		12	17.2	15.9	0.003	0.008	0.009	0.009	150	

Table 3. Results of calculation of stress values for the base depth

No	z, m	σ_{zg}, kPa	α	$\sigma_{zp} = \alpha \cdot p_{II}$, kPa	$\sigma_{zi} = \sigma_{zg} + \sigma_{zp}$, kPa	$\sigma_{zi.m}$, kPa	$\varepsilon_{Sl.i}$	$K_{Sl.i}$
0	0	29.9	1.0	150	179.1	-	-	-
1	0.5	39.8	0.907	136.1	175.9	177.9	0.068	0.88
2	1.0	48.8	0.703	105.5	154.3	165.1	0.052	0.88
3	1.5	58.5	0.488	73.2	131.7	143.0	0.047	0.88
4	2.0	63.7	0.366	54.9	118.6	125.2	0.033	0.88
5	2.5	72.8	0.243	36.5	109.3	114.0	0.028	0.88
6	3.0	79.7	0.180	27.0	106.7	108.0	0.024	0.88
7	3.5	88.5	0.138	20.7	109.2	108.0	0.021	0.88
8	4.0	95.7	0.108	16.2	111.9	110.6	0.024	0.88
9	4.5	104.4	0.087	13.1	117.5	114.7	0.015	0.88
10	5.0	111.8	0.071	10.7	122.5	120.0	0.014	0.88
11	5.5	120.4	0.060	9.0	129.4	126.0	0.017	0.88

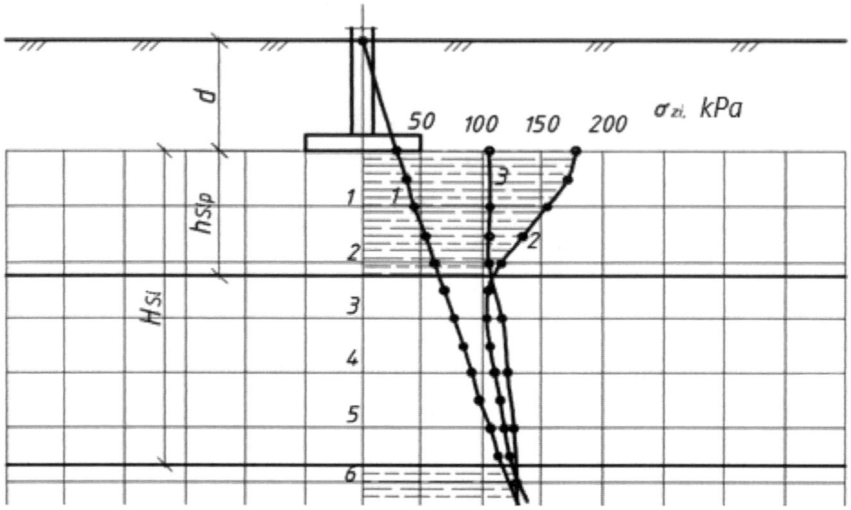

Figure 2. Schemes for calculation of the basement subsidence: 1 – vertical load from the own weight σ_{zg}; 2 – total vertical stress from the external load and own weight $\sigma_z = \sigma_{zp} + \sigma_{zg}$; 3 – change with a depth of initial drawdown pressure P_{Sl}

The total amount of settlement and complete subsidence is $S + S_{Sl} = 2.07 + 15.11 = 17.18cm$, that is more than the permissible value equal to S_u = 12 cm. Therefore, to ensure operational suitability of the building, it is necessary to eliminate subsidence properties of soils and reduce the amount of settlement.

Example 3. Calculation of the fully soaked (water-saturated) base.

As a result of analysis of the calculations, we have:

at S_r=0.31: S=0.66 cm, H_c=5.1m;

at S_{rSl}=0.6: S=1.66 cm, H_c=4.7m;

at S_{rwat}>0.8: S=4.16 cm, H_c=4.7m.

When calculating settlement, the compressible strata is divided into three layers according to the degree of humidity. As a result of the calculation, the settlement turned out to be less than at S_r> 0.8, and the depth of the compressible thickness did not change.

S=3.84 cm, H_c=4.7 m

Calculation of the settlement at S_r=0.31 (d_{fn}=1.5m))

Calculation results:

Settlement – 0.66 cm. The depth of the compressible stratum is 5.10 cm

Calculation of settlement at S_r=0.60 (d_{fn}=1.5 m)

Calculation results:

The settlement value is 1.66 cm. The depth of the compressible stratum is 4.70 cm

Calculation of settlement without P_b at d_{fn}=1.5 m

Calculation results:

The settlement value is 3.84 cm. The depth of the compressible stratum is 4.70 cm

For the first foundation, we calculated the settlement using the formula by V.P. Dyba and O.N. Osipova [9] for soils with the structural strength. The settlement value was determined by formula

$$S = \beta \sum_{i=1}^{n} \frac{(\sigma_{zpi} - P_{stri}) \cdot h_i}{E_i}$$

Example 4. The administrative and housing building in Klykov street, Elista.

The settlement value was calculated in Settlement 1.1 developed in the South SRPU (NPI) by G.M. Skibin, S.I. Evtushenko, and E.Yu. Anischenko.

Calculation of the settlement value ignoring and taking into account the structural strength of the soil:

Calculation of the settlement value without P_b at d_{fn}=1.5m

Calculation results:

The settlement value – 1.43 cm. The depth of the compressible stratum is 5.10 cm

Calculation of the settlement value accounting for P_b at d_{fn}=1.5 m

Calculation results:

The settlement value – 0.89 cm. The depth of the compressible stratum is 4.3 cm

Calculation of the settlement value without P_b at d_{fn}=1.5 m

Calculation results:

The settlement value – 1.07 cm. The depth of the compressible stratum is 4.2 cm

Calculation of the settlement value accounting for P_b at d_{fn}=1.5 m

Calculation results:

The settlement value – 0.69 cm. The depth of the compressible stratum is 3.9 cm

When taking into account R_b, the sediment value decreases 1.6 times, and the depth of the compressible depth - 1.2 times.

4. For the second foundation, we determined R and F when the moisture regime of the base changes. The forecast of the design resistance of the soil and the bearing capacity of the foundations

$$K_{\varphi 1} = \frac{\varphi_{wet}}{\varphi_{lowmoist}};\ K_{\varphi 2} = \frac{\varphi\ watsat}{\varphi_{loimoist}};\ K_{c1} = \frac{c\ wet}{c_{lowmoist}};\ K_{c2} = \frac{c\ watsat}{c_{lowmoist}}$$

Table 4. The values of the strength parameters of loam c, φ and reduction factors K_φ and K_c when changing the moisture regime of the soil base

Indicators	Loam			Reduction factors c and φ	
	Low moisture (S_r<0.5)	Wet (0.5≤ S_r<0.8)	Water saturated (0.5≤ S_r<1.0)	$K_{1\varphi}$ / K_{1c}	$K_{2\varphi}$ / K_{2c}
φ, degree	26.4	22.7	17.9	0.86	0.68
c, kPa	42	28	27	0.67	0.64

The results show that after soaking and water saturation, specific cohesion c decreases 1.5 and 1.6 times, and the angle of internal friction decreases 1.2 and 1.5 times, respectively.

For quantitative assessment of the calculated soil resistance of the base in low-moisture, wet and water-saturated states, let us consider examples of calculation.

Example 5. The foundation has a size l = b = 2 m and a depth of d_l = 1.5 m; the base is composed of low-moisture loess loam. The specific weight of soils under and above the base in the low-moisture state are 17. 5, and the wet and water-saturated state are equal to $\gamma_p = \gamma_p{}'$ =19. 4 кN/m^3. The building has no basement, therefore, $d_в$= 0. We determined that γ_{c1} = 1.1, γ_{c2} = 1, K = 1.1, K_z = 1.

1. Let us determine the design resistance of the soil in the low-moisture state. As a result of experiments, we determined the angle of internal friction φ = 26.4 °, and specific adhesion c = 42 kPa, Mg=0.85, Mg=4.44, Mc = 7,01. Then we calculated the soil resistance [1].

$$R_1 = \frac{1.1 \cdot 1}{1.1}(0.85 \cdot 1 \cdot 2 \cdot 17.5 + 4.44 \cdot 1.5 \cdot 17.5 + 7.01 \cdot 42) = 440.72кPa$$

2. When determining the design resistance for the soaked base, we conducted experiments with soil samples in the wet state and determined the angle of internal friction φ = 22.7 ° and specific adhesion at c = 28 kPa, dimensionless coefficients Mγ = 0.65, Mg = 3. 60, MS = 6.16. Then we calculated the soil resistance

$$R_2 = \frac{1.1 \cdot 1}{1.1}(0.65 \cdot 1 \cdot 2 \cdot 19.4 + 3.60 \cdot 1.5 \cdot 19.4 + 6.16 \cdot 28) = 302.46кPa$$

3. Resistance in the conditions of water saturation of the soil is calculated after determining strength parameters of the soil M_γ=0.42, M_g=2.71, M_c=5.27.

The resistance is calculated by formula:

$$R_3 = \frac{1.1 \cdot 1}{1}(0.42 \cdot 1 \cdot 2 \cdot 19.4 + 2.71 \cdot 1.5 \cdot 19.4 + 5.27 \cdot 27) = 237.45 \text{ kPa}$$

The results show that the soil resistance decreases 1.46 and 1.86 times, respectively, compared to the resistance of slightly wet soil. Thus, resistance values and reduction coefficients of resistance for base loams in low-moisture, wet and water-saturated states were determined experimentally (Table 5).

Table 5. Values of the soil resistance (R_1, R_2, R_3) and reduction coefficients of the resistance (K_1, K_2)

Soil	Soil condition			Coefficients	
	slightly wet (S_r<0.5) R_1, kPa	wet (0.5≤S_r<0.8) R_2, kPa	water saturated (0.8≤S_r<1.0) R_3, kPa	K_{1R}	K_{2R}
Loam	440.7	302.5	237.4	0.69	0.54

$$K_{1R} = \frac{R_{wet}}{R_{lowmoist}};\ K_{2R} = \frac{R_{watsat}}{R_{lowmoist}},$$

where $R_{lowmoist}$, R_{wet}, R_{watsat} – calculated resistance values for the soil in low wet, wet and water-saturated states. The basis for the carrying capacity is calculated taking into account the following conditions:

$$F \leq \gamma_c \cdot F_u/\gamma_n,$$

where F – base load; F_u – strength of base resistance; γ – coefficient of working conditions taken for silty-clay soils in a stabilized state; γ_c=0.9, for silty clay soils in unstabilized condition γ_c=0.85; γ_n – reliability coefficient equal to 1.2; 1.15 and 1.10 for buildings and facilities of the first, second and third levels of responsibility.

The vertical component of forces of ultimate resistance Nu of the base composed of non-rocky soils in a stabilized state can be determined by formula (16) of CR 22.13330.2011.

$$N_u = b^1 l^1 (N_\gamma \xi_\gamma b^1 \gamma_1 + N_g \xi_g d \gamma_1{}^1 + N_c \xi_c C_1)$$

Designations are the same as in formula (16) CR 22.13330.2011.

Example 6. Calculation parameters from example 5.

Table 6. Values of the bearing capacity of base soil (F_1, F_2 and F_3 in кN) coefficients of reducing the bearing capacity of the soil (K_{1F}, K_{2F})

Soil	Soil condition			Coefficients	
	slightly wet (S_r<0.5) F_1, kN	wet (0.5≤S_r<0.8) F_2, kN	water saturated (0.8≤S_r<1.0) F_3, kN	K_1	K_2
Loam	6585	3659	2956	0.56	0.45

$$K_{1F} = \frac{F_{wet}}{Flowmoist};\ K_{2F} = \frac{Fwatsat}{F_{lowmoist}},$$

where $F_{lowmoist}$, F_{wet}, F_{watsat} – the bearing capacity of the soil in slightly wet, wet and water-saturated states. The obtained results show that the bearing capacity of base soils composed of loess subsidence loams in wet and water-saturated states decreases 1.36 and 1.9 times compared to the bearing capacity of the slightly wet soil of natural density-humidity.

Example 7. Determination of the size of the bottom of the columnar foundation taking into account ultimate stress states. The foundations are located on loams with the following design characteristics: IL = 0.3; γ1 = 17.5 kN / m3; φ = 26 °; c = 0.042 MPa; E = 14.8 MPa. The foundation depth is d1 = 1.5 m. The foundation transmits a load of N = 1.5 MN to the base. Maximum allowable sediment is s_u=12 cm.

According to the CR method, γc_1=1.1; γc_2=1,0; κ=1.1.

$$R = \frac{1.1 \cdot 1.0}{1}[0.85 \cdot 1 \cdot 2 \cdot 17.5 + 4.44 \cdot 1.5 \cdot 17.5 + 7.01 \cdot 42] = 440.72\text{KPa}$$

Determining the sizes of the base from $P \leq R$, we have $b=l$=2x2 m, where $P=R$=441 kPa.

The calculation of the foundation settlement by the method of layer-by-layer summation (Settlement software) gives the depth of the compressible thickness H_c=5.1 m and settlement S=1.43 cm. Since the calculated settlement is still far from the maximum allowable one, we will consider the foundation of 1.5 x 1.5 m with an average pressure in the base P_{av} = 0.382 MPa.

For the foundation of 1.5x1.5 m in size, the resistance is

$$R = \frac{1.1 \cdot 1.0}{1}[0.85 \cdot 1 \cdot 1.5 \cdot 17.5 + 4.44 \cdot 1.5 \cdot 17.5 + 7.01 \cdot 42] = 433.3\text{KPa}$$

$$R=0.433 \text{ MPa and } R_0=0.200 \text{ MPa}$$

By the method of layer-by-layer summation (Settlement)

$$H_c=4.2 \text{ m}, S=S_y=1.17 \text{ cm}$$

The pressure limit is

$$p_{\lim} = N_u/b^2 = 1.5 \cdot 1.5(6.19 \cdot 0.75 \cdot 1.5 \cdot 1.75 + 11.26 \cdot 2.5 \cdot 15 \cdot 1.5 + 21.88 \cdot 1.3 \cdot 43)/2.25 = 1.98\text{MPa}$$

Let us determine the following values

$$K^p = \frac{0.382}{0.200} = 1.91; K^p = \frac{1.98}{0.200} = 9.9; x = \frac{1.91-1}{9.9-1} = 0.10$$

By x=0.10 and φ=26° using the CR method, we determine $K_{pl}^S = 1.08$

AT p=0.382<R=0.433, the base settlement is

$$S_c = S_y \cdot K_{pl}^S = 0.82 \cdot 1.17 = 1.26cm$$

With the values of the calculated load N=1.5 MH and the bearing capacity Nu=1.98·2=3.96 MH.

The reliability coefficient is

$$\gamma_n = \gamma_c \frac{N_u}{N} = \frac{3.96 \cdot 0.9}{1.5} = 2.64 > \gamma_n = 1.2 \text{ (by the CR method)}$$

The foundation with dimensions of 1.5x1.5 m can be used instead of the foundation with dimensions of 2.0x2.0 m. The example confirms that transition to the design of foundations for deformations of the foundations makes it possible to reduce the cost of foundations while maintaining the required bearing capacity.

4. For most buildings, uneven deformations (cracks) are dangerous, since all structures located on subsiding soils subside. The calculation must be carried out according to the second group of ultimate states, by deformations.

Example 8. Determination of uneven deformations

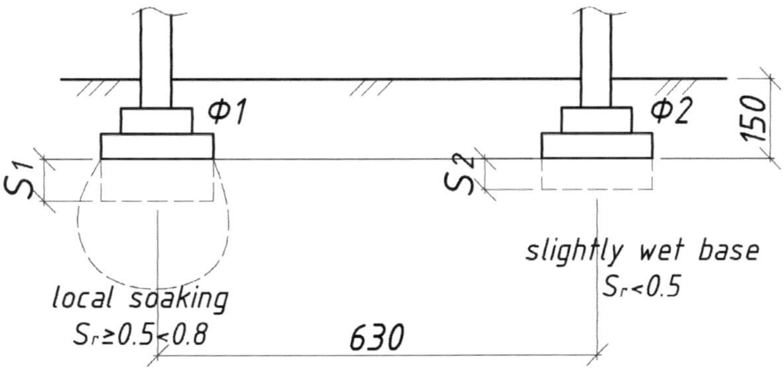

Figure 3. Determination of non-uniform deformations with local soaking

1. CR calculation:

$$\Delta S/L = (S_1 - S_2)/L = (15.11 - 2.07)/630 = 0.021$$

2. Caclualtion by the Dyba and Osipova's formula [9]:

$$\Delta S/L = (S_1 - S_{2str})/L = (15.11 - 0.89)/630 = 0.023$$

Therefore, $S_{2CR} > S_{2str} \Rightarrow \Delta S_{str}/L > \Delta S_{CR}/L$,

2.07>0.89⇒0.023>0.021

where S_2 – foundation settlement f2 at the slighltly wet base; S_1 – foundation settlement f1 at local soaking; S_{str} – settlement taking into account the structural strength of soils; S_{CR} – settlement calculated by the CR method.

4. Conclusion

1) Depending on the soil moisture, the settlement of the foundation increases from 0.66 cm at S_r=0.3 to 1.66 cm at S_{rSl}=0.6 and 3.84 cm at S_r>0.8; depth of the compressible thickness decreases from 5.10 m to 4.70 m;

2) Full water saturation of the whole subsiding strata does not occur.

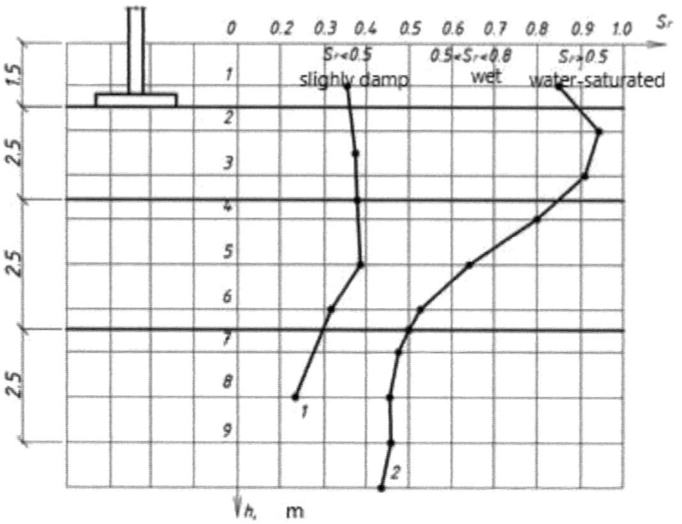

Figure 4. The degree of soil moisture base during: 1) slightly damp condition; 2) water-saturated state

3) According to the Dyba and Osipova's formula, uneven deformation of the foundations is more than acccording to the CR formula as the former takes into account the structural strength of soils.

4) The method can be used to predict uneven deformations of regional loess soils.

References

[1] *The code of rules for design and construction. Foundations of buildings and structures. Updated version of SNiP* 2.02.01-83 SP 22.13330.2011 2011 (Moscow: Ministry of Regional Development of Russia)

[2] Borlikov G M and Dordzhiev A G 1989 *Humid regime of foundation soils as a result of prolonged soaking* vol 3 (Moscow: VNIIS)

[3] Dyba V P and Shiryaeva M P 2006 Simulation of the subsidence of foundations under strip foundations *Izvestiya of universities. North Caucasian region. Techical sci.* **12** 18–25

[4] Dordzhiev A G 2006 *Engineering-geological properties of carbonate-containing subsiding soils of the North-Western Caspian Sea* (Elista: Dzhangar Scientific and Production Enterprise)

[5] Ilyichev V A and Mangushev P A 2014 *Reference geotechnics. Foundations, foundations and underground structures* (Moscow: DIA)

[6] Drucker D C, Prager W and Greenberg H J 1951 Extended limit design theorems for continuous media *Quart. Appl. Math.* **9** 381

[7] Dordzhiev A A and Dordzhiev A G 2014 Deformation characteristics of loess soils *Scientific and technical bulletin of the Volga region* **2** 130–2

[8] Terzagi K 1947 *Theoretical Soil Mechanics* (New York: Wiley)

[9] Osipova O N and Dyba V P 2009 Ground settlement taking into account the structural strength of soils. Low-rise construction in the framework of the national project Affordable and Comfortable Housing for Russian Citizens *Technologies and materials, problems and development prospects in Volgograd Region. Materials of the Int. scientific-practical conf.* (Volgograd) pp 216–9

IOP Conf. Series: Earth and Environmental Science **350** (2019) 012032 doi:10.1088/1755-1315/350/1/012032

Green building as a tool of energy saving

A M Gladkih, V Yu Konyuhov, I I Galyautdinov and E I Shchadova

Irkutsk National Research Technical University, 83 Lermontov Street, Irkutsk, 664074, Russia

E-mail: konyuhovvy@ex.istu.edu

Abstract. The article considers main aspects of green building, energy saving criteria, international certification of LEED and BREEAM, and the legislation of the Russian Federation in this area. The analytics of enterprises and office buildings of the Russian Federation with international certificates was carried out. The reasons for the weak development of green building in Russia, especially in Siberia, have been assessed. The criteria necessary for the development of this direction are revealed.

1. Introduction

Green building is a type of construction and operation of buildings whose impact on the environment is minimal. It aims to reduce the level of consumption of energy and material resources throughout the entire life cycle of a building: from site selection to design, construction, operation, repair and demolition [1-11].

Along with high-tech solutions [13-17], the relevance of environmental projects [18-22] is indisputable.

Since Russia is replete with the mineral raw material and hydrocarbons, for a long time there has been no need for the energy-effective building. Due to this, green building in Russia is only at the stage of formation.

Since 2009, the federal law No. 261 “On Energy Saving and Improving Energy Efficiency” has been in effect, providing for a number of steps to improve the energy efficiency of buildings and structures. The energy consumption of the country's GDP by 2020 should be reduced by 40 %. Meanwhile, construction facilities use 40 % of all primary energy consumed and 67% of electricity. Studies show that over 25 % of the potential savings accounted for the energy consumption of residential and office buildings [10, 21-24].

“The fundamentals of the state policy in the field of environmental development of the Russian Federation for the period until 2030” (approved in April 2012) provide for an increase in the volume of construction of buildings and structures certified in the system of voluntary environmental certification of real estate [2].

2. Methods and materials

The following international certification systems are used in Russia: LEED (Leadership in Energy and Environmental Design, USA) and BREEAM (Building Research Establishment Environmental Assessment Method from the Construction Research Institute, UK) [3].

Some things are similar in both standards. Both of them assess the estate property from the point of view of environmental safety, energy efficiency and the introduction of modern "green" technologies that provide a comfortable and conveniently controlled internal environment. The advantages of certificates are also similar. They are an image, a competitive advantage, raising awareness of the object in the professional community, attracting investment, energy efficiency and savings in operation, safety, quality and reliability, as well as the possibility of increasing the rental rate and improving conditions for employees or residents of the estate property.

The US Green Building Council USGBC developed the LEED standard (Leadership in Energy and Environmental Design).

BREEAM (BREEnvironmental Assessment Method) is a standard, or method, of assessing the efficiency and environmental performance of buildings, developed by the British company BREGlobal.

3. Results and discussion

Standards exist only in English: 817 pages of LEEDv4 (BD + C) and 454 pages of BREEAM International New Construction 2016. Both LEED and BREEAM regulate the creation of a quality environment, energy and water conservation, the use of regional materials, recycled materials, safe finishing materials, separate collection of waste, management of environmental and indoor parameters and other parameters, but prescribe requirements in different ways.

Both systems are intended for certification of estate properties, among which there may be a designed, reconstructed and operated building for any purpose, as well as part of a building – for example, an office or a store in which only finishing works are performed.

Both standards allow working with different functional buildings: residential, office, retail, warehouse, industrial, as well as with schools, hotels, exhibition halls and other types of estate properties [22-24].

Both LEED and BREEAM are guided by various international norms and rules: ASHRAE, ISO, EN and others.

For the certification on any of the systems it is required to pass three stages and the pay contributions for them in the currency:

Registration of a project as an aspirant to the certification.

Passing of an intermediate stage (examination of the "Project" stage, Design Review / Interim Certification).

Progress of the final examination of the constructed estate property (Construction Review/Final Certification).

The second and third steps can be unified.

Let us consider the differences between these standards.

First, by the fact that BREEAM gives the possibility to obtain intermediate certificate immediately after the design documentation is assessed, and in LEED the certificate is given only to the already constructed estate property.

In the second place, they differ by the sections of requirements. The requirements themselves are quite similar, and evaluate equivalent aspects, however, it is impossible to call them identical.

Thirdly, mandatory requirements are vital difference. There are several of them in LEED, and they are united for all levels from Certified to Platinum. Mandatory requirements are more complex in LEED than in BREEAM (for example, energy simulation) and, due to this fact, the cost of certification grows.

In BREEAM the mandatory requirements differ for the different levels of certificate – there are only several base requirements provided for Pass, and for Very Good or Excellent there are additional ones. The higher the level, the greater the requirements.

BREEAM allows you to more flexibly and freely demonstrate the requirements, it has more opportunities for a creative approach in a good sense. BREEAM-certified buildings can differ significantly. LEED is more strict and formal, more "engineering"-like.

In LEED, documentation is provided on the USGBC / GBCI online platform by filling out special forms, uploading drawings, calculations, specifications. In BREEAM, the appraiser checks all documents: he prepares an official report and sends it through his personal account to the BRE for verification with the entire documentation package.

LEED uses a transparent rating system in which points are awarded for each criterion. BREEAM is a bit more complicated: the scores obtained by category are multiplied by weightings, reflecting the significance of each of the requirements and the relevance of design solutions for the region where the estate property is located. The final rating in BREEAM is expressed as a percentage.

Texts of BREEAM standards are available on the net for free, LEED standards require paying a fee.

Table 1. Sections of LEED and BREEAM

LEED sections	BREEAM sections
• Integration process (design complexity)	• Process control
• Location and transport	• Health and comfort
• Construction site (building site)	• Energy
• Consumptive water use efficiency	• Transport
• Energy consumption and atmospheric parameters	• Water
	• Materials
• Consumption of materials and resources	• Waste disposal
• Environmental quality indoors	• Use of the land plot
• Innovations in the design	• Pollution

4. Important difference between standards

An important difference between standards is who can certify an estate property. In LEED, any company can register on the site, add its estate property and be certified, despite the fact that the probability of a successful outcome without experience is very small. The presence of an accredited LEED professional in the project team is an advantage and an additional point. In BREEAM, there is no choice like that, the presence of an accredited specialist - BREEAMAssessor - is mandatory. Without this specialist, the registration and certification of the project will fail [23].

Several Russian systems have been developed, including GOST R 54964–2012 "Conformity assessment. Ecological requirements for estate properties", recognized by the state as the national standard for green building. The national standard STO NOSTROI 2.35.4–2011 "Green building" is also of great importance for the market. [4, 5].

Let us consider in more detail the criteria for energy efficiency in green building. According to the standard STO NOSTROI 2.35.4–2011 "Green building", the category of assessment of environmental sustainability "Energy saving and energy efficiency" includes the following criteria:

- heat consumption for heating and ventilation systems;
- heat consumption for hot water system;
- electric energy consumption (for lighting, engineering systems, building air conditioning);
- electric energy consumption of energy consuming equipment and electrical products.

Special attention is paid to the category "Application of alternative and renewable energy", where the share of secondary and renewable energy in the annual energy balance of the facility is estimated [6].

Let us consider the solution of energy saving problems by the example of construction objects certified according to LEED or BREEAM.

Freedom passive straw house, built in the Istrinsky District of the Moscow Region, requires approximately 40 kWh/m^2 per year for heating. This 8-10 times less than for the buildings built of brick. Energy savings are achieved here through effective thermal insulation of exterior walls, floor coverings and the roof, made of blocks of timber, filled with straw. Outside, the blocks are covered with heat-resistant wood panels, and inside - with a clay plaster on a fiber flooring. Heating is done using a wood pellet stove. In the summer, a solar collector is used to heat the water; the house is equipped with photovoltaic panels.

The SKF plant in Tver (Figure 1) is the first industrial facility in Russia to be assigned the LEED standard certification. It uses energy-efficient coolers, ventilation system with heat recovery, fans with adjustable speed, and automated control system. 75 % of the building area is illuminated naturally. Due to these measures, energy savings amount up to 40 % and the fan efficiency increases by 80 % compared with traditional equipment.

Figure 1. The SKF plant in Tver.

Water consumption is reduced through the use of a closed-loop system and sanitary equipment with low water flow. In addition, rainwater is collected and used.

In addition, the plant sorts and recycles waste. Construction waste was used to build roads in the industrial zone, which saved money that would otherwise be used to purchase materials and transport waste to special landfills [7].

If we consider an office building, then an example is the Jones Lang LaSalle office in Moscow, which received LEED and BREEAM certificates. Energy saving is carried out here through the use of metering devices that allow one to control the energy consumption for lighting (light and motion sensors), the operation of fan coils, the heating system, which takes into account the perception of all employees.

Water is saved through the use of plumbing appliances and other devices that reduce water consumption by up to 50 % compared to a regular office. Waste is collected separately and materials are reused.

CO_2 remission is reduced due to bicycle parkings. In addition, due to this, the employees keep a healthy and active lifestyle.

According to the estimates of the savings, it turned out that the payment for electricity decreased on average by 43 %, while the additional costs of "greening" the office amounted to 4.8 % of the project's construction budget, and the payback period for these investments was about 6 years [8].

5. Conclusion

In conclusion, it is worth noting the factors constraining the development of green building in our country. The principles of energy-saving construction should be based on foreign experience, but always taking into account Russian specifics, especially when it comes to the implementation of projects in Siberia with its climatic features.

The current lack of awareness of broad audience of the potential of green building is also a factor hindering the development of the industry. Moreover, the main reason is, above all, Russia's being replete with cheap and affordable traditional sources of energy. There was simply no need for alternative energy. Therefore, energy-efficient construction has long developed exclusively in theory. This problem will not be solved without ideological, legal and financial support from the Government.

Acknowledgments
This paper has been prepared using the equipment kindly provided by the Joint Instrumentation Centre "Baikal Center of Nanotechnologies".

References

[1] Yan Ji and Stellios Plainiotis 2006 *Design for Sustainability* (Beijing: China Architecture and Building Press)
[2] Gozbenko V E, Kargapoltsev S K, Karlina Yu I, Karlina A I and Artyunin A I 2018 *Advances and Applications in Discrete Mathematics* **19(3)** 289-297
[3] Suslov K V, Stepanov V S and Solonina N N 2013 *IEEE International Symposium on Electromagnetic Compatibility* pp 841-845
[4] Stotzer M, Styczynski Z, Gronstedt P, Buchholz B M, Suslov K V and Glaunsinger W 2012 *IEEE Power and Energy Society General Meeting* 6345090
[5] "Green" standards are now in Russia, Available at: http://www.abok.ru/for_spec/articles.php?nid=5369/
[6] STO NOSTROI 2.35.4–2011 Green Building
[7] Jones Lang LaSalle Green Office, Available at: http://zvt.abok.ru/articles/104/Zelyonii_ofis_Jones_Lang_LaSalle/
[8] Ilyushkina E S and Konyukhov V Yu 2012 *Bulletin of Irkutsk State Technical University* **7** 181-187
[9] Gozbenko V E, Kargapoltsev S K, Kornilov D N, Minaev N V and Karlina A I 2016 *International Journal of Applied Engineering Research* **11(20)** 10367-10373
[10] Khomenko A P, Gozbenko V E, Kargapoltsev S K, Minaev N V and Karlina A I 2017 *International Journal of Applied Engineering Research* **12(23)** 13773-13778
[11] Sivtsov A V, Sheshukov O Y, Tsymbalist M M, Nekrasov I V, Makhnutin A V, Egiazar'yan D K and Orlov P P 2018 *Russian Metallurgy (Metally)* **12** 1108-1113
[12] Nekrasov I V, Sheshukov O Yu, Sivtsov A V, Tsymbalist M M and Metelkin A A 2016 *Steel in Translation* **46(6)** 435-442
[13] Sivtsov A V, Sheshukov O Y, Tsymbalist M M, Nekrasov I V and Egiazar'yan D K 2015 *Metallurgist* **59(5-6)** 380-385
[14] Gozbenko V E, Kargapoltsev S K, Kornilov D N, Minaev N V and Karlina A I 2016 *International Journal of Applied Engineering Research* **11(23)** 11132-11136
[15] Kargapoltsev S K, Khomenko A P, Gozbenko V E, Minaev N V and Karlina A I 2017 *International Journal of Applied Engineering Research* **12(22)** 12362-12368
[16] Gozbenko V E, Khomenko A P, Kargapoltsev S K, Minaev N V and Karlina A I 2017 *International Journal of Applied Engineering Research* **12(22)** 12369-12372
[17] Balanovsky A E, Shtaiger M G, Grechneva M V, Kondratev V V and Karlina A I 2018 *IOP Conference Series: Materials Science and Engineering* **411** 012012
[18] Balanovsky A E, Shtayger M G, Kondratev V V, Huy Vu V and Karlina A I 2018 *IOP Conference Series: Materials Science and Engineering* **411** 012013
[19] Ivanchik N N, Balanovsky A E, Shtayger M G, Sysoev I A and Karlina A I 2018 *IOP Conference Series: Materials Science and Engineering* **411** 012035
[20] Balanovsky A E, Shtaiger M G, Kondratev V V, Karlina A I and Govorkov A S 2018 *Journal of physics: conference series* 012006
[21] Nekrasov I V, Sheshukov O Y, Sivtsov A V, Tsymbalist M M, Egiazar'yan D K and Metelkin A A 2015 *Metallurgist* **59(3-4)** 300-304
[22] Suslov K, Solonina N and Stepanov V 2015 *Proceedings - 2015 International Symposium on Smart Electric Distribution Systems and Technologies, EDST* **7315218** 260-264
[23] Rassokhin A S, Ponomarev A N and Figovsky O L 2018 *Engineering and Construction Journal* **3(79)** 132-139
[24] Ponomarev A N and Rassokhin A S 2016 *Engineering and Construction Journal* **8(68)** 45-57

Special aspects of assessing hazard to public health while monitoring the negative impact of solid municipal waste landfills

I O Kirilchuk, A V Besedin, G P Timofeev and A V Iordanova

Southwest state University, 94, 50 October st., Kursk, 305040, Russia

E-mail: iraida585@mail.ru

Abstract. The present study considers the negative impact of solid municipal waste landfills on the environment and public health. It also assesses hazard to public health caused by the impact of solid municipal waste landfill emissions (by the example of a solid household landfill in the city of Kursk). We have carried out four interrelated stages of research work during the implementation of public health hazard assessment of the municipal waste landfill under study: hazard identification, assessment of dose-response relationship, exposure assessment, and risk characterization. At each stage, we assessed uncertainties that might affect the completeness and accuracy of final conclusions and recommendations. A particular attention was paid to some specific uncertainties that are due to the peculiarities of waste landfills' operation. These uncertainties can have a significant impact on the results of the study, especially in terms of emissions' quantitative composition. Thus, despite the fact that the risk level of the solid municipal waste landfill under study was assessed as low, the authors propose the introduction of a degassing system at the solid municipal waste landfill in Kursk. The application of this system will eliminate the release of biogas into the atmosphere, reduce the load on the air and reliably decrease hazard to public health caused by the impact of the landfill.

1. Introduction

Currently, the problem of solid municipal waste formation and disposal is one of the main environmental objectives in ensuring ecological safety in the Russian Federation and abroad. Every year the amount of solid municipal waste (hereinafter referred to as SMW) is steadily increasing. In Russia, about 130 million m^3 of SMW is formed annually [1].

At present, the first place among the methods of waste disposal belongs to solid municipal waste landfills. About 90-95% of waste is deposited at these landfills. In Russia, the problem of landfill depositing is not paid due attention to. Most SMW landfills do not meet sanitary standards [2]. Existing SMW landfills can pose a significant environmental hazard which will exist for decades. In this regard, hazard assessment of landfill impact on the environment and public health comes to the fore [3].

Nowadays, in almost all world countries and international organizations the concept of hazard assessment is considered as the main mechanism for the development and management decision-making at the international, national or regional levels, and at the level of individual production or other potential source of environmental pollution [4].

2. Materials and methods

At present, the task of assessing hazard to public health caused by the negative impact of pollutants released from the body of SMW landfill is particularly relevant [5]. The complete (basic) scheme of health hazard assessment provides for four interrelated stages: hazard identification, dose-response relationship assessment, exposure assessment, risk characterization [6]. At each stage, uncertainties affecting the completeness and validity of final conclusions and recommen-dations are assessed.

In estimating hazard to public health when being exposed to chemicals which pollute the environment, most of the uncertainties are related to various aspects of such substances' impact on human health. However, uncertainties related to the nature of environmental pollution sources such as SMW landfills might be critical.

Within this study, we consider the procedure for assessing hazard to public health from the impact of SMW landfill's emissions (by the example of a solid domestic waste landfill in Kursk) with the identification of specific uncertainties due to the peculiarities of the waste land-fill operation as a source of negative impact.

3. Results and discussion

In order to substantiate the list of chemicals for the subsequent assessment of health risk to the population, the ranking of emitted pollutants by the indices of comparative non-carcinogenic and carcinogenic danger was carried out. When ranking chemicals we have taken the population living in the affected area of landfill emissions. The total number is 160 people.

The greatest contribution to the comparative carcinogenic and non-carcinogenic hazard indices was made by the following substances: formaldehyde, ethyl benzene, ammonia, methane, xylene, hydrogen sulfide, nitrogen dioxide, and toluene.

Based on the results of calculations, a list of chemicals has been formed which includes all carcinogenic substances and substances with the highest rank of comparative non-carcinogenic hazard index. There are also substances included in the list of priority substances for the Russian Federation. These substances: Formaldehyde, Ethyl benzene, Soot, Benzapyrene, Ammonia, Methane, Xylene, Hydrogen sulfide, Nitrogen dioxide, Toluene, Sulfur dioxide, Carbon monoxide.

As part of the dose-response relationship assessment, the hygienic and toxicological specification of chemical substances emitted by the Kursk solid domestic waste landfill under study has been performed. The analysis has showed that the respiratory system is adversely af-fected by 7 substances: sulfur dioxide, nitrogen dioxide, ammonia, xylene, formaldehyde, hy-drogen sulfide, soot. The development is influenced by 3 substances: carbon monoxide, ethyl benzene, benzapyrene.

The immune system is negatively affected by 3 substances: formalde-hyde, benzapyrene, sulfur dioxide. The change in blood composition can be influenced by 2 substances: nitrogen dioxide, carbon monoxide. Carbon monoxide and sulfur dioxide have a negative impact on the cardiovascular system; xylene, ethylbenzene and toluene have a negative influence on kidneys and liver. Soot has systemic effect and a negative impact on teeth, and formaldehyde has an effect on eyes. Ethylbenzene has a negative impact on the endocrine sys-tem.

Thus, respiratory organs, central nervous system, blood, development, cardiovascular system, immune system, kidneys, liver, endocrine system, teeth, eyes are expected to be vulner-able to the impact of emissions from the landfill under study.

Exposure assessment characterizes levels, duration, frequency and methods of factors' impact on the estimated population. Taking into consideration the purpose of our study, the resi-dential area scenario was adopted as the basis of the exposure scenario, which considers chronic (lifetime) exposure.

On the basis of findings made in the hazard identification, atmospheric air was considered as the analyzed environment, and inhalation root was a priority one. At the same time, surface air concentrations of the relevant chemicals at the locations under consideration are required to characterize the risks.

The calculation of both maximum one-time and average concentrations has been carried out using computer programs that created computer simulation of aerial effluents dispersion on the basis of special mathematical dependencies. As a result of simulation, the values of maximum annual concentrations of priority substances obtained are significantly lower than the established Russian maximum allowable

IOP Conf. Series: Earth and Environmental Science **350** (2019) 012033 doi:10.1088/1755-1315/350/1/012033

concentrations for these substances.

Risk characterization is the final stage of public health hazard assessment. At this stage the results of the previous three stages are summarized, the risks and their comparative characteristics are calculated, and all the uncertainties are analyzed that can affect the validity and reliability of final conclusions and recommendations being necessary for risk management.

The probability of carcinogenic effects for the population living in the affected area of the landfill (CR) was estimated as a result of formaldehyde, ethylbenzene, soot and benzapyrene impact (Table 1).

The analysis has shown that on the border of the estimated sanitary protection zone of the landfill, in the territory of residential development and in the territory of garden plots in the zone of the landfill's emissions impact, the levels of individual carcinogenic risk without regard to the background from the effects of formaldehyde, ethylbenzene, soot and benzapyrene correspond to the first risk range (minimum level). Such risks do not require any additional measures to reduce them and their levels are subject only to periodic control. The total carcinogenic risk also corresponds to the first risk range.

Table 1. Carcinogenic health hazard due to the impact of substances emitted by the landfill

No.	Code	Substance	CAS	CR		
				min	max	average
estimated sanitary protection zone of the landfill						
1	1325	Formaldehyde	50-00-0	9.77E-07	3.25E-06	2.19E-06
2	627	Ethyl benzene	100-41-4	8.09E-08	2.69E-07	1.81E-07
3	328	Soot	-	8.13E-09	1.21E-07	3.69E-08
4	703	Benzapyrene	50-32-8	2.01E-13	5.46E-13	3.68E-13
Total				1.07E-06	3.64E-06	2.41E-06
residential space in the affected area						
1	1325	Formaldehyde	50-00-0	1.01E-06	2.61E-06	1.50E-06
2	627	Ethyl benzene	100-41-4	8.38E-08	2.16E-07	1.24E-07
3	328	Soot	-	6.68E-09	6.42E-08	1.42E-08
4	703	Benzapyrene	50-32-8	1.67E-13	4.01E-13	2.51E-13
Total				1.10E-06	2.89E-06	1.64E-06
garden plots in the affected area						
1	1325	Formaldehyde	50-00-0	9.33E-07	3.31E-06	1.82E-06
2	627	Ethyl benzene	100-41-4	7.70E-08	2.75E-07	1.51E-07
3	328	Soot	-	5.34E-09	1.19E-07	2.36E-08
4	703	Benzapyrene	50-32-8	1.78E-13	6.35E-13	3.38E-13
Total				1.02E-06	3.70E-06	1.99E-06

The results of ranking the levels of non-carcinogenic health hazard have showed that substances with the strongest non-carcinogenic effects are: formaldehyde, hydrogen sulfide, ammonia, xylene. The values of danger coefficients (HQ) excluding the background at all points of impact/receptor points are significantly below acceptable (less than 1).

This level of risk is assessed as low one, as shown in Table 2. The maximum concentrations of the priority substances do not exceed 0.0661 in the residential area and 0.0825 at the boundary of the estimated sanitary protection zone.

The total impact of pollutant emissions on critical organs and systems is presented in Table 3.

Table 2. Non-carcinogenic risks from landfill emissions without regard to the background

Substance	at the boundaries of the estimated sanitary protection territories		residential space in the affected area		garden plots	
	HQmin	HQmax	HQmin	HQmax	HQmin	HQmax
Formaldehyde	2.48E-02	8.25E-02	2.57E-02	6.61E-02	2.36E-02	8.41E-02
Hydrogen sulfide	1.01E-02	3.35E-02	1.04E-02	2.69E-02	9.61E-03	3.41E-02
Ammonia	4.13E-03	1.37E-02	4.28E-03	1.10E-02	3.94E-03	1.40E-02
Xylene	3.43E-03	1.14E-02	3.55E-03	9.15E-03	3.27E-03	1.16E-02
Sulfur dioxide	2.85E-03	1.13E-02	2.95E-03	8.45E-03	2.70E-03	1.11E-02
Nitrogen dioxide	2.57E-03	1.34E-02	2.57E-03	9.01E-03	2.32E-03	1.31E-02
Methane	8.19E-04	2.73E-03	8.49E-04	2.19E-03	7.82E-04	2.78E-03
Toluene	1.12E-04	3.73E-04	1.16E-04	2.99E-04	1.07E-04	3.80E-04
Carbon monoxide	7.60E-05	3.75E-04	7.60E-05	2.57E-04	6.90E-05	3.68E-04
Soot	3.67E-05	5.47E-04	3.02E-05	2.90E-04	2.41E-05	5.37E-04
Ethyl benzene	7.40E-05	2.45E-04	7.62E-05	1.96E-04	7.00E-05	2.50E-04
Benzapyrene	1.80E-07	4.90E-07	1.50E-07	3.60E-07	1.60E-07	5.70E-07

The data obtained indicate that at the boundaries of the estimated sanitary protection territories of the landfill, in the residential space and in the territory of garden plots located in the zone of the landfill's influence with the joint action of chemicals by their non-carcinogenic effects (HI), the negative impact on the respiratory system, development, central nervous system, blood, heart vascular system, immune system, kidneys and liver due to the landfill's emissions excluding the background are at an acceptable level.

4. Conclusion

At each stage of hazard assessment, some uncertainties were identified. Also, each stage has determined how these uncertainties affected the completeness and validity of final conclusions and whether any adjustments to the study were required. Some of the identified uncertainties can be attributed to almost any study objects – the lack of Roshydromet (Federal Service of Russia on Hydrometeorology and Environmental Monitoring) data on the background average annual concentrations at the site where study was being undertaken, the suitability of techniques for modeling chemical substances' dispersion in the atmosphere in order to assess hazard, etc. However, in the course of the work, special attention was paid to some specific uncertainties that are due to the peculiarities of the waste landfill operation under study:

1. An uncertainty concerning the volume of waste deposited at the landfill. Information on the amount of waste is usually provided by the owner of the landfill according to accounting data, which is not always true.
2. The volume of released biogas largely depends on compositional analysis of solid municipal wastes. Such analysis has never been undertaken in Kursk.
3. The components of the landfill body wastes can significantly vary over time depending on some factors, for example, from the combustion of waste at the landfill, which has been repeatedly revealed at the Kursk SMW landfill.

These uncertainties can significantly influence the results of the study, especially in terms of quantitative composition of emissions. Thus, despite the fact that the level of non-carcinogenic risk was assessed as low, and the total carcinogenic risk was attributed to the first risk range (minimum level), we propose the introduction of a degassing system at the landfill in Kursk, in which biogas released from the landfill body is to be collected and used to generate heat or electricity [7]. The use of a degassing system will make it possible almost completely to eliminate biogas release into the atmosphere, decrease load on the air and reliably reduce the level of hazard to public health from the impact of SMW landfill to minimum values.

IOP Conf. Series: Earth and Environmental Science **350** (2019) 012033 doi:10.1088/1755-1315/350/1/012033

Table 3. Total hazard indices with simultaneous input of chemical substances depending on their influence on critical organs and systems excluding the background

territories	organs	substances	HI min	HImax
estimated sanitary protection zone of industrial site	breathing organs	ammonia, xylene, nitrogen dioxide, formaldehyde, sulfur dioxide, hydrogen sulfide, soot	4,78E-02	1,66E-01
	development	carbon monoxide, ethyl benzene, benzapyrene	1,50E-04	6,20E-04
	central nervous system	toluene, xylene, carbon monoxide	4,91E-03	1,65E-02
	blood	carbon monoxide, nitrogen dioxide	2,64E-03	1,37E-02
	immune system	formaldehyde, benzapyrene, sulfur dioxide	2,76E-02	9,38E-02
	kidneys	xylene, ethyl benzene, toluene	3,62E-03	1,20E-02
	liver	xylene, ethyl benzene, toluene	3,62E-03	1,20E-02
	heart vascular system	carbon monoxide, sulfur dioxide	2,93E-03	1,17E-02
	HI total		**9,33E-02**	**3,26E-01**
residential space in the affected area	breathing organs	ammonia, xylene, nitrogen dioxide, formaldehyde, sulfur dioxide, hydrogen sulfide, soot	4,95E-02	1,31E-01
	development	carbon monoxide, ethyl benzene, benzapyrene	1,52E-04	4,53E-04
	central nervous system	toluene, xylene, carbon monoxide	5,08E-03	1,31E-02
	blood	carbon monoxide, nitrogen dioxide	2,64E-03	9,27E-03
	immune system	formaldehyde, benzapyrene, sulfur dioxide	2,87E-02	7,46E-02
	kidneys	xylene, ethyl benzene, toluene	3,75E-03	9,65E-03
	liver	xylene, ethyl benzene, toluene	3,75E-03	9,65E-03
	heart vascular system	carbon monoxide, sulfur dioxide	3,03E-03	8,71E-03

Table 3. Continuation

	HI total		**9,66E-02**	**2,56E-01**
garden plots	breathing organs	ammonia, xylene, nitrogen dioxide, formaldehyde, sulfur dioxide, hydrogen sulfide, soot	4,55E-02	1,68E-01
	development	carbon monoxide, ethyl benzene, benzapyrene	1,39E-04	6,09E-04
	central nervous system	toluene, xylene, carbon monoxide	4,68E-03	1,67E-02
	blood	carbon monoxide, nitrogen dioxide	2,39E-03	1,35E-02
	immune system	formaldehyde, benzapyrene, sulfur dioxide	2,63E-02	9,52E-02
	kidneys	xylene, ethyl benzene, toluene	3,45E-03	1,23E-02
	liver	xylene, ethyl benzene, toluene	3,45E-03	1,23E-02
	heart vascular system	carbon monoxide, sulfur dioxide	2,77E-03	1,15E-02
	HI total		**8,87E-02**	**3,30E-01**

5. Acknowledgments

This work was carried out according to a research grant of the President of the Russian Federation, MK-941.2019.5, for the support of young Russian scientists.

References

[1] Eroshina D M and Hodin V V 2009 *J. Solid Municipal Waste* **8** 31-35
[2] Gribanova L P and Gudkova V N 2008 *J. Solid Municipal Waste* **9** 18-23
[3] Caniato M, Tudor T and Vaccari M 2015 *J. of Environmental Management* **153** 93-107
[4] Kirilchuk I and Rykunova V 2015 *International multidisciplinary scientific conference on social sciences and arts (SGEM 2015)* **1(1)** 583-588
[5] Giubilato E, Zabeo A, Critto A, Giove S, Bierkens J, Hond E D and Marcomini A 2014 *J. Environment International* **65** 41-53
[6] P.2.1.10.1920-04 2004 *Guidelines for the assessment of public health risks from exposure to chemicals that pollute the environment* (Moscow: Ministry of Health of the Russian Federation) p 143
[7] Marinenko E E 2003 *Bases of biofuel production and use for the solution of energy saving and environmental protection issues in housing and communal services and agriculture* (Volgograd: VolgGASA) p 100

IOP Conf. Series: Earth and Environmental Science **350** (2019) 012034 doi:10.1088/1755-1315/350/1/012034

Efficiency of field-protective forest plantations in the first agro-climatic zone of the Stavropol territory of the Russian Federation

P V Klyushin[1], V N Khlystun[1], A V Loshakov[2], S V Savinova[1] and T A Ivanova[1]

[1]State University of Land Use Planning, 15, Kazakova str., Moscow, 105064, Russia
[2]Stavropol agrarian University, Stavropol, 12, Zootekhnicheskiy lane, 355017, Russia

E-mail: klyushinpv@gmail.ru

Abstract. This paper presents studies on the state of agricultural land in the first agro-climatic Stavropol territory, and on the basis of this forest reclamation measures for effective and rational use have been developed. The specialization of agriculture is sheep breeding and cultivation of winter wheat and winter and spring barley. For this agro-climatic zone, only one type of crop rotation is suitable – grain-steam. In the structure of acreage, pure steam reaches 40%, which has a very negative impact on the quality of arable land. Protective forest stands play an important role in protecting agricultural land from the negative effects of water and wind erosion. At present, the total area of protective plantations on the lands of the I agro-climatic zone is 22730 hectares or 1.81% of the area of agricultural land, at least 2.5% is necessary.

1. Introduction

Stavropol region is one of the largest agricultural regions of the country, producing more than 3% of gross output. It occupies the Central part of the Caucasus and its area is 6616 thousand hectares. The composition of the Stavropol territory includes 26 administrative districts and 10 cities of regional subordination. 92% belong to the territory of the Stavropol territory. The plowed land in some areas is more than 80% of their land fund. Long-term intensive use of land has led to the deterioration of the environmental situation in the region and the development of many degradation processes that significantly reduce the productivity of agricultural land [1–3].

The total land area of the I agro-climatic zone is 1806348 hectares and it includes five administrative districts of the Stavropol territory: Apanasenkovsky, Arzgir, Levokumsky, Neftekumsky and Turkmen. The main limiting factor of agricultural production is the availability of moisture. The soil cover is represented by light chestnut and chestnut soils, as well as a significant spread of saline and saline soils. These soils are very poorly provided with humus, but contain a large number of trace elements.

The main specialization of agriculture in this area is sheep breeding and production of winter wheat and barley. For this agro-climatic zone, only one type of crop rotation is suitable – steam grain. In the structure of acreage, pure steam reaches 40%, which has a very negative impact on the quality of arable land. There is a steady increase in the area of eroded and deflated agricultural land, as well as significant losses of humus on them. In rare cases, the crop rotation includes spring crops such as millet and barley.

This paper presents the study and analysis of agricultural land in the first agro-climatic zone, and on this basis – the material for the development of forest reclamation measures to protect and preserve land from degradation processes, primarily from deflation.

2. Materials and methods

Work on the identification of degraded lands of the Stavropol territory was carried out in 2000-2016, and the methodological basis for the surveys was "Guidelines for the identification of degraded and contaminated land, published in 1995".

In our work, we calculated the ecological stability of agricultural lands in the Stavropol region according to the data from 2000 to 2016, because during the period of our research there have been significant changes in the areas of land types, and therefore in the stable and unstable elements of agricultural landscapes. Based on the coefficient of environmental stabilization, we see that it is necessary to develop a set of measures to improve their stability for agricultural landscapes of the Stavropol territory, as the area of unstable elements is much larger than the area of stable ones. Thus the cheapest, but very effective means for the protection of agricultural landscapes are forest reclamation activities, which is why this work is dedicated to this topic [2–4].

3. Study of agroforestry activities in the first agro-climatic zone

The territory of the I agro-climatic zone is more than 95% represented by agricultural lands and during the period of research the area of these lands increased by 27906 hectares. This increase was mainly due to reserve lands (table 1).

Table 1. Distribution of agricultural land of the I agro-climatic zone by areas

Area	Year 2000		Year 2016	
	ha	%	ha	%
Apanasenkovsky	341635	95.32	341675	95.33
Arzgirsky	310433	91.75	313026	92.51
Levokumsky	439655	93.80	453205	96.69
Neftekumsk	349676	92.09	356997	94.02
Turkmen	247486	94.75	251888	96.43
Total for zone I	1688885	93.49	1716791	95.04

During the analyzed period, there was an increase in the area of agricultural land in all administrative areas, but the greatest growth was observed in the Levokumsky district, where this category of land increased by more than 13 thousand hectares. As observations showed, there was involvement in agricultural land turnover, which initially was not in a very good quality condition (solontsy, slope land and already degraded to varying degrees). The same situation exists on the territory of the Neftekumsk district, where 7321 ha of land was involved for sixteen years in agricultural use. The smallest increase in the area of agricultural land is noted in Apanasenkovsky area, within which this category increased by 40 ha.

In composition of agricultural lands, the majority of the area is agricultural land. Currently, 93.2% of agricultural lands of the I agro-climatic zone are involved and intensively used for agricultural production. The share of arable land in agricultural land shows the degree of anthropogenic load on agricultural landscapes. The maximum percentage of ploughing of the territory is noted in the Turkmen region (71.47%), but during the period of research it decreased by 0.67%. This increases the area of pastures in both physical and percentage terms. The area of perennial plantations has decreased slightly, such lands as hayfields and deposits are absent within the boundaries of the Turkmen district.

There are only two types of land on the territory of the Arzgir district – arable land, which occupies 62.12% of the agricultural land area, and pastures, which occupy 33%. During the 16-year period, the area of arable land increased by 128 hectares, but in relative terms its share decreased; respectively, the share of pastures increased from 32.54% to 33%.

Farmlands of the Apanasenkovsky district is represented by four types of land, the maximum percentage of which is arable land (55.62%) and pastures (of 35.94%). Hayfields and perennial plantations occupy a total of 0.9% of the land. The dynamics of areas for all types of land remains stable for the entire period of research.

The least ploughed area is observed in Neftekumsk (27.1%) and Levokumsky (38.82%) areas, and for the analyzed period, this percentage has not significantly decreased. This increases the area of grassland, especially pastures (+ 8969 hectares in the Levokumsky and + 4007 hectares in Neftekumsk district). Also in these areas, the largest areas of perennial plantations, which are represented by vineyards during the research period, a tendency to reduce their area is revealed. For a more detailed description of the use of agricultural land, we conducted surveys of forest plantations outside the state forest fund, presented mainly by protective forest plantations (table 2).

Table 2. Dynamics of the area of forest plantations not included in the state forest fund by areas

Area	Year 2000		Year 2006		Year 2012		Year 2016	
	ha	%	ha	%	ha	%	ha	%
Apanasenkovsky	4732	1.38	4744	1.39	4744	1.39	4744	1.39
Arzgirsky	3923	1.26	4104	1.31	4104	1.31	4104	1.31
Levokumsky	4365	0.99	4536	1.02	4536	1.01	4536	1.01
Neftekumsk	4561	3.52	4639	3.46	4638	3.45	4638	3.46
Turkmen	4708	1.90	4708	1.88	4708	1.88	4708	1.88
Total for zone I	22289	1.81	22731	1.81	22730	1.81	22730	1.81

The presence in the composition of agricultural land protective forest plantations and their condition plays an important role in protecting the land from the negative effects of water and wind. The total area of protective plantations on the lands of the I agro-climatic zone increased by 441 hectares in 16 years and in 2016 it was 22730 hectares or 1.81% of the agricultural land area. The highest percentage of land occupied by forest plantations is noted in the Neftekumsky district (3,46%) and the smallest within the boundaries of the Levokumsky district (1%). During the period of research in all administrative areas of the zone there is an increase in the area of protective forest plantations [5–7].

Table 3. Quantitative characteristics of field-protective forest stands of the I agro-climatic zone

Area	Area of agricultural land, thousands of hectares	The percentage of afforestation	Area of protected forest plantations, ha	Number of protective forest strips, pieces
Apanasenkovsky	315.9	1.2	3965	1706
Arzgirsky	297.7	1.0	2944	1405
Levokumsky	416.5	0.4	1653	872
Neftekumsk	326.9	1.0	2550	1474
Turkmen	239.2	1.7	4061	2220
Total for zone I	1596.2	1.1	15173	7677

Protected forest areas play an important environmental role, being part of the ecological framework of the agricultural landscape, including agricultural land, contribute to reducing soil erosion, improve the overall microclimate and hydrological regime of the territory, perform a number of useful functions. Thus, in the Stavropol region, the area of forest plantations outside the forest fund is 143.7 thousand hectares, of which 22730 hectares (15.8%) fall on the land of the I agro-climatic zone. This area differs in its importance from the materials of the inventory of trees and shrubs, which was held in the Stavropol territory in 2004. According to the materials of the "Inventory of wood and shrub vegetation of protective value and roads on agricultural land" the area of protective forest strips in the zone was 15173 hectares. The table below shows the area of protective forest plantations in areas of the I agro-climatic zone of the Stavropol territory (table 3).

The percentage of afforestation of agricultural land throughout the agro-climatic zone is critically low, especially in the Levokumsky district, where it is 0.4%. The number of forest shelter belts and the area is more than two times less than that in other zones, while the Levokumsky district is the largest in the Stavropol region [8–10].

4. Conclusion

According to the results of the inventory, it can be concluded that 5,524 hectares (26%) of forest belts are in good condition, 6,561 hectares require logging and maintenance, 1526 hectares need to be supplemented with planting, 885 hectares – reforestation, 677 hectares need to be decommissioned. The proposed activities for the administrative areas of the I agro-climatic zone are presented in table 4.

Table 4. Forest belts and proposed activities in administrative areas of the I agro-climatic zone

Area	Number of protective forest strips, pieces	Area, hectares	The area of the proposed activities acre			
			landing	care cabins	reforestation	stubbing
Apanasenkovsky	1706	3965	199	2192	23	199
Arzgirsky	1405	2944	219	535	–	104
Levokumsky	872	1653	542	176	27	75
Neftekumsk	1474	2550	548	551	30	188
Turkmen	2220	4061	18	3107	805	111
Total for zone I	7677	15173	1526	6561	885	677

As can be seen from table 4 protective forest strips require a large amount of various works aimed at improving their quality. Small afforestation of agricultural land and especially arable land, non-compliant with the rules of rational land use and disregarding the specifics of the terrain, as well as a large area of clean vapors, contribute to the development of various degradation processes, especially water erosion and deflation. And natural factors determine the presence of saline, flooded and saline lands.

The maximum area to complement existing forest identified in Neftekumsk (548 ha) and Levokumsky (542 ha) areas, and the lowest in Turkmenistan (18 ha). Logging work is required in a large volume of Turkmen (3107 ha) and Apanasenkovsky (2192 ha) areas. The Turkmen district will need to make reforestation on the area of 805 hectares, the largest area of forest belts that require stubbing detected in Apanasenkovsky (199 ha) and Neftekumsk (188 ha) areas. At the same time, the quantitative composition and condition of forest belts cannot but cause concern among agricultural specialists. Afforestation of arable land by forest belts is extremely low – 2.5%.

5. Acknowledgments

The research was carried out according to the program "About the Federal target program "the South of Russia (2014-2020)" adopted by the Order of the Government of the Russian Federation on December 26, 2013 N 1297 in Moscow.

References

[1] Ryabov E I 1996 *Wind erosion (deflation) and measures to prevent it* (Stavropol: Kniga)
[2] Poluektov E V 2003 *Erosion and deflation of agricultural landscapes of the North Caucasus* (Novocherkassk: NGMA)
[3] Zhuchenko A A, Trukhachev V I, Penchukov V M et al 2011 *Systems of agriculture of the Stavropol region* (Stavropol: AGRUS)
[4] Hasai N Yu 2009 *Improving the efficiency of environmental management on the basis of ecologically safe use of land resources for agricultural production,* abstract of candidate thesis (Stavropol)

[5] Loshakov A V, Klyushin P V, Shirokova V A, Khutorova A O and Savinova S V 2019 Environmental problems of treatment of saline lands for agricultural needs in the first agro-climatic zone of the Stavropol territory *South of Russia: ecology, development* [in Russian – Yug Rossii: Ekologija, razvitie] **14(1)** 105-116
[6] Gostishchev D P and Khutorova O A 2011 Agriculture the effect of the introduction of shelterbelts *Environmental engineering* **5** 36-41
[7] Shapovalov D A, Klyushin P V, Shirokova V A, Khutorova A O and Savinova S V 2018 Problems and efficiency of land use in the North Caucasian federal district *18th Int. Multidisciplinary Scientific GeoConf. SGEM 2018* **18(5.1)** 667-674
[8] Zelepugin A 2010 Scientific basis of land monitoring classification *Agricultural journal* **1** 50-51
[9] Larina G E 2010 Ecological control of agroecosystems state *Actual problems of humanities and natural sciences* [in Russian – Aktualnye problemy gumanitarnyh i estestevennyh nauk] **10** 39-41
[10] Yurtaev A A 2011 Agrolandscape researches: theory and practice *Scientific sheets of BelSU. Series Natural Sciences* [in Russian – Nauchnye vedomosti of Belgorod State University. Serija Estestvennye nauki] **15** 217-221

Ecological architecture: the green roofs

V Yu Konyuhov, A M Gladkih, I I Galyautdinov and T Yu Kiseleva

Irkutsk National Research Technical University, 83 Lermontov Street, Irkutsk, 664074, Russia

E-mail: konyuhovvy@ex.istu.edu

Abstract. Ecological architecture is not just a new-fashioned tradition and an attempt to keep up with the times. If you think globally, then ecological architecture is a new way of life and thinking, the opposite of what is generally accepted. For several centuries, man has been using nature as he pleases. The peak of this trend, perhaps, happened in Russia, when the Soviet authorities changed the river beds for their convenience, planted the fields with crops unacceptable for the climate and soil. In the twenty-first century, the world community understood the value of nature and took up preserving it sensibly. The problem of environmental pollution is very relevant at present. Many technologies are being created to protect our planet, and architecture is one of their main components. In this article, we looked at the new ecological type of roofing - the “green roof”, examined its pros and cons and summed it up at the end.

1. Introduction

The question of environmental protection often arises in modern society. People strive for a good, comfortable life through the development of technology [1-5], but it is also very important that it does not harm the environment [6-10]. In this case, architecture plays a key role, because the construction material is directly related to natural resources. Today we consider roofings as one of the construction parts. There are many types of roofs, such as metal, galvanized steel, slate, etc., but has anyone heard a term like "green roofs"?

"Green roofs" means "planting greenery on the roofs" in English. This phrase means partial or full filling of a roof with live plants. Plants are planted directly into the ground, and a waterproof membrane layer is installed between it and the roof. This technique appeared relatively recently, but the "green roofs" already have a history of their own, as they appeared several centuries ago. This design has excellent thermal insulation qualities. In the cold countries of Scandinavia or Iceland, “green roofs” helped to keep the houses warm, and in hot countries, such as Tanzania, for example, they kept the room cool [11-16]. Modern planting technologies appeared in the 1960s in Germany due to the deteriorating environmental situation in the city. In the following decades, this application of planting of greenery spread throughout the world.

Today this kind of technique is mostly popular in megalopolises: on the roofs of shopping centers, office buildings, residential high-rises. Just recently "green roofs" began to be made for country houses and household buildings. But one can agree that it has long been thought out how to heat houses in the modern world, and an air conditioning system was created to make rooms cool. But then why is the demand for this technology increasing every day? Let us find it out.

The traditional arrangement of the green "turf" roof.

Not every resident of a large city has the opportunity to relax in the forest or breathe the fresh air in mountain meadows. Most often, it is not enough time to take a nature hike. Multi-meter buildings, stuffy streets and scorching asphalt under their feet are depressing every day. Therefore, many happy owners of private homes seek to achieve their desire to communicate with nature in the construction of housing from natural environmentally friendly materials.

Russian experts have learned many technical solutions for the installation, weatherization, decoration and reconstruction of houses from the experience of the Scandinavian countries. Finnish, Norwegian and Swedish engineers have long been using a number of technologies in the construction that are becoming popular in our country just now.

Among them is the very "turf roof", which is being actively built on residential buildings not only in Scandinavian countries, but also in some European ones.

The general principle of building a green roof in the old Norwegian version was as follows: a clay bedding material was placed under the sod layer. Its thickness was about 10 cm. Clay served simultaneously as a heat and water proof. Using that kind of system, it was necessary to create very strong supporting structures. Not only did the dirt cake weigh a lot, in the winter a mass of snow was added to this load. If the supporting pillars were not weighty enough, the inhabitants would wake up one morning with a pile of earth and snow on their heads. Perhaps in the summer the building looked very picturesque, but the arrangement of that kind of roof was not cheap. For some time, they completely stopped doing a turf roofing.

Perhaps an interesting constructive solution would have remained a relic of the past, were it not for the new manufacturers who recalled the well-forgotten old Scandinavian customs.

Modern installation of green roofs.

The challenge for modern specialists was to reduce the overall weight of the roof structure. With this technology, it was necessary to preserve its decorative appearance. The main problem was to create a durable layer of waterproofing. It was supposed to be dense enough so as not to let in moisture and plant roots, and light, so as not to give a serious load on the roof structures [17-20].

A supporting flooring is laid as the first layer. The material in this case is unimportant. It can be boards or sheets made of pressed chips. Linings are is also suitable, but then the cost of building materials increases slightly.

The next layer is insulation. This part of the roof structure should protect the roofing rooms from rain and melt water penetration. In addition, it is the insulation that prevents the germination of plant roots. The building stores offer a lot of rubber or bitumen-based roll waterproofing. There are plenty of choices.

Figure 1. "Green roofs".

2. Methods and materials

The work is carried out using a set of modern methods of theoretical and experimental research: modeling of the formation of layers; design of the experimental batch; laboratory experiments on the selection of the optimal composition and thickness of soil layers; the use of methods of mathematical statistics and computational software package Microsoft Office.

3. Draining systems for green roofs. Drainage options

Once the way into the house for the water is blocked, it means that it must find a way out in another place. For this, roof drainage is provided. Without this layer, water will accumulate on the waterproofing surface and cause acidification of the soil. This, in turn, will cause rotting of the roots and stems of the green covering, which will spoil the whole idea of creating a green roof. A thin layer of sand or rubble can be poured as a drainage. If possible, one can buy lighter coatings such as geotextiles [17, 21-23].

Ready-made solutions are often used as drainage systems for green roofs.

1. Drainage Plates

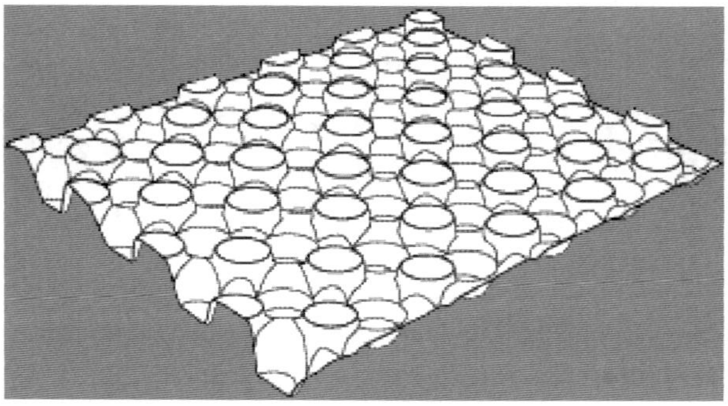

Figure 2. Drainage plates.

Drainage plates are profiled plastic panels that are interconnected and laid over the entire roof area. Water is kept in reservoirs of the plates, and its excess amount flows to the bottom plates. Thus, the water is drained away to the lower part of the roof and further to the outer drainage system.

2. Drainage pads

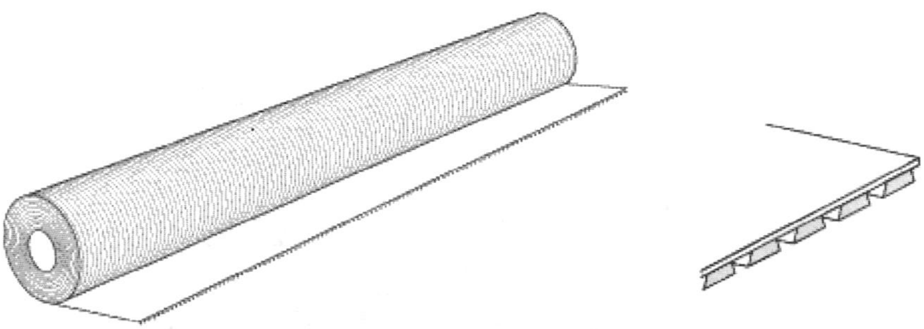

Figure 3. Drainage pads.

Drainage pads work on the same principle as the drainage plates. They are fairly quick and easy to install. Plus, they function as a waterproofing material. However, water disposal in this way is limited by the roof slope. This method will not work for too shallow or flat structures.

3. Drainage boxes

Not only properly selected material, but also the slope of the roof affects good drainage. If there is no slope, another problem arises. It is necessary to take measures to forcibly remove water from the roof or use drainage boxes. Water enters the funnel, and then flows down the sewer or drainpipe.

4. The cost of installing the "green roof"

As stated by the company "Green Roofs for Healthy Cities", installing a square meter of green roof costs about 120-180 US dollars. This price does not include a waterproof coating. The installation cost is directly dependent on the type of landscaping, the material of the coatings and the complexity of the installation of supporting structures [16-17].

It is also worth remembering that it will be necessary to spend money on annual weeding of the vegetation, and fertilization of the soil.

In some German municipalities, the installation of a green roof has already been made mandatory under penalty of fine for all owners of flat roofs. Authorities say that, as a result, there is less dirt, it becomes cooler in summer, and the air is cleared. In the vicinity of Stuttgart, even huge warehouses are covered with grass, otherwise the buildings simply will not be put into operation.

The owners of green roofs point out how interesting it is to watch how the roof's appearance changes depending on the time of year.

Figure 4. A drainage box.

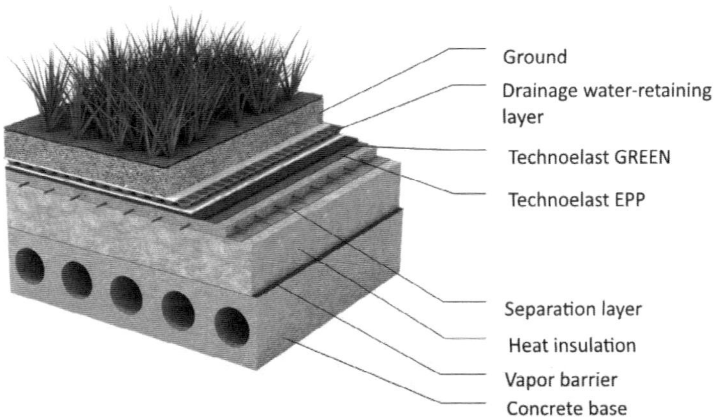

Figure 5. "Installing landscaping".

Green roofs in Russia only start appearing in the largest cities such as Moscow, St. Petersburg, Yekaterinburg, Kaliningrad, Tver, Voronezh, and some others. A prominent example is the green roof of Crowne Plaza Business Center in Pulkovo airport in St. Petersburg. In the project, implemented in 2011, a roof was created with extensive landscaping, covering an area of more than 2000 sq. m.

Green roofs are also used in the construction of international-level facilities for the 2014 Olympics in Sochi and the 2018 World Football Cup. There is an increase in the needs of real estate market to design eco-roofs in residential buildings, since the economic effect of their presence in the implementation of development projects has the prospect to exceed the costs of construction by more than a third.

Meanwhile, there are some factors that slow down the development of this idea in our country, and they cannot go unnoticed. These factors are associated not only with economic conditions, but also with established stereotypes.

Figure 6. "A green roof in Moscow in Bolshoy Gnezdnikovsky lane".

So, here are the advantages of "green roofs":

1. Improving the environment, because green plantings purify the air, retaining about 20% of harmful impurities.
2. As we have said, it is the increase of the insulation level. Plants help regulate the heat transfer between the building and the environment.
3. The increase of noise reduction. This roof is perfect for buildings that are located near railway tracks, airports, and highways.
4. Long service life. This method involves protecting the roof from negative influences, such as moisture, snow and sunlight.
5. These roofs are undoubtedly a stunning additional recreation space.
6. And of course, green roofs are the aesthetic components of modern urban planning. They are original and always memorable.

It should also be noted that, except the positive sides, there are a number of disadvantages.

1. First of all, this is a high initial cost.
2. It is very difficult to establish that kind of structure in earthquake-prone areas.
3. The problem of the permanent preservation of roof humidity is relevant for many plant species.

5. Conclusion

In addition to the obvious strengths (decorative, environmental and fire safety), the green roof has another advantage. The roofing pie is considered heavy enough to provide high-quality and uniform shrinkage of the house log. In summer, the eco-roof protects the building from sun rays inside the attic and creates a comfortable, cool indoor temperature.

Summing up, we can conclude that this type of roof has great prospects. People just need to think seriously that everyone switched to this technology, because this is our health and our clean air. It is clear that to implement this kind of project is not an easy task. But if you correctly calculate everything, the investments will be exactly justified. It will not take many years to repair that kind of roof, and over the years the appearance of the structure will be still pleasing the eye with its elegance and accuracy.

Acknowledgments

This paper has been prepared using the equipment kindly provided by the Joint Instrumentation Centre "Baikal Center of Nanotechnologies".

References

[1] Gozbenko V E, Kargapoltsev S K, Kornilov D N, Minaev N V and Karlina A I 2016 *International Journal of Applied Engineering Research* **11(20)** 10367-10373
[2] Khomenko A P, Gozbenko V E, Kargapoltsev S K, Minaev N V and Karlina A I 2017 *International Journal of Applied Engineering Research* **12(23)** 13773-13778
[3] Sivtsov A V, Sheshukov O Y, Tsymbalist M M, Nekrasov I V, Makhnutin A V, Egiazar'yan D K and Orlov P P 2018 *Russian Metallurgy (Metally)* **12** 1108-1113
[4] Nekrasov I V, Sheshukov O Yu, Sivtsov A V, Tsymbalist M M and Metelkin A A 2016 *Steel in Translation* [in Russian – Stal] **46(6)** 435-442
[5] Sivtsov A V, Sheshukov O Y, Tsymbalist M M, Nekrasov I V and Egiazar'yan D K 2015 *Metallurgist* **59(5-6)** 380-385
[6] Solodusha S, Suslov K and Gerasimov D 2016 *Stability and Oscillations of Nonlinear Control Systems* 7541227
[7] Suslov K V, Solonina N N and Smirnov A S 2011 *IEEE PES Innovative Smart Grid Technologies Conference Europe* 6162654
[8] Suslov K V, Solonina N N and Smirnov A S 2010 *Modern Electric Power Systems* 6007274
[9] Gozbenko V E, Kargapoltsev S K, Minaev N V and Karlina A I 2016 *International Journal of Applied Engineering Research* **11(23)** 11132-11136
[10] Kargapoltsev S K, Khomenko A P, Gozbenko V E, Minaev N V and Karlina A I 2017 *International Journal of Applied Engineering Research* **12(22)** 12362-12368
[11] Gozbenko V E, Khomenko A P, Kargapoltsev S K, Minaev N V and Karlina A I 2017 *International Journal of Applied Engineering Research* **12(22)** 12369-12372
[12] Gozbenko V E, Kargapoltsev S K and Karlina A I 2019 *IOP Conf. Series: Earth and Environmental Science* **229** 012020
[13] Gozbenko V E, Kargapoltsev S K and Karlina A I 2019 *IOP Conf. Series: Earth and Environmental Science* **229** 012021
[14] Sheshukov O Y, Nekrasov I V, Sivtsov A V, Tsymbalist M M, Egiazar'yan D K and Metelkin A A 2014 *Steel in Translation* [in Russian – Stal] **44(1)** 43-46
[15] Sivtsov A V, Tsimbalist M M, Fadeev V I, Ostrovskii Y I, Veselovskii I A and Afanas'ev V I 2012 *Russian Metallurgy (Metally)* **2012(12)** 1036-1040
[16] Balanovskiy A E 2017 *Welding International* **31(6)** 467-476
[17] Green Roof: Grass Roof Technology. Available at: https://krovportal.ru/krovlya/naturalnye-materialy/zelenaya-krysha-texnologiya-ustrojstva-travyanoj-krovli/
[18] Ivanov N A, Balanovskii A E, Kalashnikov Y D, Ryazantsev V V and Skripnichenko A S 2016 *High Temperature* **54(2)** 289-293
[19] Rassokhin A S, Ponomarev A N and Figovsky O L 2018 *Engineering and Construction Journal* **2(78)** 151-160
[20] Gozbenko V E, Kargapoltsev S K, Karlina Yu I, Karlina A I and Artyunin A I 2018 *Advances and Applications in Discrete Mathematics* **19(3)** 289-297
[21] Gozbenko V E, Korchevin N A, Kargapol'tsev S K, Karlina A I and Karlina Yu I 2019 *Advances in Engineering Research* **182** 120-125
[22] Gozbenko V E, Korchevin N A, Kargapol'tsev S K, Karlina A I and Karlina Yu I 2019 *Advances in Engineering Research* **182** 126-131
[23] Ponomarev A N and Rassokhin A S 2016 *Engineering and Construction Journal* **8(68)** 45-57

IOP Conf. Series: Earth and Environmental Science **350** (2019) 012036 doi:10.1088/1755-1315/350/1/012036

Economic aspects of green technologies

V Yu Konyuhov, A M Gladkih, I I Galyautdinov and Ya D Severina

Irkutsk National Research Technical University, 83 Lermontov Street, Irkutsk, 664074, Russia

E-mail: konyuhovvy@ex.istu.edu

Abstract. The article reveals the characteristics and principles of "green" technologies from an economic point of view and discusses the key points of implementation and operation of "green" technologies. The authors have analyzed the development of "green" innovations in different countries and the level of their economic development. They have considered a number of countries using innovative technologies and the process of implementing such technologies in practice in the case of water-purifying product "BINGSTI".

1. Introduction

In the modern world, the main hopes for solving urgent environmental problems are rested in tech-related scientific breakthroughs. Today, developed countries are reorienting their development towards the implementation of an environmentally-oriented growth strategy, one of the main and priority components of which is "green" technologies. Emerging economies are increasingly involved in the process of introducing that kind of technologies. This process indicates a fundamental transformation of the global economy.

Developed countries are trying to accelerate the development of "green" technology. According to the classification of Organization for Economic Cooperation and Development (OECD), "green" technologies comprise the following areas:

- general environmental management (waste management, water and air pollution abatement, land restoration, etc.);
- producing energy from renewable sources (solar energy, biofuels, etc.), mitigating the effects of climate change, reducing harmful emissions into the atmosphere, improving fuel efficiency and energy efficiency in buildings and lighting devices.

"Green" technology involves almost all sectors of the economy: energy, industry, transport, construction, agriculture, etc. They are currently being introduced into the activities of companies, including (in addition to production) consumption, management and methods of organizing production [1-5]. Among the main areas of development of "green" technology, energy is the key one. The main directions of its "environmentalization" are improving energy efficiency and the development of new energy sources, primarily renewable ones [6-10].

Analyzing the processes of development and implementation of "green technologies", we can say that they are aimed at:

- sustainable functioning and development that meets all modern needs of society;
- reproduction of products that can be fully restored or reused;
- a significant reduction in waste and in environmental pollution by changing the production system and consumption patterns;

- innovations that develop alternative solutions to technologies that had been previously associated with causing damage and harm to the environment;
- the formation of economic development centers around fundamentally new technologies and products that create fundamentally new jobs and raise the country's economy in general.

A serious issue is the economic efficiency of the development and application of "green" technologies, which depends on the specific field of activity. At the initial stage, the development of "green" technologies can be very costly, but later, as studies show, "greening" can provide not only an increase in natural capital, but also a higher level of GDP of the country which uses the technology [11-13].

This largely depends on the government's policy aimed at smoothing the differences in prices for technologies that pollute the environment and "green" technologies. Thus, the high price of "green" electric cars still is one of the main constraints for the development of this sector of innovative technologies.

Experts at various levels on the issue of "green" technologies are virtually united in their opinion that the next technological paradigm of economic development will be based on "green" innovations. Technologies for improving the energy efficiency of all sectors of the economy, as well as technologies for the production of materials and closed-cycle structures with a high degree of recycling will be greatly developed. That is, "green" technologies are considered as one of the key factors in the functioning of the future technological paradigm [14-18]. That is why, after Germany, the United States actively joined in the race for first place in the market for green technologies, proposing the development of alternative energy as the main way to overcome the economic crisis, and China is the most dynamically developing country among the major economies.

"Green" technologies will allow economic activities in which the production, distribution and consumption of goods from an economic point of view does not result in negative environmental consequences and at the same time improve human well-being. The introduction of "green" technology is considered cost-effective. For example, "green" technologies are closely connected with the global trend of transition to a new type of economy - a resource efficient, safe for the planet's ecosystem. While this field of knowledge is being developed and implemented, such situations occur as the emergence of additional work positions, improving the life quality and reducing risks in the economy [19-23].

2. Methods and materials

The work is carried out using a set of modern methods of theoretical and experimental research: the use of methods of mathematical statistics and computational software package Microsoft Office; modeling of economic processes; comparative tests of innovative energy structures; laboratory experiments on the selection of optimal modes.

3. Results and discussion

The use of green technologies begins with the construction of passive eco-houses, ending with the production of cars on clean fuel. In the global aspect, there is an active search for solutions to combat global climate change and environmental pollution. Numerous UN programs in the field of the environment intend to introduce green technologies, to gradually eliminate hazardous industries and to replace sources of anthropogenic greenhouse gas emissions, which also has a positive impact on the economy.

The concept of "green" technology is paid the most attention in the business environment. Financial foundations, venture capitalists, governments of advanced countries, businessmen and consumers are already building systems of "green" technologies, in which investments in energy-efficient technologies and natural infrastructure yield favourable results.

It is worth paying attention to the intensity with which *Denmark* invests in growth of green technologies. Since 1980, GDP grew by 78 % with only a minimal increase in energy consumption. Even in the midst of economic boom *Poland* have managed to reduce emissions by a third in the last 17 years. Such savings turn into profit for entrepreneurs. Today, European companies from the green technology sector enjoy significant advantages as pioneers: they account for a third of the green technology market that is thriving in the world.

An example of stimulating the economy with green technologies is *South Korea*, the host country of the forum, which became the only state that chose green growth as a national strategy. According to of the Organization for Economic Cooperation and Development, in this area South Korean investments amounted to 9.3 billion euros. They went to the development of "green" types of transport, alternative sources of fresh water, industrial waste processing technologies. Its green growth strategy focuses on three elements: industry, energy and investment. The strategy is aimed at preserving the range of productive economic activities with the least expenditure of energy resources, as well as minimizing the pressure on the environment of all types of energy and resources used and taking measures to transform investments into environmental protection activities and the driving force of economic growth.

A successful example of the practical application of "green" technologies is the development of Alexander Prigodin, a scientist from Novocherkassk. So, wastewater and its sediments, as a product of human life, contain a large number of pathogens of parasitic infections. The concentration of parasites in sediments has a higher degree than in wastewater, which makes them even more dangerous to the environment and all of humanity. The innovation proposed by the company is a modern disinvasive formulation "BINGSTI" that can solve this problem. The drug is intended for the purification of waste water and their sediments. It naturally kills infections without having an effect on human health.

From an economic point of view, BINGSTI it is the advantageous solution, since one liter of liquid is sufficient for cleaning 10 of thousand m^3 of effluents. Instead of cubic meters or even tons of other means, a consumer can use only a few liters of BINGSTI.

It is also important that the drug does not adversely affect the environment and humans. BINGSTI is easily integrated into any existing treatment technology. The introduction of the drug into the treatment technology allows for guaranteed purification of the entire volume of wastewater and sediments with an efficiency of 99 %.

The formulation includes such components as dried potato stalks up to 20 cm high, disinfectant “SSD-200”, drinking water, sodium benzoate (E211) and sorbic acid (E200) according to the current technical documentation.

"BINGSTI" is most advanced in use. In its introduction to wastewater treatment technology, the costs are insignificant. Also, this development can be used as the means of the additional purification of the unpurified wastewater while applying other methods of processing. The ambient temperature does not affect the effectiveness of the drug, so it can be used for disinvasion of environmental objects in any climate zone. The introduction of the environmental technology with the use of innovative drug "BINGSTI" allows you to:

- remove highly toxic chemical reagents from the technology or significantly reduce their dosages by replacing them with an environmentally friendly biological product;
- at the same time, process wastewater and sediment;
- significantly reduce the level of energy consumption for wastewater treatment;
- lower purification costs by 6-12 times;
- remove restrictions in the technological modes of disinvasion;
- maintain a high level of disinvasion efficiency in extreme technological modes;
- start disinvasion of 100 % of wastewater with repeated cost reduction and reduction in the prime cost of disinvasion throughout the wastewater treatment cycle.

4. Conclusion

Thus, the development and introduction of a new environmentally friendly product is hygienically relevant. It makes it possible, while respecting the environment, to more effectively carry out disinfection from pathogens of parasitic diseases of various environmental objects and reduce the risk of infecting the population, as well as expanding the possibilities of using wastewater sludge as an organic fertilizer.

Thus, today developed and developing countries are reorienting their policies towards "green" growth, stimulating the development and introduction of environmental technologies, aiming educational programs towards the formation of environmental awareness, and conducting extensive information campaigns. All these measures increase not only the ecological state of the countries, but also raise their economy, helping to overcome the crisis.

Despite the complexity of introducing innovative technologies, the future of the planet remains with them. The priority was chosen to protect nature, not destroy it, which increases the strength not only of the economy, but also of the state itself. Globalization is forcing many countries to reconsider the existing line in economics and politics, increasingly paying attention to "green" improvements and the introduction of modern innovations.

Acknowledgments

This paper has been prepared using the equipment kindly provided by the Joint Instrumentation Centre "Baikal Center of Nanotechnologies".

References

[1] Shamarova N A, Krupenev D S and Konyukhov V Yu 2016 *Improving the efficiency of energy production and use in Siberia* pp 423-424
[2] Kargapoltsev S K, Khomenko A P, Gozbenko V E, Minaev N V and Karlina A I 2017 *International Journal of Applied Engineering Research* **12(22)** 12362-12368
[3] Gozbenko V E, Khomenko A P, Kargapoltsev S K, Minaev N V and Karlina A I 2017 *International Journal of Applied Engineering Research* **12(22)** 12369-12372
[4] Nekrasov I V, Sheshukov O Yu, Sivtsov A V, Tsymbalist M M and Metelkin A A 2016 *Steel in Translation* **46(6)** 435-442
[5] Sivtsov A V, Sheshukov O Y, Tsymbalist M M, Nekrasov I V and Egiazar'yan D K 2015 *Metallurgist* **59(5-6)** 380-385
[6] Gozbenko V E, Kargapoltsev S K, Kornilov D N, Minaev N V and Karlina A I 2016 *International Journal of Applied Engineering Research* **11(20)** 10367-10373
[7] Khomenko A P, Gozbenko V E, Kargapoltsev S K, Minaev N V and Karlina A I 2017 *International Journal of Applied Engineering Research* **12(23)** 13773-13778
[8] Lombardi P, Sokolnikova T, Suslov K V and Styczynski Z 2012 *IEEE PES Innovative Smart Grid Technologies Conference Europe* 6465706
[9] Prigodin A V, Malysheva N S, Samofalova N A and Suslov N V 2013 *Technologies and prospects in the disinvasion of drains of livestock breeding complexes (Theory and practice of parasitic animal diseases)* pp 21-22
[10] *The Russian Council on the International Affairs,* Available at: http://russiancouncil.ru/inner/?id_4=508#top-content
[11] Balanovsky A E, Kondratev V V, Shtayger M G, Nebogin S A and Karlina A I 2018 *IOP Conference Series: Materials Science and Engineering* **411** 012014
[12] Balanovsky A E, Shtaiger M G, Grechneva M V, Kondratev V V and Karlina A I 2018 *IOP Conference Series: Materials Science and Engineering* **411** 012012
[13] Balanovsky A E, Shtayger M G, Kondratev V V, Huy Vu V and Karlina A I 2018 *IOP Conference Series: Materials Science and Engineering* **411** 012013
[14] Bulatov Y N, Kryukov A V and Suslov K V 2017 *Far East Journal of Electronics and Communications* **17(5)** 1197-121
[15] Sivtsov A V, Sheshukov O Y, Tsymbalist M M, Nekrasov I V, Makhnutin A V, Egiazar'yan D K and Orlov P P 2018 *Russian Metallurgy (Metally)* **12** 1108-1113
[16] Nekrasov I V, Sheshukov O Y, Metelkin A A, Sivtsov A V, Tsymbalist M M and Egiazar'yan D K 2015 *Metallurgist* **59(3-4)** 300-304
[17] Ivanchik N N, Balanovsky A E, Shtayger M G, Sysoev I A and Karlina A I 2018 *IOP Conference Series: Materials Science and Engineering* **411** 012035

[18] Balanovsky A E, Shtaiger M G, Kondratev V V, Karlina A I and Govorkov A S 2018 *Journal of physics: conference series* 012006

[19] Sivtsov A V, Tsimbalist M M, Fadeev V I, Ostrovskii Y I, Veselovskii I A and Afanas'ev V I 2012 *Russian Metallurgy (Metally)* **2012(12)** 1036-1040

[20] Gozbenko V E, Kargapoltsev S K, Karlina Yu I, Karlina A I and Artyunin A I 2018 *Advances and Applications in Discrete Mathematics* **19(3)** 289-297

[21] Gozbenko V E, Korchevin N A, Kargapol'tsev S K, Karlina A I and Karlina Yu I 2019 *Advances in Engineering Research* **182** 120-125

[22] Gozbenko V E, Korchevin N A, Kargapol'tsev S K, Karlina A I and Karlina Yu I 2019 *Advances in Engineering Research* **182** 126-131

[23] Suslov K V, Stepanov V S and Solonina N N 2013 *IEEE International Symposium on Electromagnetic Compatibility* 841-845

IOP Conf. Series: Earth and Environmental Science **350** (2019) 012037 doi:10.1088/1755-1315/350/1/012037

Assessing the ecological state of agricultural irrigated fields of the Orenburg gas processing complex with dumping sewage water for crop irrigation

I P Kruzhilin[1], T A Gamm[2], A A Mushinskiy[3] and S D Fomin[1,4]

[1]Federal State Budget Scientific Institution of the All-Russian Research Institute of Irrigated Agriculture, 9, Timiryazeva str., Volgograd, 400002, Russia
[2]Orenburg State University, 13, Victory ave., Orenburg, 460018, Russia
[3]Federal Scientific Centre of Biological systems and Agrotechnologies of Russian Academy of Sciences, 29, 9 Yanvarya str., 460000, Orenburg, Russia
[4]Volgograd State Agrarian University, 26, University ave.,Volgograd, 400002, Russia

E-mail: fsd_58@mail.ru hammtam@mail.ru

Abstract. The article is devoted to using wastewater of the Orenburg Gas Chemical Complex in the amount of up to 7.0 million m^3 for irrigating agricultural crops with the purpose of additional cleaning of wastewater and increasing the yield of agricultural crops. Analyzing the obtained data, it is shown that in the arid climate with rainfall deficiency, the use of wastewater for irrigation of forage crops is the best variant in providing the necessary measures for preserving the ecological status of the agricultural sewage farms (ASF). As a result of own research over years 1982 – 2015, the influence on the ecological state of ASF soils was detected in the form of deterioration of the water-physical properties and increased content of easily soluble salts in the soil; with that, the content of gross and mobile forms of heavy metals is significantly lower than MPC, but slightly above the background values, which does not affect the quality of sorghum harvest. Pollutants accumulate in the top layer of soil; therefore, the impact on surface water bodies has not been established. The authors suppose that maintaining the ecological state of ASF requires some measures like the introduction of calcium-containing substances and optimization of the irrigation conditions. The practical importance of scientific research is that based on the obtained results, a project of ASF soil rehabilitation has been developed and put into production.

1. Introduction

Extraction and processing of fossil fuels have a significant effect on flora and fauna, agriculture, and landscape aesthetics [1, 2].

Additional treatment of large volumes of wastewater with low concentrations of contaminants to the standards that would allow discharging into water bodies is a problem. The problem of sewage and sediment disposal is also being solved by foreign scientists [3–7]. Therefore, the studies of gas-processing industry sewage disposal are relevant. The relevance of the research studies lies in the fact that extraction and cleaning of associated gas from sulfur-containing components at the Orenburg Gas Chemical Complex (OGCC) requires significant consumption of water, which simultaneously results in an impact on the components of the environment [8–10].

Discharging sewage in the natural-technical system of OGCC is associated with two problems: improving the quality of wastewater based on tertiary treatment in the natural subsystem of the agricultural sewage farm (ASF), and creating safe conditions for their return into the environment and increasing the yield of forage crops upon irrigation.

OGCC ASF was designed as an ecological measure. They have been functioning since 1976, and are an integral part, the last step of the existing post-treatment of wastewater from three enterprises: the Orenburg Gas Processing Plant, the Orenburg Helium Plant, and the Kargalinskaya CHP [9, 10]. After biological treatment, up to 7.0 million m^3 of wastewater are sent to the setting pond (SP).

In the technical subsystem, wastewater is discharged by Fregat sprinkling machines. The existing closed type irrigation network is implemented according to the dead-end system and consists of the main and the distribution pipelines.

The scientific novelty of the research is in the fact that it was the first long-term study with the aim of assessing the effect on the environment during the disposal of wastewater from gas processing enterprises.

The object of the research was using sewage from gas processing enterprises for irrigating fodder crops.

The subject of the research was the impact of wastewater irrigation on the environment.

2. Materials and methods

The research methods were planned in accordance with RD 52.18.595-96 "Federal List of Measurement Procedures allowed for environmental pollution monitoring", MU 2.1.7.730-99 "Hygienic assessment of soil quality in residential areas", SP 11-102-97 "Engineering-Ecological Surveys for Construction"; the ecological state of the soils was assessed by the total chemical pollution index. The ecological-geochemical state of the soils was assessed based on the background characteristics and the maximum (approximate) permissible concentrations of harmful substances in the soil. Using the methods of mathematical statistics, the significance of controlled parameters was promptly assessed in the "Statistics" software. Monitoring of the soils chemical contamination was developed according to GOST 27593-88. The chemical composition of the soils, the surface runoff and vegetation was analyzed in certified and accredited laboratories.

The effect in the natural-technical system was studied in laboratory and field experiments. In a field experiment, the formation of the technical parameters of the subsystem was studied with the wetted soil layer depth of 0.5 m. For studying heavy metals behavior in the soil, 30 points were allocated on the territory of the ASF. The total of 120 soil samples were taken annually. The water-physical properties of the soil were determined according to [11].

Samples of water were taken in accordance with GOST R 51592 – 2000 "Water. General requirements to sampling"; chemical analysis of water samples were performed in accordance with RD 52.24.309-92 “Protection of nature. Hydrosphere. Organization and observation of the conditions of surface runoff pollution with the use of the Roshydromet network. St.-Petersburg. 1992". The accuracy of measurement was according to GOST 27384-82 "Water. Rates of measurement error of characteristics of composition and properties" and GOST 27384-87. Requirements to instruments and devices for taking samples, primary processing and storage of samples were according to GOST 17.1.5.04-81.

The yield was accounted for using the quadrant sampling method by mowings within the boundaries of the affected area and in the reference plot. The number of samples per plot was three. The results were processed by the analysis-of-variance method [12].

3. Results

In the natural subsystem of ASF, the climate of the Orenburg region is characterized by hot summers and frosty winters, and low year-round rainfall [13].

Before wastewater discharge, the soils of ASF were presented by southern black soils with the humus content of about 4 % and the thickness of the humus layer of 35 – 55 cm with the predominance of physical clay by more than 50 %. The natural conditions of the OGCC ASF location are shown in Table 1.

Table 1. ASF natural conditions

Parameters	Orenburg district, the zone affected by OGP
Geological structure	the area belongs to the Paleozoic group of the Permian system; the composition of the parent rocks includes sandstones, limestones, dolomites, and clays
Soil texture	fine- and medium-loamy
Soil type	ordinary black soil
Zonal type of vegetation	fescue-feather grass steppe
The average depth of soil freezing (end of February), cm	100 – 120
Wetting area	Arid
Hydrothermal coefficient (HTC)	0.6
The sum of temperatures above +10 °C	2,700
Duration of the frost-free period, days	136
Heat provision area	warm
Snow depth, cm	30 – 40
Average annual rainfall, mm	350 – 400
Air temperature, °C:	
– July	+21
– January	-15
The absolute maximum of temperatures, °C	+42
The absolute minimum of temperatures, °C	-43
Snow cover melting	April 4
The absolute value of potential water resources, l/s×km^2	0.5 – 1
Ecological state of the natural environment	unfavorable with three factors, anthropogenically disturbed environment
Index of environmental factors	12
Elements concentration in the rocks	8 MPC
The thickness of the aeration zone, mm	4,000 – 36,000

The natural environment in the area of the gas processing plant is experiencing significant pressure from technical systems: the environment is anthropogenically affected, the index of ecological factors is 12, and elements concentration in the rocks reaches 8 MPC. Irrigation has a significant effect on the soils of ASF.

For studying the dynamics of changes in the water-physical properties of soil over the many-years period, the authors performed a comparative analysis of their own average data on the ASF and the data for the year 1988 (Table 2).

These data show the pattern of soil compaction on the entire ASF massif, compared to the values of this parameter in 1988. Soil humidity at full and minimum water capacity during ASF operation reduces. The soils in ASF are not only greatly over-compacted, but the water capacity decreases steadily throughout the entire territory of the massif. Soil water capacity in the spring is higher, which clearly indicates their seasonal over-wetting.

Accumulation of sulfur-containing compounds may be considered as the main negative process in the soils of ASF, despite the fact that the overall mineralization of wastewater does not exceed 1 g/l. The concentration of sulfate ions in the top soil layer in the autumn is notably higher than in the spring, their coefficient of local concentrations is in the range between 1.89 and 2.50. The content of hydrosulphide ion across the entire massif exceeds the background value by 5 – 10 %.

Table 2. The dynamics of changes in the water-physical properties of ASF soil (average for the wetted soil layer)

Indicator / Date and location of taking soil samples		Humidity, %	Bulk density, g/cm³	Humidity at the lowest water holding capacity, %	Humidity at full water holding capacity, %
Background		21.50	1.17	35.29	54.80
Average values for ASF	autumn 1988	-	1.13	24.80	48.72
	autumn 2014	26.20	1.22	21.53	46.51
	spring 1988	-	1.18	36.33	54.08
	spring 2015	26.90	1.36	32.78	51.85

Comparative analysis of changes in the soil chemical composition in the years 1982 through 2015 is shown in Table 3. In 1982, the concentration of chlorides and sulfates in the soil exceeded the background values only 1.2 times, in 2015 — 2.18 times.

Table 3. The content of chemicals in the arable soil layer, the average for the ASF, 1982 through 2014, mg/kg

Period of time	pH	Cl^-	HCO_3^-	Ca^{2+}	Mg^{2+}	HS^-	NH^{4-}	SO_4^{2-}	Zn^{2+}
Background	7.5	69.90	742.90	64.50	32.70	38.50	35.60	41.00	0.046
1982 – 1984	8.2	84.007	279.76	56.14	26.99	-	-	51.307	-
2014 (autumn)	7.3	95.35	441.38	104.02	25.49	39.55	57.08	89.24	0.106

The content of mobile forms of heavy metals in ASF soils is much less than MPC, Table 4. Compared to the background, the content of mobile forms of manganese, chromium, nickel, lead, and cadmium is elevated.

The highest concentration of heavy metals and the indicator of their total pollution are observed in the vicinity of hydrant No. 7 where the amount of discharged water is greater than in other analyzed areas. Here, the content of mobile forms of manganese is three times higher than the background value, of lead — 2.5 times higher, and of cadmium — 2.2 times higher than the background value.

Table 4. The content of mobile forms of heavy metals in ASF soil (0.1 – 0.25 m horizon depth), mg/kg, October 2014

Location of taking soil samples	The content of mobile forms of heavy metals					Chemical Pollution Indicator
	Mn	Cr	Ni	Pb	Cd	
Background, mg/kg	4.40	0.58	0.61	0.44	0.05	
Hydrant No. 7	11.90	0.51	0.80	1.14	0.11	9.68
Hydrant No. 11	5.00	0.37	0.67	0.43	0.07	5.26
Hydrant No. 15	9.30	0.49	0.60	0.84	0.10	7.84
Average for ASF	8.73	0.46	0.69	0.80	0.09	7.59
MPC	140.0	6.00	4.00	6.00	1.00	

The Chernaya River nearby is the area of groundwater discharge. Therefore, the chemical composition of the waters in the river upstream and downstream of the ASF was studied, Table 5.

The excess of MPC upstream of the ASF was noted for copper — 8.5 times, for iron — 7.1 times, for chlorides — 1.44 times, for sulfates — 2.7 times, for COD — 3.3 times, for BOD_5 — 2.14 times, and for suspended solids — two times. In the water downstream of the ASF, the concentration of these substances exceeds the MPC, but the content is much lower than at the cross-section close to the tributary.

A clear decrease is visible in all analyzed indicators in the river below the ASF, compared to the indicators in the midstream. This is explained by the fact that in the aeration zone, a complete final treatment of sewage occurs.

Table 5. Chemical composition of the waters in the Chernaya River upstream and downstream of the ASF

Date and location of taking samples / Indicator	Unit of meas.	MPC (fishery)	Source		05.2015 Upstream of the ASF (close to the tributary)		Downstream of the ASF		MPC (potable)
			Concentration	MPC multiplicity	Concentration	MPC multiplicity	Concentration	MPC multiplicity	
Cl^-	mg/l	300.00	353.90	1.18	432.20	1.44	197.30	0.66	350.00
HCO_3^-	mg/l	not rated	826.60	-	732.00	-	594.70	-	1,000.00
SO_4^{2-}	mg/l	100.00	230.40	2.30	268.80	2.69	153.60	1.54	500.00
HS^-	mg/l	absent	0.16	-	0.15	-	0.16	-	3.00
Ca^{2+}	mg/l	180.00	119.60	0.66	102.20	0.57	100.80	0.56	200.00
Mg^{2+}	mg/l	40.00	60.00	1.50	53.60	1.34	85.00	2.13	100.00
Total hardness	mg-eq/l	not rated	10.98	-	9.58	-	8.60	-	7.00
Zn^{2+}	mg/l	0.01	0.008	0.80	0.005	0.50	0.004	0.40	1.00
Fe	mg/l	0.10	0.19	1.90	0.71	7.10	0.34	3.40	0.30
Cu	mg/l	0.01	0.056	5.60	0.085	8.50	0.061	6.10	1.00
Ammonium nitrogen	mg N/l	0.40	not found	-	0.20	0.50	not found	-	2.00
Suspended matter	mg/l	10.00	23.50	2.35	20.00	2.00	5.26	0.53	40.00
COD	mg O_2/l	15.00	33.30	2.22	50.00	3.33	25.00	1.67	30.00
BOD_5	mg O_2/l	2.00	-	-	4.28	2.14	4.04	2.02	4.00
Dissolved O_2	mg/l	not lower than 4.00	-	-	6.96	-	7.17	-	not lower than 4.00
Oil products	mg/l	0.05	5.00	100.0	not found	-	not found	-	0.10
pH	LgcH	6.5 – 8.5	8.0	-	7.6	-	8.1	-	6.5 – 8.5
Dry residue	mg/l	not rated	820.00	-	1140.00	-	600.00	-	1000.00
Settling substances	cm^3/l	not rated	0.20	-	0.20	-	0.20	-	not rated

Being an indicator of the ecological status of the agroecosystem, the quality of the obtained products is of great importance. Therefore, the authors determined the mineral content and crude protein in sorghum cultivated in the ASF, and compared the results to similar indicators in the non-irrigated reference plot (Agricultural Farm Kolkhoz Imeni Lenina in the Orenburg district). The results are shown in Table 6.

The content of sulfur, sodium, calcium, nitrogen, and protein in the ASF massif was lower than in the reference plot; with that, irrigation provides insufficient yield increase. Lower sulfur content in the sorghum cultivated in the ASF, compared to the plants cultivated in the reference plot, showed that the soil wastewater treatment in the studied massif had been efficient.

Table 6. Chemical composition of sorghum green mass

Indicators	OGCC ASF	Agricultural Farm Kolkhoz imeni Lenina (reference)
Yield, t/ha	156.00	148.00
Content, %:		
– nitrogen	1.68	3.00
– crude protein	10.50	18.70
– sodium	0.05	0.07
– calcium	0.54	0.62
– sulphur	0.49	0.50
– lead	0.11	0.12
– cadmium	0.013	0.014
– copper	35.9	40.0
– zinc	13.5	15.0

4. Discussion

As a result of the authors' own studies, it has been found that over the years of wastewater irrigation, soils were significantly dispersed (the percentage of annealed residue increased, the fraction of silt increased, bulk density increased, and porosity decreased). Relative predominance of sodium over calcium is observed; with that, the content of mineral forms of carbonates has been steadily declining, being contributed to by the process of sodium chloride content reduction [1].

An increased concentration of soluble salts was noted in the soil as a result of wastewater discharge, compared to the indicators for the period between 1982 and 1984. The maximum value of this indicator was observed in the autumn of 2014. The content of gross and mobile forms of heavy metals is significantly below the MPC, but slightly above the background values.

The content of nitrogen (56 %) and crude protein (56.15 %) is below the background values, which confirms the scarcity of nutrients in the wastewater and, consequently, in the soil, and almost complete absence of measures aimed at increasing ASF soil fertility; therefore, mineral and organic fertilizers are to be introduced into the ASF soil. The sorghum yield increase in the ASF, compared to the non-irrigated reference plot, is explained by three mowings as a result of irrigation, vs. single irrigation at the agricultural farm Kolkhoz Imeni Lenina.

Wastewater from OGCC does not contain nutrients in the amounts sufficient for fertilizing plants, thus, the discharge conditions are determined by water consumption by the cultivated crops. Due to the fact that in the recent years, only sorghum has been cultivated for green fodder, it is necessary to present new requirements to the quality of wastewater, and to the design of ASF and watering conditions. The level of technogeneous load on the existing natural-technical system with the seasonal discharge of wastewater is determined from the prerequisites for creating optimal conditions for growth and development of the plants in the agrocoenosis [14].

5. Conclusion

OGCC wastewater has unstable chemical composition, is represented by large volumes, has low concentrations of pollutants; it is, therefore, expedient to discharge it to ASF for irrigating fodder crops in a natural-technical system with the aim of returning into the environment after additional purification in the natural subsystem and obtaining high yields. The wastewater discharge conditions are determined by water consumption by the crops. As a result of years of research studies, it has been found that wastewater irrigation deteriorates the water-physical properties of soils; therefore, additional agrotechnical measures are needed for their stabilization. Easily soluble salts, including sulfates, accumulate in the soil during the period of irrigation. It is, therefore, necessary to perform salt flushing from the soil in the autumn. This prevents the accumulation of heavy metals in the soil. The effect of wastewater on the groundwater and surface water has not been established. In order to increase sorghum yields in the ASF, the introduction of organic and mineral fertilizers is required.

6. Acknowledgments

The studies were performed in accordance with the project: "Development of theoretical foundations and practical methods of increasing the efficiency of crop production using nanotechnology solutions" in 2018-2020 Federal Scientific Centre of Biological Systems and Agrotechnologies of Russian Academy of Sciences (No. 18-8-9-18).

References

[1] Gamm T A and Klunov A G 2015 The effect of the wastewater from the Orenburg Gas Chemical Complex on the chemical composition of the soil cover in ASF *Bulletin of the Orenburg Agricultural University* **6(56)** 124–126
[2] Stepanova M I 1980 The effect of wastewater from the Orenburg Gas Processing Plant on the ameliorative condition of land *Collection of scientific works* 62–69
[3] Vogel C 1977 *Klarschlammverwertung in der Landwirtschaft. ErgebnisseeinesmehriahrigenFeldversuches auf einemBasaltwiherungsboden* (Landwirt: Forsch) p 30
[4] Xanthaclis D 1978 *Utilisation des Baiesresiduarlsengrandes Cultures* (AnuSembloux) p 245
[5] Baver H D 1974 Landtreatment of wastewater *Adv. Agronvor.* 194
[6] Pescod M B 1992 *Wastewater treatment and use in agriculture* (Rome: Food and agriculture organization of the united nations) p 169
[7] Wiesmann U 2007 *Fundamentals of Biological Wastewater Treatment* (Weinheim, WILEY-VCH Verlag GmbH & Co. KGaA) p 364
[8] Kaliev A Z and Damrin A G 2016 Assessment of the effect of emissions from a gas chemical complex on the ecological state of irrigated fields *Bulletin of the Orenburg State Agricultural University* **4(60)** 176–179
[9] Kaliev A Z 1981 *The effect of irrigation with waste waters of the Orenburg Gas Complex on the ameliorative condition of ASF* (Saratov) p 22
[10] 2009 *Assessment of the level of anthropogenic impact on the environmental components in the area of facilities of LLC Gazprom Dobycha Orenburg* **BOOK 1** (Orenburg) p 121
[11] Kachinsky N A 1970 *Soil physics* Part 2 *Water-physical regimes and soil properties* (Moscow: Higher School) p 359
[12] Dospekhov B A 2012 *Field experiment methodology (with the basics of statistical processing of results research)* (Moscow) p 352
[13] Chibilev A A 1999 *Geographical Atlas of the Orenburg region* (Moscow: DIK) p 96
[14] Gamm T A, Shabanov S V, Gamm A A and Sagitov R F 2016 Plants growth and development upon recultivation of soils polluted with oil using waste water *Bulletin of the Orenburg Agricultural University* **1(57)** 127–129

Phytomonitoring of woody plants in the urban agglomeration

L G Seraya[1], G E Larina[1,2], O G Griboedova[1], A V Petrov[1] and F F Zhukov[1]

[1]All-Russian Research Institute of Phytopathology, 5, Institute st., Bolshie Vyazemy, Moscow Region, 143050, Russia

[2]State University of Land Management, 15, Kazakov st., Moscow, 105064, Russia

E-mail: lgseraya@gmail.com, galara7@yandex.ru

Abstract. Phytomonitoring including mycological studies with information about plant pathogens, etc., is required to develop strategic directions for maintaining green spaces in urban agglomeration. Mycological analysis of the aerial part and roots of urban tree plants can be used to determine the composition and structures of microbial-plant associations in visually healthy and affected plants. The relevance of these works is due to rapid development of diseases and pests and their active distribution in close proximity to city residents. In coniferous and deciduous plants with symptomatic lesions (*Picea abies* (L.) H. Karst., *Pinus* spp., *Quercus robur* L., Tilia spp.), the species diversity of micromycetes in the phylloplane and rhizosphere decreased by 21–38% compared to visually healthy plants. It was suggested using the following micromycetes as marker organisms for assessing the soil fungistasis of artificial bicenoses and the state of coniferous trees: D*icoccum sp., Fusarium moniliforme J. Sheld., Fusarium culmorum.* Perennial urban plants require improvement of the above-ground part using foliar fertilizers and phytostimulants, bioprotection tools, humates, mycorrhizal preparations, destructors of organic compounds of roots and adjacent lawns. The prospects of phytomonitoring of old-growing trees were emphasized. Reconstruction and enhancement of their resistance to aggressive environments require complex works.

1. Introduction

Moscow agglomeration includes satellite cities closely located to the Moscow Ring Road (within a radius of 15 km) as well as cities having a "specialization" scientific cluster, industrial parks, etc. Geographical boundaries of Moscow agglomeration cover the vegetation of public gardens, boulevards, yard and street plantings, and lawns. They exist under anthropogenic pressure. The influence of aggressive conditions of the urbanized environment is leveled by agrotechnical measures: soil loosening, fertilization, moderate watering, preventing mechanical damage to roots and branches.

In the green areas of Moscow and satellite cities, active development of infectious diseases of woody plants occurs regularly. They cannot be treated by agrotechnical methods. These diseases are as follows: infectious drying of linden branches (thyrostromosis, *Thyrostroma compactum* (Sacc.) Hoehn.), Dutch elm disease (graphism, *Ophiostoma ulmi* (Buisman) Nannf.), etc.. Under the influence of various harmful factors, green plantings weaken and suffer from pathogens and pests. During the entomo-phytopathological examination of young plantings, 18% of plants had water shoots; among infectious diseases, 32% are due to branch drying (thyrostromosis, etc.), 77% are damaged by leaf-eating insects, 20% - by sucking insects (including ticks) [1]. Entomo-phytopathological monitoring of planting material, soil and substrates is mandatory for newly created landscape gardening areas and street plantings within the urban agglomeration. Phytosanitary monitoring involves the use of modern methods to identify and account for the number of pests, structure of artificial biocenosis and abundance of its individual components taking into account weather conditions and cultivation methods [2].

Collection and analysis of multi-year information are relevant for growing and keeping trees. Therefore, phytosanitary monitoring of plants, including old 150-year old trees, should be carried out at all stages of their development and growth. Right decisions and balanced practical actions begin with systematic, scientifically grounded on-site surveys (visual diagnostics) in combination with instrumental techniques and scientific analysis.

To carry out phytosanitary monitoring, an integrated approach accompanied by data grouping is required:

- *meteorological* - information about the climate, including data from regions. This information has several levels: average long-term indicators, retrospective assessment of main indicators of previous years and the current year. The following weather conditions (climatic) have the greatest value in phytomonitoring: air temperature (minimum/maximum, °C), soil temperature (°C), precipitation (mm), relative air humidity (%) ;
- *agrotechnical* - data on the species composition of plantings, growth characteristics, location, phenology of the plant and its reaction to external factors. The following agrotechnical indicators have the greatest value: soil-agrochemical conditions (mechanical composition, particle size distribution, moisture and air permeability, organic matter content, pH, acidity, alkalinity), elements of mineral nutrition; ecological and biological features of growth, plant development and longevity associated with vital processes (thermal regime, insolation, medium reaction, soil microorganisms, moisture and soil aeration, the ratio of macro and microelements in the substrate);
- *biocenotic* - data on plant pests and pathogens (biology, phenology), structure and species composition of phytophages, entomophages, entomopathogenic microorganisms, etc. This unit of phytosanitary monitoring is extremely important for developing common strategic directions: control, protection and planting.

The research purpose is to study prospects of phytonomonitoring of woody plants in the urban agglomeration, including mycological research, in order to make timely optimal management decisions.

2. Materials and methods

The basis of phytomonitoring data in the urban environment was made up of observations of the state of greenery (route surveys of 2014-2018) in garden and park systems in satellite cities of the megalopolis. The state of old plants was accounted by assessing categories of the state of trees used in forestry [3]. The studies assessed the spread and development of diseases, pests and non-infectious damage. Information was systematized and analyzed using MS Excell, MS Access and Internet resources.

For an instrumental analysis, at the beginning of the growing season, samples of soil and plants (branches, shoots) were taken. The general scheme of soil sampling from a depth of 0–10 cm for a mycological analysis included mixed samples from the root zone at a distance of 10–15 cm from the trunk and background samples taken above 50 cm from the trunk. Each sample was placed in an individual plastic bag. The samples were taken three times. The samples were analyzed by traditional methods used in microbiology: a wet chamber, sowing in the nutrient media. Isolated species were identified according to modern determinants of fungi [4, 5]. The terminology is used in accordance with *Index Fungorum* (http://www.indexfungorum.org).

3. The study of the structure of the modified lead-tin-base bronze

According to the results of phytomonitoring surveys, general signs (symptoms) of lesion were identified in a number of tree species in artificial biocenoses. For example, in conifers (Picea abies (L.) H. Karst., Pinus spp.), these are shortening needles and shoots (the current year increment), curvature and deformation of branches and trunks, staining of needles at the edges of shoots in rusty-brown color. The following symptoms were observed in deciduous species: pedunculate oak (Quercus robur L.) – deformed chlorotic leaves with small yellow spots on the edge of the central veib; cream or black shoots;

Linden species (Tilia spp.) - dark brown spots on leaves, bucket-like twisted leaves; shoots have brown round spots.

Mycological analysis of plants showed differences in the biodiversity of the microbial community: 34 filloplans and 21 microscopic fungi from the root zone were identified. In visually healthy plants, for example, *Quercus robur,* the following fungi were isolated in the aerial part: Alternaria alternata (Fr.) Keissl., Aspergillus glaucus (L.) Link, Aureobasidium pullulans (de Bary & Löwenthal) G. Arnaud, Chrysosporium merdarium (Ehrenb.) Jw Carmich., Penicillium chrysogenum Thom, Phyllosticta sp.. In the root zone, the following fungi were identified: Aspergillus niger Tiegh., Cunninghamella echinulata (Thaxt.) Thaxt. ex Blakeslee, Cylindrocladiella parva (P.J. Anderson) Boesew., Fusarium solani (Mart.) Sacc., Mucor sp., Trichoderma viride Pers. For deciduous trees with lesion symptoms (oak, linden), the following fungi were identified: A. alternata, A. niger, Ch. merdarium, Fusarium sporotrichioides Sherb., Fusarium culmorum (Wm.G. Sm.) Sacc., Trichoderma virens (J.H. Mill., Giddens & A.A. Foster) Arx [= Gliocladium virens J.H. Mill., Giddens & A.A. Foster.], Micelia sterilia.

For coniferous and deciduous plants with symptomatic lesions, species diversity of micromycetes in the phylloplan and rhizosphere decreased by 21–38%, compared with visually healthy plants. Representatives of 10 species were identified with a frequency of 6% and higher, with the maximum presence of the fungi Fusarium oxysporum Schltdl. and Fusarium solani. In the fungi observed in the root zone, the number of Fusarium and Clonostachys genera fungi was higher by 20% compared to Fusarium, Alternaria, Aspergillius fungi. In the root zone of all conifers (spruce, pine species), the following fungi were identified: Clonostachys rosea (Link) Schroers, Samuels, Seifert & W. Gams, Fusarium incarnatum (Desm.) Sacc., T. virens [= g. virens,] C. echinulata, Micelia sterilia, Mucor sp., Penicillium purpurogenum Flerov, T. viride.

The differences in the composition of fungi in the root zone were established. The following states were identified for the (visual damage symptoms): Norway spruce - Cylindrocladiella parva (P.J. Anderson) Boesew., Fusarium gibbosum Appel & Wollenw., Gliocephalotrichum simplex (J.A. Mey.) B.J. Wiley & E.G. Simmons, species of the genus Penicillium, Verticillium albo-atrum Reinke & Berthold; Scotch pine - Fusarium sambucinum Fuckel, Gliocladium sp.; for plants of the 3-4th categories (damage symptoms), the following states for identified: Norway spruce - Coniothyrium sp., Dicoccum sp., Fusarium moniliforme J. Sheld .; Scotch pine - Fusarium culmorum. For urban green spaces, it is important to preserve soil fungistasis, i.e. the action of the biological component of the soil in order to limit the mycelial growth of fungi and inhibit the germination of resting forms of pathogens. Therefore, it is useful to consider fungi communities that characterize the state of the 1-2nd and 3-4th categories as markers of the state of the tree and soil fungistasis of artificial bicenoses.

It is useful to supplement the data of phytomonitoring with information on the mycology of plant parts and soil. At the same time, natural processes are accompanied by succession of dominant representatives in the microbial community. For example, in the zone of the root system of woody plants, the succession of complex of micromycetes or destroyer organisms, including phytopathogens (Fusarium, Pythium, etc.) capable of invading roots of weak plants is initiated. Then they are replaced by saprotrophic fungi (Mucor, Penicillium, Aspergillus) which are able to absorb available substances – mono- and oligosaccharum, hemicellulose; then, they are replaced by cellulolytic micromycetes (*Trichoderma, Demaiium, Cladosporium, Alternaria, Ceratocystis, Cunninghamella*) and lignin-destroying fungi In the urban environment, agrotechnical measures (loosening, mulching, fertilizing) affect the development cycles of living organisms and infectious processes, formation of resistance to host plants.

Particular attention should be paid to the problem of old (150 years and older) for which the results of phytomonitoring must be accompanied by comprehensive works to reduce the overcrowding of the upper soil layers by loosening and pitting, removing weeds and lawn grass near the trunk; introduction of structural elements (sand, peat, clay, expanded clay, coarse material); mulching; removal of rotten root patches; seasonal use of potassium humate, fungicides, morphoregulators and biological preparations with active destructive organisms, etc.

It is necessary to say about the seasonality of "critical periods" of plant growth (the survival rate after transplantation, flowering, fruiting) during which the risk of disturbances in plant development, weakening vitality and loss of decorativeness is high. The following measures can be used: watering, use of foliar fertilizers, macro, micro and chelated fertilizers. Biological activity of the soil microflora can be suppressed (changed) by applying complex mineral fertilizers which worsens the soil structure, increases humus mineralization and leaches calcium, magnesium and mobile organic substances.

Therefore, it is important to observe calendar dates for maintenance works. Phytomonitoring (visual diagnostics of planting conditions) is important for plant protection, prediciton of the development of pests and assessment of the effectiveness of seasonal methods for protection and maintenance of green areas. In case of danger of development of diseases and pests, their active distribution in the human environment according to the official "List of pesticides and agrochemicals which can be used in the Russian Federation" [6], there is a limited range of pesticides with regard to environmental safety requirements which can be used in urban areas. Therefore, maintenance of plant immunity, creation of optimal conditions for their growth is crucial when designing, creating and keeping urban greenery.

4. Conclusion

The relevance of phytonomonitoring of woody plants in the urban agglomeration is associated with rapid formation of foci of diseases and pests and their active distribution in close proximity to humans which affects their health, performance, and physical endurance, etc. The development of pathogens is a predictor of events, reflects changes in abiotic factors and the influence of human activities. For perennial urban plantings, given the importance of work on their reconstruction and increasing resistance to aggressive environments, it is important to carry out seasonal works to improve aerial parts of plants using foliar top dressings with fertilizers and phytostimulants, treatment of roots with bio-protectors, humates, mycorrhizal preparations, destructors of organic compounds, etc.

References

[1] *Report on the state of green spaces in the city of Moscow in 2006* 2007 (Moscow: Department of nature management and environmental protection in Moscow)
[2] Tkachenko O B, Keldysh M A, Kashtanova O A, Mukhina L N, Seraya L G, Chervyakova O N, Traivas L Yu and Kozarzhevskaya E F 2018 *Protection of woody plants from pathogens and pests* (Moscow: Russian Academy of Sciences)
[3] Resolution of the Government of the Russian Federation dated May 20, 2017 no. 607 On the Rules of Sanitary Safety in Forests *Government of the Russian Federation website* Available at: http://government.ru/
[4] Blagoveshchenskaya E Yu 2015 *Phytopathogenic micromycetes* (Moscow: LENAND)
[5] Ainsworth G C, Kirk P M and Bisby G R 2001 Dictionary of the Fungi *CAB International*
[6] *List of pesticides and agrochemicals permitted for use in the territory of the Russian Federation* 2019 Online version of the "Directory..." Available at: https://www.agroXXII/ goshandbook

Conservation agriculture: new crop production technologies

G E Larina[1,2], L G Seraya[2], I O Ivanova[2], I N Kalembet[2] and L M Poddymkina[3]

[1]State University of Land Management, 15, Kazakov St., Moscow, 105064, Russia
[2]All-Russian Research Institute of Phytopathology, 5, Institute st., Bolshie Vyazemy, Moscow Region, 143050, Russia
[3]Russian State Agrarian University – Moscow Agricultural Academy n.a. K.A. Timiryazev, 49, Timiryazevskaya st., Moscow, 127550, Russia

E-mail: galara7@yandex.ru

Abstract. Among the arguments in favor of new crop production technologies are the need for improvement of regional ecology and biodiversity, and high-quality biological products. Methods for reducing crop losses caused by air and soil droughts in the central and southern regions of Russia were analyzed. Field observations (2010-2016) were carried out using a crop rotation scheme of experimental fields with a wide range of crop varieties: wheat, barley, corn, sunflower, peas, chickpeas, alfalfa, and flax. The purpose of the article is to assess regional characteristics of effective soil fertility using different resource-saving technologies. Biological productivity (regional, local) was used as an integral indicator of effectiveness of agricultural technologies and crop yields. High effective fertility was observed when using a zero agrotechnology in Volgograd region and minimum tillage method in Stavropol region. The need for the control of soil fungistasis as a basis of biological activity of the soil and long-term improvement of effective fertility of the arable land was emphasized.

1. Introduction

Under intensive agricultural production in the central and southern regions of Russia, traditional tillage using plowing equipment increased degradation of soil fertility, changed environmental conditions and ecosystem functions (biodiversity) of agrocenoses [1]. The potential of varieties and hybrids of field, fruit and medicinal crops, the range of plant protection products and agrochemicals are not able to compensate for destruction and drying of a fertile layer of the arable land [2]. Resource-saving technologies where the use of mechanical tools is minimal (plowing is carried out in three years or a zero tillage system with soil covering with mulch shredded crop residues is applied) are of great interest [3, 4]. It is necessary to take into account a natural (global) role of the soil as a basis of the ecosystem, including regional ecology and biodiversity. It is necessary to produce high-quality biological products and reduce losses. Among the advantages of using resource-saving agrotechnologies, one can mention improvement of physical properties of arable land which are important when growing field, fruit and medicinal plants in conditions of frequent air and soil droughts.

The purpose of the article is to assess regional characteristics of effective soil fertility using different resource-saving agricultural technologies for growing crops.

2. Materials and methods

The route surveys of experimental fields covered three regions with high risks of prolonged dry periods (Volgograd, Rostov and Stavropol). Field observations (2010-2016) were carried out in a crop rotation scheme of experimental fields with a wide range of economically significant crops: wheat, barley, maize, sunflower, peas, chickpeas, alfalfa, flax, sunflower, etc. Regional factors affecting soil formation processes and increasing the risk of losses of effective fertility during traditional tillage were identified.

Volgograd region has the following zones: a steppe, a dry steppe, a desert and a semi-desert. The most favorable conditions were formed in the steppe zone (significant soil productivity, high rainfall and favorable distribution for the vegetation of agricultural plants). The relief is hilly with elevations which increases the risks of wind and water erosion.

Rostov region has two zones: a steppe and a dry steppe. According to the moisture conditions, there are very arid (the amount of the annual rainfall is less than 400 mm) and arid (the amount of the annual rainfall is more than 400 mm). The relief is flat with gentle slopes of the southern and northern directions, but the risk of wind erosion is high.

Stavropol region belongs to the dry-steppe and semi-desert agro-climatic zone. According to the conditions of moisture supply, the territory is divided into extremely arid (the amount of the annual rainfall is less than 400 mm), arid, unstable and sufficient moisture zones. The soil cover is variegated represented by southern and ordinary carbonate chernozem, dark chestnut and chestnut soils. The terrain relief is mostly flat with gentle slopes.

In all regions, annual plowing can violate water-physical properties of the arable layer which increases the loss of humus and nutrients available to plants, and decreases crop yields (from 10 to 70%) and product quality [5].

In the field and laboratory conditions, methods of agrosobrancheology allowed for a wide range of indicators for assessing changes in effective fertility under the influence of resource-saving technologies [6]. Experimental information was grouped by the method of scientific analysis taking into account the corresponding sets of elements (characteristic parameters): <u>*Group A*</u>. Structural and water-physical properties (content of mesoaggregates, density of soil addition, resistance to penetration, content of readily soluble salts, soil porosity, soil moisture, soil temperature, etc. .) <u>*Group B*</u>. Organic matter (total humus content, active carbon content, soil reaction, cation exchange capacity, etc.); <u>*Group B*</u>. Biological activity (the number of mesofauna (earthworms), intensity of soil respiration, the total number of bacteria, enzymatic activity of the soil (dehydrogenase, catalase, invertase), etc.); <u>*Group G*</u>. Agriculture (biological productivity of the fields, the total number of fungi, the number of pathogenic fungi, etc.). Regional (Bp) and local (Bp*) biological productivity expressed in conventional feed units were used as an integral indicator of the effectiveness of applied agrotechnologies [1].

The array of experimental data was analyzed in Microsoft Excel (2013); standard statistical indicators were calculated (arithmetic average). To establish the significance of differences between the compared signs, a 5% level of significance was used.

3. The study of the structure of the modified lead-tin-base bronze

According to the statistics of the Russian Federation for 2010-2016, the value of the regional biological productivity was calculated for three regions (Table 1).

In <u>*Volgograd region*</u>, the best result was obtained for perennial grasses (Bp = 0.258) and wheat (Bp = 0.168). By the regional climatic conditions, 2014, 2015, 2016 were optimal years (average Bp = 0.173-0.215) compared with 2010 and 2012. (Bp =0.122) which is due to low moisture reserves and long periods of air and soil drought. For economically valuable sunflower crops, the best results were obtained in 2013, 2015 and 2016. (Bp = 0.145-0.162). In <u>*Stavropol region*</u>, Bp = 0.414 was obtained for sunflower. According to the response to regional climatic conditions, 2016 was optimal (Bp = 0.357), in 2012, Bp was 0.190. According to the response to the agricultural technology and the choice of culture, 2016 with Bp = 0.223-0.513 was the best year. The lowest yield results were recorded in 2012 (Bp = 0.171-0.350) which is due to long periods of air and soil drought. In <u>*Rostov region*</u>, high biological productivity was observed for perennial grasses and wheat (Bp = 0.248-0.339). By the response to regional climatic conditions, 2011, 2014 and 2016 were optimal years (Bp = 0.242-0.325); in 2012 and 2013, Bp was 0.193-0.198. The lowest yield results were obtained in 2012 and 2013 due to long periods of air and soil drought.

Table 1. Regional crop yields in feed units in different years of observation (BP)

Plant	2010	2011	2012	2013	2014	2015	2016	Average
				Volgograd region				
wheat	0. 111	0. 153	0. 154	0. 163	0. 200	0. 165	0. 231	0. 168
barley	0. 056	0. 120	0. 079	0. 080	0. 105	0. 095	0. 137	0. 096
sunflower	0. 083	0. 124	0. 098	0. 149	0. 134	0. 145	0. 162	0. 128
perennial herbs (hay)	0. 237	0. 221	0. 219	0. 234	0. 278	0. 285	0. 329	0. 258
linen	-	0. 113	0. 059	0. 094	-	-	-	0. 089
average	0. 122	0. 146	0. 122	0. 144	0. 179	0. 173	0. 215	0. 148
				Stavropol region				
wheat	0. 284	0. 323	0. 190	0. 255	0. 328	0. 329	0. 357	0. 295
barley	0. 217	0. 284	0. 173	0. 261	0. 269	0. 298	0. 321	0. 260
corn (grain)	0. 151	0. 180	0. 171	0. 181	0. 183	0. 176	0. 223	0. 180
sunflower	0. 412	0. 423	0. 350	0. 391	0. 421	0. 386	0. 513	0. 414
perennial herbs (on hay)	0. 266	0. 303	0. 221	0. 272	0. 300	0. 297	0. 353	0. 288
average	0. 284	0. 323	0. 190	0. 255	0. 328	0. 329	0. 357	0. 295
				Rostov region				
wheat	0. 227	0. 243	0. 198	0. 198	0. 278	0. 270	0. 324	0. 248
barley	0. 140	0. 165	0. 139	0. 126	0. 177	0. 183	0. 208	0. 163
corn (grain)	0. 121	0. 218	0. 162	0. 209	0. 203	0. 213	0. 280	0. 201
sunflower	0. 108	0. 133	0. 143	0. 161	0. 159	0. 175	0. 240	0. 160
perennial herbs (hay)	0. 345	0. 342	0. 274	0. 260	0. 354	0. 306	0. 488	0. 339
average	0. 208	0. 242	0. 193	0. 198	0. 253	0. 243	0. 325	0. 238

In comparison with the regional level, analysis of local biological productivity (Bp*) was carried out using different resource-saving agricultural technologies (Table 2). On experimental fields N1-N3 (Volgograd region, Mikhailovsk district), for winter wheat, Bp* was 0.102-0.323, for sunflower - 0.106-0.306, for flax - 0.144. Taking into account soil and climatic conditions, the best results were observed on N1-N3 in 2014 and 2016 (Bp* = 0.236-0.269). Low yields of sunflower were obtained in 2012 and 2015 (Bp * = 0.106-0.175) due to weather conditions and fertility levels. Using the zero agrotechnology, a good level of fertility was formed. It allowed for levelling negative weather conditions (soil and air drought) in 2012. In 2015, in Volgograd region, the yield of sunflower was high in contrast to the results for N1-N3 due to insufficient agricutural measures using a zero technology.

On S1-S3 experimental fields (Stavropol region, Ipatov District), for winter wheat, Bp* was 0.461, for sunflower - 0.575, and for pea - 0.404. There is an excess of the average Bp* for the main crops on the S1-S3 fields compared with the average data for the region (Bp) which characterize the use of minimum tillage to maintain effective fertility in the conditions of Stavropol region. Taking into account the local soil and climatic conditions, the best Bp* was observed in 2012 and 2014. (0.480-0.499). On the C1-C4 experimental fields (Rostov region, Salsk district), for winter wheat, Bp was 0.291, for sunflower - 0.193, and for pea - 0.234. Compared with the regional results, there is an excess of average Bp for major crops at the local level, i.e. on the C1-C4 fields where the application of a no-tillage technology characterizes an optimal combination of agricultural methods with fertility of the field and the choice of crop rotation.

Table 2. Local yield of crops in the feed units accounting for field conditions in different periods of observation (Bp*)

Field number	Agrotechnology	Years of observation					Average
		2012	2013	2014	2015	2016	
		Mikhailovsk district, Volgograd region					
N1	zero	-	0. 323	0. 305	0. 168	0. 306	0. 276
N2	zero	0. 106	0. 106	0. 152	0. 175	0. 188	0. 145
N3	zero	0. 133	0. 102	0. 251	0. 087	0. 311	0. 177
	average	0. 120	0. 177	0. 236	0. 144	0. 269	0. 189
		Ipatov District, Stavropol region					
S 1	minimal	0. 404	0. 461	0. 461	0. 117	-	0. 361
S 2	minimal	0. 461	0. 461	0. 575	-	0. 377	0. 468
S 3	minimal	0. 575	-	0. 461	0. 273	0. 404	0. 428
	average	0. 480	0. 461	0. 499	0. 195	0. 390	0. 405
		Salsk district, Rostov region					
C1	zero	-	0. 212	-	0. 263	-	-
C2	zero	0. 379	0. 174	0. 204	0. 188	0. 353	0. 158
C3	zero	0. 368	0. 203	0. 204	0. 274	0. 357	0. 260
C4	zero	0. 290	0. 203	0. 204	0. 239	0. 363	0. 281
	average	0. 259	0. 198	0. 204	0. 241	0. 358	0. 260

Let us discuss adequacy of the assessment of resource-saving technologies for changing a moisture regime and plant nutrition. For example, in Rostov region, in different years of observations, on the C4 field the crop yield was lower (28.5 centners per hectare) as compared with the C2 field (43.4 centners per hectare) which confirms the Bp = 0.158 for the C4 field. Having analyzed the yield of the experimental fields of tilled and leguminous crops (peas, sunflower), it was concluded that the C3 field (Bp = 0.098) is characterized by "worse" local fertility conditions compared to C1, C2, C4 fileds (Bp = 0.21-0.24). This is due to long-term disturbances in local soil processes under conditions of moisture deficiency (soil and air drought) and deterioration of the ecosystem functions of agricultural landscapes which, when regulating water-physical properties and humus formation, are insufficient to restore the biological activity of the arable layer. Therefore, soil fungistasis is weakened and the risk of phytopathogen accumulation is high.

4. Conclusion

Effective soil fertility and biodiversity are a basis for sustainable development of the agroecosystem and intensive agricultural production. Regional and local biological productivity is an integral indicator of the effectiveness of resource-saving agricultural technologies used in combination with the crop yield of the crop rotation. High effective fertility was observed when using a no-till agrotechnology in Volgograd region (Bp - 0.09-0.26, Bp* - 0.14-0.28); the minimum-till agrotechnology was used in Stavropol region (Bp - 0.18-0.41 and Bp* - 0.36-0.47); in Rostov region, a zero agrotechnology was used (Bp - 0.16-0.34 and Bp* - 0.15-0.28).

References

[1] *The State report On the state and protection of the environment of the Russian Federation in 2015* 2016 (Moscow: Ministry of Environment of Russia, NIA-Nature)

[2] Alliaume W, Rossing A H and Tittonell P 2017 Modelling soil tillage and mulhing effects on soil water dynamics in raised bed vegetable rotations *European J. of Agronomi* **82** 281–6

[3] Chauhan Bhagirath, Singh Ravi and Mahajan Gulshan 2012 Ecology and management of weeds under conservation agriculture *Crop Protection* **38** 57–65 10.1016/j.cropro.2012.03.010

[4] Lili G, Becker E and Guopend L 2017 Effect of different tillage systems on aggregate structure and inner distribution of organic carbon *Geoderma* **288** 97–104
[5] Fedotov V A, Kadyrov S V and Schedrina D I 2011 *Agrotechnologies of field crops in the Central Black Soil Region* (Voronezh)
[6] *Agrochemical methods of soil analysis* 1976 (Moscow: Nauka)

Ecological organization of the territory

D V Novikov and A V Novikov

State University of Land Management, 15, Kazakova st., 105064, Moscow, Russia

E-mail: prof.novikov@hotmail.com

Abstract. The article describes the reasons for deterioration of agricultural lands which affects the environmental management system violating the landscape and biosphere compatibility. It is necessary to implement the ecological-landscape organization of the territory at various levels of development. The nature of the ecological landscape organization of the territory was described. Sustainable agro-landscape systems should be formed while designing land categories. Territories of the agricultural land should be formed taking into account zonal features of landscapes. The article provides land management recommendations on the ecological-landscape basis. Its advantages in comparison with traditional methods were revealed.

1. Introduction

The world has realized that reproduction of natural systems, their integrity and restoration contribute to economic development and preservation of vital interests of future generations.

The "Concept for the Transition of the Russian Federation to Sustainable Development" adopted in 1996 provides for

- ecologization of economic activities;
- preservation and restoration of the biosphere focusing on the needs of future generations;
- formation of the noosphere. [1]

The starting point on the path to a sustainable development model is ecologization of agriculture, rational use of natural resources, including land.

Land resources of the Russian Federation are a driving force of economic development. Their rational use and protection are primary tasks of the government.

At the same time, the analysis of land resources used in Russia identified serious problems.

Over the past years, reconstruction of irrigation and drainage systems has stopped, the volume of land reclamation works have been reduced by 5-8 times. The volume of anti-erosion, forest reclamation, and agrochemical works has been reduced.

Significantly increased the area of land not used in agricultural production.

Productive lands that are not used in agriculture become abandoned.

One of the reasons for deterioration of the phytosanitary condition of agricultural lands is almost complete elimination of crop rotations.

In most of the agricultural enterprises of the Russian Federation, they were destroyed during the land redistribution.

Destruction of crop rotations causes an irreparable damage to the ecology of agricultural landscapes, fertility of fields and the economy of agricultural enterprises.

The analysis of the state of land resources of the Russian Federation shows that its quality is deteriorating, and in some regions the influence of negative factors on land is catastrophic.

The presence of these serious problems is evidence that the entire environmental management system does not meet requirements of landscape and biosphere compatibility, principles of adaptive agricultural land management and rational economic development.

The crisis of the agrarian sector dispelled illusions of preferential regulation of soil fertility by man-made means. Modern industrial technologies have reached their environmental, energy and production limits.

The conceptual basis for combating land degradation is their environmental quality and adaptive landscape arrangement of agricultural areas.

With regard to the agricultural sector, many ecologists see the solution to these problems in the transition to alternative farming systems, primarily to adaptive-landscape systems and their regional modifications.

As applied to the use of land resources, this is a transition to the ecological-landscape organization of the territory in the system of land management associated with farming systems.

2. Methods and materials

Theoretically, ecological landscape management is a natural continuation of the development of scientific knowledge in the field of rational organization of the territory, use and protection of land resources. It is based on the recognition of classical provisions of land management science, its definition as a socio-economic and environmental-economic process and a set of measures to form objectively determined systems of land tenure, land management and land relations, territorial organization of agricultural and industrial production, environmental protection. Its distinctive features are versatile consideration of ecological-landscape, ecological-economic and agro-ecological conditions and properties of the land, consumer demand for the results of land management, crop and livestock production.

The goal of ecological-landscape land management is to create conditions for mobilization of the natural resources and adaptive potential of agricultural crops aimed at economically efficient, socially-oriented and environmentally safe production and nature management stabilization by creating environmentally sustainable agrolandscape systems and eliminating effects of land degradation and pollution [2]

The categories of land funds granted for agricultural purposes, human settlements, specially protected areas have defined ecological-landscape dimension. The ecological-landscape content is embedded in the concept "land" when viewed from the location and systematic use or suitability for specific uses and differences in natural historical features.

Optimization of the land composition is a necessary task at the economic and regional levels. A stable structure of agricultural land should be ecologically substantiated.

Elements of the ecological-landscape organization of the territory must be present at all territorial levels of land management documentation development. They can differ in the degree of specification, but assessment and justification methods for landscape, environmental and economic indicators should be identical.

The tasks of ecological-landscape organization of the territory are determined by the territorial level of development, the ecological state of the territory, prospects for its development, availability of an information base, etc.

When developing top-level land management schemes, data on the agro-ecological state of lands are summarized, factors, sources and results of anthropogenic impact are established, their relationship and localization with respect to structural elements and landscape components, groups of agricultural landscapes are identified, and recommendations for their use in order to ensure their sustainability and effective management are provided.

Recommendations are developed to improve agricultural enterprises and areas of economic activities; environmental protection measures are developed.

The territorial organization should be differentiated according to the types of landscapes based on their zonal-provincial characteristics.

As a rule, the landscape design should be part of design works on land management, preparation of medium-scale landscape maps.

In order to determine the permissible economic burden on certain landscapes when developing land management schemes for the territories of the Russian regions, it is advisable to make regional landscape-ecological forecasts.

Continuing and not always controlled redistribution of land during the creation of farms, granting of land shares require special attention to the integrity of the landscape structure.

3. Results

Lands of collective farms should be as compact as possible without interband sections and other territorial inconveniences, which will contribute to efficient environmental management, development and implementation of environment-stabilizing measures.

The organizational and territorial structure of agricultural enterprises is formed during inter-farm (territorial) and on-farm land management. At the same time, characteristic properties of the landscape are used to achieve ecological well-being. The organization of agricultural landscapes is changed. Production and its territorial organization is adapted to peculiarities of the landscape structure which transforms the natural environment.

Ecological and landscape land management which has great advantages over traditional methods of territory organization is complex due to the unity, integrity and compactness of tasks and measures for organization of land management and protection, and increased costs of its implementation. At the same time, it responds more quickly to changes in the economic development, land and environmental management, and takes into account the prospective dynamics of a land-based facility and conditions for its operation.

Landscape conditions and agro-ecological quality of land should determine the volume, specialization and intensity of production, land management parameters, composition and costs of environmental measures taking into account the balance and consistency of the balance of natural and economic resources, land, working-age population, technical means, financial flows, gross and marketable products, etc.

Improvement of land management should serve the interests of agricultural enterprises and improve management activities of municipal and state bodies and living standards [3]

Theoretical and methodological aspects of the ecological-landscape organization of the territory are considered at the level of on-farm land management.

For the levels of the General scheme of land management of the territory of the Russian Federation, land management schemes of the territory of the Russian regions, land management schemes of municipal districts, ecological landscape requirements are not taken into account.

Existing methodological recommendations on the development of land management schemes and the schemes themselves have no measures for land protection.

4. Discussion

One of the documents that can be used to address the issue of land management and protection on the landscape basis is the "Landscape-ecological zoning of the territory" (basics of methods and zoning schemes). [4]

Suggestions made by E.V. Ponomarenko and S.V. Ponomarenko in their work "Land reformation in Russia: Negative Consequences and Possibilities for Sustainable Development" are of interest [5]

Organization of the territory on an ecological-landscape basis is discussed in the practical guide "Land management scheme of the administrative district". [6]

To solve the problem of the optimal ratio of natural and economic lands, the Concept highlights three important tasks:

1. determination of the optimal share of land for the landscape region (province);
2. identification of the minimum required area of a plot with natural vegetation;
3. planning of the optimal territorial structure of natural land plots.

The main territorial object for which the ratio of natural and economic land plots should be determined is an AL-district (agrolandscape); auxiliary objects are an AL-terrain and an AL-district.

The need to develop land management on the ecological-landscape basis is determined by both the negative state of land resources and the need to form an organizational-territorial structure for land management and crop production on the basis of land management. This structure has been formed over many centuries. It is based on differences in the landscape quality and land suitability. The conditions for implementing the methods of ecological-landscape organization of the territory are the scientific support and scientific and technical achievements. The issues of ecological and landscape substantiation of the organization of agricultural production are theoretically developed. Only on the basis of environmental laws and a healthy economy, one can create an optimal environmental management system.

5. Conclusion

The main task of the ecological and landscape organization of the territory in the process of land management is to create conditions that ensure the mobilization of the territory's resources, maintaining the stability of landscapes and, first of all, agrolandscapes, or increasing their stability on degraded agricultural lands, provided that they fulfill their socio-economic functions within the given limits.

The final result of the ecological and landscape organization of the territory should be the preservation of the parameters of the landscape structure and the properties of the individual components, ensuring its relative stability and their expedient optimal productivity.

The solution of these issues in the process of land management is associated with the presence of a number of unsolved problems of a theoretical and methodological nature, which does not allow for the ecological and landscape organization of the territory at the proper level.

Unresolved issues include the following:

– lack of information on the patterns of differentiation of landscapes, the nature of the relationships and the relative position of their morphological parts, necessary for designing at different territorial levels of the organization of the territory, ensuring stability of landscapes;

– lack of criteria and parameters characterizing and ensuring the stability of landscapes;

– lack of recommendations for establishing an environmentally and economically sound structure of agrolandscapes in various climatic zones;

– lack of development of methods for determining permissible anthropogenic loads on landscapes;

– lack of ecological-landscape zoning of the land fund at the local level;

– lack of a working methodology of ecological and landscape organization of the territory;

– lack of experimental-production base for designing of ecological and landscape organization of the territory [8]. This information is not enough for developing conditions that ensure the conservation and increase of stability of agrolandscapes. Information is needed on the boundaries and structure of agrolandscapes, their actual state, characteristics of threshold values of the state of components of agrolandscapes, as well as information on the detailed assessments of the optimization capabilities of technologies and agrotechnical techniques, the specifics of local soil, climatic, microclimatic and weather conditions, features of adaptive responses to their effect of cultivated plant species and varieties.

References

[1] Decree of the President of the Russian Federation 1996 *Concept of the transition of the Russian Federation to the sustainable development model*

[2] Volkov S N and Wolves S N 2000 *The concept of modern land management and methodological foundations of land management during the transition to new land relations* (Moscow)

[3] Barsukova G N 2016 (Krasnodar)

[4] Kartsev G A et al 2002 *Land management plan of the administrative region. Practical Guide* (Moscow: Uni-press)

[5] Novikov D V and Novikov A V 2017 *Zoning of the territory and the establishment of zones with a special mode of land use: Tutorial* (Moscow: Neolithic)

[6] Ponomarenko E V and Ponomarenko S V 1996 *Land reform in Russia: negative consequences and opportunities for sustainable development, environmental design laboratory* (Moscow)
[7] Andriyshyn and Koltunov N M 1993 *Landscape-ecological zoning of the territory. Fundamentals of the methodology and zoning scheme* (Moscow: RAAS)
[8] Novikov D V 2009 *The organization of rational use of agricultural land on an ecological-landscape basis on the example of the Southern Federal District of Russia – territory, economy, organization* (Moscow)
[9] Saprin S V 2016 Estimation of anthropogenic load on agricultural landscapes of Voronezh region *Bull. of the Voronezh State Agrarian University* **2** 236–41
[10] Chursin A I 2008 *Landscape organization of the territory of the forest-steppe zone of the middle Volga region* (Penza: PGUAS)

Development of the methods of environmental and sanitary-hygienic assessment of urbanized lands (on the example of the Ob-Tom interfluve)

E Yu Pasechnik and L N Chilinger

Division for Geology, School of Earth Sciences & Engineering, Tomsk Polytechnic University, 2 building 5, Lenina Avenue, Tomsk, 634028, Russia

E-mail: paseyu@yandex.ru; lilichilinger@gmail.com

Abstract. Modern transformations to land legislation draw the government to establishing the boundaries of use-restricted zones throughout the territory of the Russian Federation. First of all, it is connected with the goal of the state to provide safety and provide a favorable background for preserving a person's life and ensure his/her life activity. The Land Code of the Russian Federation has defined a list of use-restricted zones, according to which, the procedure for establishing and approving their boundaries for each type of zone to put data into the Unified State Register of Immovable Property should be defined. Consequently, there is a need to develop an ecological and sanitary-hygienic assessment of the territory, which takes into account all these factors. However, the analysis of the approved regulations for flooding zones shows that the procedure does not stipulate for an integrated analysis of the territory as a single hydraulically connected system, which consists of hydraulic structures, surface, and groundwater. The lack of an integrated geo-information analysis in the current regulations for the corresponding territories induces intensive economic development, the consequence of which could increase human impact. Thus, unreliable information will be put into the Unified State Register of Immovable Property. Consequently, considering all these factors, it is necessary to develop an ecological and sanitary-hygienic assessment of the territory.

1. Introduction

Water is the most important natural resource for all living organisms including humans. However, the development of a large number of enterprises in cities significantly deteriorates the quality of surface waters. In this regard, many localities are switching to groundwater use.

The environmental and sanitary-hygienic condition of the lands of the Ob-Tom interfluve (OTI) is essential since there is the Tomsk underground water intake supplying the population of the city of Tomsk and the Tomsk region since 1973 with potable water.

Despite the fact that water intake is a strategically important object, the necessity to consider the borders in the registry emerged due to of the Federal Law "On cadastral activities" No. 221-FZin 2007 [1]. Thus, it is a long period from the calculation of the boundaries of the protective sanitary zone (PSZ) [2] to their establishment that leads to the human impact change being a key parameter in the calculation of the PSZ. Such a problem exists in the (OTI) territory. This territory is attractive for the development of residential buildings due to its proximity to the administrative center of the city, and in recent years is intensively being developed, although OTI was planned to be used for recreation when building water intake due to inefficient protection of groundwater.

The last reevaluation of groundwater reserves in this area to consider the water demands of Tomsk [3] and the human impact was carried out in 2003 for a period of 27.4 years and the boundaries of the protective sanitary zone were established. After, the areas of localities within the sanitary protective zones have significantly changed, and the human impact on landscapes in general and on groundwater has increased. In this regard, it is necessary to assess and predict changes in the components of the natural environment due to human impact to obtain reliable information about the state of the land and ways of sustainable land use in the operating conditions of the Tomsk underground water intake.

2. Materials and research methods

From 1992 to the present, the study of economic activities in the selected area was conducted by analyzing land resources use. The system, statistical and correlation-regression analyses were used to estimate and predict the population in the work. To assess the dynamics in the residential area change, the landscape analysis, and system mapping, cartometric analysis using geo-information technologies were used.

In the study, the following source materials were used: topographic maps with 1: 10000 scale (release of 1992 (not updated)), which are the map of the Tomsk Region Administration, the development master plan of the Tomsk administration and rural settlements, the Esri maps, cadastral plans of the Federal Registration Service, cadastral register and cartography of the Tomsk region, information of forest management regulations for Timiryazevsky forestry area in the Tomsk region.

The Federal State Statistics Service data, project declarations and commissioning certificate of the residential district « Severnyj Park» were used to calculate and predict the population size. Statistical analysis and population forecast for the period from 2002 to 2040 were conducted using the Microsoft Excel spreadsheets by constructing diagrams with a trend line.

3. Algorithm ecological and sanitary and hygienic assessment of urbanized lands

The Tomsk underground water intake refers to the water intakes of the linear type and consists of 198 production wells. At the present time, 95 wells are in operation in the normal load mode, 14 of which belong to the control wells to sample the quality of the water supplied. The study subject is the territory of the third zone of the SPZ of the first stage of water intake (wells 1э-69э) launched on December 13, 1973 [4]. Complete construction of the first stage was over in October 1974 [5].

The area of the SPZ of the Tomsk underground water intake is 26000 hectares. The study area within the third zone of the first stage is 12359 hectares.

The territory choice for the developed technique testing is due to the following main factors: proximity to the city of Tomsk (the administrative centre of the Tomsk region), a large number of settlements and the maximum intensity of economic development human impact on the territory of the Ob-Tom interfluve.

The methods for ecological and sanitary-hygienic assessment of urbanized lands are based on the integrated analysis of the territory with population dynamics analysis, change of the boundaries of urbanized territories, identification of reasons of human impact, evaluation of the landscape and the ecological and economic balance of the territory. The technique consists of the following successive stages:

3.1. 1st stage. Analysis and prediction of population dynamics

The Federal State Statistics Service data on the number of residents registered in urban areas with low-rise buildings (the total population in 2015 was 13756 people) was used to estimate the population dynamics within 2 and 3 zones of the SPZ of the underground water intake. There has been a sharp increase in the population over the period from 2002 to 2008. This is associated with the incorporation of some settlements into the municipal entity "Gorod of Tomsk" in November 2004 (v. Timiryazevskoe, v. Dzerzhinskoe, etc.). In addition, for the period from 2008 to 2015 there is an insignificant stable increase in the population by 11.5%.

To analyze the further possible change in population growth, a forecast was given based on the trend and fluctuations of the statistical series (extrapolations of statistical series) for 2030 and 2040 (fig. 1). The analysis showed that with the current trends in population size within the PSZ intake, by 2030 the theoretical value of this indicator may increase by 700 people, and by 2040 it may increase by 860 people. Considering the error of the method and the confidence band, according to Student's criterion, in 2030 the actual population value will be from 13579 to 15741, and in 2040 it will be from 12887 to 15487.

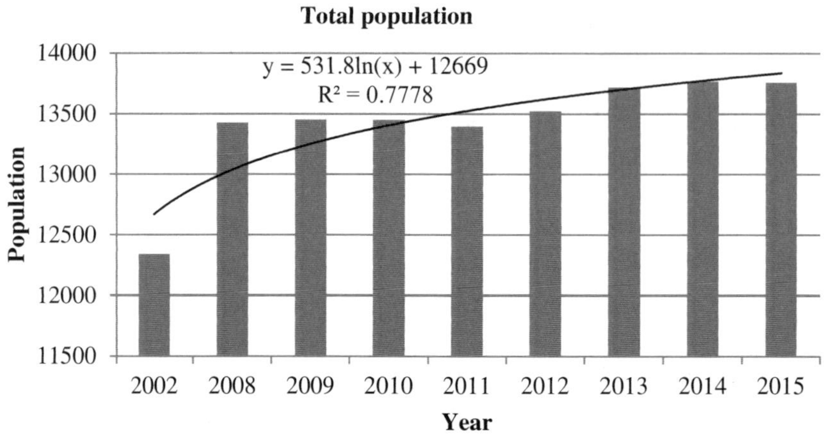

Figure 1. Type of a function for population projection in the populated areas under consideration by extrapolation

However, in recent years, a multi-storeyed residential building has been developing on the left bank of the Tom River. At the moment the area of the residential district Severnyj Park is 35.5 hectares. It is located on the border of 3d zone of the SPZ of the water intake. In this area, according to project statements and commissioning certificates, there are 19 five-storeyed houses (2077 apartments), an eight-storeyed house (76 apartments) and an 11-15 storeyed house (158 apartments), except public amenities. Moreover, in the area of the residential district Severnyj Park there are 12 land plots for low-rise buildings and a large one, not divided land mass (approximately 12 land plots). It can be proposed that the population number of the residential district Severnyj Park after developing the whole territory will be 9340 people (2311 apartments and 24 land plots, providing that the average family composition is 4 people). If we assume that these areas are built similarly to the project documentation of the residential district Severnyj Park, and the number of families is calculated, the population number of the left bank areas will increase by more than 42 thousand people.

3.2. Stage II. Analysis of changes in the boundaries of urban areas

An analysis of the areas occupied by settlements within the territory using archival cartographic materials (tablets) of 1992 and spatial data Esri maps of 2018 led to the conclusion that the territories occupied by settlements increased by 1317.3 hectares, or slightly more than double. Thus, the area reached 2415.7 ha in 2018. For comparison, in 1992 it was 1098.4 ha. It should be noted, that the increase in the area of settlements within the sanitary protective zones falls on individual housing and country house constructions, and not on apartment buildings.

It is important to emphasize that the boundaries of some settlements have increased severalfold (v. Zorkaltsevo; v. Kaidalovka) due to the construction of new the residential districts, which are located in the in neighborhoods of villages and settlements ("Zorkaltsevskie Usadby", "Sloboda Volnaya", " Solnechnyj Plyus ", " Snegiri ", " Serebryanyj Bor ") [6].

3.3. Stage III. Source identification of human impact and boundaries determination of sanitary protective zones

For this study, the classification of factors and impacts of technogenic disturbances of the hydrosphere of the Ob-Tom interfluve [7] was used, on the basis of which a map of sources of human impact was created.

In this area, various farms specializing in livestock production are actively developing. The operation of these farms implies the availability of animal waste, animal burial sites, and manure storage facilities. The growth of settlements leads to an increase in the number of different domestic waste landfills. Within the SPZ [2], there are also industrial enterprises, such as a concrete-asphalt plant, gas stations, as well as warehouses of fuel and lubricants and mineral fertilizers, and other objects that directly exert human impact, and are unacceptable in compliance with sanitary standards (fig. 2).

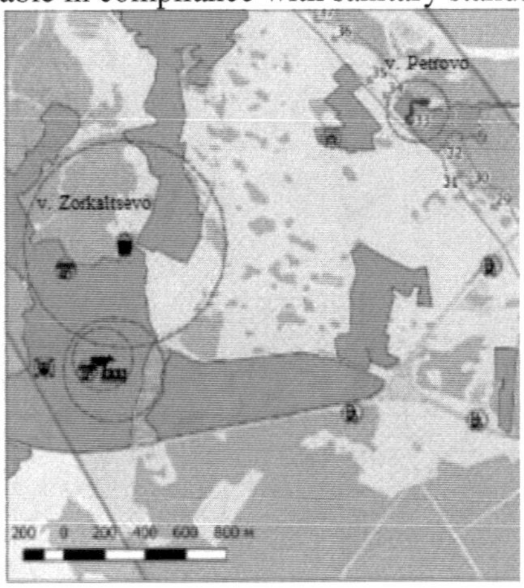

Legend		**Sources of human impact**	
Underground water well		Livestock complexes	Food industry objects
Second belt of sanitary protection zones		Slaughterhouses	Waterhouses mineral fertilizers
Third belt of sanitary protection zones		Reservoirs	Concrete-asphalt plant
Boundaries of settlements		Cemeteries	Boiler rooms
		Gas station	Solid waste landfills
		Waterhouses of fuel and lubricants	Landfills

Figure 2. Fragment of the map of sources of anthropogenic impact

According to the sanitary rules and norms, sanitary protective zones of sources of human impact were determined. Based on the cadastral plans of the territories, there is no record of the majority of the boundaries of such objects in the Unified State Register of Immovable Property, so the corresponding sanitary conditions are not ensured and human activity is not limited. Moreover, most of the objects are located within the boundaries of the SPZ water intake, which is unpermitted.

3.4. Stage IV. Landscape analysis to assess the possibility of migration of pollutants with surface runoff

To determine the surface runoff direction, a digital model of the landscape [8] was constructed, analyzing which it was concluded, that the eastern part of the OTI is characterized by low and slightly sloping forms compared to the western part (fig. 3). There are many bayou lakes of the river Tom. Thus, the landscape is a cause of the flow of pollutants from the drain to the water wells.

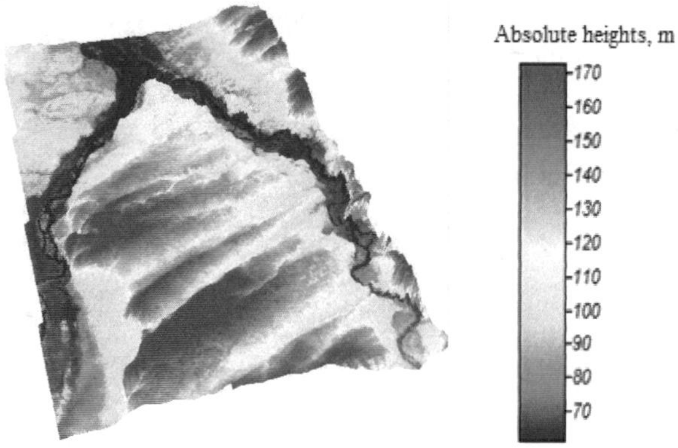

Figure 3. Digital elevation model of the territory of the Ob-Tomsk interfluve

3.5. Stage V. Assessment of the ecological and economic balance of the territory

The ecological and economic balance (EEB) of lands was carried out on the basis of a site map of the territory. The presented types of landscapes were ranked in accordance with the degrees of human impact by means of an expert method (table 1).

Table 1. The grading of Tomsk central part of the underground water intake by degrees of anthropogenic load.

Degree of anthropogenic load	Score	Types of land	Area, (ha)
Higher	6	Residential areas	6 112.4
Very high	5	Agricultural land for comprehensive use	7 502.5
		Land used for growing crops: irrigated, arable	10 398.5
High	4	Agricultural land used for livestock	5 417.0
Average	3	Forests, perennial plantations	5 206.9
Low	2	Forest Park Zones	9 517.9
		Green areas	5 018.6
Very low	1	Conservation lands	479.0
		Unused land (reserve land)	45.7
Total area			49 698.4

The assessment of the condition of the land use structure of the study area was carried out based on the expert assessment method [9] and the classification [10] of land use. To assess the EEB, the absolute tension ratio of the territory was used, which characterizes the ratio of severely disturbed areas (the development of urbanized areas leads to not only the activities of industrial enterprises but also traffic congestion) and lands of poorly disturbed or uncultivated.

As a result of the ecological and economic balance assessment of the territory, it was discovered an imbalance between human impact and reproduction potential of natural ecosystems on the territory of the Ob-Tom interfluve within the SPZ of Tomsk underground water intake with the most intensive human impact caused by the active urbanization processes.

4. Conclusion

The proposed technique for the environmental and sanitary-hygienic assessment of the urbanized lands is a comprehensive approach to the study of the interaction of natural and human factors.

The Ob-Tom interfluve is considered to be a favorable territory for the formation and preservation of groundwater due to hydrogeological, physiographic, and ecological conditions. However, when developing this territory, it is necessary to take into account the possibility of infiltration of groundwater in some areas. The application of the developed technique for the environmental and sanitary assessment of urbanized lands allowed us to conclude that during a period of the Tomsk underground water intake development, the boundaries of the populated areas within the PSZ boundaries significantly extended, and human impact in the landscape increased. It may lead to groundwater pollution.

Moreover, the ecological situation in the study area is becoming worse due to noncompliance with technological regulations at the located industrial facilities and agricultural enterprises, deforestation, the development of garden co-operatives and cottage development within the boundaries of the SPZ.

Special attention should be paid to the districts that are being built in close proximity to the SPZ boundaries, in which, a large number of people and vehicles will be concentrated in a small area unlike in low-rise areas. In the event of an increase in the boundaries of the SPZ with the next revaluation of groundwater reserves, this territory may fall into the border of 3 zones. Thus, the residential district Severnyj Park is an example of the Ob-Tom interfluve. During the construction and operation of such facilities, special attention should be paid to the creation of measures that minimize the impact on the environment.

The economic development of territories can be considered successful [11], and rational only taking into account all the natural factors that determine the condition of the landscape as a whole. It is impossible to choose a rational direction of land use, predict the further development of the territory, as well as achieve environmental compatibility neglecting natural factors and the ecological situation of the territory.

Thus, currently, land management (in the use and protection of the land reserves of the Russian Federation) should take into consideration increase in anthropogenic activities: intensive agriculture, urbanization [12], technogenesis effect [13], climate change.

To organize the sustainable development of land use in such areas, the following measures are proposed (for example, OTI) within the framework of land-use planning:

1. To introduce the boundaries of the PSZ water intake on the scheme of land-use planning of the Tomsk region.
2. To establish the industrial object of the SPZ and put such boundaries into the Unified State Register of Immovable Property.
3. To revise the Rules for Land Use and Development and town planning regulations in terms of the territory use restrictions, taking into account the SPZ.
4. To review and change the territories of perspective development within the boundaries of the SPZ.

Regarding the sources of human impact: liquidation and conservation of objects, restriction of activities, land restoration, environmental protection measures: landscaping, creation of natural parks, organizational and technical measures for waste disposal, discharge treatment.

5. Acknowledgments

The authors express their appreciation and deep gratitude to Victor Konstantinovich Popov, professor of the Division for Geology of the NI TPU, for advice and valuable comments when working on these studies.

6. References

[1] Federal Law N 221-FZ 2007 On cadastral activities, Available at: http://www. consultant. ru/

[2] Sanitary-hygienic standards No. 2. 1. 4. 1110-02 2002 Sanitary protection zones of water supply sources and drinking water supply systems, Available at: http://www. consultant. ru/

[3] Makushin Yu V 2005 *Revaluation of operational groundwater reserves of the Tomsk field* (Tomsk: OAO «Tomskgeomonitoring»)

[4] Tomskvodokanal 2018 Official website of Tomsk water and sewer service, Available at: http://www. vodokanal. tomsk. ru/.

[5] Popov V K, Kozina M V, Levak Y Y and Shvagrukova E V 2016 Ecologization of water-land property matters on the territory of the Tom lower course *IOP Conference Series: Earth and Environmental Science: Contemporary Issues of Hydrogeology, Engineering Geology and Hydrogeoecology in Eurasia* **33** 1–5

[6] Chilinger L N 2018 Formation of new microdistricts area within the sanitary protection zone of the Tomsk underground water *Proc. of Problems of geology and subsoil development: works of the XXII Int. Symp. named after academician M. A. Usov of students and young scientists* (vol 2) ed Boev A S (Tomsk: National Research Tomsk Polytechnic University) pp 846–847

[7] Popov V K, Lukashevich O D and Korobkin V A 2003 *Ecological and economic aspects of groundwater exploitation of the Ob-Tomsk interfluve* (Tomsk: TGASU)

[8] Redkina V I and Chilinger L N 2017 Analysis of the territorial development of the Ob-Tomsk interfluve using geographic information systems *Materials of the I Int. scientific and practical conf. "Actual problems of geodesy, cadastre, rational land and environmental management"* (vol 2) ed A M Oleinik (Tyumen: TIU) pp 94–98

[9] Kochurov B I 2003 *Eco-diagnostics and balanced development* (Smolensk: Madzhenta)

[10] Erofeev A A, Kopysov S G and Nikiforov A N 2018 Landscape-geophysical approach to zoning the structure of land use in the small catchment area of Sakhalin *Bulletin of the Tomsk polytechnic university-geo assets engineering* **329(6)** 39–47

[11] Melkiy V A, Verkhoturov A A and Popova Y P 2018 Monitoring of land status and assessment of dynamics of anthropogenic impact on the territory of Sakhalin *Bulletin of the Tomsk polytechnic university-geo assets engineering* **329(6)** 48–57

[12] Petina M A, Petin A N, Chendev Y G, Lebedeva M G and Belevantsev V G 2015 Drainage degradation influenced by anthropogenic load in different landscapes of the forest-steppe zone (in terms of the region of Belgorod) *Research journal of pharmaceutical, biological and chemical sciences* **6(4)** 61–65

[13] Qiu Z, Kennen J G and Giri S 2019 Reassessing the relationship between landscape alteration and aquatic ecosystem degradation from a hydrologically sensitive area perspective *Science of the total environment* **650** 2850–2862

Transboundary protected areas

L E Petrova, O A Sorokina, I V Fomkin and E E Mamedova

State University of Land Use Planning, 15, Kazakova street, Moscow, 105064, Russia

Abstract. The degradation of ecosystems on sizable land and water area necessitates the protection and restoration of ecosystems, their efficient use, cessation and prevention of the degradation processes, and biodiversity preservation. Those activities are integral for The Concept of Sustainable Development currently accepted and used by the global community. One of the universally accepted mechanisms of ecological problem solving is the establishment and development of protected area (PA) systems. The complex, multifaceted and transboundary nature of ecological problems necessitates international cooperation. As a result of this cooperation, transboundary protected areas (TPAs) are formed according to international treaties. *Research methods:* abstract-logical, monographic, statistical, cartographic. The data base for the research consists of statistical and analytical materials of federal and regional executive bodies and the guides and reports written by scientific organizations in the field of environment preservation and land use planning. *Results:* analytical scrutiny of evidence and legal documents reveals the necessity of improvement of legislative and procedural guidelines for TPA formation.

1. Introduction

At the current stage of development of the society, human transformational activity has the irreversible negative consequences, and measures should be taken [1, 2] to prevent them. The sustainable development agenda set by The United Nations in September 2015 for a period until 2030 has 17 goals, and one of them is the protection and restoration of land ecosystems, the efficient use of them, cessation and prevention of degradation processes, and biodiversity preservation. The transboundary properties of natural processes necessitate the interregional and international cooperation in the field of territorial nature preservation. As a result of this cooperation, the bilateral and tripartite agreements are signed and the transboundary protected areas (TPAs) are formed [3]. This problem is especially important for the Russian Federation because of the great length of its national border. In Russia, the first TPA was formed only in 1989. The nomenclature and the statutory regulation of TPA formation and functioning need to be refined.

2. Methods

The data base for the research consists of statistical and analytical materials of federal and regional executive bodies of the Russian Federation (Russian Federal State Statistics Service, Ministry of Natural Resources and Environment, etc.), the guides and reports written by scientific organizations in the field of environment preservation and land use planning. Those guides and reports contain the systematized data about the actual state of environment in Russia, including the information about the state of specific environment components, ecosystems, current processes and occurrences, natural and anthropogenic factors influencing the state of environment, analysis, trends, and the forecast of their impact on the environment. The research object necessitates the use of diverse methods of gathering, processing and analyzing of scientific information: abstract-logical, monographic, statistical analysis (including the analysis of government statistics data and state informational systems), analysis of the key Russian and international forecasting reports.

3. Results

The complex, multifaceted and transboundary nature, the global scale of ecological problems necessitate the international cooperation and formation of TPAs [4–6]. The analysis of statistical and cartographic data and Environmental Law in Russia shows that there are no regulations in place for a number of TPA formation questions. The first TPA in Russia was formed in 1989, and now there are 5 TPAs in Russia. But until now, the term TPA has not been used in the Russian legislation.

Table 1. Transboundary Specially Protected Natural Areas in Russia

№	TPA name	Name(s) of the Russian PAs	Name(s) of other countries PAs	The international treaty regulating the TPA's activity
1	Russian-Finnish Nature Reserve "Friendship"	Kostomuksha Nature Reserve	5 PA in Finland	The Agreement between the Government of the USSR and the Government of Finland of 26.10.1989
2	China-Mongolia-Russia International Protected Area «Dauria»	Daursky Nature Reserve	Mongol Daguur (Mongolia) and Dalainor Biosphere Reserve (China)	The Agreement between the national environmental agencies of the Russian Federation, Mongolia, and the People's Republic of China of 29.03.1994
3	Russian-Chinese Nature Reserve "Lake Khanka"	Khanka Nature Reserve	Sinkai-Hu Reserve (China)	The Agreement between the Government of the Russian Federation and the Government of the People's Republic of China of 26.04.1996
4	Transboundary reserve "Grand Altay"	Katunskiy Reserve	Katon-Karagay Reserve of Kazakhstan	The Agreement between the Government of the Russian Federation and Kazakhstan of 15.09.2011
5	Pasvik–Inari Trilateral Park	Nature Reserve "Pasvik"	Vätsäri Wilderness Areain Finland and Øvre Pasvik National Park and Øvre Pasvik Landscape Protection Area in Norway	The Joint draft completed in 2005-2008

One of the results of the developing integration processes between countries is the formation of transboundary biospheric reserves. These reserves are part of the UNESCO World Network of

Biosphere Reserves. In 2016, the Network consisted of 669 territories in 120 countries, and 16 of them were transboundary.

The first transboundary biospheric reserve in the Russian Federation, the Grand Altay, is composed of the Katunskiy biosphere reserve (Russian Federation) and the Katon-Karagay biosphere reserve (Kazakhstan) formed in 2017.

The regulating documents defining the environmental policy are the Government program «Environment Preservation for 2012 – 2020» (Enactment of the Government of the Russian Federation of 15 April 2014 № 326) and the implementation strategy for «Basic Principles of State Policy of the Russian Federation in the field of environmental development for a period until 2030» (approved by the Government Resolution of 18.12.2012 № 2423). These regulating documents provide for preparation and implementation of projects of international treaties on the formation of TPAs between the Russian Federation and adjoining countries: the People's Republic of China, Mongolia, the Republic of Abkhazia, the Kingdom of Norway, the Republic of Finland, Ukraine, the Republic of Azerbaijan, the United States of America. The aim of these treaties is the landscape and biodiversity preservation in the border zones adjoining the national border.

To create optimal territorial conditions and achieve the objectives of PAs, it is necessary to implement land use planning. All activities concerning the formation of new PAs or the reorganization of the existing ones should be carried out only on the ground of land use planning layouts and projects. Such projects are important because not all of the existing PAs create the conditions to maintain viable population of organisms, even though the PAs were formed to preserve these organisms. The land use planning implementation will result in proper organization and achieving the objectives of PAs [7, 8].

The disconnexion of PAs and haphazard use of adjoining territories eventually lead to loss in environmental value and gradual degradation of ecosystems. Thus, it is necessary to form the interstate networks of PAs.

4. Discussion

The world's first TPA, the Waterton-Glacier International Peace Park, was formed in 1932 as a result of a merger between two national parks, Waterton Lakes National Park (Canada) and Glacier National Park (USA).

The history of TPAs in Russia starts in 1989, when the Agreement between the Government of the USSR and the Government of Finland of 26 October 1989 was signed and the first international zapovednik, the Russian-Finnish Nature Reserve Friendship, was formed. It consists of Kostomuksha Nature Reserve (Republic of Karelia, Russia) and 5 PAs of Finland: Juortanansalo mire reserve, Iso-Palonen - Maariansärkät nature reserve, Lentua nature reserve, Elimyssalo nature reserve in the province of Kainuu, and Ulvinsalo strict nature reserve.

The representatives of the Russian Federation and Mongolia set up the project of the Agreement between the two countries to form the Daursky international reserve. The project included the schematic map of the prospective international reserve, its objectives, process of cooperation and funding, and the monitoring of TPA's working. Later, the People's Republic of China acceded to the Agreement as well.

As the result of this cooperation, in 1994, the Daursky Nature Reserve (Russia), Mongol Daguur (Mongolia) and Dalainor Biosphere Reserve (China) merged into the tripartite Dauria International Protected Area (DIPA).

A prerequisite for the formation of the transboundary biospheric reserve was the fact that by 2006, all three reserves had been included in the MAB program («Man and Biosphere»).

The environmental protection of separated parts of the ecosystem is inefficient, as disappearance of animals in one part of it leads to overall reduction of the population. The need to preserve the spatial and genetic connection between animals living on the territory of three countries necessitated the formation of the ecoregion. Because of that, in December 2009, a territory of about 2 900 000 ha in steppe and forest-steppe regions of Zabaykalsky Krai adjoining the border of Mongolia and China became a model zone of cooperation. The main objectives of its formation were preservation and restoration of threatened species; ecological education; providing for efficient working of PAs of regional and federal importance; improving the cooperation between government bodies, educational, social, commercial, and other organizations in the field of environment preservation.

The official name of the TPA is the China-Mongolia-Russia International Protected Area «Dauria» (CMR DIPA), and its activity is coordinated by the Joint Commission headed by deputy heads of national environmental agencies. The current activities are carried out by the working group led by directors of the reserves. Later, those reserves achieved various statuses of international importance and participated in scientific and environmental projects, confirming their significance.

In 1996, the Russian-Chinese Nature Reserve Lake Khanka was formed according to the Agreement between the Government of the Russian Federation and the Government of People's Pepublic of China. This TPA consists of two reserves: Khanka Nature Reserve (Russia) and Sinkai-Hu Reserve (China). The Agreement of 31 May 2011 between the Government of the Russian Federation and the Government of Mongolia formed another transboundary reserve. It consists of the Uvs Lake Basin (Russia) and Uvs Lake (Mongolia). To date, this Agreement has not come into force. Therefore, there are 5 currently functioning transboundary protected areas in the Russian Federation. They are managed by the joint commissions of representatives of the participating countries.

The global community pays a lot of attention to TPA formation. This topic is mandatorily examined at IUCN World Parks Congresses: the recommendations of the Vth IUCN World Parks Congress (Durban, 2003) support the global initiative to form TPAs on every continent and in the World Ocean, and the VIth IUCN World Parks Congress (Sydney, 2014) also emphasizes formation and provision of sustainably functioning TPAs.

5. Conclusion

The analytical scrutiny of evidence and legal documents reveals that around the world and in the Russian Federation, the trend of TPA formation is likely to persist. Thus, the improvement of legislative and procedural guidelines for transboundary reserves formation is necessary to achieve the TPA objectives.

References

[1] Volkov S N, Shapovalov D A and Klyushin P V 2017 Solutions of problems in defining indicators of agricultural land within the framework of activities for the implementation of the concept of development monitoring in the Russian Federation *GeoConference SGEM* **17(52)** 819–28

[2] Shapovalov D A, Klyushin P V and Shirokova V A 2018 Problems and efficiency of land use in the North Caucasian federal district *GeoConference SGEM* **18(5.1)** 667–74

[3] Klyushin P V, Lukyanova T S and Shirokova V A 2018 The current state of boggy territories in the Meshchyora national park *GeoConference SGEM* **18(2.1)** 775–82

[4] Semenova E P and Shirokova V A 2018 Assessment of the adverse effect objects as a starting point for ecological reform *Current aspects of natural resource management problem solving* (Moscow: State University of farming) pp 431–7

[5] Shapovalov D A, Klyushin P V, Shirokova V A, Khutorova A O and Savinova S V 2018 Problems and efficiency of land use in the North Caucasian federal district *18th Int. Multidisciplinary Sci. GeoConf. SGEM–2018* **18(5.1)** 667–74

[6] Shirokova V A, Khutorova A O, Gurov A F, Shirokov R S and Yurova Y D 2018 Environmental Assessment of Anthropogenic Effects on Recreation Zones Osetr River in the Moscow Region *Baltic Geodetic Congress (BGC Geomatics)* (Olsztyn) pp 181–5

[7] Petrova L E 2016 Land use planning is main mechanism of realization of ecological doctrine in Russian Federation *Int. Sci. Techn. and Industrial Electronic J. GeoScience* **1** 71–6

[8] Volkov S N, ed 2017 *Land use planning in state programs and high-priority national projects in development of the agroindustrial complex and other fields of the economy* (Moscow: State University of farming)

Dependence of health conditions of population on the level of environment pollution

L P Stepanova[1], A V Pisareva[2], Nechushkin Yu V[2] and A I Myshkin[3]

[1]Orel State Agrarian University, 69, General Rodin street, Orel, 302019, Russia
[2]Bauman Moscow State Technical University, 5, 2-nd Baumanskaya, Moscow, 105005, Russia
[3]Orel State University named after Turgenev, 95, Komsomolskaya street, Orel, 302026, Russia

Abstract. This article presents a comprehensive assessment of the ecological condition of landscapes of the Orel region. Factors affecting the state of contaminated territories, in turn, affect human health. The effect of industrial waste and motor vehicle emissions on the geochemical features and environmental sustainability of soil, plants, surface water and the atmosphere in the study area is shown. The dependence of the level of exposure to toxicants on the health of the population of the Oryol region is established. Based on the assessment of the parameters of ecosystem pollution, we identified areas of ecological disadvantage in the Oryol region. It is shown that human activities and anthropogenic impact on the environment lead to loss of properties, providing a normal life and safety of human health. The degree of influence of soil degradation factors on the biogeochemical characteristics of the landscapes of the Oryol region is shown. The influence of anthropogenic factors on the concentration and form of chemical compounds of ecosystem components in the studied territories is established. The anthropogenic influence of sources of soil pollution by heavy metals and radionuclides is proved. Special zones of environmental catastrophe in the Oryol region, where the anthropogenic impact leads to a violation of human health safety, have been identified.

1. Introduction

Territories with a tense environmental situation or with a special regime of environmental management and economic activity are distinguished in connection with the real technological transition of most of Russia. Oryol is a city located 368 km southwest from Moscow in the Oryol region on the Central Russian Upland on both sides of the Oka River and a tributary of the Orlik River. The city is a part of the Central Federal District, the Central Economic Region and the administrative center of the Oryol region. The Oryol region was formed on September 27, 1937. Currently, the Oryol region is a constituent entity of the Russian Federation. It is also a part of the Central Federal District and the Central Economic Region of Russia. The population is 754,816 people; the share of urban population is 66.72%. On the territory of the Oryol region, there is 24,652 km^2. Oryol is the administrative center. The whole region is divided into 24 districts. The region is located in the central part of the Central Russian Upland in the forest-steppe zone in the south-west of the European part of Russia [1]. The climate of the Oryol region is moderately continental. In the eastern part of the region, the river Sosna flows with its tributaries. To the west of the city of Orel, the rivers Nerussa and Navlya flow [2].

The region is located in the zone of soil movement from sod-podzolic to predominantly leached and podzolized chernozemic soil [3, 4]. In addition, there are various types of soils in the region in composition, ranging from light gray forest to alkaline and typical chernozems [4, 5]. The industrial sectors of the region are: mechanical engineering (this part of the industry branch makes up 30% of the total production), the food industry in the Oryol region is more than 25%, and the share of the construction industry is only 13% [6]. In addition to the city of Orel, where most of the enterprises are concentrated, there are separate factories that are located in Livny and Mtsensk, as well as in other towns of the Oryol region [7]. The region's ferrous metallurgy is only 7%, which began to develop in the region in the second half of the 20th century [8]. In the city of Oryol, the Oryol Steel Rolling Plant open joint stock company is located, and in the city of Mtsensk, there are an aluminum plant and a plant for the production of metal products.

Relevance: the article discusses the factors of anthropogenic and man-made origin that determine the formation of disturbed areas, and also identifies areas of environmental problems among the population of the Oryol region [9]. As a result, the daily impact of the population on nature becomes significant, and the surrounding biosphere becomes not only the habitat of people, but also the sphere of their activity. In Russia, the concept of demographic policy for the period up to 2025 defines the main task of the government, which speaks of the need to reduce mortality, maintain and improve the health of the population, increase fertility [10, 11]. As a result, there is a need for scientific forecasting of environmental reactions to the impact of human activity on natural processes and the creation of anthropobiogeochemical zones representing an environmental hazard to humans [11].

Purpose: to determine the influence of the ecological situation on the health of the population in the administrative districts of the Oryol region and in the city of Orel.

Task: to identify the factors of anthropogenic and man-made origin which determine the formation of anthropogenic territories. To achieve this goal, we have chosen areas that are affected by various types of degradation changes.

2. Materials and methods

We selected as an object of the research pilot areas subjected to prolonged exposure to the waste slag of Mtsensk aluminum foundry plant. Area of continuous observation (PPN) was located in the village of Great Domino at different distances from the subject land pollution [12]. Soil type: light-grey forest, medium loamy soil on loess-like loam [13]. To determine the content of heavy metals in the soil, we took soil samples from the layers 0-20 cm and 1 kg each. In addition, for the analysis of 1 kg from the layer of 0-20 cm in 5-fold repetition, we took an individual mixed sample. Chemical analysis of soils was conducted on the following parameters: soil reaction (pH), gross and mobile forms of heavy metals (lead, copper, Nickel, zinc, cobalt, cadmium). The accumulation of heavy metals in soil and plants was determined by atomic absorption, fiery photometric method according to the method of the all-Union scientific-research Institute of mineral raw materials and Arinushkina E. V. [14, 15]. Granulometric analysis of soil was conducted using N. A. Kachinzynski method [16]. Total soil analysis was performed according to the method by N. I. Savvinov [17, 18]. Determination of the quantitative indicator of the humus was made using the method by I. V. Tyurin [19]. Preparation of salt extract and determination of its pH according to the method of TsINAO [20]. Determination of the amount of absorbed bases and hydrolytic acidity was made using the method by Capena Helikovice [21]. Definition of mobile forms of phosphorus and potassium was performed using Kirsanov method [20]. To determine the level of influence and nature of anthropogenic impact on the spatial variability of incidence rates, we used statistical data on the administrative districts of the Orel region over a 15-year period [22].

3. Results

The reaction of any human body to the ecological situation in ecologically unfavorable areas is manifested in the form of such diseases as: microelementosis, cancer, pathology of the human cardiovascular system, as well as other common and orphan diseases. A multi-component assessment of ecosystems makes it possible to identify zones of ecological disaster [23]. Continuous monitoring of the ecological situation provides for a multi-component assessment of the state of ecosystems, with human health being the most important indicator reflecting the state of the human environment [24]. It should be noted that in many cases it has become more convenient to divide the disease into infectious and non-infectious, the occurrence of which may be due to the influence of any physical or chemical factor.

As a result of our research, it was established that diseases of the circulatory system, oncology, and diseases of the respiratory organs, the musculoskeletal system and the genitourinary system prevail among the diseases of the population of the Oryol region. The influence of natural and anthropogenic and technogenic factors of the Oryol region increases the incidence. For example, diseases of the sensory organs, the state of the nervous system, respiratory organs, digestion, blood, skin diseases, the urogenital system and other organs in the human body can be directly associated with the pollution by aerotechnogenic toxicants; nitrogen compounds, sulfur dioxide, carbon oxides and other suspended particles prevail in their composition.

Thus, according to the regression analysis, the value of the correlation coefficient between pollution levels and environmental indicators of the territory and morbidity indicators was set in the range of 0.8-0.9. At the same time, there is no direct correlation between the phenomena studied; this is due to a different combination of natural, man-made, and social factors. It should be said that in the Oryol region, regular monitoring, continuous environmental studies and the creation of maps of the environmental situation are necessary, where the environmental situation can vary significantly within certain limits. It has been established that for the most effective point of assessing the ecological status of contaminated areas on the incidence of the population, it is necessary to take as an integral indicator of environmental impact and human health average annual indicators for the last five to ten years for various classes of diseases, which reflect the degree of radioactive and chemical influence of the environment on the level of health, for example, diseases of the respiratory and cardiovascular systems, as well as the digestive organs.

Overall, an analysis of the incidence rate on a medical-geographical basis in the Oryol region showed heterogeneity of the studied territories of the region. For example, in the Oryol region, the incidence of respiratory diseases is 1.1 times higher than the average for Russia. At the same time, in the Oryol region, this class of diseases is growing by 1.6 times, and in the city of Orel - by 1.4 times. Respiratory diseases in the region increased by 1.3 times over the fifteen-year period, with a predominance of certain nosological forms, such as chronic bronchitis and asthma. In the first place is the incidence of the circulatory system in the region. Over the fifteen-year period, this disease increased by 2.2 times and exceeded by 1.1 times the incidence level of the circulatory system in the whole country. The increase in the incidence of blood circulation in the city of Orel and the Oryol region was 1.4 times larger.

In the Bolkhovsky district of the Oryol region, there is a stabilization of the dynamics of diseases of the respiratory and circulatory organs, and in the Mtsensk district of the Oryol region, there is a slight decrease in such diseases. According to the research data, the largest share of skin diseases and subcutaneous tissue accounts for the Orlovsky district of the Oryol region; the share of diseases diagnosed with Acute Respiratory Viral Infection (ARVI) in the Bolkhovsky district of the Oryol region is 84.7%, and in the city of Orel, the share of diseases diagnosed with SARS is only 23.7%.

In the fourth place in the region are disfunctions in the endocrine system of the population, while this figure increased by 2.3 times, due to the deterioration of the ecological situation in this period. On the territory of the Mtsensk District, the situation has somewhat stabilized due to the incidence of the endocrine system and remains at the level of 1.65-1.7 thousand people.

The urogenital system of the population also suffers and exceeds in 1.17 times the level in the country, while over the fifteen-year period, the growth of this group of diseases among the population was noted by 2.3 times, including 2.5 times in the city of Orel. However, if for the Mtsensk and Bolkhov raions, there is a slight decrease in the incidence of the genitourinary system, then for such areas as Krasnozorensk, Zalegoschensk, Znamensk and Livensk in the Oryol region, there is a significant increase in this group of diseases among the population.

Analysis of the incidence of the digestive organs of the population shows that the increase in the level of diseases twice during the study period is noted in the territory of the Oryol region, the Oryol district and the city of Oryol.

Since annually the pollution of the territory of the Oryol region with radionuclides, heavy metals and harmful emissions of gases increases and becomes the cause of oncology diseases of the population, over the fifteen-year period, an increase was noted of 1.5 times in the population diseases of this group of diseases in the Oryol region, including 1.8 times in the city of Oryol, in the Oryol district – 2.4 times, in the Mtsensk district – 1.4 times, and in the Bolkhovsky district – 1.3 times.

Over the past three years, the following situation has persisted in the Oryol region: the environment for the spread of the population diseases: (18.2–18.8%) of the circulatory system, (14.3–14.6%) of the respiratory organs, (9.5–10.5%) of the musculoskeletal system, (8.9–9.2%) of the urogenital system, (6.1–6.4%) of the digestive organs, (3.2–3.9%) of the endocrine system.

We would especially like to note that diseases of the circulatory system make up a significant proportion of the total number of diseases of the population that require expensive treatment with the help of high-tech medical equipment. For the main causes of death, 62–64% are diseases of the circulatory system, 11.7–12.3% are causes of oncology, 4.1–4.9% are respiratory diseases, and 3.4–3.9% are diseases of the digestive organs. In the system of diseases of the respiratory organs, the main share is bronchitis and bronchial asthma.

An analysis of the incidence of the population among children is carried out, which is associated with the medical and environmental status of the administrative districts of the Oryol region. Statistics show an increase in the incidence of children over the past fifteen years. The highest increase in the incidence of children can be identified in the following areas: Novosilsky, Bolkhovsky, Mtsensky, Livensky, Soskovsky, Dmitrovsky, Kolpnyansky, Shablykinsky, Orlovsky and Hotynetsky. On average, in rural areas of the region, the incidence rate of children increased 1.9 times, and the incidence rate of adolescents in the city of Oryol increased 2.5 times over the observed period.

The territory of the Oryol region can be divided into four groups according to the incidence of the population of the region in comparison with the excess of the incidence of the population of the Central Federal District. The first group includes areas with an excess of the incidence rate in 1.5 - 2 times (Bolkhovsky, Dmitrovsky, Krasnozorensky, Trosnyansky areas), which are considered relatively safe in medical and environmental terms. The second group includes the districts of the region with an increase in the incidence of the population relative to the average level of the Central Federal District in 2-2.5 times. This group includes the territories of the following districts: Verkhovsky, Glazunovsky, Dolzhansky, Korsovsky, Kromsky, Maloarkhangelsky, Novosilsky, Orlovsky, Uritsky, Shablykinsky. Districts of the region included in the third group with an excess in 2.5 - 3 times the incidence rate in the Central Federal District, are environmentally tense. These are Mtsensk, Livensky, Znamensky, Khotynetsky, Novoderevenkovsky, Soskovsky districts of the region. In the fourth group, there are two districts of the region - Zalegoshensky and Kolpnyansky, where the incidence rate is more than three times higher than the average level in the Central Federal District. At the same time, our studies have shown that in the areas with relative ecological well-being, loci of high environmental tension can be distinguished. For example, on the territory of the Mtsensk District, this is the village of Bolshoye Dumchino, where the incidence rate is four times higher; on the territory of the Oryol District, it is the city of Oryol, where the incidence rate exceeds the incidence rate of the Central Federal District three times, and on the territory of the Bolkhovsky District of the Oryol region with relative environmental well-being, estimated at 1.5-2 points, areas are allocated with a high level of radioactive contamination (more than 5 Ku/km^2) and population morbidity exceeding the average federal level by more than four times.

4. Conclusion

Analysis of medical and environmental indicators found that the incidence in the areas of the Oryol region, both children and adults, has a close relationship with the current environmental situation, as evidenced by the increase in indicators for the studied diseases. As a result, the most thorough approaches may be required, both to monitoring the health status of the population, and to organizing their forms and rebuilding methods to maintain a healthy lifestyle. Studies have shown that the incidence of the population of the village of Bolshoe Dumchino, Mtsensk District, as an area of ecological risk, has a correlation dependency between the level of anthropogenic and technogenic pollution of the soil, air basin, water resources of the region and indicators of health of the people living in the area.

It has been established that a change in geochemical conditions at a narrow regional level, in the presence of a large source of environmental pollution, occurs so rapidly that the human body cannot adapt to extreme conditions in a timely manner. As a result, the natural biochemical processes that occur in the human body are disrupted. This manifests itself in the form of diseases of various systems of the body, and requires careful development of a system of measures for the recovery and adaptation of humans to extreme changes in ecosystems. The results of the study indicate that the health status of the population of the Oryol region reflects a change in ecosystems and may well be an indicator of the degree of anthropogenic and man-made pollution of the natural environment.

In our opinion, in order to improve the environmental situation, in addition to taking measures on environmental safety, changes in the structure of nature management are necessary. At the same time, greater attention should be paid to recreation as an alternative type of economic activity in the territory under consideration. In general, the conducted studies have shown that the geoecological situation in some districts of the Oryol region and in the city of Oryol is tense, and the medical and ecological situation is typical for the urbanized territories of the Russian Federation.

References

[1] Schepaschenko D, Moltchanova E, Shvidenko A, Blyshchyk V, Dmitriev E, Martynenko O, See L and Kraxner F 2018 Improved estimates of biomass expansion factors for Russian forests *Forests* **9(6)** 312

[2] Bayas J C et al 2016 Crowdsourcing in-situ data on land cover and land use using gamification and mobile technology *Remote Sensing* **8(11)** 905

[3] Ksenofontov B S, Antonova E S, Ivanov M V, Kozodaev A S and Taranov R A 2015 The influence of oil contaminated soil on the quality of surface waste water *Water Practice and Technology* **10(4)** 814-822

[4] Ksenofontov B S, Butorova I A, Kozodaev A S, Afonin A V and Taranov R A 2017 Problems of toxicity of ash and slag waste *Ecology and Industry of Russia* [in Russian – Ekologija i promyshlennost Rossii] **21(2)** 4-9

[5] Belotserkovsky M Yu, Belyakova T M and Berkovich K M 1993 *Environmental tensions of Russia (factors, their assessment and zoning). Problems of assessing the environmental stress of the territory of Russia: factors, zoning* (Moscow: Moscow State University) pp 93-100

[6] Arkharov I A, Simakova E N and Navasardyan E S 2016 Landfill Gas as Feedstock for Energy and Industrial Processes *Chemical and Petroleum Engineering* **52(7-8)** 547-551

[7] Bednova O V, Kuznetsov V A, and Tarasova N P 2018 Eutrophication of an Urban Forest Ecosystem: Causes and Effects *Doklady of Russian Academy of Sciences. Earth Sciences* **478(1)** 124-128

[8] Bednova O V, Kuznetsov V A and Tarasova N P 2015 Transformation of urban forest ecosystems: Indication and integral assessment *Doklady of Russian Academy of Sciences. Earth Sciences* **463(2)** 868-872

[9] Ksenofontov B S and Goncharenko E E 2016 Intensification of Purification of Surface Sewage by use a Bioflocculant *Herald of the Bauman Moscow State Technical University, Series Natural Sciences* **3** 118-127

[10] Fritz S et al 2017 A global dataset of crowdsourced land cover and land use reference data *Scientific Data* 4,170075

[11] Stepanova L P, Yakovleva E V and Pisareva A V 2019 Spatio-temporal dynamics of soil geochemical anomalies in the zone of impact of slag residuals *Ecology and Industry of Russia* [in Russian – Ekologija i promyshlennost Rossii] **23(3)** 44-48

[12] Ksenofontov B S, Kozodaev A S and Taranov R A 2016 The Problems of Rare Earth Metals Extraction from Coal Ash in the Form of Solvents and its Concentrating *Ecology and Industry of Russia* [in Russian – Ekologija i promyshlennost Rossii] **20(4)** 12-15

[13] Lavrentyeva G V, Mirzeabasov O A and Synzynys B I 2017 Monitoring of soil contamination in the zone of impact of a regional near-surface radioactive waste repository in the decommissioning stage *Radiation biology. Radioecology* [in Russian – Radiatsionnaja biologija. Radioekologija] (Moscow) **57(3)** 279-285

[14] Yakubovich A L and Ryabkin V K 2007 *Nuclear-physical methods of analysis and quality control of mineral raw materials* (Moscow: VIMC) p 263

[15] Arinushkina E V 1970 *Chemical soil analysis guide* (Moscow: Moscow State University) p 488

[16] Kachinsky N A 1958 *Mechanical and microaggregate composition of the soil, methods for its study* (Moscow: USSR Academy of Sciences) p 191

[17] Nesmeyanova M A 2015 Structural-aggregate composition and water resistance of the soil under the influence of perennial leguminous plants *Perm Agrarian Bulletin* **1(9)** 50-55

[18] Valkov V F, Kazeev K Sh and Kolesnikov 2004 *SI Soil science: Textbook for universities* (Rostov: Mart) p 490

[19] Tyurin I V 1965 *Organic matter of the soil and its role in fertility* (Moscow: Nauka) p 320

[20] GOST 26483-85 1985 *Soil. Determination of pH of salt extract, metabolic acidity, metabolic cations, nitrate, metabolic ammonium and mobile sulfur methods TsINAO* (Moscow: Publishing house of standards)

[21] Kaurichev I S 1980 *Workshop on soil science* (Moscow: Kolos) p 273

[22] Stepanova L P, Myshkin A I and Korenkova E A 2011 Assessment of the state of the environment to identify areas of ecological distress *Bulletin of Peoples' Friendship University of Russia. Series: Ecology and life safety* **1** 42-48

[23] Stepanova L P, Yakovleva E V and Pisareva A V 2016 Ecological characteristics of anthropogenically transformed soils contaminated with heavy metals *Agrochemistry* [in Russian – Agrokhimija] **12** 60-67

[24] Voronov A G 1986 *Medical geography. Anthroponoses: study guide* (Moscow: Moscow State University) p 82

Geo-ecological factor of water influence on the relief of steppe zones and semi-deserts of Kalmykia

M M Sangadzhiev, L M Rubeko, V A Onkaev, K G Kazakov and O Sh Kedeeva

Kalmyk State University named after B.B. Gorodovikov, 11, 1A building, Pushkina str., Elista, 358000, Republic of Kalmykia, Russia

Email: smm54724@yandex.ru; nauka.kalmsu@gmail.com; vik.onkaev@yandex.ru; koka68@inbox.ru; kedeeva.osh08@yandex.ru

Abstract. The Republic of Kalmykia is the only region in Russia with up to 10% of deserts and semi-deserts. Geographically, it is located in the western part of the Caspian lowland. One of the problems of the residents is the lack of fresh water. The aim of the work is to show the influence of the geo-ecological factor on the relief and its connection with surface and groundwater, human health. To solve the problem, results of the expedition routes in Kalmykia were analyzed. Expeditionary and laboratory studies of water resources, express analysis of water, soil and plants was carried out. The decision-making hypothesis adopted a synergistic algorithm for the interconnection system of the water system, the steppe relief and human health. This approach identified that values of the environmental impact are interrelated. Changes in one or several parameters cause instability of the entropy of equilibrium in environmental management. The results can be used to assess the relationship between the relief, water quality, the ecological situation and human health.

1. Introduction

In Kalmykia, one of the areas with minimal rainfall and snowfall, there are its own peculiar desert-steppe relief. The aim of the work is to analyze changes in the local relief caused by the influence of water resources.

The northern and western parts of the Caspian lowland are represented by plains with a rugged ravine-girder system, some of which are located below the sea level (up to -28 m in Lagan) [1]. The northern part of the Republic of Kalmykia (RK) is represented by the Yergenin Upland. It was formed in the Proterozoic and served as a shore of the Caspian Sea. In the eastern part, due to the influence of the Caspian Sea, spits are formed. Their height reaches 60 m. The length of the spit is several kilometers (https://youtu.be/azj1O1noV54).

Ground and surface waters are fed by melting snow which accumulates in the ravine-gully system. Climatic conditions (rains, amounts of snow) are an influencing factor as well [2].

The horizon line of rainfall is not uniform in Ergeny and the Pre-Caspian region. This is due to the air flow, both in the lower part of the horizon and at the heights where the clouds are located.

One more important factor is wind influencing on the surface and altitudes of more than 50 meters. Frequent winds form sand tornadoes, especially in the second half of summer. Dust rising from the eastern and southeastern parts of the territory reaches Elista within a few hours (about 250 km). The same dust with dry and strong winds has a strong effect on ravines and beams. There is a pool-type system; instead of water, a strong wind influences the relief (wind speeds is more than 10 m/s; at a height of more than 10-15 meters, it is 15 m/s and more). There are winds with a speed of 30 m/s (Figure 1).

Figure 1. Dust storm in Kalmykia

Wind and dust adversely affect the flora and fauna, the food chain and human health.

The relief is influenced by such factors as climate, drought, sand and dry winds. The role of the influence of tectonic factors is not great. But for the entire period of well operation and drilling in Chernozemel and Lagan district, probability of dig formation is high. This is due to the formation of voids in the lenses of water wells used mainly for technical needs. They form voids at different depths. The lithospheric composition is composed of a sedimentary cover with a depth of 2-2.5 km in the south-east of Kalmykia.

Due to tectonic phenomena, seismic waves with a force of 2-3 points reach Kalmykia. They can serve as an impetus for destruction in the hollows of the earth.

2. Analysis of literary sources

Scientific publications on climate changes were analyzed. The global warming is directly related to the economic potential of the regions. The climate has undergone several changes. As an example, one can analyze descriptions by Gumilev and other philosophers [2–4]. Employees of the Faculty of Engineering and Technology of the Kalmyk University published an article on climatic features of the region based on long-term observations and measurement data [5].

The influence of anthropogenic and natural factors on the formation and development of the ravine network is ambiguous. A positive aspect is accumulation of snow and rain. As a rule, small lakes are formed in the lower part of the ravines; they feed small rivers [6].

The influence of the human factor and regional sources on environmental management, ecology and subsoil management was described in “Peculiarities of subsoil management in the territory of the RK” [7].

The data on geological and geographical characteristics of the region were presented in [7, 8].

The interaction of ground and surface waters was studied by V.A. Onkaev, M.M. Sangadzhiev. and others. [5]

Frequent dust storms, dry winds carry sand over long distances. The focus is deserts located in the Chernozemel, Yustin and Yashkul districts. They are formed due to the influence of man on the nature, violations of the soil cover by animals, the anthropogenic impact of man, and land reclamation.

In the republic, there are more than 40 small rivers which form a wide ravine-gully system and channels of a complex geometry. A network of meanders is directly related to the destruction of the coastline and steep formation [5].

Groundwater located in the lithospheric zone at depths of 100-150 m is mineralized. Surface water is not suitable for household and drinking supply as well [5]. The theoretical data on groundwater are presented in [9].

The geotechnical characteristics of the soil, its salt composition and changes during long-term water filtration are described in the work by KalmGU workers [8].

3. Research methods and area. Data on water bodies in Kalmykia

Small rivers and intermittent lakes create a peculiar flat terrain. It is carved with beams, ravines and river beds.

The university staff and researchers of the Geographical faculty of Moscow State University study the areas of Kalmykia. The purpose of their expeditions was to study channels of small rivers, their coastal part. Heights on the slopes of the Yergenin Upland were measured. Some data were presented as informative on the websites for students studying general and regional geology and geography of the republic. Data can be interesting for schoolchildren studying the nature of the RK.

During the expeditionary routes, water was taken from various sources: streams, rivers, lakes, springs and wells. The data were presented to the laboratory. In addition to water and soil, the vegetable layer was analyzed.

A one-time sampling of water and soil was used. The dynamics of water changes was not studied due to the fact that main indicators of the water composition do not change. We did not carry out a detailed analysis of water for chemical compounds and elements.

Before the expedition, cartographic materials of the expedition area, photo and video were studied. We used a Garmin navigator with a built-in map of the region. With the help of the navigator, geometric measurements of the shoreline of rivers and lakes, steep heights, slopes were made. Quadcopters were used. The areas are presented on the Internet sites (https://youtu.be/O_HbiROOpEM, https://youtu.be/aSKEqLrezwE).

4. Results and discussion

Kalmykia is cut up by a network of ravines and beams, especially at the foot of the Yergenin Upland. This is due to the fact that thawing and rainwater flowing from the surface of the hill wash the upper ground layer. This layer consists of loams, sandy loams, and clay.

Therefore, the influence of surface and underground waters on the lithospheric part of the earth forms a ravine-ravine system. This system violates the cover, changes the relief, reduces areas suitable for agriculture. The growth of ravines and beams is linear.

The economic damage is due to the loss of agricultural land areas, expansion of ravines and land plots and water reservoirs with ravines. To assess and determine the possible damage from ravines, the choice of anti-vegetation facilities and their rational placement, data on ravine growth rates are necessary.

Growth rates at different stages of ravine development vary considerably. In the initial stage of gully formation, the intensive linear growth is observed; the growth rate is more than 100-150 m/year. For the territory of the Republic of Kalmykia, the maximum linear growth rates vary from a few meters to 10 m/year. The average annual growth ranges from 0.7 to 10 m/year. Growth rates vary in individual years and seasons. Therefore, the use of short-term observations to estimate average growth rates can lead to erroneous conclusions. In addition, the ravines at different stages of their development are compared; the measured speed rates are atypical for evaluating the process as a whole. Cartographic materials that provide reliable data on the growth of ravines make it impossible to determine a seasonal increase and identify differences in speed rates at different stages of ravine development.

Various external factors influence the rate of changes in gullying conditions. But the growth of ravines changes under the influence of changes in the conditions of ravine formation produced by ravine erosion and transformations of the channel bed. This process is directional and goes through a series of stages characterized by significant changes in trends (acceleration, deceleration) and growth intensity in different dimensions (linear, area, volume) and directions (depth, width, length). Growth rates can change. As a result, at the late stages of development with relatively favorable conditions for the ravine growth, the growth is slower than at the first stages with less favorable environmental conditions.

One of the most characteristic features of the dynamics of the ravine development is the uneven growth of all its morphometric parameters.

In the initial period of gullying, an increase in intensity of erosion during the formation of a channel stream is important. When small slope streams merge into the mainstream, a significant increase in flow velocity is observed. It is due to the decrease in energy losses spent on the movement of water masses within the channel compared with the movement of small stratified streams on the surface of the slope.

Development of suffusion and gravitational processes predetermines the occurrence of a primary differential stimulating a sharp increase in flow energy. The latter causes a rapid increase in the drop height and speed of its regressive movement (the linear growth of the ravine), as well as the rate of deepening of the channel below the drop (deep growth). At the same time, an increase in the relative height of the slopes under the influence of deep erosion intensifies gravity processes. A decrease in the speed of ravine growth which marks the transition from the first phase to the second one is caused by two factors. The linear growth slows down as a result of a gradual reduction in the mass of water entering the top of the ravine due to the linear growth of the ravine caused by a decrease in the apex of the catchment area. Deepening and lateral growth of the ravine slow down due to a drop in the flow energy caused by a decrease in its speed due to a decrease in the average slope. Figure 2 shows a photo taken by D. Badmaeva in the vicinity of Sadovoe (Sarpin District, Republic of Kalmykia).

Figure 2. A ravine near Sadovoe (Republic of Kalmykia)

The deformation of slopes and the ravine top is not always associated with the flood. Gravitational processes are regulated by the critical height and steepness of the slopes which can reach significant values at any runoff value. Filtering water filling side cracks plays a crucial role. This process is delayed. Dry periods contribute to the formation of slope debris; accumulation of slope destruction products at the bottom of the ravine is completely unrelated to the runoff.

The second feature of the dynamics of ravine growth is changes in the parameters of the ravine over time. The maximum frequency of oscillation of rates of all parameters of the ravine is observed during the first 10-20% of the period of its existence.

When assessing the growth rate, growth trends are of great importance, since at different stages of development at the same average speed, the ravine growth can be characterized by general acceleration and deceleration.

5. Influence of the water collection morphology on the ravine morphometry

The relief accelerates or slows down the transition of ravines from one stage to another one or ceases their growth at different stages. The main changes in the development of the ravine parameters under the influence of geomorphological factors (depth of the erosion basis, length and steepness of the slope, shape and area of the water collection area) are as follows: 1) frequencies and amplitudes of fluctuations in growth rates and ravine parameters; 2) limiting sizes and time.

Large morphological parameters of the water collection are in compliance with large maximum and average growth rates of the ravine in its different dimensions, as well as the frequency and amplitude of their oscillations. The size of the gully watershed has a major effect on the volume growth and intensity of the linear growth. The volume of the ravine and its growth rate at all are higher with large values of the water collection. The exception is a more active volume growth at the first stage of development on short steep slopes in comparison with more gentle long ones. At all the stages, the ratio changes and remains in this form until a longitudinal profile is developed by the ravine.

Changes in the morphology change absolute values of dimensions of the gully forms and the time of their development. In all cases, t large sizes of the ravines correspond to the large morphological parameters of the water collection. However, maximum dimensions of any parameter depending on the morphological characteristics can reach different values causing a different width and area in the watershed zone of the “undetected” gully erosion, a different percentage of gully lands destruction, etc.

At large depths of the erosion basis when the length of the ravine is 40% of the maximum one, its volume does not exceed 5% of the final one. On a slope with a ledge, the ravine develops much earlier, and its volume (with the same length of 10-15%) exceeds the volume of the ravine developing on a straight slope.

With a large steepness and a base depth of the erosion basis, tops of the ravine can be so close to the line of the watershed and the width of the zone of "non-emerging" erosion becomes small.

The different shape of a slope affects the flow speed and its eroding ability at the early stages of development determining intensity of linear and volumetric growth of the ravine. The shape of the slope affects the marginal volume of the ravine: on a convex slope (ceteris paribus), it is significantly higher than on a straight one. At the same time, it was revealed that the shape of the slope has no effect on the shape of the “developed” longitudinal profile and the maximum length of the ravine.

6. Conclusion

Geological activity and other factors influencing the relief in the desert and semi-desert territories of the Republic of Kalmykia change the landscape. The main geological unit that changes the relief of Kalmykia is the Yergenin Upland with its numerous ravines and beams. The waters flowing from the summits of the hills in the Western and North-Western part of the Caspian lowland form a network of small rivers. They collect water for numerous lakes and ponds used for land reclamation, household and drinking purposes (for livestock).

Water is not used rationally. In the Soviet era, reservoirs, dams and other water-accumulating facilities were built. Currently, they are not used due to the collapse of the republic’s economy; settlements that used water from reservoirs and rivers ceased to exist or became very small (the number of inhabitants is 300 people per village). The lack of funding, reduction of the car and tractor fleet do not allow them to repair dams and monitor their performance.

References

IOP Conf. Series: Earth and Environmental Science **350** (2019) 012044 doi:10.1088/1755-1315/350/1/012044

[1] Ovchinnikov A S, Borodychev V V, Dedova E B and Ivanova V I 2015 Peculiarities of the ecosystems of the salt water bodies of Kalmykia *News of the Nizhnevolzhsky agrouniversity complex: science and higher professional education* **4(40)** 10–21

[2] Berg L S 1922 *Climate and Life* (Moscow: GosIzdat)

[3] Bobylev S N and Gritsevich I G 2005 *Global Climate Change and Economic Development* (Moscow)

[4] Bolshakov V A 2003 *The new concept of the orbital theory of paleoclimate* (Moscow)

[5] *Climate database* Available at: http://ru.climate-data.org/region/686/

[6] *Gully erosion* 1989 ed R S Chalov (Moscow: Moscow State University)

[7] Sangadzhiev M M 2015 *Features of subsoil use in the Republic of Kalmykia* (Elista: Kalmyk University)

[8] Dordzhiev A A, Dordzhiev A G, Sangadzhiev M M, Rubeko L M and Onkaev V A 2018 Salt Composition of Clay Soils and Its Variation with Long-term Water Filtration *J. of Environmental Management and Tourism* **IX 1(25) Spring** Available at: http://dx.doi.org/10.14505/jemt 130–5 ISSN 2068–7729

[9] Kene V 1932 *Groundwater Doctrine* (Moscow; Leningrad: NKTP USSR; State scientific and technical publishing house of the construction industry and shipbuilding; Gosstroyizdat)

Heavy metals in the soil of Kalmykia arid territories

L H Sangadzhieva[1], Ts D Davaeva[1], B V Tsombueva[1], O S Sangadzhieva[1] and M S Esenamanova[2]

[1]Kalmyk State University named after B.B. Gorogovikov, 11, 1A building, Pushkina str., Elista, 358000, Russia

[2]Atyrau State University named after Kh. Dosmukhamedov, 1, Studencheskiy ave., Atyrau, 060011, Republic of Kazakhstan

Abstract. In the composition of soils of the Republic of Kalmykia a large diversity of the gross content of heavy metals was found, which is explained by different genesis of top sediments and the leading role of fine particles in them. Several natural soil and geochemical provinces as well as technologically polluted territories were identified where an imbalance of chemical elements in food chains is observed. It shows promising use of the relationship between the gross amount of individual metals to study the genesis of soil-forming rocks and soils.The limits of the content of 23 elements were revealed, of which 5 elements are absent in the soil of the Republic of Kalmykiaor are within the sensitivity of the instruments. In the south of the Republic of Kalmykiaa soil and geochemical province with a high total As content was found. Its core is located in the southeastern part of the Black Lands. The arsenic content here ranges from 15 to 35 mg/kg. The province is polluted with oil. Besides, there is a uranium and phosphorite region, where, along with the phosphorus the arsenic was concentrated. The arsenic-rich rock destruction products were transferred to the neighboring plain and were part of sediments. However, toxicosis associated with a high content of arsenic was not observed in the described area, since soil As is immobile and its supply to food chains is limited. This can be judged by the number of elements in plant products: it remains environmentally friendly.

1. Introduction

The study of chemical elements in the natural objects is currently of great importance, since it helps to understand many biogeochemical and ecological events and, if they are negative, develop methods for eliminating them or mitigating their actions [1–5].

The soil cover of the Republic of Kalmykia (RK) has been formed for a long time and in different parts of it (in different environmental conditions). As our studies have shown all this is reflected in the regional and local specificity of the microelement composition of soil-forming rocks, as well as in the soils formed on them.

These circumstances make it possible to addsome important details to the ideas about the history of the soil cover and the genesis of soils. It has to be admitted that the soil cover of RK has been insufficiently studied. New data may increase interest in the topic considered. Some approaches aimed to clarify the origin of light and loose sediments, on which most of the soil cover of RK is formed, are not purely regional.

The purpose and the objectives of this work is to study the gross content of trace elements in the soils of the arid territories of Kalmykia to identify their biogeochemical distribution in natural landscapes, to identify the deficiency or excess of trace elements.

2. Materials and methods

The Republic of Kalmykia is located in the south-east of the European territory of the Russian Federation between 44°40′ and 47°35′ north latitude and 40°10′ and 40°50′ east longitude. Kalmykia is the driest region of the European part of Russia. In terms of climate aridity this territory is the second after the deserts of Central Asia. In summer and winter, there is a small amount of precipitation here, which is explained not so much by the remoteness of the region from the ocean, as by the relatively rare penetration of cyclones. Due to its southern position the territory of Kalmykiareceives a lot of heat. The sunshine duration here is 2180-2250 hours/year. The amount of total solar radiation entering the area varies from 115 kcal/cm^2 in the north and west and is up to 120 in the central and southeast regions [5–7]. The territory of Kalmykia belongs to the low-water-supplied regions of the country, as the hydrographic network here is poorly developed. Only the extreme northern section, bordering the Volgograd region, is characterized by a relative water content of about 0.5–1.0 l/s per 1 km^2. In the rest of the territory, it is even lower with a tendency to decrease in the direction from the north-west to the south-east and in the south is almost close to zero.

The Republic is located in the semi-desert zone, the distinctive feature of which is the complexity of the soil and vegetation cover, which manifests itself in a mosaic combination of steppe and desert areas. This combination is due to the abundance of salt and salt marshes that are widespread throughout the country [6]. The vegetation of the Caspian lowland repeats the variegated complexity of the soil cover. It is common to attribute associations and formations with the dominance of herbaceous perennial xerophytes, mainly turf grass and forbs to the steppe type of vegetation, and the dominance of halophytic and xerophytic semi-shrubs and half-shrubs to the desert type [7–9].

The average annual rainfall is small. In different parts of Kalmykia, it falls from 200 to 300 mm, and only on the western outskirts of the Manych valley it reaches 400 mm and more. The precipitation on the territory of the Black Lands is relatively little, i.e. 200–240 mm per year. More than half of the annual precipitation falls during the warm season, when the air temperature is high, and therefore a significant part of the precipitated rainfall evaporates, not having time to penetrate the soil.

The study of the geological and morphological conditions of the Caspian lowland proves the fact that the underlying rocks of the quaternary system are represented by the ancient Caspian sediments of the Khvalynsky stage. The lithological characteristic shows that the quaternary system is represented by clays and sands. The brownish gray and chocolate clay with thin lense sands. The gray and gray-green sands, fine-grained, slightly clayey, micaceous sands with the thickness from 30 to 130 m.

The soil survey of the North-West Caspian of the Russian territory and the Republic of Kalmykia was carried out. The features of the distribution of 25 elements in the soil and ecological Pre-Caspian District in the semi-desert soil-climatic region, where two districts (Chernozemelsky and Pri-Caspian) are studied. There are three administrative districts of Kalmykia and one district of Kazakhstan in this territory.

The semi-desert and soil-bioclimatic region is located on the flat territory of the southeast and the east of the republic. The Caspian province is characterized as a zone of light chestnut and brown soils, sodic complexes, sandy massifs and salt marshes.

The selection of soil samples for physicochemical and chemical analyzes and their subsequent cameral treatment was carried out according to the recommendations [10]. In accordance with the generally accepted agrochemical methods, the main physicochemical properties of soils were determined (the particle size distribution of soil, the content of organic matter, carbonates, exchangeable Ca+2 and Mg+2, pH, dry residue of aqueous extract and anion-cation composition).

The assessment of the ecological state of the studied soils was carried out according to the total content of chemical elements in the soil and in terms of metal retention by soils. The gross content of Mn, Cr, Ni, Cu, Zn, Pb, Cd, As in soils was determined by the atomic absorption method on an MGA-100 spectrometer with argon atomization.

An environmental load rationing was carried out by various methods: 1) while monitoring the content of heavy metals in soils, the soil pollution levels were compared over the natural background by MPC, clarke in the lithosphere and the background level; 2) biogeochemical load rationing. The concentration coefficient was calculated by the ratio of the values of trace elements to clarke in the lithosphere according to A.P. Vinogradov [1] (Kk). The coefficient of background concentration of heavy metals in soils (anomaly coefficient) is calculated by the ratio of the values found to the background level (Кф).

3. Results

Chernozemelsky district is a district of brown, mostly solonizedand slightly saline soils of light grain size: sandy,sabulousand loamy soils, as well as their irregularly spotted complexes with solonetzonthe sea and alluvium and marine sediments.

Brown semi-desert soils lack nutrients and humus. The alkaline reaction of the medium is up to pH 8.5–9.0. The absorption capacity islow. The alkalinity processesare ubiquitous. All the soils are complex with solonetz and with salt marshes.

The limits of the content of heavy metals in the soils of the Chernozemelsky district are as follows: Al 1400–7107; Fe 1965.2 - 3015.6; Ti 17 - 45.0; V 32.3 - 90.5; Cr 16.2–37.3; Mn 30.4–174.1; Co 6.88–14.5; Ni 6.6–16.08; Cu 13.58 - 36.04; Zn 57.61 - 100.39; Ga 5.2–34.8; As 11.9 - 17.66; Sr 7.5– 12.7; Mo 0.05 - 0.19; Cd 0.01 - 0.16; Sn 0.10 - 0.42; Hg 0.001 - 0.013; Pb 1.85 - 8.84 mg/kg. Ba, Be, Li are absent.

The greatest variability is observed in Ti, Cr, V, Cu, Zn, As, Mo, Cd, Pb. The profile distribution of heavy metals differs insignificantly, since the soils have a light particle size distribution, vegetation is sparse, and the projective cover is less than 30%.

As far is the concentration range is concerned the picture is as follows: $n*10^3$ – Al, Fe; $n*10^2$ – Ti, Mn; n*10 – Cr, Co, Ni, Cu, Zn; n – Pb; 0.1*n – As, Mo, Cd, Sn; 0.01*n – Hg. According to the concentration row the elements are arranged in a row: Al> Fe> Zn>Mn> Cu> Cr> Ni>Sr>Pb>Sn> Mo> Cd> As> Hg. The concentration coefficient indicates equal or close to 1 value of clarke - Cd (0.95), the excess of clarke for Co (1.6), Zn (4,2), Cu (1.2), below clarke - Fe (0.1), Ti (0.5) V (0.5), Cr (02), Mn (0.6), Ni (0.3), Sr (0.01), Mo (0.1), Sn (0.03), Pb (0.3), Al (0.5), Hg (0.003). For toxic metals the levels for Cd below clarke are 5 times, for Hg are 5–7 times, for Pbare 4-8 times.

Comparison with the background level of heavy metals in soils (Kf): below the background:Mn (0.2), Cd (0.15), Mn (0.2), Hg (0.05), Pb (0.4), As (0.4); equal to the background: Co (0.95), Mo (0.98); exceeds the background: Cr (1.1)Cu (1.6), Zn (1.7). Maximum permissible concentration is exceeded for Co (1.6), Zn (1.4).

The Primorsk part of the Pre-Caspian lowland of Kalmykia territory is a province of brown soils, sable complexes, sandy massifs and salt marsh spots. The studied soils have a light particle size up to marching sands on the marine sediments.

The limits of the content of heavy metals in the soils of the Lagansky district are as follows: Al 400–1107; Fe 756.2-1895.6; Ti 24–147.1; V 30.0–64.5; Cr 7.5– 20.3; Mn 75.1–206.0; Co 1.3-11.5; Ni 0.7–18.6; Cu 6.2–34.12; Zn 17.5–42., 9; Ga 12.0–29.8; As 0.8–3.49; Sr 3.0–11.2; Mo 0.01–0.19; Cd 0.002–0.34; Sn 0.05–0.35; Hg 0.001–0.013; Pb 0.5–5.4 mg/kg. Ba, Be, Li are absent.

The greatest variability is observed in Ti, Cr, V, Cu, Zn, As, Mo, Cd, Pb. The large variability of such toxic heavy metals as Cd and Pb is due to confinement to sources of pollution, i.e. gas stations and heavy traffic. The whole territory of Laganicity is divided into microdistricts with canals. The sewage is mostly brought into canals, therefore pollutants are distributed throughout the city. This distribution is especially dangerous for toxic metals. The profile distribution of heavy metals differs insignificantly, since soils have a light particle size distribution, vegetation is sparse, and the projective cover is from 25 to 35%.

On the concentration range: $n*10^3$ –Al, Fe, $n*10^2$ – Ti, V, Mn, n*10 – Cr, Co, Ni, Cu, Zn, Sr, n – As, Pb, 0.1*n - Mo, Sn, 0.01*n – Cd, Hg.

According to the concentration row the elements are arranged in a row: Al> Fe>Mn> V> Ti> Zn> Cu> Cr> Ni>Sr> Co>Pb> As> Cd> Hg.

The concentration factor (according to A.P. Vinogradov) indicates the levels: equal to clarke or with a slight excess of clarke - As (1.1), for Zn (1.5), Cu (0.98), Cd (1.8), other elements below clarke- Al (0.5), Fe (0.2), Ti (0.2), V (0.45), Cr (0.06), Mn (0.15), Co (0.5), Ni (0.25), Sr 0.03, Mo (0.06), Sn (0.2

), Pb (0.3), Hg (0.02). For toxic metals, the levels of Cd are 2–10 times below Clarke; for Hg - 100 times; for Pb - 5–7 times.

Comparison with the background levels of heavy metals (Kf): below the background -Sr (0.03), Mn (0.3), Mo (0.6), Hg (0.07), As (0.7), Cu (0.6); equal to the background - Mo (0.89), Zn (0.8); exceed the background - Cd (2.5). The MPC is exceeded for As (1.2).

In the Zhylyoi region of the RKthe brown semi-desert soils of light grain size are highly saline due to their development on the marine sediments.

The limits of the content of heavy metals in the soils of the Zhylyoi region of the Republic of Kazakhstan are as follows: Al - 1425; Fe –1959; Ti 26.6 - 147.6; V 16.9 - 45.7; Cr 3.1 - 8.5; Mn 86 - 182.5; Co 2.15 - 8.11; Ni 2.3 - 7.65; Cu 59.2 - 213.7; Zn 0.68 - 11.04; As; Sr 9.6 - 31.4; Mo 0.02 - 0.12; Cd 0.003 - 0.025; Sn 0.03 - 0.09; Hg 0.002 - 0.05; Pb mg/kg. Ba, Be, Li are absent.

As far as the concentration range is concerned the picture is as follows: $n*10^3$ – Al, Fe; $n*10^2$ – Mn, Zn, Cu; n*10 – Ti, V, Co, Pb; n – Co, Ni, Zn, Sr; 0,1*n Mo, Sn; 0.01*n – Cd. The concentration row: Al> Fe>Mn> Cu> Ti> Ni>Sr> V> Zn> Cr>Pb> As> Hg. As far as the coefficient of concentrationis concerned the picture is as follows: no equal to clarke, exceedance of clarke- Ni, Cu, below clarke- Al, Fe, Ti, V, Cr, Mn, Co, Ni, Zn, Mo, Cd, As, Pb, Hg. For Cd 5–7 times lower than clarke, 2–4 times for Hg, 5 times for Pb. The MPC exceedancefor Ni, Cu.

4. Discussion

The studies of the soil cover in the recent years have given grounds to assume that there are natural habitats with a natural deficit or an excess,and other poorly studied chemical elements on the Caspian Lowland.

The content of heavy metals is partially inherited by soils from parent rocks, as evidenced by the concentration ratio above 1 or close to it; however, it turned out that there are areas on the territory of the RK with significant differences, i.e. from very low to very high levels. A low gross amount is typical of soils formed on sands and sandy loams, and a high amount is typical for soils of heavy particle size distribution. The main factor of this difference is the unequal saturation of sandy, silty and clay particles with heavy metals, and high saturation with many heavy metals of the main carriers of silt particles is not so much modern as relic: it already took place in the sediments that became parent rocks [2, 3, 11, 12].

If we proceed from what has been said, it is easy to foresee that low and dropped down concentrations of metals should be confined to the soils of high terraces of rivers and valleys of ancient runoff, which are found in the southeast of the RK on the Black Lands.

In the south of the region there are vast territories, where redistribution of water-soluble salts between geomorphological structures and relief elements has been and is currently taking place, which has become one of the causes of soil salinity in low spaces of the steppe zone [5, 8, 9].

Compared to the natural soil-geochemical provinces, the polluted urban lands are not so extensive in the area and compact in configuration. Within such an area, the content of pollutant elements in soils is represented by a complex mosaic of contours, where along with the citywide pollutants the local ones (producers of individual industrial enterprises) may be present.Compared to natural soil-geochemical provinces, polluted urban lands are not so extensive in area and compact in configuration. Within such an area, the content of pollutant elements in soils is represented by a complex mosaic of contours, in which, along with citywide pollutants, local (producers of individual industrial enterprises) may be present.

In the south of the RK a soil-geochemical province with a high concentration of As was found. Its main part is located in the southeastern part of the Black Lands. The arsenic content here ranges from 15 to 35 mg/kg. One may correctly estimate these data if to take into account the fact that in the soils of the European part of the country the background content of arsenic varies between 0.8–8.6 mg/kg, and the domestic hygienic standard (MAC) for light soils is 2; for heavy soils with neutral medium –10 mg/kg. The products of rock destruction rich in As were transferred to the neighboring plain andwere part of the loose sediments. However, the toxicosis associated with a high content of arsenic was not observed in the described area. The fact is that the soil As is immobile and its supply to food chains is limited. This can be judged, in particular, by the amount of the element in plant production: it remains environmentally friendly.

Thus, we are dealing with a natural soil-geochemical province, enriched with the hygienic As standard, which, however, has no negative manifestation for the objective reasons. Similar cases when soils contain a large amount of inactive chemical element are possible in other places. They should be considered as potential biogeochemical provinces where the transition of sedentary forms of the element into mobile ones will occur when the properties of the soil change. In relation to As the alkalinity in soil must increase to pH of 9–10 in order for the element to become more mobile. The alkalization of soil often occurs in case of itstechnogenic pollution.For example, now the pH is 8.5–9.0in some soils of Elista. However, at the indicated values, the alkalinity becomes very toxic for plants, which makes it difficult or impossible to cultivate agricultural crops. In alkaline medium the mobility of Mo is increased. An increased gross quantity of this element was found in the soil cover in the south of the region. It is especially high (exceeds the MPC) in the soils in the south of the northwest of the Caspian Lowland, which are influenced by phosphorus-bearing rare-metal mineralization of rocks, of which Mo is a component. High concentrations of metal were found in marsh-saline and solonetz-saline vegetation on pastures and haymakings of the Caspian Sea [5, 8, 9].

Such (potentially dangerous) biogeochemical provincescan change their ecological status during other types of human activities. Therefore, the search, delineation, study of soil and geochemical features of such areas should be presented in the regional monitoring.

5. Conclusion

The study of heavy metals in the soil cover on the territory of Kalmykia has revealed a great diversity of their gross content due to the genesis of the surface sediments on which modern soils have been formed, as well as the unequal content in the soil-forming rocks of fine particles. The content of heavy metals in the air and soils of the cities and their suburbs is significant, which worsened the living conditions of the population and even made them dangerous.

1. The set of the excessive chemical elements contains not only well-known "priority" Pb, Cd, Zn, Cu, but also, judging by the soil cover in the city of Elista, not well studied Hg, Ga, Sn, Se, Ni , Mo and others.
2. The obtained information on the content and ratio of some elements in the soil cover made it possible to clarify the genesis of soil-forming rocks and soils. The role of aquatic migration and exogenous processes has been enhanced.
3. The aforementioned is reflected in the peculiarities of the microelement composition of soils in certain territories, which we consider as soil geochemical provinces. The deficiency or excess of chemical elements in soils typical for such provinces can negatively affect the life of plants, animals and humans through food chains.

The study was supported by the Kalmyk State University named after B.B. Gorodovikov in the framework of the research project 1096 Comprehensive analysis of accumulation and emission of polycyclic aromatic hydrocarbons and radionuclides in natural and man-made landscapes of arid territories.

References

[1] Vinogradov A P 1963 Biogeochemical provinces and their role in organic evolution *J. Geochemistry* **2**199–213
[2] Glazovskaya M A 1988 *Geochemistry of natural and technogenic landscapes of the USSR* (Moscow: High school)
[3] Ilyin V B 1988 Biogenic and technogenic accumulation of chemical elements in soil *J. Soil sci.* **7** 110–21
[4] Kabata-Pendias A and Pendias H 1980 *Trace elements in soils and plants* (Moscow)

[5] Sangadzhieva L H, Erdniev O G, Davaeva Ts D and Rubeco L M 2008 Distribution of macro– and microelements in plants of oil-contaminated territories *J. of Natural science* **1(22)** 32–6
[6] Bakanova T I, Vorobyeva N P and Zelensky E A 1994 *Soils of The Republic of Kalmykia* (Elista: Dzhangar)
[7] Bananova V A and Gorbachev B N 1990 *Natural grasslands of the Kalmyk Autonomous Soviet socialist Republic and their rational use* (Elista)
[8] Davaeva Ts D 2006 The current state of ecosystems in the oil fields in the South of Kalmykia Republic *Scientific thought of Caucasus SKNTS VSH Special iss.* **5** 64–7 (Rostov-on-don)
[9] Ildzhinova M I, Tsombueva B V, Davaeva Ts D, Sangadzhieva O S and Sangadzhieva L H 2010 Estimation of negative influence of oil pollution of the reserve Black Earth *J. of Natural sci.* **3** 57–64
[10] *Guidance on sanitary-chemical study of soil* 1993 (Moscow)
[11] Perelman A I 1989 *Geochemistry* (Moscow: High school)
[12] Solntseva N P and Pikovsky Yu I 1980 *Migration of pollutants in soils and adjacent environments* (Leningrad: Hydrometeoizdat) p 25

IOP Conf. Series: Earth and Environmental Science **350** (2019) 012046 doi:10.1088/1755-1315/350/1/012046

Assessment of total chlorine and bromine content in semi-volatile organic substances in water

M Yu Vozhdaeva[1,2], A R Holova[1], N V Trukhanova[1], I A Melnitskiy[1,2] and E A Kantor[2]

[1]Center of Analytical Control of Water Quality, Municipal unitary enterprise "Ufavodokanal", Rossyiskaya str., 157/2, 450098, Ufa, Russia
[2]Ufa State Petroleum Technological University, Kosmonavtov str., 1, 450062, Ufa, Russia

E-mail: vozhdaeva@mail.ru

Abstract. The integral indicators for assessment of total concentration of chlorine and bromine in semi-volatile organic substances (SVOS) of water are presented. The method is based on liquid/liquid extraction of SVOS from a sample of water (natural, potable, industrial, purified sewage water etc.), concentrating the extract and without derivatization or conversion analyzed by the method of gas chromatography, coupled with atomic emission detecting (GC-AED). In this case the AED operates in two modes: the chlorine selective channel with detection at a wavelength of 479 nm and the bromine selective channel – at 478 nm. Under the given conditions of sample preparation and the chromatographic procedure the components are registered with Kovats retention index on non-polar phases from 900 to 3200. The range of the analytical procedure covers classes of SVOS, such as halogen-containing pesticides, aromatic and aliphatic hydrocarbons, fatty acids, ethers, polyaromatic hydrocarbons, phenols and their halogen derivatives. The total concentration of chlorine and bromine is calculated by component-independent calibration which is available by using an AED. The indexes are named Technogenic Organic Chlorine and Technogenic Organic Bromine, because analyzed compounds have technogenic or secondary technogenic origin. The sensitivity of determination is 1×10^{-6} mg/dm^3 for the specified elements.

1. Introduction

The control of water contaminates is usually reduced to the definition of a specific list of compounds: global pollutants that have become ubiquitous and some site-specific emerging contaminant groups, which may be present in the water. The number of such compounds is not large, and it is defined on practical grounds taking into account the possible availability of laboratories and necessary rapidness of the analysis. Under this approach the informative value of the received data is not sufficient to assess the total water pollution because of the fact that a significant amount of the contaminants including chlorinated by-products has not been estimated [1].

In heavily industrial regions and megapolises the reservoir water usually has a complex matrix of organic pollutants. Especially in view of possible rain-storm flows from the territory of river watershed which can be agricultural lands or industrial sites, atmospheric transfers and other antropogenic sources of contamination. Chemical plant wastewaters also pose risk to normal functioning of water resources [2]. First of all it depends on their complex chemical composition. The water treatment technologies available at the plant not always cope with sewage treatment. The toxicity of the sewage increases at the disinfection stage, special by using chlorine-containing agents. Eventually, organic compounds including toxic halogenated xenobiotics, which are irrelevant to the biosphere, enter the water body [3, 4]. Also the formation of a large number of halogenated chlorination by-products in potable water is a known problem. Many of these compounds have not been identified and not quantitatively determined by routine control [5].

Most of the natural water bodies are receivers of treated wastewater and used as centralized drinking water resources simultaneously [6]. Therefor monitoring of them is relevant. Despite of the fact that only traces of halogen-containing organic compounds are often present in the purified effluent, they can accumulate in the water bodies not undergoing biotransformation and photooxidation due to their stability.

So, organization of a reliable system for monitoring the quality of water from natural water resources and potable water after chlorinated stage is an important task. Availability of information about growing general level of pollution provides for measures of strategic water management and control of the compliance with the standards for water protective areas.

2. Existing Assessment Methodology

In order to address the task of informational monitoring management, it is practical to use survey methods of analysis which allow for monitoring the general background pollution level and incidental emissions [7, 8], combined with control of the most toxic indicator compounds which are characteristic of the given object under observation [2].

In general, the survey methods of analysis can be subdivided into three categories: methods which allow for assessing the general group composition of pollutants, general ultimate composition and methods assessing the possibility of the pollutants to interact (to oxidize under the action of strong oxidizers, to be extracted, sublimated with steam and reduced, etc.). These are some examples of integral water quality indicators: Total Organic Carbon (TOC), Dissolved Organic Carbon (DOC), Ultraviolet Absorption 254 nm, Steam-volatile Phenols, Permanganate index, Chemical oxygen demand, Total Organic Halogen (TOX), Adsorbable Organic Halogen (AOX) and Extractable Organic Halogen (EOX) [9-11].

The increased focus on the total content of halogen-containing organic water pollutants is due to their high toxicity, mutagenic and carcinogenic activity [12, 13].

According to the method for determination of extractable halogen-containing organic compounds in water (EOX), analytes are extracted from water and placed into the reactor where chlorine organic compounds are converted to HCl under oxygen atmosphere, bromine organic compounds – to HBr, etc. Then the conversion products are absorbed by water. The absorbate solution is analyzed by ion chromatography [11]. The degree of conversion is about 100%, the minimum detected level is 10^{-8}-10^{-6} g/mm^3 depending on the element.

3. Results and discussion

This paper presents a new approach to assessing the total chlorine, bromine content in semi-volatile organic substances (SVOS) of water without derivatization or conversion stages. The new approach methodology is as follows. The sample of natural or drinking water is extracted with an organic solvent (e.g. methylene chloride), the extract is boiled down and analyzed by gas chromatography with atomic emission detection (GH-AED), the detector being adjusted for chlorine (479 nm) or bromine (478 nm) emission line.

Under the given conditions of sample preparation and analysis, the chromatograms record the components with Kovats retention index on non-polar phases from 900 to 3200, i.e. boiling out in the range of 150-500°C. This range covers such classes of limited-volatile organic compounds as pesticides (including highly molecular decil), aromatic and aliphatic hydrocarbons, fatty acids, ethers, phenols, polyaromatic hydrocarbons and also halogen derivative organic compounds produced upon chlorination of water [14]. The resulting element selective chromatograms show the total area of chlorine- and bromine-containing compounds (figure). The total concentration of chlorine and bromine in the analyzed compounds is calculated by component independent calibration available when using the AED [15]. As far as the compounds are mostly of technogenic or secondary technogenic origin, the values are named Technogenic Organic Chlorine (TgOCl) and Technogenic Organic Bromine (TgOBr). The sensitivity of determination is 1×10^{-6} mg/dm^3 on the given elements.

Definition of such values does not require a close-cut component separation of the sample, still the pollutants can be identified, if necessary, by the elements content in the molecule of the component or in combination with the results of chromatography mass-spectrometry. Such data cannot be obtained with the known composite indices.

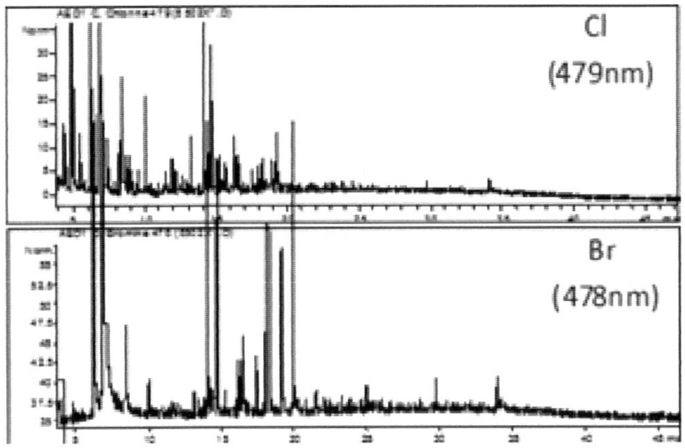

Figure 1. The element selective chromatograms of natural water extract after chlorination, registered on the characteristic wave lengths: CL (479 nm), Br (478 nm).

The integral indicators TgOCl and TgOBr give much information when assessing not only the background pollution of water bodies with halogen compounds, but also when assessing the efficiency of water purification and water treatment at any stage of production and control. The using of these indicators is especially informative in selecting the type and dose of chlorinating agent during the water treatment to minimize the formation of halogenated derivatives.

4. Conclusion

The increased focus on the total content of halogen-containing organic water pollutants is due to their high toxicity, mutagenic and carcinogenic activity. Most of them belong to semi-volatile organic substances. A new approach to assessing the total chlorine and bromine content in semi-volatile organic substances of different type water without derivatization or conversion stages is present. The approach allows one to evaluate the total concentration of these halogens. The sensitivity of determination is 1×10^{-6} mg/dm^3 on the given elements. The proposed indicators are named Technogenic Organic Chlorine and Technogenic Organic Bromine because analyzed compounds have technogenic or secondary technogenic origin.

5. Acknowledgments

The work was performed within the framework of State Task No. 5.12863.2018 / 8.9 "Development of a system for identification and quantitative analysis of environmental risks arising from the water supply of a large urban agglomeration".

References

[1] Lepoma P, Brown B, Hanke G, Loos R, Quevauviller P and Wollgast J 2009 Needs for reliable analytical methods for monitoring chemical pollutants in surface water under the European Water Framework Directive *J. Chromatogr. A* **1216** 302–15

[2] Dickenson E, Snyder S, Sedlak D and Drewes J 2011 Indicator compounds for assessment of wastewater effluent contributions to flow and water quality *Water Res.* **45** 1199–212

[3] Sun Y-X, Wu Q-Y, Hu H-Y and Tian J 2009 Effect of bromide on the formation of disinfection by-products during wastewater chlorination *Water Res.* **43** 2391–8

[4] Shan J, Hu J, Kaplan-Bekaroglu S, Song H and Karanfil T 2012 The effects of pH, bromide and nitrite on halonitromethane and trihalomethane formation from amino acids and amino sugars *Chemosphere* **86** 323–8

[5] Simmons J, Richardson S, Speth Th and Miltner R 2002 Development of a research strategy for integrated technology-based toxicological and evaluation of complex mixtures of drinking water disinfection byproducts *Environ. Health Perspect.* **110** 1013–24

[6] Swayne M, Boone G and Bauer D 1980 *Wastewater in Receiving Waters at Water Supply Abstraction Points* (Washington, D.C.: U.S. Environmental Protection Agency)

[7] Ademollo N, Patrolecco L, Polesello S, Valsecchi S, Wollgast J, Mariani G and Hanke G 2012 The analytical problem of measuring total concentrations of organic pollutants in whole water *Trends Anal. Chem.* **36** 71–81

[8] Ricci M, Kourtchev I and Emons H 2012 Chemical water monitoring under the Water Frame-work Directive and Certified Reference Materials *Trends Anal. Chem.* **36** 47–57

[9] Pereira J, Moreira C, Albers C, Jacobsen O and Flores E 2011 Determination of total organic halogen (TOX) in humic acids after microwave-induced combustion *Chemosphere* **83** 281–6

[10] Muller G 2003 Sense or *no*-sense of the sum parameter for water soluble ''adsorbable organic halogens'' (AOX) and ''adsorbed organic halogens'' (AOX-S18) for the assessment of organohalogens in sludges and sediments *Chemosphere* **52** 371–9

[11] Virius E, Kapinus E, Revelsky I and Borzenko A 2003 Determination of total organochlorine content in water, based on microfluid extraction and microcoulometric analysis of the extract *Zav. Lab* **69** 3–6

[12] Vozhdaeva M, Holova A, Wagner E, Kantor E, Kantor L, Trukhanova N and Melnitsky I 2018 The use of results of expanded monitoring research for the integrated assessment of drinking water according to indices of chemical harmlessness *Hygiene and sanitary* [in Russian – Gigiena i Sanitaria] **97** 117–24

[13] Bulla R, Reckhow D, Li X, Humpaged A, Joll C and Hrudeyc S 2001 Potential carcinogenic hazards of non-regulated disinfection by-products: Haloquinones, halo-cyclopentene and cyclohexene derivatives, N-halamines, halonitriles, and heterocyclic amines *Toxicology* **286** 1–19

[14] Vozhdaeva M, Safarova V, Kantor L, Paramonov E, Teplova G, Tsypysheva L and Kantor E 2004 Basic classes of organic compounds of technogenic nature in water sources of Ufa city *Chemistry and Technology of Water* [in Russian – Khimiya i Tekhnologiya Vody] **1** 78–81

[15] Vozhdaeva M, Tsypysheva L, Kantor L and Kantor E 2007 Using gas chromatography with atomic emission detection for determining organic pollutants in water *Journal of Analitical Chemistry* **11** 1186-91

IOP Conf. Series: Earth and Environmental Science **350** (2019) 012047 doi:10.1088/1755-1315/350/1/012047

Compositional change of solid and liquid phases of soil suspension depending on pH value

S P Zamana[1], T G Fedorovsky[1], N A Golubkina[2] and A V Sokolov[3]

[1]State University of Land Use Planning, 15, Kazakova str., Moscow, 105064, Russia
[2]Federal Scientific Center of Vegetable Production, 14, Selectsionnaya str., VNIISSOK, Moscow region, 143072, Russia
[3]Russian state agrarian correspondence university, 1, Fuchika str., Balashikha, Moscow region, 143907, Russia

E-mail: svetlana.zamana@mail.com

Abstract. The content of mobile forms of chemical elements in soils is affected by many factors including soil pH, grading and humus concentration. The most pronounced influence on elements behavior and their absorption levels by plants is produced by pH, which gives an opportunity to characterize quantitatively the ratio between elements forms in the solution and the equation between liquid, solid and gaseous forms of soil. Thus, the element composition of solid and liquid soil phases depending on pH may become of special interest. The article presents the results of a model experiment to determine the changes occurring in different pH ranges in the solid and liquid phases of aqueous soil suspensions taken from the organogenic horizon of the peat-podzolic soil of the Moscow Region. Potentiometric titration and atomic absorption methods for determination of chemical elements (Ca, Mg, K, Mn, Fe, Al, Si) were used. The element composition of solid and liquid phases of peat-podzolic soil suspension is considered at different pH. The results demonstrate that pH of the soil suspension correlates with element composition changes of both liquid and solid phases, thus indicating soil bioavailability of elements.

1. Introduction

Mobile forms of chemical elements in soils are known to depend on soil pH, grading and humus content. Among the above factors, soil pH influences especially intensively the elements behavior in soils and their accumulation by plants [1-3]. Only at neutral pH (6.5-7.5) most of biogenic elements (N, P, K, S, Ca, Mg, Fe, Mn, B, Cu, Zn, Mo) become bioavailable for plants [4]. Absorption of cations and anions by plant decreases their concentration and activity in soil solution thus decreasing gradually the efficiency of plant nutrition. New ion portions maintaining the activity at the same level enter the solution from the solid phases as a result of dissolution of the sparingly soluble salts present in the solid phases, exchange reactions or aluminosilicate and organic matter destruction. pH value may characterize quantitative ratios between different elements forms in the solution and also characterize conditions of equilibrium of soil solid-liquid-gaseous phases.

Different buffer systems may exist at different soil pH [5]: 1) at pH 3.0 – 4.5 – permanent charge of clay minerals, silicates destruction and formation, organic aluminum complexes; 2) at pH 4.5 – 6.0 – amino groups, carbonates equilibrum, hydroxo complexes of aluminum; 3) at pH 6.0 – 7.5 – hydroxo complexes of aluminum and carbonate equilibrum; 4) at pH 7.5 – 9.0 – hydroxo complexes of aluminum, phenol hydroxyls, carbonate equilibrum. Ca^{2+}, Mg^{2+}, K^{+}, Al^{3+} are released during destruction of alumosilicates [6, 7]. In case of soil liquid phase pH, less than 4 significant amounts of Al may be released from the silicate grid.

The following reactions may take place as a result of interaction of organogenic soil horizons with an acid: dissociation of organo-mineral complexes, containing Al, Fe and Mn; protonation of funcional groups of specific and nonspecific organic acids; cation exchange reactions; solubilization of strong bases salts and week organic acids with the appropriate protonation of acid anion. Al, Fe and Mn appear in the solution in case of mobilized ions excess above the capacity of their complex bounding by organic compounds.

During the base titration of aqueous suspensions of organogenic soil horizons up to pH 7, carboxyl (COOH) groups of humic (GA) and fulvic acids (FA) can be titrated. At pH>9, deprotonation of GA and FA phenolic and NH_2 groups takes place [8]. Additional reactions of iron and aluminum hydroxo complexes with organic ligands may occur. Basic titration of GA and FA is consistent with deprotonation (at pH 3-7) of the first of the most acidic carboxyl groups and at pH > 8 – phenolic and NH_2 groups. Carboxyl groups are titrated in a large pH range (from 3 to 7), but COOH groups with pKa from 4 to 5 predominate. At high pH levels, several aluminum organic compounds are destructed, which results in a conversion of $A1(OH)^{4-}$ and $A1(OH)_5^{2-}$ ions into a solution.

Taking into account a great influence of soil pH on chemical elements absorption by plants, studies on the element composition of solid and liquid soil phases are of great importance.

The work presents the results of element composition changes of solid and liquid phases of aqueous soil suspensions depending on pH.

2. Materials and methods

We carried out a model experiment to determine changes occurring in different pH ranges in the solid and liquid phases of aqueous soil suspensions from the organic horizon of the peat-podzolic soil of the Moscow Region.

3 g of soil were diluted with water at a ratio of 1:25; pH equilibrium values were measured and continuous titration with the base was achieved up to pH 5, pH 6, pH 7, pH 8, pH 9, pH 10 and also with an acid up to pH 4 and pH 3.

The applied method of potentiometric titration consists in registering (using a pH meter) changes in the pH values of the soil suspension or extraction from the soil when a certain amount of acid or base is added to the system [9, 10].

The resulting solution was centrifuged during 10 minutes at 4000 r/min, then the liquid phase was separated from the solid one. The solid phase was dried and used for the determination of: 1) Ca, Mg, K, Mn, Fe, Si, Al in NH_4CI extracts using the AAS method; 2) organic carbon was determined via calculating mass losses during calcination. The same parameters were determined in the liquid phase using AAS, and Turin method was used for determination of water- soluble organic carbon (Cp).

3. Results and discussion

Results of Ca, Mg, K, Fe, Al and Si determination in liquid phase of the organogenic peat-podzol soil horizon suspension in different pH ranges (3-4; 4-4.2; 4.2-5; 5-6; 6-7; 7-8; 8-9; 9-10) are presented in Table 1.

The calcium content at the initial pH value reached 8.0 mM-eq/kg; the lowest value (0.9 mM-eq/kg) was detected at pH 5-6, and the highest (22.2 mM-eq/kg) – at pH 8-9. At pH 5-7, calcium was transported to the soil absorbing complex. Due to the increase of the cation exchange capacity at higher levels of pH. Increase in the soluble Ca content at pH 8-9 is caused by dissolving the Ca-containing organic matter that may explain the highest increase in the Ca concentration and organic matter at pH 8-9. Decrease of the Ca level at pH 9-10 (up to 12.9 mM-eq/kg) is associated with a decrease of the soluble organic matter (C_p= 6.7 g/kg).

The lowest concentration of soluble Mg was manifested at the initial pH equal to 4.2 (2.0 mM-eq/kg); the highest (5.3 mM-eq/kg) – in the pH range of 8-9, which correlate, to a certain extent, with change in the organic matter content at different pH values.

Table 1. The composition of the liquid phase of an aqueous suspension of peat-podzolic soil (mM-eq/kg).

Indicators	Content of		chemical	elements at different values of pH ranges					
	3-4	4-4.2	initial pH	4.2-5	5-6	6-7	7-8	8-9	9-10
Ca	14.5	10.3	8.0	5.8	0.9	4.5	8.7	22.2	12.9
Mg	4.8	3.0	2.0	2.4	2.8	2.6	2.9	5.3	5.0
K	4.6	3.1	3.0	22.8	36.8	40.8	54.8	94.8	114.8
Mn	1.2	0.4	0.3	0	0	0	0	0	0
Fe	0.5	0.4	0.4	0.9	1.1	1.5	2.8	4.2	7.0
Al	6.9	5.5	5.6	6.0	5.6	4.0	6.8	11.2	18.8
Si	1.8	1.4	1.5	2.0	2.8	4.4	4.7	7.7	11.1
C_p, g/kg	0.8	0.9	1.1	4.3	5.5	5.8	6.3	9.6	6.7

The lowest concentration of soluble K was shown at the initial pH level of 4.2 (3.0 mM-eq/kg). With the increase of pH from 4.2 to 10, the potassium content in the solution increased dramatically, parallel to the organic matter increase, reaching 114.8 mM-eq/kg at pH 9-10. Potassium goes into solution in higher quantities than those of Ca and Mg, which is connected with greater solubility of its derivatives.

The iron concentration at pH 4.2 reaches 0.4 mM-eq/kg. By adding alkali, its concentration increased up to 7.0 mM-eq/kg at pH 9-10 due to dissolution of Fe-organic complexes.

The aluminum content in the solution at the initial pH value was equal to 5.6 mM-eq/kg. pH increase caused the elevation of the Al level up to 6.0 mM-eq/kg at pH 4.2-5, and then a decrease at pH 5-6 (5.6 mM-eq/kg) and pH 6-7 (4.0 mM-eq/kg). At pH 8-10 its concentration again increased (18.8 mM-eq/kg). The observed changes of soluble Al with pH increase may be connected with the fact that at lower pH there is a partial displacement of the exchangeable aluminum in the solution; at pH 5-6 and pH 6-7 Al precipitates in a form of hydroxide, and at the higher pH hydroxide dissolution is observed with the formation of aluminate ions.

The silicon concentration in a solution at initial pH was equal to 1.5 mM-eq/kg. Increase of pH resulted in a gradual Si level increase up to 11.1 mM-eq/kg at pH 9-10 due to dissolution of organic residues.

Decrease of pH from the initial value to 3 resulted in significant increase of all the above elements in the solution: Ca – up to 14.5 mM-eq/kg, Mg – to 4.8 mM-eq/kg, K– up to 4.6 mM-eq/kg, Fe – up to 0.5 mM-eq/kg, Al – up to 6.9 mM-eq/kg, Si – up to 1.8 mM-eq/kg.

The soluble organic carbon content in a liquid phase of the water suspension investigated at initial pH was equal to 1.1 g/kg. During titration from initial pH to 5 and from 5 to 6 pH, a noticeable increase of the soluble organic carbon content occurred up to 4.3 and 5.5 g/kg, accordingly. At pH 6-7 and 7-8, the soluble organic carbon content changed insignificantly reaching 5.8 and 6.3 g/kg, respectively, and at pH 8-9 the value again increased significantly up to 9.6 g/kg. Increase of the soluble organic carbon content at lower pH is achieved due to predominantly nonspecific organic acids; while at higher pH – by participation of specific humus compounds. During a dissolution of the organic matter as a result of alkali supplementation, a hydrogen bonds destruction in a specific organic matter takes place. The phenomenon may be enhanced with the pH increase.

The content of chemical elements in the solid phase of the organogenic peat-podzol soil horizon suspension investigated is presented in Table 2.

The exchangeable Ca content at initial pH was 98 mM-eq/kg. pH increase up to 9-10 caused its decrease up to 58 mM-eq/kg. The exchangeable Mg content at initial pH was 25 mM-eq/kg, and elevation of pH resulted in its increase differing for various pH ranges, reaching 27 mM-eq/kg at pH 7-8. Alkali titration was accompanied by similar to Ca changes in the Mg content , i.e. at pH 4.2-5 and pH 7- 8 it increases, while at pH 5-7 and pH 8-10 - decreases. The phenomenon may be connected with the facts that during titration with alkali a simultaneous displacement of Mg and Ca from the soil absorbing complex takes place with the appropriate solubilization of Ca, Mg containing organic compounds, and the mobility increase of solid phase Ca and Mg.

Table 2. The composition of the solid phase of an aqueous suspension of peat-podzolic soil (mM-eq/kg).

Indicators	Content of		chemical	elements at different values of pH					
	3-4	4-4.2	initial pH	4.2-5	5-6	6-7	7-8	8-9	9-10
Ca	73	86	98	96	93	87	89	74	58
Mg	19.5	23.9	25	26.7	26.3	25.1	27.0	26.5	25.3
K	3.7	3.8	5.0	20.3	26.1	36.9	38.5	68.5	96.1
Mn	1.3	1.5	1.9	3.1	3.1	2.7	2.8	3.0	2.5
Fe	0.5	1.7	2.9	9.0	3.0	3.0	3.3	3.1	2.0
C_{org}, g/kg	88.1	88.0	88.6	89.1	88.6	92.7	91.0	88.7	89.1

The content of exchangeable potassium in the solid phase at initial pH was 5.0 mM-eq/kg and the value increased significantly at higher pH up to 20.3 mM-eq/kg (pH 4.2-5) and up to 96.1 mM-eq/kg (at pH 9-10), which may be connected with the increase of its mobility.

Exchangeable Mn at pH=4.2 was equal to 1.9 mM-eq/kg; at pH 4.2-5 the value increased to 3.1 mM-eq/kg, while at pH 6-10 it was equal to 2.5-3.0 mM-eq/kg. Exchangeable Fe at initial pH was 2.9 mM-eq/kg. Alkali treatment from initial pH to 5 caused the value increased due to the increase of mobility of complexes with organic ligands of lower acidity. At pH 5-7 the value decreased up to 3.0 mM-eq/kg; at pH 7-8 and pH 8-9 it increased a little and became lower 2.0 mM-eq/kg at pH 9-10.

Acidification up to pH 3 resulted in a decrease of all exchangeable cations: Ca – up to 73 mM-eq/kg, Mg – to 19.5 mM-eq/kg, K – to 3.7 mM-eq/kg, Mn – to 1.3 mM-eq/kg, Fe – to 0.5 mM-eq/kg.

4. Conclusion

The model experiment with the organogenic horizon of peat-podzol soil allows revealing that pH of the soil suspension affects changes in the element composition of both liquid and solid phases, thus influencing the chemical elements bioavailability of soil for plants. Acidification of the organogenic horizon up to pH 3 increases the Ca, Mg, K, Mn, Fe, Al, Si content in the liquid phase from 0.2 mM-eq/kg (Fe) up to 6.7 mM-eq/kg (Ca). The pH increase up to pH 10 results in Ca, Mg and Al increase at a certain pH range and the decrease at other pH ranges, while the concentration of K, Fe and Si increases in all pH ranges.

The carbon content of the soluble organic matter in the solid phase of water suspension during pH elevation from 4.2 to 10 increases from 1.1 to 6.7 g/kg, while during acidification from pH 4.2 to pH 3 the value decreases from 1.1 g/kg to 0.8 g/kg.

References

[1] Yagodin B A 1985 *Agrochemistry* [in Russian – Agrochimiya] **11** 117-127
[2] Barber C A 1988 *Biological availability of nutrients in the soil. Mechanistic approach* (Moscow: Agropromizdat)
[3] Kabata-Pendias A and Pendias X 1989 *Trace Elements in Soils and Plants* (Moscow: Mir)
[4] Thompson L M and Throw F P 1982 *Soils and their fertility* (Moscow: Kolos)
[5] Filepp D and Radley M 1989 *Soil Science* **12** 48-59
[6] Dronova T I 2000 *Soil Science* **5** 61-65
[7] Tolpesta I I and Sokolova T A 2013 *Soil Science* **9** 81-90

[8] Gugenberger et al 1989 *Science of the Total Environment* **81-82** 447-457
[9] Sokolova T A, Pakhomov A P and Terekhin V G 1993 *Soil Science* **7** 97-106
[10] Feodorovsky T G, Ladonin D V and Sokolova T A 2005 *Vestnik of MSU. Soil Sci. Series* **1** 22-28

Layout for drilling seismic wells by vibration method

L M Zaripova, M S Gabdrakhimov

Ufa State Petroleum Technological University, Branch of the University in the City of Oktyabrsky, 54a, Devonskaya St., Oktyabrsky, Republic of Bashkortostan, 452607, Russian Federation

E-mail: Lilyabert31@mail.ru

Abstract. More and more common methods are used to reduce the cost of drilling wells and extracting oil and gas. One of these methods is the increase of energy capacity of drilling rigs through the use of dynamic methods of drill bits loading. Various vibro-mechanisms are used as vibration exciters with helpful use of vibration. Vibro-mechanisms by type of drive are divided into mechanical, electromechanical and magnetostrictive. All of them are intended for the excitation of forced oscillations in the elements attached to them that provide the desired technological effect. The authors have carried out the work to solve the problem of studying the transfer of dynamic load on the drill string. In this case the effectiveness of bottomhole vibration is largely determined by layout parameters of the drilling tool. Oscillations of the drill string are described by a wave equation in partial derivatives. With the solution of the problem, conditions for the transfer of bottomhole vibration load are being determined. The solution shows that dynamic bottomhole load can be effectively transferred from the mouth up to 100 m; the frequency may be greater at lower striker weights.

1. Introduction

The rotary percussion drilling method began to develop and investigated by the following scientists G.I. Neudachnyi, L.E. Graf, F.F. Voskresenskii, D.D. Barkan, V.M. Slavskii, O.I. Tagiev, B.Z. Sultanov, and others who created various hydraulic hammers of both simple and double action, mechanical vibrators, vibratory hammers, which increase the mechanical speed of the drilling process of both soft and hard rocks [1-6].

Formulation and solution of the problem of studying the dynamic load transfer along the drill string. In this case, the effectiveness of bottomhole vibration effect is largely determined by the drilling tool layout parameters [7-10].

Oscillations of the drill string are described by a wave equation in partial derivatives:

$$\frac{d^2U}{dt^2} + b\frac{dU}{dt} + a^2\frac{d^2U}{dx^2} = 0 \qquad (1)$$

where U=U(x,t) – cross section offset of x at the moment of t time from static equilibrium position; b – parameter characterizing the resistance forces acting on drill string; a – distribution speed of oscillations in the drill string.

A vibrator of mass M acts on upper end of the drill string with frequency ω and power P (amplitude value):

$$EF\frac{dU}{dt} = M\frac{d^2U}{dt^2} - P\cos\omega t \qquad (2)$$

where E – Young's modulus of pipe material; F – cross-sectional area of pipe material.

Lower end of the drill string is of length L, rests on the bottomhole with stiffness and viscous resistance of μ.

$$EF\frac{dU}{dx} = cU - \mu\frac{dU}{dt} \quad (3)$$

Equation (3) for integration is more convenient to present in the following form:

$$EF\frac{dU}{dx} = M\frac{d^2U}{dt^2} - Pe^{i\omega t} \quad (4)$$

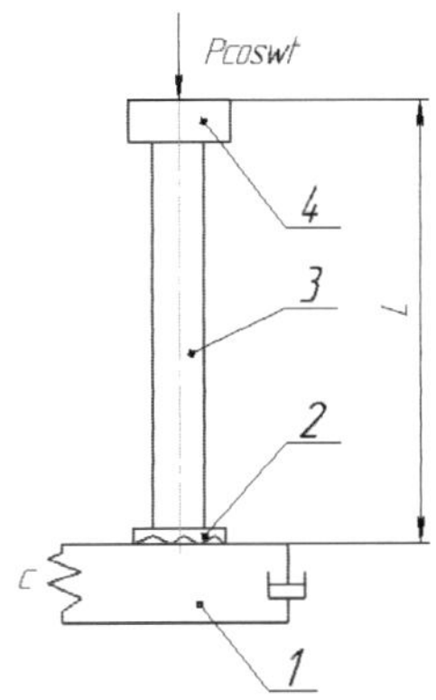

Figure 1 Design layout of the drilling tool equipped with a ground hydraulic hammer:1 - bottomhole of a well; 2 - bit; 3 - drill pipe string;4 - hydraulic hammer

We will seek a forced solution in the following form:

$$U(x,t) = X(x)e^{i\omega t} \quad (5)$$

After substituting equation (5) into wave equation (1), we obtain:

$$-\omega^2X(x)e^{i\omega t} + ib\omega X(x)e^{i\omega t} - a^2\frac{d^2X}{dx^2}e^{i\omega t} = 0 \quad (6)$$

$$\frac{d^2X}{dx^2} + \frac{\omega^2 - ib\omega}{a^2} = 0 \quad (7)$$

General solution of equation (7) is as follows:

$$X = Ae^{kx} + Be^{-kx} \quad (8)$$

where

$$k = \sqrt{\frac{ib\omega - \omega^2}{a^2}} \quad (9)$$

From boundary conditions (2), (3) after substitution (5) is followed:

$$EF\frac{dX}{dx} = -M\omega^2X - P \quad (10)$$

$$EF\frac{dX}{dx} = -(c + i\omega\mu)X \quad (11)$$

Solving the equations of (10) and (11) after a series of transformations and substitutions, we determine the complex value of dynamic effort at the bottomhole:

According to the well-known decision the complex value of dynamic effort at the bottomhole is determined:

$$P_{з.к} = EF\frac{dU}{dx} = EF\left(Ake^{kL} - Bke^{-kL}\right)E^{i\omega t}. \quad (12)$$

where A and B – constant integrations, k – root of the characteristic equation.

The dynamic force amplitude is determined through the modulus of the force complex value:

$$Q(\omega) = |P_{з.к}| = EF|Ake^{kL} - Bke^{-kL}|. \quad (13)$$

Decisively, the dynamic bottomhole load amplitude is determined by the formula:

$$Q(\omega) = |EF(A(\omega)e^{k(\omega)L} - B(\omega)k(\omega)e^{k(\omega L)}| \quad (14)$$

Equation (14) was solved with the following data: B = 10; 30; 90; 270; M=250; 500; 1000; 2000 kg; L=25; 100; 250 m; E=2.1·10^{11} H/m^2; D=0.146 m; d=0.126 m; P_i=3.1415926; M=250; 500; 1000 kg;

IOP Conf. Series: Earth and Environmental Science **350** (2019) 012048 doi:10.1088/1755-1315/350/1/012048

F=8.9723884656·10^8; ω=1…1000 s^{-1}; i=$\sqrt{(-1)}$; b=10; 30; 90; 270; k(ω)=$\frac{-(\omega)^2}{a^2}+i\cdot\omega\cdot\frac{b}{a^2}$; P= 10000 H; a=5.1887452166·10^3 m/s; s= 100000000 H/м; μ= 100000Pa·s.

Results of the decisions are shown in Figures 2, 3 [5].

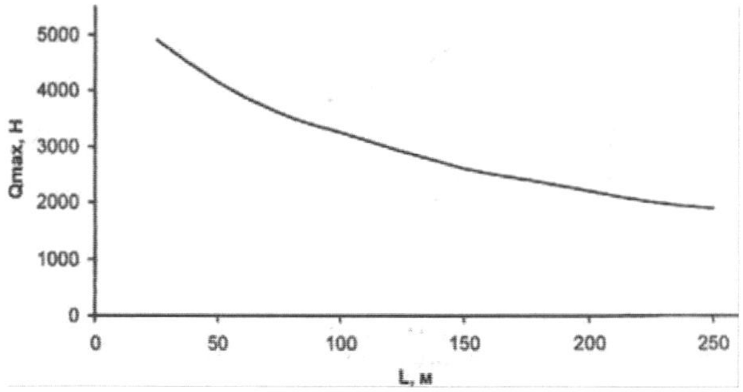

Figure 2. Dependence of maximum dynamic bottomhole load on the depth at M = 1000kg, b=30.

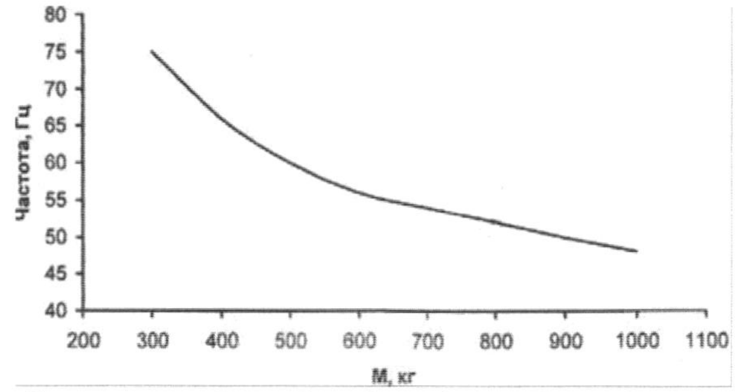

Figure 3. Effect of striker mass on the purity of maximum dynamic bottomhole load L=100 m, b=30.

Частота, Гц – frequency, Hz

The solution shows that dynamic bottomhole load can be effectively transferred from the mouth up to 100 m; the frequency may be greater at lower striker weights; parameter b obstructs the transfer of dynamic bottomhole load.

Theoretical studies of technical parameters influence of technological process, on the transfer of dynamic load from a hydraulic hammer installed on the drilling tool, on the bottomhole of a well have been conducted.

The research results indicate that with increasing the depth of the well, frequency of oscillations transmitted to the bit increases; maximum load decreases with increasing the drill string length, and with increasing the mass of the striker, oscillation frequency decreases.

2. Results and Discussion

A schematic diagram of the ground hydraulic hammer is shown in Figure 4 to study the effect of a hydraulic hammer on drilling efficiency.

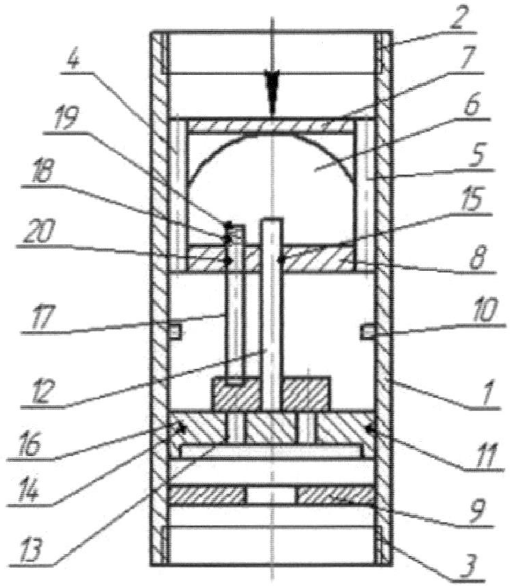

Figure 4. Schematic diagram.

A constructive scheme of a ground hydraulic hammer-pulsator is developed. The striker is a two-stage piston. In this case, the piston is connected to the atmosphere, the large piston is equipped with a valve, which closes channels of the large piston in the upper position, and opens in the lower one.

The hydraulic hammer consists of a container 1 fitted with an upper adapter sub 2 and a lower 3, lateral drains outlets 4 and 5, slot 6, partitions 7 and 8, anvil 9 and a guard 10. Piston-hammer 11 is installed inside container 1 and has a shank adapter 12 and port-holes 13. The piston and shank adapter are equipped with congestions 14 and 15. A disk 16 is hafted on the shank adapter and is connected by a pin 17, which interacts with a partition 8 by means of a spring 18. The pin 17 is sealed in the partition by means of congestion 20.

The device during operation is installed under the swivel. At initial moment, the spring 18 is expanded; the disk is in upper position. As the flushing fluid flows, the piston-hammer 11 and elements connected to it move upward due to the pressure difference under the piston and in the annulus.

Movement occurs until the closure by disk 16 of port-holes 13 of the piston-hammer 11. As a result, a hydraulic shock happens, pressure above the piston sharply increases, the disk moves down and strikes the anvil. Before strike, the pin opens the disk, which, due to expansion of the spring, occupies the initial position. The piston moves upwards due to hydraulic forces. The cycle is repeated. The increase of indicators in rotary drilling during the drilling of seismic wells by this device is carried out at the expense of creating a dynamic load on the bit and pulsating flushing of bottomhole wells. With the use of a ground hydraulic hammer in the layout of the drilling tool, the well No. 61 was drilled with the creation of vibration.

To study the dynamics of a hydraulic hammer, a differential equation of the piston movement together with the valve is compiled (down):

$$M\frac{d^2x}{dt^2} = \sum F_{ix} \tag{15}$$

$$M\frac{d^2x}{dt^2} = G + P_{at} + P_{Hyb} - F_{fr_1} - F_{fr_2} - F_{el} - cx \tag{16}$$

where $M=m_1+m_2$ – total mass of both the piston and valve; x=x(t) – current coordinate defining piston position; G – total piston gravity with valve; P_{at} – atmospheric pressure force; P_{HyT} (hydraulic top) – vertical (top) hydraulic pressure; P_{Hyb} (hydraulic bottom) – vertical (bottom) hydraulic pressure; F_{fr_1} –

piston friction force against container walls; F_{fr_2} – shank adapter friction force against container walls; F_{el} – elastic force of the valve spring; c – spring stiffness.

Here the total constant forces acting when the piston moves down are equal to

$$F_{Hy} = P_{HyT} - P_{HyB} \quad (17)$$

$$F_{fr} = F_{fr_1} + F_{fr_2} \quad (18)$$

Then initial equations (16) can be written in the following form:

$$\frac{d^2x}{d\,t^2} = g + a + f_{\Gamma} - \frac{cx}{M} \quad (19)$$

All forces (right side of the equation) refer to the mass unit of a moving piston with a valve, that is

$$g = \frac{G}{M}; \quad a = \frac{P_{at}}{M}; \ f_{Hy} = \frac{F_{Hy}}{M}; f_{fr} = \frac{F_{fr}}{M}, \quad (20)$$

We combine the constant members of sum in equation (19) g + a + f_r +f_{fr}=P-Const.

Considering, that $\frac{c}{M}$=k^2(const), differential equation (19) takes the final form

$$\frac{d^2x}{dt^2} + k^2x = P \quad (21)$$

The differential equation (20), which determines the downward movement of the piston, is a linear non-uniform second-order differential equation with constant coefficients. To integrate this differential equation, initial and boundary conditions are written: i.e at t=0

$$X_0 = 0, \quad V_{x.0} = 0 \quad (22)$$

This means (based on nature of the work of the piston with valve) that the piston was initially static and was located at the top guard.

The solution of equations (20), (21) determines the speed of the piston movement, which has the form:

$$V = \frac{P}{k^2}\sin(kT), \quad (23)$$

where time value T is found by the formula

$$T = \frac{1}{k}\arccos(1 - \frac{lk^2}{P}), \quad (24)$$

where 1 – piston stroke

Knowing the speed of V piston at the moment of strike with the anvil, we determine the total impact pulse by the formula

$$S_{sh} = m_1(l + k_{col})\frac{P}{k^2}\sin(kT) \quad (25)$$

where k_{col} – collision coefficient at strike, $k_{col} = \frac{u}{u_0}$; u – piston speed after strike.

Calculation of the dependence of the impact pulse on the piston diameter is carried out according to the following parameters: piston diameters D range from 0.1m to 0.2m with a range of 0.025m; piston thickness b=0.04m; shank adapter length l =0.1m.

As a result of the calculation, the dependence of impact pulse on piston diameter (Figure 5) is obtained, which shows that the impact pulse increases from 33 H·s to 101 H·s with an increase in the piston diameter from 0.1 m to 0.2 m with different spring stiffness.

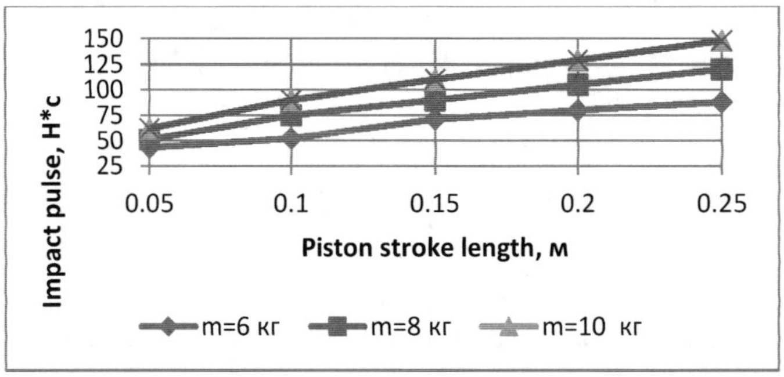

Figure 5 Dependence of impact pulse on piston diameter with different stiffness of spring C

Calculation of the dependence of the impact pulse on the piston stroke length is carried out according to the following parameters: piston stroke length L varies from 0.05 to 0.25 m; piston thickness b=0.04 m; collision coefficient k_{col} =1; spring stiffness s=300 H/m; shank adapter diameter d=0.04 m; shank adapter length l =0.1m.

As a result of the calculation, the dependence of impact pulse on piston stroke length (Figure 6) is obtained, which shows that the impact pulse increases from 40 to 147 H·s with an increase in the piston stroke length from 0.05 to 0.25 m with different piston mass.

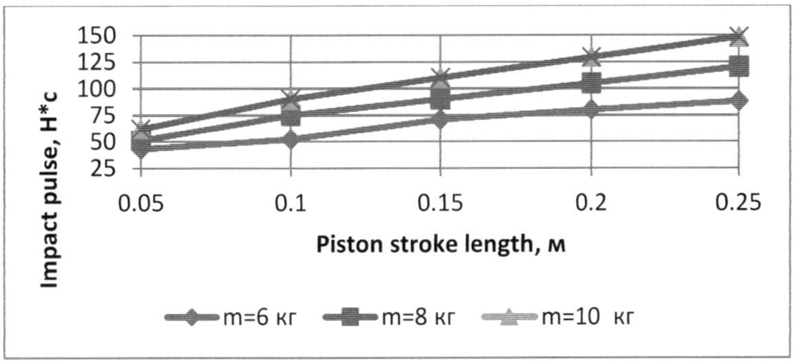

Figure 6. Dependence of impact pulse on piston stroke length at different values of piston mass

The results of laboratory studies of the hydraulic hammer model, the main elements of which were a cylinder, a piston and a pin, selected on the basis of geometric modelling. The test model was carried out on the stand, which had a circulating system, instruments for monitoring the parameters of fluid under and above the piston, a device for measuring the axial force of the reverse stroke of the rod.

When testing a model of a hydraulic hammer, the following questions were investigated: working efficiency of the hydraulic hammer model; selection of optimal diametric dimensions of the piston bores; ratio of the diameter of the piston and rod; effect of piston and rod friction force on model working efficiency; study of the dependence of the rod stroke return force on the piston cross-section area. In the study, the number of experiments required for static processing in order to obtain results was carried out. For this, the values of arithmetic mean and root-mean-square deviations were determined. Main geometrical dimensions of the hydraulic hammer were specified according to the results of tests of the hydraulic hammer model: adopted piston diameter - 145 mm; rod diameter - 30 mm; rod length - 40 mm; the ratio of areas of the piston and holes - 7.

Theoretical studies of technical parameters influence of technological process, on the transfer of dynamic load from a hydraulic hammer installed on the drilling tool, on the bottomhole of a well have been conducted.

The research results indicate that with increasing the depth of the well, frequency of oscillations transmitted to the bit increases; maximum load decreases with increasing the drill string length, and with increasing the mass of the striker, oscillation frequency decreases.

By solving the problem, we determine how the bottomhole load is transmitted.

Experimental drilling conditions are given below. The test was carried out while drilling a well with a depth of 40m No. 61, profile No.1240 of Sorochenskaia Square. The drilling was carried out by the drilling rig УББ -2А-2, drilling pump НБ-50, flushing fluid and technical water, pressure on the riser 2 .. 5 MPa.

Tool layout: drill pipes with a diameter of 60.3 mm, a two-blade bit with a diameter of 190 mm, and a ground hydraulic hammer was installed between the swivel and the drill string.

The installation was equipped with a combined rotor-swivel, rotor speed of 140 rpm (r/min). The load on the bit was created by a hydraulic vibration device 1.2 ... 2.6 t.f. Geological section of the well was composed of sediments of quaternary age, represented by loams and sands.

3. Conclusion

As a result of drilling a well by a vibration layout, 40 m of rocks for 1 spudding drilling is drilled. Drilling time is 20 minutes; mechanical speed is 2 m/min. We have compared mechanical speeds of neighboring wells drilled in the same area to assess the results of experienced drilling with the use of a ground hydraulic hammer. The average mechanical speed of the estimated wells is 1.64 m/min.

An increase in the mechanical drilling rate by 18% has been obtained as a result of experienced drilling, established when drilling the well No. 61 using a ground hydraulic hammer.

References

[1] Zhou X, Lü X, Song X, Bai Z, Zhang J 2017 Petroleum source of oil seepages in the Kalpin area, NW Tarim Basin *China Journal of Petroleum Science and Engineering* 83-96

[2] Duan J, Liu H, Guan J , Hua W, Jiao G, Gong J. 2016. Wax deposition modeling of oil/gas stratified smooth pipe flow. *AICHE Journal* **62(7)** 2550–62

[3] Zhang Z, Zhang Y, Zhu G, Chi L, Han J 2016 Impacts of Thermochemical Sulfate Reduction, Oil Cracking, and Gas Mixing on the Petroleum Fluid Phase in the Tazhong Area, Tarim Basin, *China Energy and Fuels* **33(2)** 968-978

[4] Fournanty S, Guer Y L, Omari K E, Dejean J-P 2008 Laminar flow emulsification process to control the viscosity reduction of heavy crude oils. *J. Disper. Sci. Technol* **29(10)** 1355–66

[5] Ghannam M T, Esmail N 2006 Flow enhancement of medium-viscosity crude oil. *Petrol. Sci. Technol.* **24(8)** 985–99

[6] Suleimanov R I, Zainagalina L Z, Khabibullin M Ya, Zaripova L M, Kovalev N O 2018 Studying heat-affected zone deformations of electric arc welding IOP *Conference Series: Materials Science and Engineering* 327-3 032053 DOI: 10.1088/1757-899X/327/3/032053

[7] Suleimanov R I, Gabdrakhimov M S, Khabibullin M Y, Zaripova L M, Vasilyeva E R 2018 The study of hydraulic hammer device in drilling tool assembly in hydraulical rotary drilling *International Journal of Engineering and Technology* **7-2** 28-30

[8] Khabibullin M Ya, Suleimanov R I 2018 Selection of optimal design of a universal device for nonstationary pulse pumping of liquid in a reservoir pressure maintenance system *Chemical and Petroleum Engineering* **54(3-4)** 225-232

[9] Khabibullin M Ya, Suleimanov R I, Sidorkin D I, Arslanov I G 2017 Parameters of damping of vibrations of tubing string in the operation of bottomhole pulse devices *Chemical and Petroleum Engineering* **53, 5-6** 378-384

[10] Davydov A Yu 2018 Utilization of expert systems for screw pump sets with surface drive management *Advances in Engineering Research (AER)* **157** 129-133

IOP Conf. Series: Earth and Environmental Science **350** (2019) 012049 doi:10.1088/1755-1315/350/1/012049

Territory monitoring and suggestions on oil filling station arrangement according to ecological standards

E E Zhelonkina, L I Boitsenyuk, V S Gruzdev, J S Valiev and E G Pafnutova

State University of Land Use Planning, 15, Kazakova street, Moscow, 105064, Russia

E-mail: pecypc-86@mail.ru

Abstract: The article studies the ecological state of the territory of Novoselki OFS and adjacent territories. The situational analysis of the territory was carried out. Its results will be applied in the section "Ecology and development of an ecological passport of the industrial enterprise". Practical recommendations on greening and reducing the fire risk were provided.

1. Introduction

The current unfavorable situation is connected with the environment and its protection. Having learned how to mine and use natural resources, human does not care about territories where industrial and capital enterprises are located. The predominant amount of extracted natural resources is directly related to the ecosystems. In addition, qualitative and quantitative characteristics of natural resources (animal and plant life, water, soil, etc.), and anthropogenic resources (buildings, structures, production wastes, etc.), as well as possibilities of their use depend on the characteristics of land plots and their use. Economic development of territories, the construction of industrial infrastructure facilities, in particular, oil and gas complex facilities, can danger the environment [1, 2]. It is necessary to form sanitary and protective zones with a special legal status. Safe operation of the facilities and the state of the environment depend on their efficiency. [3]. In this regard, it is necessary to investigate the ecological state of the territory of industrial enterprises (e.g., an oil loading station) and adjacent sanitary protection zones [4, 5].

2. Research methods

The geobotanical description of the territory and sanitary zones of the enterprise was performed. Soil cuts were described; soil and vegetation samples were taken for chemical analysis. Chemical analysis of soil and vegetation samples for the presence of heavy metals of organic pollutants was carried out using standard techniques.

3. Results and Discussion

Novoselki oil filling station (OFS) of the ring oil pipeline around Moscow is located in Podolsk district. The industrial site is located at a distance of 400 m from Hryvna, on the land plot of the forest fund. There is no hydrographic network with permanent watercourses on the territory surrounding the industrial site, except for a small pond and a boggy stream hydrologically connected with the Rogozhka river. In accordance with technical requirements, the OFS has a protection zone 100 m in width. The soils are sod-podzolic, formed on homogeneous loess-like loams, have an average particle size distribution. Significant areas show degradation of the soil cover [6, 7]. This is plane erosion caused by light erosion of loess loams widespread in the territory due to the disturbance of vegetation cover during production. Pollution with heavy metals and acidification with sulfur dioxide was observed [8] (Table 1).

Table 1. The chemical composition of the soil test site of the southern part of the sanitary protection zone (fragment)

Horizon designation	Sampling depth, cm	pH	P2O5, mg per 100g of soil	K2O mg per 100g of soil	Humus, %	Oil products mg/kg	Zn mg/kg	Cu mg/kg	Mn mg/kg
A_p	0-24	6	15	2.5	1.8	22.6±2.3	21.2	4.8	388
B_1	30-40	4.3	8.8	0.5	-	14.3±1.4	6.5	-	-

As a result of geobotanical surveys of the territory, key herbaceous areas were identified, and herbaceous species of ruderal plants, including tall stems capable of accumulating dry plant residues, which can cause fire, were described (Table 2).

On the territory of the oil filling station, samples of the upper soil layer were taken and chemical analysis was carried out [9, 10] (Table 3).

Table 2. Characteristics of plants growing on the key plot (fragment)

Name of plants	Projective cover, %	Height, cm	Phenophase
Barbarea	2.2	60	Fruits
Taráxacum	25	25	Fruits
Tussilágo	10	15	Vegetates
Tanacétum vulgáre	0.2	20	Vegetates
Plantágo	0.2	25	Vegetates
Atriplex	5	70	Vegetates
Trifolium pratense L.	25	20	Vegetates
Achilléa	10	50	Vegetates
Árctium	24	150	Vegetates

Based on the results of the survey of the OFS territory, it was proposed to cut grass near the underground tanks and within the protected zones in early June, i.e. before insemination of dandelions and other ripening herbs.

The discrepancy in the sanitary and fire regulations of the forests adjacent to the OFS protection zone and the presence of damaged trees and shrubs within the sanitary protection zone were identified.

In the western part of the protection zone at a distance of 10 meters from the enterprise's border, there is deadwood. In the southern part of the protection zone, in the 10-meter zone from the enterprise, there is a large amount of dead birch, aspen, willow. Technical works have to be performed. To renew protective strips, it is necessary to clean the dead wood and carry out drainage works. Based on the soil and climatic conditions, the most suitable woody plants for the restoration of protective bands will be spruce, pine, larch, or oak. They are well adapted to these conditions.

To the east of the enterprise and beyond its boundary, the tree layer is represented by conifers – pine and spruce providing a sufficient level of soil moistening. On the northern side of the enterprise territory, there is an asphalted parking for cars; the distance to the forest stand corresponds to the standards.

The survey of self-seeding and undergrowth revealed self-seeding oaks in the vicinity of tanks with oil products.

To improve the working space, it is recommended to equip lawns and flower beds near the buildings. Plants were selected. A plan for planting green spaces was developed. Chamomile can be planted on

IOP Conf. Series: Earth and Environmental Science **350** (2019) 012049 doi:10.1088/1755-1315/350/1/012049

lawns to reduce the number of dandelions in the area. Flowers for beds were selected: Bergenia cordifolia - Pulmonalis, Aquilégia, Asteraceae alpinus, etc.

Table 3. List of pollutants emitted into the atmosphere

Code	Name of pollutants	Quality criteria for the atmospheric air MPC m. (mg/m3)	MPC pp (mg/m3)	RSIL (mg/m3)	Hazard Class
123	Dioxide Iron trioxide, Iron oxide	-	0.0400000	-	3
143	Manganese and its compounds (IV oxide)	0.0100000	0.0010000	-	2
301	Nitrogen dioxide; (NO2)	0.2000000	0.0400000	-	3
304	Nitric oxide (II)	0.4000000	0.0600000	-	3
328	Carbon soot	0.1500000	0.0500000	-	3
330	*Sulfur dioxide and sulfurous nhydride (SO 2)*	0.5000000	0.0500000	-	3
337	Carbon oxide (CO)	5.0000000	3.0000000	-	4
344	Inorganic Fluorides	0.2000000	0.0300000	-	2
405	Hydrocarbons C1-5 (pentane)	100.0000000	25.0000000	-	4
501	Pentilenes, Amylenes (mixture of isomers)	1.5000000	-	-	4
602	Benzene C6H6	0.3000000	0.1000000	-	2
616	Dimethylbenzene, Xylene (mixture of isomers o-, m-, p-)	0.2000000	-	-	3
621	Methylbenzene, Toluene	0.6000000	-	-	3
627	Ethylbenzene	0.0200000	-	-	3
703	Benz [a] pyrene; 3, 4-benzpyrene	-	0.0000010	-	1
1042	Butan-1-ol; N-butyl alcohol	0.1000000	-	-	3
1061	Ethanol; Ethanol	5.0000000	-	-	4
1210	Butyl acetate	0.1000000	-	-	4
1240	Ethyl acetate	0.1000000	-	-	4
1401	Propan-2-one; Acetone	0.3500000	-	-	4
2704	Hydrocarbons (gasoline)	5.0000000	1.5000000	-	4
2732	Hydrocarbons (kerosene)	-	-	1.2000000	
2735	Oil mineral e oil (spindle, machine) Oil aerosol	-	-	0.0500000	
2750	Solvent naphtha	-	-	0.2000000	
2754	Alkanes C12-C19; Carbonwater limit C12-C19	1.0000000	-	-	4
2902	Suspended substances Paint spray	0.5000000	0.5000000	-	3
2908	Inorganic dust	0.3000000	0.1000000	-	3
342	Hydrogen fluoride (H2F2)	0.0200000	0.0050000	-	2

It was recommended to plant trees and shrubs (linden, viburnum, lilac, bird cherry, yellow acacia, barberry, etc.) that will trap dust and create comfort and aesthetics along the paths where the administrative building is located. The collected data and recommendations will be included in the section "Environmental certification of enterprises" that will justify the requirements for the operation of oil filing stations and their territories, buildings, facilities and equipment.

4. Conclusion

The survey revealed an insignificant level of anthropogenic pollution of the territory and sanitary protection zones. Soil trampling has the greatest negative impact. It causes the spread of ruderal plant species. Near the underground oil reservoirs, self-seeding and undergrowing trees were found. They have to be removed in accordance with sanitary standards. In the ten-meter sanitary zone, there are dead wood and shrub species which have to be cut. It is recommended to perform sanitary cutting, remove self-seeding tree species and mow the territory. It is necessary to conduct environmental monitoring of the territory and develop an environmental passport of the enterprise which have to be updated every five years.

References

[1] Zhelonkina E E, Boytsenyuk L I and Valiyev D S 2014 Analysis of the ecological condition of lands in the Nefteyugansk region of the Khanty-Mansiysk Autonomous District *Land Management, Cadastre and Monitoring of Lands* [in Russian – Zemleustrojstvo, kadastr i monitoring zemel] **12(120)** 42–4

[2] Udie J, Bhattacharyya S and Ozawa-Meida L 2018 Vulnerability assessment of climate change impact on critical oil/Gas infrastructure A decision-maker's perception in the Niger Delta *Int. J. of Climate Change Impacts and Responses* **10(4)** 25–39

[3] Smith J W N and Kerrison G 2013 Benchmarking of decision-support tools used for tiered sustainable remediation appraisal topical collection on remediation of site contamination *Water Air and Soil Pollution* **224(12)** 224

[4] Saatsaz M, Monsef I, Rahmani M and Ghods A 2018 Site suitability evaluation of an old operating landfill using AHP and GIS techniques and integrated hydrogeological and geophysical surveys *Environmental Monitoring and Assessment* **190(3)** 144

[5] Shapovalov D A and Gruzdev V S 2008 The impact of technogenic emissions on soil and vegetation as in the case with OJSC Severstal *Ecology and Industry of Russia* [in Russian – Ekologiya i promyshlennost Rossii] **7** 32–5

[6] Boytsenyuk L I, Zhelonkina E E and Pafnutova E G 2019 Estimation of the condition of the natural and recreational zone of the Novoseltsevo village, Mytishchi district *Int. Scientific Practical Conf. "Current issues of land management and cadasters at the present stage"* pp 48–54

[7] Danilov V F, Shurygin, V Y and Alexandrovich D A 2017 *System of the automated leak detection of oil products* (Astra Salvensis) pp 287–95

[8] Xie D, He Q, Zhang X, Wang J and Zhou K 2017 Study on the Pollutant Emissions Characteristics of Typical Petrochemical Enterprises in Hainan Province *IOP Conf. ser. Earth and Environmental Sci.* **233(5)**

[9] Marvin C H, Burniston D A, Shen L, Helm P A and Reiner E J 2018 Polychlorinated dibenzo-p-dioxins and polychlorinated dibenzofurans associated with settling particles in Lake Ontario *Chemosphere* **212** pp 983–93

[10] Heibati B, Godri Pollitt K J, Charati J Y, Karimi A and Mohammadyan M 2018 Biomonitoring-based exposure assessment of benzene, toluene, ethylbenzene and xylene among workers at petroleum distribution facilities *Ecotoxicology and Environmental Safety* **149** 19–25

Assessment of monitoring efficiency for the state of sanitary and epidemiological well-being of the population of metallurgical monocities

E G Zinovyeva, S V Koptyakova, N R Balynskaya and P Y Stepashkov

Magnitogorsk State Technical University n.a. Nosov, 38, Lenin ave., Magnitogorsk, 455000, Russia

E-mail: ekaterina_7707@mail.ru

Abstract. The relevance of the study is due to the importance of the health status and living standards of the population under negative impacts of a number of factors (the environment, large enterprises, nutrition, natural resources). The article presents the results of assessment of the main areas characterizing the sanitary and epidemiological well-being of the population of metallurgical monocities (the environment and its impact on human health, occupational morbidity). At industrial enterprises of Magnitogorsk, the number of workers working under harmful and dangerous conditions is 42.95%. In 2016, the share of workplaces that do not meet sanitary and hygienic noise requirements increased; the share of workplaces that do not meet vibration, microclimate, and illumination standards decreased; there were no violations of EMF standards. In 2016, chronic occupational diseases were caused by imperfect technological processes (36.65% of cases), imperfect workplaces (30.0%), constructive drawbacks of labor means (16.65%) and sanitary equipment (6.65%), and professional contacts with infectious agents (10.0%). The results of the study can be used to improve programs at the federal and regional levels.

1. Introduction

The sanitary-epidemiological well-being of the population includes public health, the environment without harmful environmental factors and favorable conditions for human activities.

Sanitary and epidemiological monitoring conducted by the Russian Federal Consumer Rights Protection and Human Health Control Service is an important element the system of well-being monitoring.

The monitoring system identifies factors that have a harmful effect on humans, assesses them, predicts the health status of the population and the environment, develops proposals for decision-making in the field of sanitary and epidemiological well-being of the population, identifies measures to prevent and eliminate the impact of harmful environmental factors on the human health. This allows for efficient identification of health indicators and living standards. The government can promptly respond to identified deviations.

The purpose of the article is to study the efficiency of monitoring of the sanitary and epidemiological wel-being of the population of metallurgical monocities on the example of Magnitogorsk. To achieve this goal, a comprehensive study of the sanitary and epidemiological well-being of the population in Magnitogorsk was carried out.

2. Materials and methods

Issues of sanitary and epidemiological well-being of the population are studied by Russian and foreign researchers.

Theoretical and methodological issues of sanitary and epidemiological well-being of the population are analyzed in the works by Aksenov [1], Ivanov, Belyaev [2], Volkov [3], Butaev, Reshetnikov, Kucherenko [4], Kuzmin [5], Zaitsev [6], Steblyanko, Zinovyev [7, 8], etc.

Theoretical and methodological issues of the quality of atmospheric air, soil and water, the level of occupational morbidity, emissions of pollutants are discussed in the works by Alferov [9], Yakovenko, Avaliani [10], Butorin [11], Onishchenko [12], Mayorova [13, 14], etc.

The use of foreign experience of sanitary-epidemiological well-being of the population was studied by Smolensky. [15], Toporkov, Leaves [16], Herbert, Jaakkola [17] etc.

The information base for the study of performance indicators are documents of the legislative authorities, regulatory acts of the Federal Service for Supervision of Consumer Rights Protection and Human Welfare of the Russian Federation [18, 19], data provided by the Magnitogorsk Department of the Federal Service for Supervision of Consumer Rights Protection and Human Welfare for 2014–2016 [20].

3. Analysis of efficiency of the sanitary-epidemiological well-being of the population of Magnitogorsk

The sanitary-epidemiological well-being of the population of metallurgical monocities of the Russian Federation is studied in the following areas: the human environment and its impact on public health; the medical-demographic situation and development of massive non-communicable diseases (poisoning) due to harmful environmental factors; occupational morbidity.

The article analyzes the level of occupational morbidity.

In Magnitogorsk, the level of occupational morbidity is affected by working conditions as one of the main risk factors for the development of professional and professionally determined diseases. At the industrial enterprises of Magnitogorsk (controlled by the territorial department), the number of employees working in harmful and dangerous conditions is 42.95%. Over the past three years, there has been no downward trend in the share of workers working in harmful and hazardous conditions; even a slight increase in the indicator has been observed: 2014 – 42.6%, 2015 - 42.8%, 2016 - 42.95%. In 2016, the state of industrial workplaces changed in comparison with 2015. In 2016, the share of industrial jobs that did not meet noise requirements increased; the share of working places that did not meet vibration, microclimate and illumination standards decreased (Figure 1).

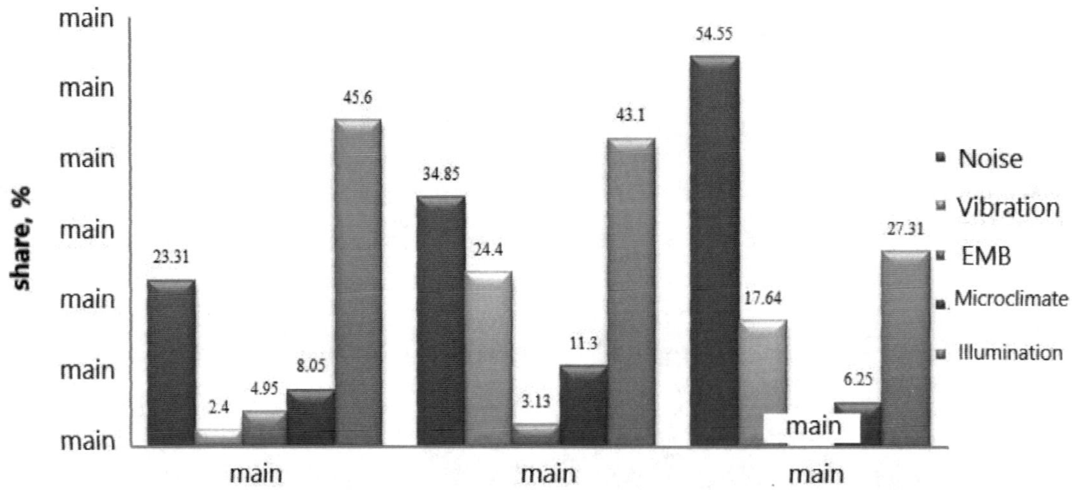

Figure 1. The share of industrial workplaces that do not meet sanitary and hygienic requirements, %

The highest rate of occupational morbidity per 10 thousand workers was recorded at manufacturing enterprises - 3.93 (Table 1).

Table 1. Occupational morbidity indicators by economic activities

Economic activities	Indicator per 10,000 employees		
	2014	2015	2016
All economic activities	1.72	1.81	3.19
Manufacturing industries	2.54	2.12	3.93
Production of other non-metallic mineral products	4.31	-	-
Metallurgy	2.71	3.11	6.60
Production of finished metal products	-	4.54	1.13
Manufacture of machines and equipment	1.64	0.55	2.91
Recycling	11.16	11.16	-
Construction	1.99	1.33	2.24
Health and social services	-	1.37	4.11

In 2016, chronic occupational diseases were caused by imperfect technological processes (36.65% of cases), imperfect workplaces (30.0%), constructive drawbacks of labor means (16.65%) and sanitary equipment (6.65%), professional contacts with infectious agent (10.0%) (Figure 2).

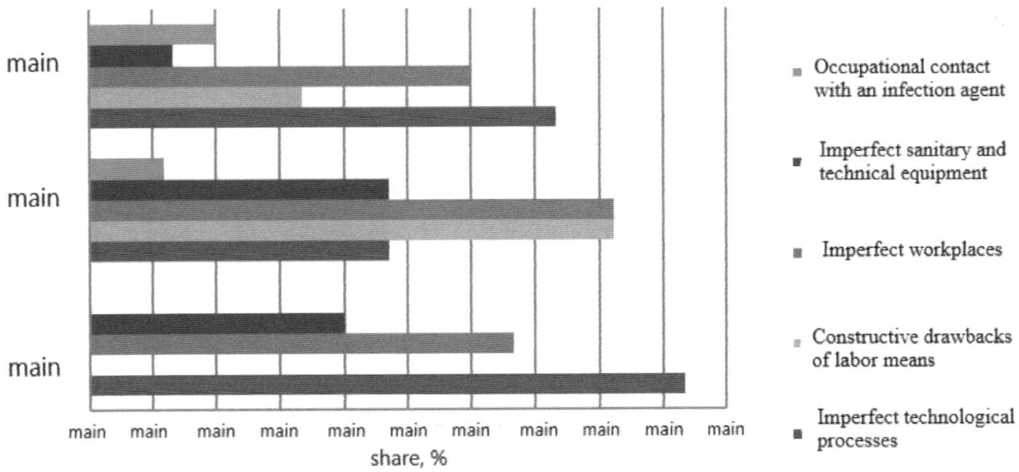

Figure 2. Circumstances and conditions of chronic occupational diseases, %

The level of detection of chronic occupational diseases depending on the type of medical institutions and their specialization did not change much. In 2016, Magnitogorsk Center for Occupational Pathology identified 96.67% of occupational diseases (in 2015, 95.24% of the total number of diseases; in 2014 - 86.67%). Analysis of the dependence of the level of occupational morbidity on jobs, contacts with harmful production factors and the age identified individual professional groups who are subject to occupational diseases.

Analysis of occupational diseases allowed us to conclude about the relationship of the length of service, contacts with harmful production factors and the level of occupational morbidity; the maximum risk of occupational diseases is characteristic of male workers affected by harmful production factors for over 30 years and female workers affected by harmful production factors for over 20 years. In the group of male workers with a time record of 20–30 years, the share of registered occupational diseases is 31.82%; with a time record of 30–40 years - 63.64%. In female workers with a time period of 20-25 years, the share of occupational diseases is 66.7%.

Workers aged 50-60 years are most subject to occupational diseases: occupational diseases in men are 90.91%, in women - 66.67% of all occupational diseases. The following jobs are most subject to occupational diseases: electric and gas welders, coke machine operators, choppers. In 2014–2016, the share of occupational diseases was 36.21%.

Most of the occupational diseases were identified during regular medical examinations. All chronic forms were identified in workers aged over 50 who have been dealing with harmful production factors for more than 30 years. Despite the decline in the number of injured people, severity of occupational diseases and the degree of loss of professional fitness, the indicator remains higher than the national average due to low-level identification of occupational diseases during medical examinations of workers.

4. Conclusion

Thus, according to the results of the study, it can be argued that there are a lot of factors affecting the sanitary and epidemiological welfare of the population (the environment, food quality, working conditions). To reduce negative consequences, it is advisable to implement the following measures:

1) air protection:
- construction of bypass roads in order to remove some vehicles from the city center;
- ban on the construction of facilities that are sources of air pollution.

2) food quality and safety:
- development of a city security program for the consumer market;
- expansion of the range of local products enriched with iodine, other macro and micronutrients;
- explanatory work in the mass media aimed at the prevention of diseases caused by insufficient intake of micronutrients;
- promotion of the healthy lifestyle and nutrition.

3) healthy working conditions:
- development of the target program "Improving the working conditions and medical support for the working population of Magnitogorsk";
- production control over sanitary regulations, hygienic standards and implementation of preventive measures;
- regular medical examination of employees affected by harmful production factors;
- implementation of planned measures aimed at improving working conditions, developing additional measures to improve working conditions according to hygienic standards (improving the efficiency of ventilation systems, replacing obsolete equipment, upgrading process equipment, and repair technologies for metallurgical furnaces and aggregates).

The primary task of any municipal body is to control the environmental situation, food quality, labor protection, and medical examination of workers.

References

[1] Aksenova O I 2015 *Hygienic aspects of providing the population with safe drinking water: problems and solutions* (Moscow: Forum, SIC INFRA-M)

[2] Belyaev E N, Fokin M V, Novikov S M, Prusakov V M, Shashina T A and Shayakhmetov S F 2013 Actual problems of improving the health risk assessment of the population to ensure sanitary and epidemiological well-being *Hygiene and Sanitation* [in Russian – Gigiena i Sanitariya] **5** 53–5

[3] Volkova N A 2015 Organization of supervision of nutrition of the population *Ecology of Russia* **2** 34–42

[4] Kucherenko V Z 2014 *Organizational and legal framework of the Federal Service for Supervision of Consumer Rights Protection and Human Welfare* (Moscow: GEOTAR-Media)

[5] Kuzmin S V, Gurvich V B, Dikonskaya O V, Malykh O L and Yarushin S V 2016 Methodology of assessing and managing risk to public health in the system of legislative regulation of the sanitary and epidemiological welfare of the population *Occupational medicine and industrial ecology* **1** 4–8
[6] Zaitseva N V, Popova A Yu, May I V and Shur P Z 2015 Methods and technologies of health risk analysis in the public administration system while ensuring the sanitary and epidemiological well-being of the population *Hygiene and Sanitation* [in Russian – Gigiena i Sanitariya] **94(2)** 93–8
[7] Steblyanko V L, Zinovyeva E G and Koptyakova S V 2019 Efficiency of monitoring the sanitary-epidemiological state of Magnitogorsk *Steel* [in Russian – Stal] **4** 80–3
[8] Zinovyeva E G and Koptyakova S V 2019 Integrated assessment of sanitary and epidemiological safety management efficiency of the population in Industry–based cities in the Russian Federation *Advances in economics, business and management research: proc. of the Int. Sci. Conf. Far East Con (ISCFEC 2018)* vol 47 pp 512–6
[9] Alferov I N and Yakovenko N V 2016 The problem of providing high-quality drinking water to the population of the water-deficient region of the Orenburg region *Human Ecology* [in Russian – Ekologiya cheloveka] **4** 3–8
[10] Avaliani S L, Novikov S M, Shashina T A, Skvortsova N S and Mishina A L 2012 Problems of harmonization of standards in atmospheric pollution and ways to solve them *Hygiene and Sanitation* [in Russian – Gigiena i Sanitariya] **5** 75–80
[11] Butorina N N, Bychkov A A, Samutin N M, Orlova I G and Pozdnyakov S A 2017 Modern approaches to the regulation of waste management of non-canalized facilities at the regional level *Ecology and Industry of Russia* [in Russian – Ekologiya i promyshlennost Rossii] **10** 68–71
[12] Onishchenko G G 2012 Results and prospects of ensuring the sanitary and epidemiological welfare of the population of the Russian Federation *Hygiene and Sanitation* [in Russian – Gigiena i Sanitariya] **4**
[13] Kolokoltsev V M, Vdovin K N, Maiorova T V and Ponomareva O S 2017 Ecological indicators in the system of non-financial reporting at industrial enterprises *CIS Iron and Steel Review* **13** 4–10
[14] Mayorova T V and Ponomareva O S 2015 Methodology for assessing the economic efficiency of environmental management of metallurgical enterprises *Bulletin of Magnitogorsk State Technical University named after G I Nosov* **4(52)** 112–6
[15] Smolensky V Yu, Toporkov V P, Karnaukhov I G, Toporkov A V and Kutyrev V V 2012 Organizational and methodological aspects of international cooperation in ensuring the sanitary and epidemiological well-being of the population *Problems of especially dangerous infections* [in Russian – Problemy osobo opasnyh infektsij] **2(113)** 5–13
[16] Leaves G D and Herbert R D 2002 Economic and environmental impacts of pollution control in a system of environment and economic interdependence *Chaos, Solitons & Fractals* **13(4)** 693–700
[17] Jaakkola J J K et al 2000 Use of health information systems in the Russian Federation in the assessment of environmental health effects *Environmental Health Perspectives* **108(7)** 589–94
[18] Federal law of 30.03.1999 no 52-FZ On the sanitary-epidemiological welfare of the population 1999 *Collected legislation of the Russian Federation* **14** 1650
[19] Federal Law of 21.11.2011 no 323-FZ On the basis of the protection of public health in the Russian Federation 2011 *Russian newspaper* **263**
[20] *The impact of environmental factors on public health. Medico-demographic situation in the Chelyabinsk region*, Available at: http://www.mineco174.ru/files/media/doklad/2014/6-1.htm

IOP Conf. Series: Earth and Environmental Science **350** (2019) 012051 doi:10.1088/1755-1315/350/1/012051

Features of townland management

A A Varlamov, S A Galchenko, D V Antropov, S G Kuznetsova and Yu S Sinitsa

State University of Land Management, 15 Kazakov st., Moscow, Russia, 105064

E-mail: antropovzem@gmail.com

Abstract. The town is a complex social and economic system that has significant land, social and other resources. An integrated approach should be used to solve townland management problems. Improvement of the efficiency of land management should be based on the comprehensive analysis of the land management system from the planning process to the control over the use of land plots taking into account financial and investment opportunities of land management facilities. Successful development of urban and rural settlements depends on the effectiveness of land management. These issues determine the significance of the present article. The article describes problems and features of townland management in Russia. Directions of the public land policy were determined.

1. Introduction

The modern town is a complex social and economic system which consists of diverse objects and independent structures. Land management issues are studied by many scientists, research, design, survey and industrial organizations. Significant contribution to the theory and methodology of land management, including land use management is made by A.A. Varlamov, S.N. Volkov, S.A. Galchenko, N.V. Komov, N.I. Kresnikova, P.F. Loiko, V.N. Khlystun, S.A. Lipsky et al. [1–9]. In cities, various elements of production and non-production spheres interact; a large number of people live and work. All of them use urban resources, including land, buildings, structures and other real estate objects. The regulatory process is a complex task aimed at meeting needs of the urban system in resources while preserving and reproducing the natural and economic potential of its territory.

The Land Code of the Russian Federation (Chapter 15) determined the legal status of settlement lands: they are used for building and developing urban and rural settlements and divided by boundaries from other categories of land. The boundary of an urban settlement is its external border which separates its land from other categories of land. The border is established on the basis of an urban plan and land management documents.

The land of settlements may include land plots assigned to the following territorial zones in accordance with town planning regulations:

- residential, social, business, industrial;
- engineering and transport infrastructures, recreational and agricultural,
- for special purpose, of military facilities, other territorial zones.

The rules of land use and development establish town planning regulations for each territorial zone taking into account peculiarities of its location and development, as well as possible territorial combination of different types of land use. Reconstruction and expansion of existing real estate objects, construction of new real estate objects associated with land can be carried out in accordance with established urban planning regulations.

IOP Conf. Series: Earth and Environmental Science **350** (2019) 012051 doi:10.1088/1755-1315/350/1/012051

2. Methods

The land management system of settlements can be improved through studies of the research object, adoption of legal acts on land management, research on theoretical foundations of the land management system formation, organizational and functional structures as a set of interrelated elements that are in stable relationships.

The town is a complex social and economic system that has significant land, social and other resources. An integrated approach should be used to solve townland management problems. Improvement of the efficiency of land management should be based on the comprehensive analysis of the land management system from the planning process to the control over the use of land plots taking into account financial and investment opportunities of land management facilities.

At the same time, the shift of priorities towards the economic efficiency without taking into account social and environmental aspects of land management, quantitative and qualitative characteristics of land plots is unacceptable. Land resources of urban districts as an object of management have their own characteristics. Land resources are a powerful economic basis for developing towns. Compared with a number of foreign countries, the share of land taxes is small. Therefore, it is necessary to increase the land market turnover, the financial potential of land.

The subject of townland management is land management measures applied in various areas of the land management system (social, residential, industrial, etc.) and taking into account the following features:

- spatial limitations of the territory;
- territorial combination of production facilities and the habitat of the population which raises requirements for the land, safety, engineering structures and communications;
- high density of artificial real estate;
- a variety of buildings (large unique buildings and large industrial facilities, garages and transformer substations);
- negative changes in the state of land caused by man-made loads and pollution;
- cultural, historical and architectural objects of the land management system which imposes restrictions on their use.

The object of management is a land plot. In built-up areas, the object of management is land plots with different legal status as well as general use land plots. Settlement land plots provides satisfy needs of residents. Therefore, land management is characterized by a large number of management areas [1].

One of the main functions of local self-government is to solve issues of ownership, management and disposal of municipal property independently. However, the main problem is determination of the scope of authority, separate powers between the Russian Federation and municipal authorities.

The powers of local governments include withdrawal, including redemption of land plots for municipal needs, adoption of land management regulations and development of urban and rural areas taking into account legislative requirements of the Russian Federation, development and implementation of local programs for land management and protection, etc.

The objectives of subjects of management are different. Therefore, when developing a common land management system for settlements, it is necessary to work out a system of management goals.

3. Research results, problems and features of land use management

The following problems were revealed:

1. Reduction of the number of towns and villages, migration of young people to large cities.
2. "Optimization" of social facilities (hospitals, educational institutions, cultural facilities, post offices, shops).
3. Termination of production activities of ruined (or already ruined) enterprises (agricultural enterprises, timber industry enterprises, sanatoriums, clinics using local sources of mineral water, mud, etc., textile industry enterprises, shops, public catering enterprises, etc.
4. Special economic zones, technology parks and other areas with a special mode of financing. Towns located outside special zones and technoparks lose their attractiveness.

5. Lack of convenient logistics, undeveloped transport infrastructure.

6. Depopulation of villages. Villages existed owning to towns where there were hospitals, schools, and shops [1].

Successful development of urban and rural settlements depends on land use management mechanisms. The existing structure of land management is regulated by legislative and executive documents that do not solve arising problems. Difficulties arise due to the fact that many organizations and departments that take uncoordinated decisions are involved in the management process. The adopted laws allowed it to solve a number of issues, but there are unresolved problems.

The number of natural properties of the land are as follows: soil (density, permeability); groundwater (depth, chemical and bacteriological composition); surface waters; relief and dissection of the territory; vegetation; soil (pollution, salinization, quality).

Implementation of one or another method of land development is attractive to the consumer and creates demand in the real estate market. The wider the range of possible directions for using the land plot, the higher its value. Accordingly, there are requirements for the size and shape of the land plot. The most common defects of land plots are their small size, elongation, blunt or acute angles, non-linear boundaries, irregular configuration, etc. Since the territory is a limited resource, formation of territorial objects should be subject to the maximum efficient use. lands. The boundaries of the object cannot be changed. Therefore, the presence of defects impedes the turnover and development of this object.

The property has other properties. These are architectural and planning characteristics of the structure, constructive solutions which are largely determined by territorial characteristics of the land plot, including its location. In addition, the need for economic activities related to the purpose of the building may require land plots adjacent to the building. The territorial characteristics of the property object are a factor in its development, including a change in the way of development and replacement of all buildings located on the land.

A building and a specific land plot form a single land and property system which should be considered as a single object.

Based on these features, the most important areas of state land policy in small towns are as follows:

- development of land use efficiency criteria;
- development of the strategy of transformation of legal relations taking into account the specifics of various land categories, features of urban land management and development of the municipal economy;
- identification of an optimal balance of various forms of land granting taking into account investment tasks of municipalities, development of the real estate market and mortgage lending, satisfaction of state and municipal needs, and solution of social problems;
- expansion of stock market instruments when implementing land management measures as a means of increasing the efficiency of state regulation of financial flows;
- rational combination of strategic and tactical interests, i.e. careful townland management and raising local funds;
- simplification and reduction of real estate title registration fees;
- ensuring the reliability and efficiency of state land control over municipal lands;
- regulation of tourism and rental relations and other forms of land management;
- rational calculation of land tax rates of market and cadastral prices of land;
- formation and accounting of real estate objects;
- monitoring of the real estate market and its information support;
- informational support of the requests of the government authorities.

A large number of land plots, phenomena and processes makes the land management system important.

4. Conclusion

Townland management is fundamentally different from the management of territories of other administrative and territorial entities. This is due to a large number of people, industries, facilities, historical and cultural monuments which dramatically increases the importance of social, environmental and transport conditions. Therefore, the importance of logistical plan development for cities and towns is crucial, since any changes in the land management system affect these plans. These plans are developed due to the growth of cities, creation of new cities in accordance with the scheme of settlement, improvement of the existing land management system.

When granting land plots, both land management and cadastral activities are carried out; architectural and planning documents are developed.

Development of urban planning and other project documentation, construction and reconstruction of urban and rural settlements, individual buildings and structures are carried out in compliance with the requirements of environmental protection, environmental safety and sanitary rules taking into account the state of land management in rural settlements and environmental safety requirements established by territorial integrated schemes of nature protection and nature management as well as taking into account harmful effects of economic and other activities on the environment and human health.

When carrying out planning activities climatic, medical, sanitary and other favorable conditions of the territories and settlements are taken into account in order to ensure the right of citizens to rest. In the places of permanent residence, land plots for recreational facilities are granted in accordance with urban planning documents, state urban planning regulations and rules.

The documents on urban planning determine the composition and structure of social, cultural and community services taking into account complexity of the development and interests of various socio-demographic and professional groups of the population. In addition, the system of social, cultural and everyday services in urban and rural settlements is created in accordance with the planning structure of settlements, their inter-settlement value taking into account business development and competitiveness the service sector [1, 4].

References

[1] Varlamov A A and Galchenko S A 2018 Development of land use system in modern Russia *International Agricultural Journal* **6 (366)** 8-9

[2] Volkov S N and Lipsky S A 2018 Improving land legislation is a necessary condition for the effective management of land resources *Land Management, Cadastre and Monitoring of Lands* [in Russian – Zemleustroystvo, kadastr i monitoring zemel] **7(162)** 5-10

[3] Volkov S N, Komov N V and Khlystun V N 2015 How to achieve effective land management in Russia? *International Agricultural Journal* **3** 3-7

[4] Galchenko S, Varlamov A and Bogdanova O 2018 Theoretical and methodological foundations for the formation of a sustainable land management system *IOP Conference Series: Materials Science and Engineering* 012141

[5] Zhdanova R V and Rasskazova A A 2017 Basic concepts of the economic efficiency of sustainable land management *International Agricultural Journal* **1** 10-13

[6] Komov N V 2017 Effective land management is the basis of statehood and wealth of the people *Economy and Ecology of Territorial Formations* [in Russian – Economika i ecologiya territorialnykh obrazovaniy] **2** 6-14

[7] Komov N V and Cheshev A S 2018 An integrated approach to planning and rational use of land resources *Economy and Ecology of Territorial Formations* [in Russian – Economika i ecologiya territorialnykh obrazovaniy] **1** 6-21

[8] Komov N V 2019 On the creation of a system of unified state land administration in Russia *Land management, cadastre and monitoring of land* [in Russian – Zemleustroystvo, kadastr i monitoring zemel] **1(168)** 5-8

[9] Sevostyanov A V 2017 Improving the efficiency of land use of the settlement by establishing its boundary *Moscow Economic Journal* [in Russian – Moskovskiy ekonomicheskiy zhurnal] **4** Available at: http://qje.su/zemleustrojstvo-i-zemledelie/moskovskij-ekonomicheskij-zhurnal-4-2017-59/

Geomorphology of the northern part of Side Ridge Gorstian anticlinorium: orography of Shanski and Makhis-Magalinsky high-mountain massifs

V V Dotsenko[1] and T K H Bachaeva[2,3]

[1]Academy of the Chechen Republic, 13, M A Esambaeva ave., Grozny, 364001, Russia
[2]Kh. Ibragimov Complex Institute of the Russian Academy of Sciences, 21 A, Staropromyslovskoe Highway, Grozny, 364906, Russia
[3]Grozny Oil Technical University named after M D Millionshchikov, 100, Kh. Isaev Ave., Grozny, 364061, Russia

E-mail: bachaeva@bk.ru

Abstract. The article shows the orography of the Shansky and Makhis-Magalinsky high-mountain massifs, selected in the geomorphological zoning scheme with the rank of subdistricts corresponding to the fourth-order morphostructures. Geomorphological processes, types of relief, the latest tectonic movements and glaciation of mountain ranges are characterized.

1. Introduction

The high-mountainous massifs of the Side Ridge (SR) morphostructure are associated with tectonic blocks formed by transverse, diagonal and longitudinal faults to which the river valleys are confined. These massifs are fourth-order morphostructures or subregions of SR [1].

However, the massifs themselves are also dissected by faults and large gaps into blocks of smaller orders. The watersheds of the valleys are represented by the branches of SR, mostly of the northern direction. In the Ingush and Chechen Republics, there are five high-mountain massifs: Shansky; Mahis-Magalinsky; Tebulos; Komito-Kobulamsky and Donos-Diklossky. In the scheme of geomorphological zoning, they are distinguished in the rank of subdistrict SR [2].

Administratively, the study area is located within the mountainous part of the Chechen and Ingush Republics. Its geological and geomorphological structure on the stretch from the Belaya River in the Western Caucasus to the r. Chanty-Argun in the Eastern Caucasus, in general, corresponds to the morphostructural zonality of the northern wing of the Greater Caucasus megalithiclinorium.

Nevertheless, in the east direction, the complexity of the structure is gradually becoming more complex. In particular, in the Order of the geomorphological subregion and further east, to the Shakh-Dag mountain, the greatest heights of the axial part of the Greater Caucasus are confined to the Lateral ridge. The heights of the Main Caucasus (Watershed) ridge here are on average lower than 1000 m, which is due to the peculiarities of the newest tectonic activity.

2. Materials and methods

In the research, a set of methods was used: laboratory-based methods and materials for remote sensing: interpretation of satellite images and topographic maps of scales from 1: 10'000 to 1: 500'000; field methods and materials of expeditionary research of reconnaissance routes along the gorges of the Terek,

Demos-Tskali, Kistinka, Armkhi, Shandon, Salgi, Assa, Sartu, Guloykhi, Chanty-Argun, Meshehi rivers; climbing tops: Salgi, Dzeneladze; A. Japaridze, A. Chavchavadze, Tsereteli, Bachahi, Rustavi, Kuro, Shino, Kideganismagali.

The names of the ridges of their spurs, peaks and saddle saddles, as well as the heights of peaks and passes in the article contain the names indicated on topographic maps of a scale from 1: 200'000 to 1: 10'000, published by the General Staff and the General Directorate of Geodesy and Cartography, also the names given in the classifier routes to the mountain peaks [3] and the list of classified high mountain passes [4].

3. Orography of the Shan and Mahis-Magalinsky high-massifs

The Shansky the high-mountain massif is quite extensive and lies between the antecedent participants of the Tereka Valley and Assy, which are associated with the same-name deep faults. In the west, it is separated by the valley of the Terek River from the Dzhimarai-Kazbek high-mountainous massif of the Central Caucasus (Figure 1), and in the north - the Puy fault, which passes through the gorge of the r. Armhi and its right tributary r. Salgi [5].

Figure 1. The eastern wall of the Shavansky Range, with hanging glaciers, avalanche-scree couloirs, punishments at the base of the wall and the peak of Shang. In the foreground are the shale "feathers" of the ridge of the Yukrolamduk ridge dividing the gorges of the Shandon and Armkhi rivers. Photo by A. Plotnikov

In the basin of the r. Assa, it is connected with a fault by a gorge Sart river. In the south, the border is also associated with a fault (Adayk-Kazbek), which runs along the upper course of the starboard side of the Djuta-Snoskali river valley, establish the Bezhitinskoy depression. The main part of the Shan massif is located in Ingushetia, and the south-western part is in Georgia. The north-western part is bordered by North Ossetia.

Centerline morphostructure of the SR on the Shan massif, as well as on other massifs, is shifted to the south and is associated with high modern and modern tectonic activity, which is manifested by high absolute elevations of the mountain peaks: Kuro (4071.2 m); Shang (4451.8 m); Kideganismagali (4229.1 m) and Chimochis-Magali (3851 m). In addition, appears abnormally high slopes of the longitudinal profiles of rivers appear there. High tectonic activity and the height of the massif determine the intensive manifestation of exogenous processes – nival-glacial, periglacial and gravitational.

The internal part of the Shansky massif is dissected by deep glacial-erosion-tectonic gorges of the Kistinka, Shandona and Armkhi rivers. The depth of the vertical division reaches 2350 m. As a result, high spurs, Ohkkur-Shavana and Yukurulamduk, were formed within the massif, which stretched subparallel from south to north. The main link of the SR is represented by the Kuro, Kidegani ranges and nameless segments.

The ridges have sharp ridges and steep slopes, often exceeding 60°. The tops of the slopes are corroded with carats and circuses filled with small kar and hanging glaciers or snowfields (Figure 1). The ancient carats located below are filled with moraine deposits or sod. The largest are the Kidagano and Kibishi glaciers. In addition, the slopes of the ridges are dissected by margins and avalanche trays, ending with non-sodded actively developing debris. The gorges of the main rivers in the lower part of the slopes are narrow, V-shaped, and in the upper part they acquire a U-shaped or trough-shaped form and are cluttered with a moraine and crumbling deposits, which is typical of young troughs.

The Shansky massif is very popular among climbers and mountain tourists. Tens of routes were laid on its tops [3] and over 30 passes of various categories of difficulty were classified [7]. The main watershed SR on the Shan massif makes a large and complex arc. In the west, it is represented by the Kuro Range, which in Georgia divides the gorges of the Terek and Kistinka rivers. The range extends from the north-west to the south-east to the town of Kora (3631 m). It has two ridge peaks: Shino (4047 m) and Kuro (4071.2 m). Further east of the Mt. Kora in the arc of the SR, there are three bends, the height of the crests of which does not exceed 4000 m.

In the western part of the massif, the first bend from the south bends around the Kibishi glacier, located in the upper reaches of the r. Kistinka. Then the second from the north - the origins of the r. Dzhavarthorhistskali, which is a right tributary of the Juta river, and then the third bend, again from the south goes around the sources of the Shandon and Armkhi rivers. The first and third bends, formed by the nival-glacial denudation and erosional activity of the Kistinka and Shandon rivers, strongly extend into the Bezhitinsky depression, so the ridge of SR approaches here with the ridge of the Dividing Range to a distance of up to 3.6 km.

From the nodal vertex of 3943 m, towering above the active glacial rocks of the headwaters of the rivers Kistinka and Shandon and above the "dead" car in the sources of the r. Dzhavarthorhistskali in the south, Shavan Ridge branches off. After the vertex of mountain Erzi (3963.3), it changes its direction to the north-west and is already called the Ohkur Ridge. The Shavana-Okhkur Ridge divides the gorges of the Kistinka and Shandon rivers and along its ridge passes the border between Georgia and Ingushetia. This ridge (spur) has a sharply alpine-type character and a large both absolute and relative height (Figure 2).

Figure 2. The upper part of the gorge of the r. Kistinka. In the distance is the western slope of the Shavan Range (right side of the gorge). On the left is the peak of Mt. Kitsch, in the center of Mt. Shan, at the end of the ridge of the peak of the "Georgian corner". In the foreground to the right is the tip of the Kuro Range (left side of the Kistinka Gorge). View from the slopes of Mt. Kaijan (northeastern part of the Kazbek-Dzhimarai massif). Photo by E. Manukyants

On it to the south, a continuous chain of 19 peaks rise, most of which exceed 4000 m. Here is also the highest mark of this mountain range and of Ingushetia - Mt. Shan (4451.8 m). In the southeastern corner of the Shandon gorge, above the narrow trough and surrounded by punishments, there is Mt. Chimgis (3970 m). After 9 km north of it rises Kideganismagali (4229.1 m). This part of the SR is called Kidegani and has a strictly meridional direction. In the ridge of the ridge the most significant is the top of the Mt. Sakharismt (3963 m), also called Sakharismagali [6] and the Sakharisgel pass (3597 m) is located. The top of Mt. Kideganismagali is a nodal. The short, but high spur Yukrolamduk with peaks of 4031.2 m, 3600 m, 3781 m, 3663 m, 3575.4 m, which inherits the direction of the Kidegan ridge, separates the lower part of the gorge of r. Shandon and the upper trough of the river valley r. Armhi. The main ridge of the SR from the top of Kideganismagali rounds from the south a deep circus and trough valley of the Kidagano glacier, and then 5-6 km stretches north-north-east to the top of Martinismt (3898.1 m).

There are a number of passes and peaks of mountains on the north-west on the SR from the top of the town of Kideganismagali to the top of Martinismt: 3968 m; 3935 m; Mt. Chimochismagali (3851 m); Mt. Tatelismagali (3755 m); Mt. Maamgos (3869 m). In front of the top of Mt. Martinismt, the main link of the Side ridge of the Shansky massif sharply turns to the east and soon near the town of Mt. Schuachviri (3112 m) is interrupted by the valley of the r. Assa. From Mt. Martinismt to the northeast, then east to the town of Mt. Terkhkort (3209 m) there stretches a large, punishable spur to Argelam, which is also interrupted by the valley of the r. Assa. On this spur is the border between Ingushetia and Georgia.

In the upper reaches of the r. Armhi valley is the Kidagano valley glacier. It is the largest in the Shan massif. The length of its language in the last century reached 2.5 km. According to V.D. Panov [7], in the period from 1881 to 1966 years, the glacier retreated at an average speed of 20.5 m per year. As a result, four terminal moraines with a height of 60-80 m were formed in the trough valley. The fifth, most ancient terminal moraine is located in front of the transom, represented by a steep ledge 300 m high, which faces the V-shaped gorge of the r. Armhi. The tongue is surrounded by two ridges of lateral moraines up to 50 m high.

4. Orography of the Mahis-Magalinsky high-massifs

The Mahis-Magalinsky high-mountain massif is located on the border territory of the Ingush and Chechen Republics and is bounded by antecedent sections of the valleys of the Assa and Chanty-Argun rivers, laid along faults. The main link in the SR of this massif in the tourist literature is called the Zusunkort ridge [6]. It is tortuous, but generally has a latitudinal direction. The western part of the ridge extends in the direction of the peaks of Shautskhviri (3112.1 m) and Martinismt of the Shan massif. On the ridge Zusunkort from west to east there are peaks 3348.9 m, Artsivismagali (3863 m), Mahismagali (3990 m), 4007 m, 3298 m, and Zuzunkkort (3437.9 m), which are the highest marks of this massif (Figure 3).

Figure 3. Mahismagali massif from the side of the Nelkh gorge Photo by T. Agirov

From the Mt. 4007 m to the northeast, to the Gul North pass stretches the Vegilam ridge with peaks: 3773 m; 3546.8 m; Harsacort (3035.1 m); Kurelam (3074.8 m). After the peak of Kurelam, the ridge sharply decreases and continues to the north in the form of a low web connecting the lateral and rocky ridges. The border between the Chechen and Ingush Republics runs along this jumper. Gul Pass Northern divides the gorge. Guloykhi, located in the Ingush Republic and the gorge r. Geshichu, located in the Chechen Republic. These gorges form the northern boundary of the morphological structure of SR and along their southern slopes passes the regional Puy fault.

From the Mt. Kurelem (3074.8 m) to the southeast at right angles departs the eponymous spur Kurelem. Then after 4 km, also at a right angle from the ridge Vegilam, the Bastilama departs from. The spurs of Kurelam and Bastila are almost parallel to each other and are separated by a narrow gorge Omchu, which drains the r. Bastikhi. The configuration of spurs and gorges and their straightness reflect regional faults. In the valley of the r. Chanty-Argun, the ridge to the Bastylam ends with a peaks of the same name, 3179.6 meters high.

Of the snowfields located at the top of the 4007 ridge of the Zuzunkort and the springs of the Vegilam and Kurelem ridges originate in the r. Meshehi. The Vegilam and Kurelam ridges are joined at right angles on the Gul South Pass. Accordingly, in the upper reaches of the river two ravines were formed. The gorge of the northeastern direction, starting from two glacial circuses and the ancient circus on the slope of the Zusunkort ridge, is called Vasichu, and the gorge of the south-southeastern direction, beginning under the Gul South pass, is called Dudilkhi. Upper and lower reaches of the gorge. Meshehs have different morphologies — the lower part of the gorge is V-shaped and erosive, and the upper part is U-shaped with a wide bottom, since it is an ancient trog (Figure 4).

Figure 4. Gorge r. Meshehi. In the distance, the wall of the Vegilam spur and the confluence of the Vasichu gorges (left) and Dudilkhi (right). One can see the transition of the V-formed gorge to the trough-shaped valley. Photo by T. Agirova

The trough part of the valley itself is in connection with this feature. Meshikhi and its comparatively gentle slopes of the southwestern exposure were very densely populated in the past. The landmark of the gorge is also the famous necropolis Tsoi-Pede, which is located on the cape, washed by the river beds of Meshekhi and Chanty-Argun, which indicates the past population of this high-mountainous region (Figure 5).

Figure 5. Necropolis Tsoi-Pede. Photo by A. Bersaev

East of the gorge of Chanty-Argun, already in the Chechen Republic, the SR is represented by a single mountain chain. Before Mt. Tebulos-Mta, its segment is called the Mutsossky ridge. Further to the summit of the Mt. Diklos-Mta, the chain is called the Pirikitelsky or Tushetsky Range, and the short northeast segment is called the Snow Range or Boghu Range [8]. The morphostructure of SR includes only its highland part.

5. Conclusion

The relief and landscapes of the Shansky and Mahis-Magalinsky high-mountainous massifs of the Lateral Range differ from the Western and Central Caucasus, as well as the high mountains of Dagestan, in their severity and "wildness". The upper reaches of river valleys are trophies developed during the epoch of the Quaternary glaciations, and their middle and lower parts are represented by deep and narrow V-shaped gorges, sometimes not having low terraces, but also floodplains. The differences are also manifested in the fact that the geological structure and climate change here.

References

[1] Lotiev B K and Sazonov I G 1979 *To the geomorphological zoning of Chechen-Ingushetia. Problems of the physical geography of the North-Eastern Caucasus* (Grozny: Chechen State University) pp 74–84

[2] Dotsenko V V 2014 Geomorphological zoning of the Chechen and Ingush Republics *Sustainable development of the mountain regions of the North Caucasus in the context of global changes: research and practice: Mater. of the Int. Sci. and Pract. Conf.* (Grozny: Chechen State University) pp 99–106

[3] *Classifier of routes to the mountain peaks* 2013 (Moscow: Federation of Alpinism of Russia)

[4] The list of classified high mountain passes 2016 *Federation of sports tourism in Russia* Available at: www.tssr.ru

[5] Agibalova V V 1988 *On Assu through Armhi Sun Route Tourist Routes* (Grozny: Chechen-Ingush Prince) p 80

[6] Panov V D 1971 *Glaciers of the river basin Terek* (Leningrad: Gidrometeoizdat)

[7] Anokhin G I 1978 *Eastern Caucasus By native expanses According to native spaces* (Moscow: Physical Culture and Sports)

[8] Khrustalev Yu P and Panova S V 2002 *Snow lines of the Greater Caucasus* (Rostov-on-Don: RSU Publishing House)

IOP Conf. Series: Earth and Environmental Science **350** (2019) 012053 doi:10.1088/1755-1315/350/1/012053

Processing, analysis and interpretation of time-frequency series for magnetotelluric monitoring

E A Bataleva

Research Station of RAS in Bishkek, Bishkek-49, 720049, Kyrgyz Republic

E-mail: bataleva@gdirc.ru

Abstract. This work presents the results of processing and interpretation of data of magnetotelluric monitoring obtained in the territory of the Bishkek geodynamic station (Northern Tien Shan). A new approach to the analysis of magnetotelluric monitoring data is proposed. The correlation of the obtained time-frequency series of variations of electromagnetic parameters with the moon-solar tidal effects, deformations of the Earth's crust, variations of apparent resistance and the distribution of seismicity was performed. A concept, linking the change in the stress-strain state of the medium with the redistribution of fluids between fracture systems, which causes variations in electrical resistance, was confirmed.

1. Introduction

Geodynamic processes, which are monitored on the territory of the Bishkek Geodynamic Proving Ground (BGPG), have a significant impact on the electrical properties of the geological environment [1-4]. Under the conditions of submeridional compression of the intracontinental Tien Shan orogene, both changes in the electrical properties of the geological environment (the first kind effect) and changes in the intensity of the electromagnetic fields generated there (the second kind effect) are possible [5]. These phenomena make it possible to study geodynamic processes occurring in the crust and upper mantle using electromagnetic methods. Performing the monitoring studies to predict catastrophic events of natural or man-made origin by the method of magnetotelluric sounding may be the most effective one. At the same time, interest in studying of lunar-solar influences and their correlation with changes in geophysical fields is currently growing, which is due to the possible use of such correlations in studying the geological processes occurring in the Earth's crust. This becomes especially relevant when implementing complex geophysical monitoring conducted on the territory of the BGPG, the main purpose of which is to study the patterns of seismic processes under conditions of submeridional compression of the Earth's crust [3, 4].

2. Data and methods

During geophysical monitoring, along with seismological data obtained from a network of seismic stations, data of electromagnetic, geodesic, gravimetric and other types of observations are collected (accumulated) on the territory of BGPG [1, 4]. All the results of complex geophysical monitoring are stored in a database on the server of the Research Station of the Russian Academy of Sciences, with the aim of further using these data for complex interpretation of geophysical data. Electromagnetic monitoring is carried out in two versions with controlled and natural sources that have different depths of research - sounding the formation of a field in the far zone (the source is an EGS-600-2 electro impulse system) and magnetotelluric (MT) monitoring, where the natural electromagnetic field is used as a source. The method of processing and interpreting the MT monitoring data includes plotting the frequency dependences of the components of the impedance tensor and their transformants - curves of apparent resistance and impedance phases and building time-frequency series of electromagnetic parameters for monitoring observation points. The processing of MT-monitoring data has some peculiarities as compared with the conventional processing technique: phase curves were chosen as reference curves, since they are considered to be more stable [6]; the use of histograms has significantly improved the quality of phase curves, and, consequently, simplified editing; the use of the dispersion ratio of the phase curves and the curves of apparent resistances also contributed to the improvement of the quality of the resulting. In this work, we continue research related to the development of the azimuthal MT-monitoring technique, which consists in analyzing the obtained time series of electromagnetic parameters in order to determine the contribution of each of the components of the impedance tensor to the information content of monitoring studies [3, 4]. The azimuthal MT-monitoring technique [3] for assessing the study of the characteristics of variations in electrical resistance includes the following procedures: 1) calculating variations of apparent resistance and impedance phase for 12 azimuths, in the interval from 0.01 to 10 sec (the duration of time series at point observations must be at least 24 hours); 2) the calculation of the correlation of variations of the resistance and phase of the impedance with the components of the lunar-solar tides depending on the period of soundings and azimuth; 3) comparison of tidal effects with pseudo-cuts of variations in electrical resistance in order to analyze the response of electrical conductivity variations to the elastic deformations of the Earth's crust from lunar-solar tidal influences. The most compact form of representation of the time-frequency structure of the components of the electromagnetic field transfer operators is their pseudo-sections — maps with the observation time horizontally and the depth parameter (the logarithm of the period) vertically. The time series of "azimuthal" MT - monitoring consists of several time-frequency diagrams (pseudo-sections) according to the number of azimuths for which the impedance tensor is recalculated (Figure 1). The logarithm of the probing period (T) characterizing its depth is plotted along the vertical axis of the pseudosections. The pitch along the horizontal axis of the pseudo-sections (time axis t) is a unit time interval within which MT sounding is processed to calculate apparent resistivity values in the working interval of periods T.

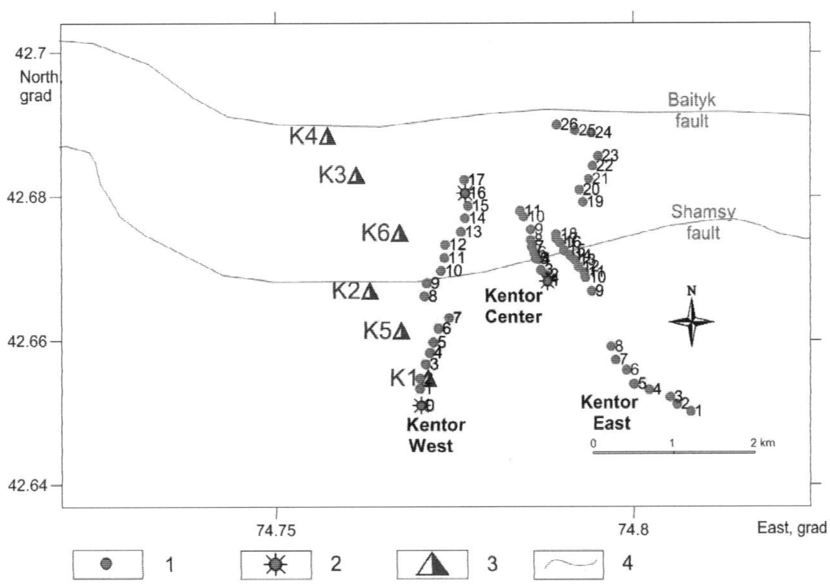

Figure 1. Scheme of points of geophysical observations on the Kentor mini-poligon. MT-soundings are shown by circles (1), monitoring observation points are marked with stars (2), line-angular observation points are marked with triangles (3), and active faults (4)

IOP Conf. Series: Earth and Environmental Science **350** (2019) 012053 doi:10.1088/1755-1315/350/1/012053

On the time-frequency diagrams for the current time interval (t) and for the period of sounding, variations of electromagnetic parameters are deposited, i.e. the difference between the average and current values, for example, the apparent resistance at the considered azimuth.

3. Results and discussion

We assume that during the compression of the Earth's crust due to the interconnectivity of fluid-filled microcracks in rocks, there is a redistribution of fluids between fracture systems, which leads to a decrease in the resistivity of rocks in the direction of compression [1, 3, 7]. When expanding, respectively, the opposite process should occur, so, the resistivity of the rocks increases in the direction of expansion. This hypothesis is confirmed by the results of experiments, including those in the territory of the Central Tien Shan [3, 4, 7]. The change in electrical conductivity and other electromagnetic parameters of the Earth's crust is not a factor itself that could directly affect the geological environment. However, this change can cause the occurrence of natural and artificial electromagnetic fields, both in the upper part of the Earth's crust and in the surface layer of the atmosphere. In addition to changes in electrical conductivity during deformations of the rocks of the earth's crust, other effects arise that can cause changes in the electric and magnetic fields, such as piezoelectric, piezomagnetic and tectomagnetic, but these effects are very difficult to isolate in the general electromagnetic background.

For all items of MT monitoring (mode, profile, stationary) in the territory of BGPG, the estimates of correlation coefficients of variations of the module of apparent resistance and impedance phases with lunar-solar tidal influences, determined using the Tide.exe program, are shown in diagrams. Analysis of the correlation relationships between variations of apparent resistance and lunar-solar tides indicates the existence of a relationship between these phenomena (Figure 2).

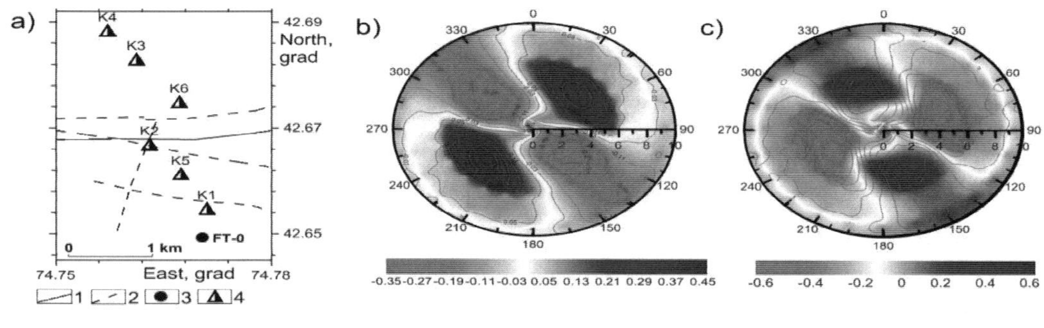

Figure 2. Correlation polar diagrams for the Kentor mini-poligon: a) - location of monitoring points and faults: 1 - Shamsinsky regional fault, 2 - local faults, 3 - MT-monitoring point, 4 - points of remote-ranging observations. Correlation diagrams for variations of apparent resistance (dRo) at FT-0 and: b) by the northern component of lunar-solar tidal effects (Av), c) by changes in the lengths of the base lines between points K5 and K2 according to the results of distance ranging observations

To compare the time-frequency series of electromagnetic parameters with gravitational tidal effects in different geological and tectonic conditions, we propose a way to present the results in the form of correlation polar diagrams. Based on the analysis of the behavior of the electromagnetic and gravitational fields in comparison with the geological and tectonic structure of the monitoring points, their strain sensitivity was assessed. The results of MT-monitoring allow us to conclude that under the influence of lunar-solar tides, the maximum variations in apparent resistivity occur in stress-sensitive zones confined to fault structures. The high level of the values of the correlation coefficients, as a rule, indicates the strain sensitivity of the zone of dynamic influence of the fault in correlation with the lunar-solar tides. In order to study the connection of deformation processes with the characteristics of variations of geophysical fields at the Kentor mini-poligon (Figure 1), the Research Station of the Russian Academy of Sciences in 2014 conducted observations on the Kentor line-angular measurement network and electromagnetic sounding using the azimuthal magnetotelluric monitoring method. The network of remote-range observations consists of six points, with the main focus on the study of deformation processes in the Shamsi-Tyundyuk fault zone. The network of MT-monitoring points in the territory of the Kentor mini-poligon, consisting of three points (Figure 1), was laid in various geoelectric conditions. Point "0" on the Kentor West profile is located on the northern slope of the Kyrgyz Range on the Paleozoic rocks overlapped with a cone, point "4" on the Kentor Central profile is located directly above the Shamsi-Tyundyuk fault zone, and point "16" on the Kentor West profile is located on Neogene-Quaternary sediments of the Chui basin. Figure 3 shows the time-frequency series of azimuthal MT-monitoring of apparent resistance from point "0" on the Kentor West profile in comparison with seismicity. In the time interval of 6–9 hours, for periods of 1–100 sec, a decrease in apparent resistance is observed at azimuths of 30°–60°, and an increase in azimuths of 120°–150°. The difference in azimuths of oppositely directed variations is about 90°, i.e. variations of apparent resistivity during this time interval may be due to the redistribution of fluid in the pore-fractured space and are "coherent in azimuths of 45° and 135°". Furthermore, along the time series in the time interval of 9-12 hours, variations coherent in azimuths of 60° and 150° are observed. In the range of 25–30 hours, a very contrasting radial variation in the range of periods $0 < \log T < 2$ is distinguished. Figure 2 presents the results of the experiment of simultaneous registration on the Kentor line-angular measurement network and MT-monitoring. In general, the characteristic of correlation dependences suggests that sources of deformation processes, apart from periodic tidal influences, are also sources of a different nature. There is a significant similarity for point 4, which is installed directly above the Shamsi-Tyundyuk fault zone (Figure 1) in the correlation diagrams of variations of apparent resistance between the components of lunar-solar tides and changes in baseline lengths. Situation is similar for points located in other geological and tectonic conditions. Although the difference in the morphology of the correlation fields for different observation points is significant, it is probably due to the presence and orientation of large geoelectric structures that characterize the electric conductivities caused by the distribution of pore fracture fluids. Variations of apparent resistivity are formed with changes in the lengths of baselines and with tidal effects on two-dimensional correlation diagrams of correlation fields of similar shape and orientation. This indicates a relationship between tidal effects, deformation processes, and variations in electrical conductivity. The level of correlation coefficients of 0.4–0.5 in the diagrams is obviously related to the low energy of tidal impacts and the insufficient accuracy of determining estimates of variations in apparent resistance and baseline lengths, but the selection of stable-form correlation fields indicates the relationship between the processes under consideration. Theoretical calculations of gravitational tidal effects are confirmed by experimental observations for the territory of the Bishkek geodynamic station in [3]. To verify the calculated data, the Research Station of the Russian Academy of Sciences performs observations using the Scintrex CG-5 Autograf gravimeter, which is installed in a gallery with a constant temperature of + 8° in the territory of the Research Station (30 km from the city). Observations are carried out at intervals of 12 s, and then during processing they are averaged at intervals that are necessary for the tasks and the required data discretization. It should be noted very high quality of the observed data (about 0.001 mGal). Interferences in this case are seismic events that are clearly recorded by a gravimeter, which is a Golitsyn pendulum. Thus, the completed studies complemented the existing database, including both magnetotelluric fields and their transformations, determined during measurements using the Phoenix MTU-5 broadband equipment, and the corresponding seismic activity parameters (KNET network) of the region under study (magnitude of seismic event magnitude, distance from the epicenter of the earthquake, clusters of seismic events). The work carried out showed the need for the use of MT-monitoring of the environment with the placement of observation points in seismically active fracture zones filled with decomposed rocks with increased water content and low resistivity at the so-called stress-sensitive points. It seems that such structures, including conducting core objects, can be combined into a single BGPG seismic monitoring system. In this case, MT-monitoring can be an effective tool for studying geodynamic processes, as well as physical mechanisms controlling the change in the geoelectric state of the environment. In the time-frequency series of apparent resistance at moderated monitoring points of the Kentor mini-poligon, anomalies of electromagnetic parameters have been identified, corresponding to the fluid redistribution model in the porous-fracture space, that is, in-phase reduction and increase in the parameter values on orthogonal azimuths (Figure 3). Thus, a phenomenological model has been tested that relates the change in the stress-strain state of the medium to the redistribution of fluids between fracture systems, which causes variations in the active and reactive components of electrical resistance.

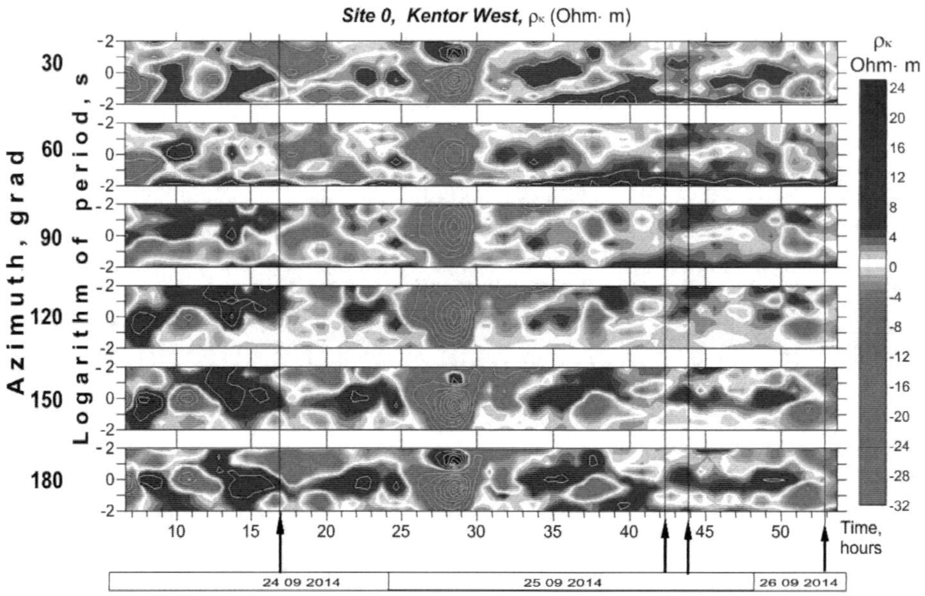

Figure 3. Time-frequency series of azimuthal MT-monitoring of apparent resistance from point "0" on the Kentor West profile in comparison with seismicity. Moments of earthquakes are shown by arrows

4. Conclusion

An interdisciplinary approach to solving the problems of complex geophysical monitoring of the territory of the Bishkek Geodynamic Proving Ground makes it possible to create new methods for studying the ongoing geodynamic processes (determining and preventing dangerous geological processes), including seismic events. The research resulted in the following:

- the depths of the geoelectric profile, which are most sensitive to the influence of geodynamic processes, have been determined;
- in the recorded time series of electromagnetic parameters, it was possible to isolate the characteristic signs of changes in the stress-strain state of the environment due to seismic events;
- it has been shown that a significant part of earthquake hypocenters is confined to gradient zones near the contacts between blocks or bodies with contrasting geoelectric parameters;
- an assessment of the informativeness of additional impedances was made, on the basis of which it was concluded that the use of this parameter is promising for evaluating the strain sensitivity of observation points;
- as a result of the correlation analysis of gravitational tidal effects with magnetotelluric parameters, it was found that the nature of the relationship depends on the geoelectric structure of the sensing point and is reflected in the distribution of stable clusters in the correlation polar diagrams;
- the concept linking the change in the stress-strain state of the medium with the redistribution of fluids between fracture systems, which causes variations in the active and reactive components of electrical resistance, was confirmed.

5. Acknowledgments

This study was supported by the Russian Foundation for Basic Research (grant 17-05-00654a).

References

[1] Bogomolov L, Bragin V, Fridman A, Makarov V, Sobolev G, Polyachenko E, Schelochkov G, Zeigarnik V and Zubovich A 2007 *Tectonophysics* **431(1–4)** 143
[2] Kuksa Y I and Svetov B S 2009 *Geomagnetism and Aeronomy* **49(6)** 816
[3] Bataleva E A, Batalev V Yu and Rybin A K 2013 *Izvestiya Physics of the Solid Earth* **49(3)** 402
[4] Bataleva E A and Mukhamadeeva V A 2018 *Geodynamics & Tectonophysics* **9(2)** 461
[5] Ivanov A G 1949 *Doklady Earth Sci.* **68(1)** 53
[6] Berdichevsky M N and Dmitriev V I 2008 *Models and Methods of Magnetotellurics* (Springer-Verlag Berlin Heidelberg)
[7] Busby J P 2000 *Geoph. Prospecting* **48** 677

IOP Conf. Series: Earth and Environmental Science **350** (2019) 012054 doi:10.1088/1755-1315/350/1/012054

Optimization of the location of production centers (slaughter points) of reindeer farms using iterative methods

R V Bazaeva

State University of Land Use Planning, 15, Kazakova str., Moscow, 105064, Russia

E-mail: r443827@yandex.ru

Abstract. The article deals with options for the most effective placement of slaughter points, the location of which would reduce the distance of deer drive to the places of slaughter from reindeer herding farms. When using different criteria, the location of the deer slaughtering production center shifts towards those organizations that have the largest number of deer and the economic efficiency of optimal placement of the slaughter points depends on the number of conditions used in the criterion.

1. Introduction

The first models of production location on the farm territory, optimization models of the sectoral structure were developed by N.Ya. Kuvshida. The most well-developed was the model of optimal combination of industries in agricultural enterprises, the formulation of this task belonged to R.G. Kravchenko and I.G. Popov, on which various modifications were subsequently created - the placement and specialization of agricultural production.

Under the leadership of M.I. Korobochkin, a complex work was conducted on the creation of mathematical models, consisting in the development of linear models of minimizing the amount of work and minimizing the costs of moving, which were later reflected in his doctoral thesis "Models and methods of optimal geodetic design of vertical planning" (1974).

The tasks had a very large dimension and in order to ensure their practical solution, well-known methods of linear programming were significantly modified; special methods of quadratic programming were developed. Based on these developments, in 1973 E.G. Larchenko the book "The use of mathematical methods in land management" was published, which served as a guide for the application of economic and mathematical methods in the design land management calculations [1].

In the period from 1974 to 1977 in the work of S.N. Volkova, "Some theoretical questions of mathematical modeling in land management" for the first time the presented classifications of mathematical methods and models for grounding land management projects were reflected [2].

When drafting the land management projects, it is often necessary to solve problems related to the placement of various industrial infrastructure facilities associated with a specific resource base in the territory. For example, slaughter points are located in such a way that they are as close as possible to reindeer herding, in order to reduce the distance of the reindeer drive to the places of slaughter.

The basic rule is that the slaughter point should be located in the center of gravity of deer pastures, affected farms. At the same time, the slaughtering point will shift towards those organizations where the most deer are located.

2. Materials and methods

To solve this problem, we suggest applying an iterative method (the method of successive approximations), which involves the gradual achievement of the initially chosen location of the slaughter point to its optimal value.

Let us determine the settlement procedure when optimizing the location of the slaughter points of several reindeer herding farms located in the territory of Taimyr (Dolgano-Nenets) municipal district. The task deals with the following: to establish the coordinates of the slaughter point location at which the total distance of the deer driving to the slaughter point will be minimal. The methodology of calculation and the problem solving sequence are reduced to the following actions.

1. Determine the need in feed for livestock of deer, intended for slaughter.

It is assumed that a deer's daily need in feed during the summer is 20 kg of green feed and 11 kg of moss during the winter. The number of days in the summer period is 122, and of the winter period - 243. Thus, the volume of feed consumed by one deer when it is transported to the slaughter point is 2440 kg in the summer period and 2673 kg in the winter season (Table 1) [3].

Table 1. Baseline data for solving the problem of optimal placement of the slaughter point

Sources of feed	Feed Type	The volume of feed in natural units, kg
Pasture	Green feed	2440
	Reindeer moss	2673

2. *Establish the number of deer, intended for slaughter, in each reindeer herding.*

For this purpose you can use the following data. In 2017, the Taimyr (Dolgan-Nenets) municipal district in 2017, there were 17 private reindeer herding organizations (Table 2) [1]. According to the methodological recommendations on the integrated survey, assessment and use of land in areas of northern reindeer herding, including the territories of traditional nature management of the indigenous peoples of the North, Siberia and the Far East of the Russian Federation (edited by Titova EA, Kurakina VI, Gavrilova M. A., Ph.D., Gladkova A.A. and others), the average allowable slaughter rate is 65-70% of the annual offspring or 22-23% of the total herd [4]. This indicator is determined on the basis of calculations of the movement of the livestock of deer in pasture seasons and the planned annual turnover of the production herd of deer of commercial and reindeer herding farms of the Nenets, Yamal-Nenets and Taimyr Autonomous District. Thus, the total number of deer for slaughter is 27.739 th. heads [5].

3. Calculate the gravity centers coordinates of the economic centers of the reindeer herding farms, from which they will drive the reindeer to the bottom. For this, you can use the local coordinate system or determine it in a conventional coordinate system graphically. After that, develop the initial matrix to optimize the placement of the bottom-hole point (tables 3 and 4).

4. Solve the problem according to a special algorithm using the iterative method (the method of successive approximations).

The amount of work Z on the movement of deer to the point of slaughter Pi from given points (reindeer enterprises) M (xi, yi) (i = 1, 2, ..., m) to point N (X, Y) can be expressed by the formula:

$$Z = \sum_{i=1}^{m} P_i \cdot R_i, \quad (1)$$

where Z is the total distance traveled by deer, thous. Km; R_i is a distance from the point of slaughter to reindeer enterprises; P_i is a number of deer, thous. heads [6]. Expressing the value of R through the coordinates, we get

$$Z = \sum_{i=1}^{m} P_i \cdot \sqrt{(X - x_i)^2 + (Y - y_i)^2} \rightarrow min, \quad (2)$$

where X и Y are the coordinates of the slaughter point; x_i and y_i are coordinates of the reindeer enterprises location in the local coordinate system [7].

Thus, it is required to determine the coordinates of the point N (X, Y) at which the value of Z will be minimal [8]. In this case, for the points of departure Mi (xi, yi), the coordinates of the settlements centers where the household centers of reindeer enterprises are located are taken.

The amount of feed consumed (Pi) is determined according to the number of deer found in reindeer enterprises placed in one locality [9]. Considering six locations of reindeer enterprises, m = 6. Then he formula takes a detailed form:

$$Z = P_1\sqrt{(X - x_1)^2 + (Y - y_1)^2} + P_2\sqrt{(X - x_2)^2 + (Y - y_2)^2} + P_3\sqrt{(X - x_3)^2 + (Y - y_3)^2}$$
$$+ P_4\sqrt{(X - x_4)^2 + (Y - y_4)^2} + P_5\sqrt{(X - x_5)^2 + (Y - y_5)^2} + P_6\sqrt{(X - x_6)^2 + (Y - y_6)^2} \to min \quad (3)$$

To reduce the number of iterations, A.A. Starkov and E.G. Larchenko suggested the following formulas for the approximate determination of the gravity center of the array, taking into account the mass of feed consumed when driving the reindeer to the slaughter point [10. p.112]:

$$X^{(0)} = \frac{\sum_{i=1}^{m} x_i \times P_i^2}{\sum_{i=1}^{m} P_i^2}; \quad (4)$$

$$Y^{(0)} = \frac{\sum_{i=1}^{m} y_i \times P_i^2}{\sum_{i=1}^{m} P_i^2}, \quad (5)$$

where $X^{(0)}, Y^{(0)}$ are approximate coordinates of the location of the slaughter point; x_i, y_i are coordinates of reindeer enterprises; P_i is the number of deer, thous. heads. We used these formulas to calculate the initial (zero) approximation.

Table 2. Characteristics of reindeer farms in the territory of the Taimyr (Dolgan-Nenets) municipal district

№	Name of farms	Location of the center of the reindeer husbandry	The total number of goals on 01.01.2016	The total number of goals on 01.01.2017	The number of goals for slaughter 01.01.2017
1	MUE "Ust-Yeniseets"	Karaul	11,983	12,956	2,915
2	RAPC "Suzun"	Tuhard	14,227	12,167	2,738
3	RAPC " Yara-Tanama"	Nosok	48,921	50,886	11,449
4	ARIPC "Tundrovik"	Khatanga	5,541	5,059	1,138
5	ACsC "Nerm Nenecha"	Dudinka	0	2,770	623
6	IE-chief Peasant farm Ilkiv L.B.	Dudinka	12,407	16,313	3,670
7	IE-chief Peasant farm Ilkiv R.P.	Dudinka	14	14	3
8	IE-chief Peasant farm Yadne G.A.	Nosok	9,926	11,043	2,485
9	IE-chief Peasant farm Sabelfeld E.V.	Dudinka	1,360	1,379	310
10	IE-chief Peasant farm Schukin G.K.	Dudinka	1,802	1,804	406
11	IE-chief Peasant farm Stepin A.V.	Dudinka	10	12	3
12	IE-chief Peasant farm Tsygankova V.A.	Dudinka	3,153	3,287	740
13	IE-chief Peasant farm Bolin S.Y.	Potapovo	5	6	1
14	IE-chief Peasant farm Vengo V.N.	Nosok	0	583	131
15	IE-chief Peasant farm Porotova A.I.	Dudinka	0	5	1
16	IE-chief Peasant farm Nikiforov P.T.	Dudinka	3,583	3,585	807
17	IE-chief Peasant farm Yamkina E.A.	Nosok	0	1,416	319
	Total:	-	1,166	1,269	27,739

3. The study of the structure

The existing slaughter point is located near Dudinka. To determine the economic efficiency of its location relative to the reindeer enterprises, let us calculate the coordinates of the slaughter point location in the conventional coordinate system in a graphical way: X = 138.30 km; Y = 165.96 km.

Table 3. Baseline matrix of optimal placement of slaughter points considering the mass of feed consumed

№	Farm location	Number of reindeer, th heads	The volume of feed consumed, th tons (Pi)	Pi^2	Coordinates of reindeer herding enterprises, km xi	yi	$xi*Pi^2$	$yi*Pi^2$	The distance from reindeer herding enterprises to the point of slaughter
1	Karaul	2.915	14.90	222.14	108.5	193.6	24,102.30	43,006.50	40.64
2	Tuhard	2.738	14.00	195.98	91.5	144.7	17,932.45	28,358.74	51.40
3	Nosok	14.384	73.55	5408.92	78.7	198.9	425,682.37	1,075,835.12	68.10
4	Khatanga	1.138	5.82	33.86	414.9	259.6	14,046.87	8,789.03	292.02
5	Dudinka	6.563	33.56	1126.05	138.3	140.4	155,732.26	158,096.95	25.56
6	Potapovo	0.0001	0.00	0.00	146.3	112.5	0.00	0.00	54.06
	Total	27.7381	141.82	6,986.95	-	-	637,496.24	1,314,086.34	
						Xo/Yo	91.24	188.08	
						Z initial		8,890.50	

To analyze the dynamics of the objective function, we consider the matrix of optimal placement of the slaughter points, when the evaluation criterion consists only of the number of deer driven to the slaughter.

Table 4. Baseline matrix of optimal placement of the slaughter points considering changes in the criterion

№	Farm location	Number of reindeer, th. heads (Pi)	Pi^2	Coordinates of reindeer herding enterprises, km xi	yi	$xi*Pi^2$	$yi*Pi^2$	The distance from reindeer herding enterprises to the point of slaughter
1	Karaul	2.915	8.50	108.5	193.6	921.95	1,645.06	40.64
2	Tuhard	2.738	7.50	91.5	144.7	685.94	1,084.76	51.40
3	Nosok	14.384	206.90	78.7	198.9	16,282.99	41,152.30	68.10
4	Khatanga	1.138	1.30	414.9	259.6	537.31	336.19	292.02
5	Dudinka	6.563	43.07	138.3	140.4	5,956.99	6,047.44	25.56
6	Potapovo	0.0001	0.00	146.3	112.5	0.00	0.00	54.06
	Total	27.7381	267.26	-	-	24,385.18	50,265.77	
					Xo/Yo	91.24	188.08	
					Z initial		1,738.80	

The method of successive approximations (iterations) was used to determine the coordinates of the optimal location of the slaughter for different criteria:

- taking into account the mass of feed consumed (table 5 and 6);
- considering only the number of deer (table 7 and 8).

Table 5. Calculation of the optimal location of the slaughter point by the method of successive approximations (1st iteration)

№	Farm location	The volume of feed consumed, th tons (Pi)	Coordinates, km			1st iteration		
			xi	yi	Ri(o)	Pi/Ri(o)	xi*Pi/Ri(o)	yi*Pi/Ri(o)
1	Karaul	14.90	108.5	193.6	18.12	0.82	89.24	159.23
2	Tuhard	14.00	91.5	144.7	43.38	0.32	29.53	46.70
3	Nosok	73.55	78.7	198.9	16.57	4.44	349.41	883.06
4	Khatanga	5.82	414.9	259.6	331.47	0.02	7.28	4.56
5	Dudinka	33.56	138.3	140.4	66.99	0.50	69.28	70.33
6	Potapovo	0.00	146.3	112.5	93.51	0.00	0.00	0.00
	Total					6.10	544.74	1,163.88
						X1/Y1	89.25	190.69
						Z1	6,272.33	

Table 6. Calculation of the optimal placement of the slaughter point considering the mass of feed consumed by the method of successive approximations (2nd iteration)

№	Farm location	The volume of feed consumed, th. tons (Pi)	Coordinates, km			2nd iteration		
			xi	yi	Ri(1)	Pi/Ri(1)	xi*Pi/Ri(1)	yi*Pi/Ri(1)
1	Karaul	14.90	108.5	193.6	19.47	0.77	83.07	148.22
2	Tuhard	14.00	91.5	144.7	46.05	0.30	27.82	43.99
3	Nosok	73.55	78.7	198.9	13.37	5.50	433.01	1,094.35
4	Khatanga	5.82	414.9	259.6	332.86	0.02	7.25	4.54
5	Dudinka	33.56	138.3	140.4	70.25	0.48	66.06	67.06
6	Potapovo	0.00	146.3	112.5	96.79	0.00	0.00	0.00
7	Total					7.07	617.21	1,358.16
						X2/Y2	87.34	192.19
						Z2	6,212.08	

Table 7. Calculation of the optimal location of the slaughter point by the method of successive approximations (1st iteration)

№	Farm location	Number of reindeer, th. heads (Pi)	Coordinates, km			1st iteration		
			xi	yi	Ri(o)	Pi/Ri(o)	xi*Pi/Ri(o)	yi*Pi/Ri(o)
1	Karaul	2.915	108.5	193.6	18.12	0.16	17.45	31.14
2	Tuhard	2.738	91.5	144.7	43.38	0.06	5.78	9.13
3	Nosok	14.384	78.7	198.9	16.57	0.87	68.34	172.71
4	Khatanga	1.138	414.9	259.6	331.47	0.00	1.42	0.89
5	Dudinka	6.563	138.3	140.4	66.99	0.10	13.55	13.75
6	Potapovo	0.0001	146.3	112.5	93.51	0.00	0.00	0.00
	Total					1.19	106.54	227.63
						X1/Y1	89.25	190.69
						Z1	1,226.74	

Table 8. Calculation of the optimal location of the bottom-hole point by the method of successive approximations (2nd iteration)

№	Farm location	Number of reindeer, th. heads (Pi)	Coordinates, km			2nd iteration		
			xi	yi	Ri(1)	Pi/Ri(1)	xi*Pi/Ri(1)	yi*Pi/Ri(1)
1	Karaul	2.915	108.5	193.6	19.47	0.15	16.25	28.99
2	Tuhard	2.738	91.5	144.7	46.05	0.06	5.44	8.60
3	Nosok	14.384	78.7	198.9	13.37	1.08	84.69	214.03
4	Khatanga	1.138	414.9	259.6	332.86	0.00	1.42	0.89
5	Dudinka	6.563	138.3	140.4	70.25	0.09	12.92	13.12
6	Potapovo	0.0001	146.3	112.5	96.79	0.00	0.00	0.00
	Total					1.38	120.71	265.63
						X2/Y2	87.34	192.19
						Z2	1,214.96	

At this stage, we finally determine the location of the bottom-hole point (Figure 1, in this example X = 87.34, Y = 192.19) within the limits of the accuracy of the calculations (up to 0.5 km). The economic efficiency of the decision on the value of the objective function of the problem (Z = 1,214.96) was further determined. Economic efficiency in terms of the objective function to determine the optimal location of the bottom-hole point is very different depending on the applied criterion: the more conditions, the lower the economic efficiency.

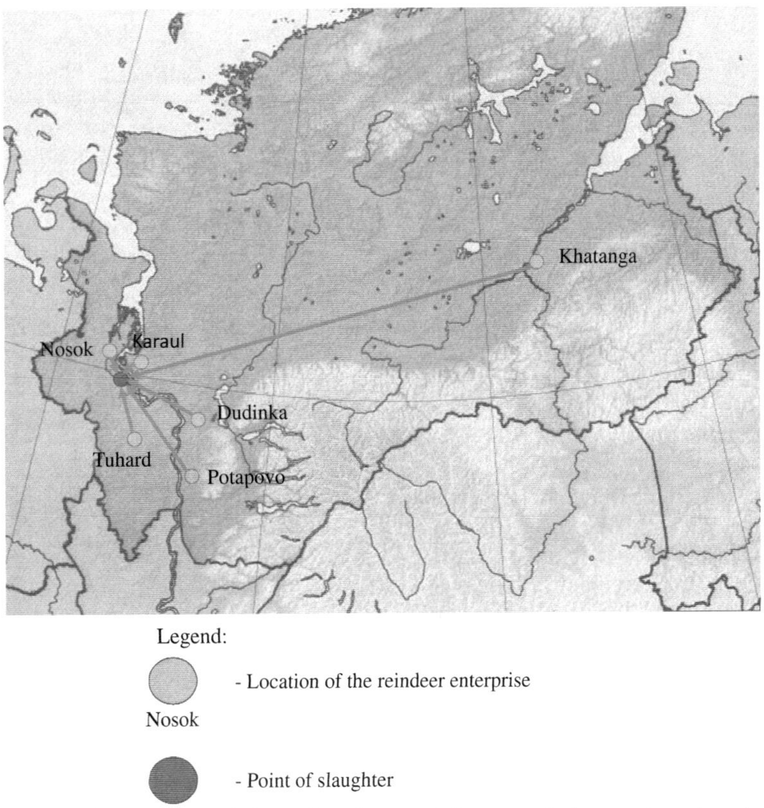

Figure 1. The layout of the point of deer slaughter relative to reindeer enterprises by the method of successive approximations.

4. Conclusion

The most rational location of deer slaughter point relative to the reindeer enterprises was determined by the method of successive approximations. The coordinates obtained do not coincide with the coordinates of the existing point of deer slaughter in Dudinka. Different criteria were used with varying the dynamics of the objective function. The result of the calculations was determined by the highest economic efficiency indicator, that is, a location that minimizes the distance that deer walk to the point of slaughter.

When conducting practical calculations using this technique, one should always compare the results obtained by economic-mathematical and traditional methods in order to bring the mathematical optimum to the real one. The method of calculation used in this paper can be applied directly to other territories, given their territorial affiliation to reindeer herding farms.

References

[1] Larchenko E G 1973 *The use of mathematical methods in land management* p 399
[2] Volkov S N 1977 *Some theoretical questions of mathematical modeling in land management* p 82
[3] Beletskaya L I 2016 *The agro-industrial complex of the Krasnoyarsk region in 2011 - 2015* pp 212-217
[4] *Methodical recommendations for integrated survey, assessment and land use of areas of northern reindeer husbandry, including the territories of traditional nature management of indigenous peoples of the North, Siberia and the Far East of the Russian Federation* 2015 p 232
[5] Volkov S N 2002 *Land Management in 9 volumes* Vol 9 *Regional Land use planning* pp 496-517
[6] Volkov S N 2013 *Land Management* p 992
[7] *Organization of events for the maintenance and development of the forage base of deer pastures in the forest-tundra zone (assessment of pasture quality, remote sensing methods) Guidelines* 2013 p 18
[8] *Methods and technological schemes for the use of perennial cereal grasses for the biological recultivation of disturbed lands* 2012 p 28
[9] Yemelyanova T A 2005 *Organization of the territories used by the communities of the small indigenous peoples of the North, Siberia and the Far East of the Russian Federation to ensure their traditional way of life (methodical and practical issues)* p 136
[10] Volkov S N 2002 *Land Management in 9 volumes* Vol 4 *Economic-mathematical methods and models* pp 109-120

Types of land massifs for essential oil crops cultivation in forest-steppe altitudinal belt of Greater Caucasus (Krasnodar territory)

E V Cherkashina[1] and K I Cherkashin[2]

[1]State University of Land Management, 15, Kazakova Street, 105064, Moscow, Russia
[2]Federal Scientific Center of Agrarian Economy and Social Development of Rural Territories, All-Russia Research and Development Institute of Agricultural Economics, Building 2, 35, Khoroshevskoe Highway, 123007, Moscow, Russia

E-mail: cherkashina@infokad.ru

Abstract. The results of studies on land suitability for cultivation of essential oil crops in terms of the forest-steppe altitudinal belt of the Greater Caucasus are presented by the example of Anapa district of the Krasnodar Territory. For the purpose of type design of land massifs the cumulative assessment of land by soils, topography, degree of erosion, and cultural condition is necessary for cultivation of ether tracts in accordance with the requirements to growing conditions. The conducted studies determined the areas of land plots for growing the perennial essential-oil-bearing plant and placing essential oil crop rotations on the land plot selected for analysis.

1. Introduction

The development of the world market of essential oils, which is an important component of food, pharmaceutical, perfumery, cosmetic, paint and varnish, textile, leather, metallurgical, printing and other industries dictates the growing need for essential oil raw materials the production of which in terms of the industrial scale was ruined in 1990-2010. Domestic enterprises operate mainly on import deliveries estimated at not less than 700 million US dollars annually [1].

The variety of natural and climatic conditions in Russia makes it possible to cultivate various types of essential oil crops, providing not only their own needs for these raw materials, but also can successfully export natural essential oils to the world market. One of the favorable regions for the cultivation of essential oil crops is the Krasnodar Territory.

This paper presents studies on the types of land massifs for cultivation of essential oil crops in the conditions of forest-steppe zone of the Greater Caucasus (Anapa region).

2. Materials and methods

The altitudinal zonality allows cultivating a wide variety of crops, including oil-bearing ether due to different temperatures, humidity, soil composition, steepness and slope exposure. The key issue of industrial cultivation of essential oil crops is the requirement for soil fertility of land plot, its technological properties, i.e. contour, configuration, relief, rockiness, and energy intensity.

To determine the suitability of land for the cultivation of essential oil crops the following indicators were identified: soil, relief, cultural and technical condition (bushes, hillocks, contamination by stones, etc.), and the degree of erosion.

The climate of the Anapa Region is of the Mediterranean type, moderately humid continental-steppe, which is a combination of three microclimates, i.e. steppe, mountain and sea foothill-steppe with a fairly steady barometric pressure. The prevailing winds in the territory are northeast in winter and southwest in summer. As a result of complex historical, geological transformations in the region, a relief was formed, represented by low mountains, ridge hills and lowland plains [2]. The soil composition of the Anapa Region is characterized as sod-calcareous shallow soil, from low to high stone, medium washed and strongly washed, weak and strongly eroded soil. The grain-size composition is marl clays, marl eluvium, talc heavy loam. The sandstone and fen clays are found. The relief is difficult with the slope steepness from 0° to 25°; the exposure is varied.

The above characteristics of climate, soil, terrain are suitable for essential oil crops, which have low requirements for soil fertility, and are capable of preventing the development of erosion processes on the land plots. On the other hand, any producer of essential oil raw materials takes into account the needs of the market. At present, lavender, rosemary, mint, wormwood, calendula, chamomile, and thyme oil are in great demand. Taking into account the long-term experience of cultivation of these crops in Russia [3-6], as well as modern foreign experience [7-10] the most suitable for the conditions of the region are lavender, rosemary, and wormwood. For example, lavender is a perennial, cold-resistant, light-loving, drought-resistant plant that grows on many types of soil, carbonate, gravelly or stony-gravelly loams. The southern and southwestern slopes are especially favorable for its cultivation. Having strong, fibrous and branched root system lavender can grow on steep slopes (over 15 degrees), protecting them from soil erosion and as a honey plant. Rosemary is a perennial light- and heat-loving plant, drought-resistant, that prefers dry calcareous, permeable soils with good aeration. It also grows on dry sandy and rubbly soils. For their placement it is necessary to have sunny areas protected from prevailing winds. Wormwood is undemanding to the conditions of growth; light-requiring and relatively thermophilic, frost hardy, when placing this plant, you must choose the southern exposure, a place protected from the winds. All of these crops are cultivated as perennial plantations. The wormwood can be cultivated in special medicinal crop rotations [1].

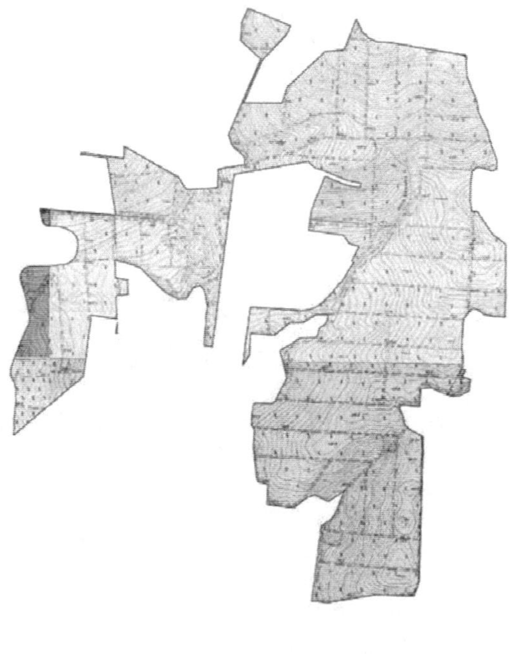

Figure 1. Topographic area scale plan 1:5,000.

The type design of land massifs was carried out on one of the land plots of the territory of the former collective farm Caucasus. The study was carried out by QGIS, the topographic terrain plan of 1:5,000 scale was used as a source material for the analysis of the terrain (Figure 1). Besides, planned and high-altitude survey of individual sections was created on the basis of which a digital model of the terrain using irregular triangulation network was developed (TIN). The soil characteristics of the land massifs were established on the basis of the 1990 soil map and the agrochemical soil survey of 2018.

3. Results

Let us consider the process of determining the suitability of land of Anapa Region in order to cultivate the essential oil crops mentioned above. To identify homogeneous land plots in terms of relief, a slope and slope exposure map was compiled based on the digital elevation model created (Figures 2, 3).

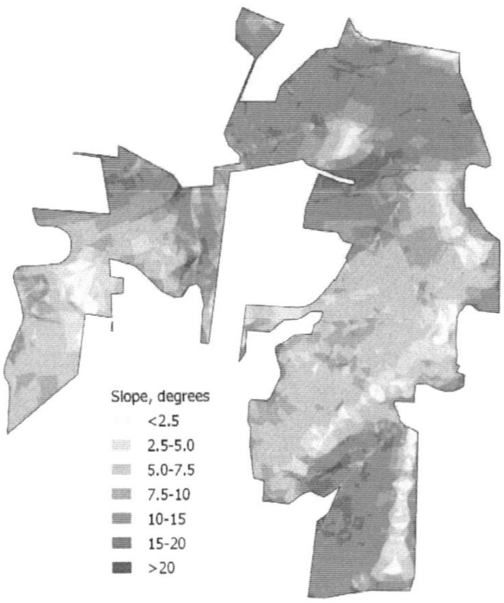

Figure 2. Map of slope steepness

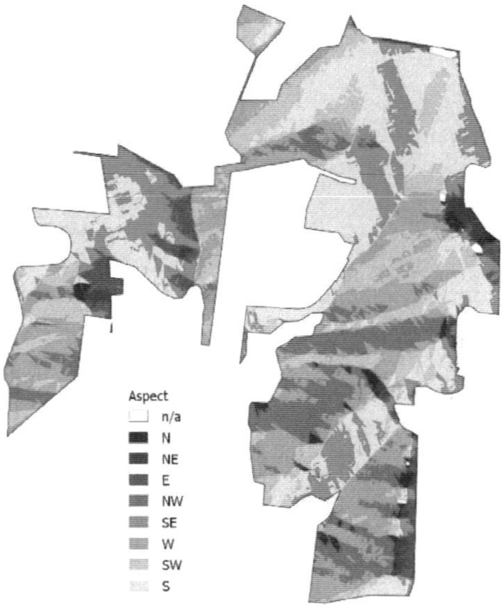

Figure 3. Map of slop exposure

The slopes of the area are mainly southern, southwestern and southeastern exposure (Table 1); the slope is in the range from 0° to 20°, the largest share is made up of the areas with a steepness of 5° -15° (Table 2).

Table 1. Characteristics of land on slope exposure

Indicators	Direction of slop exposure								
	North	Northeast	East	Southeast	South	Southwest	West	Northwest	Total
Area (ha)	1.44	2.08	4.89	17.36	32.41	20.44	27.04	15.88	121.54[1]
Specific weight (%)	1.18	1.72	4.02	14.29	26.67	16.82	22.24	13.07	100.00

The area is characterized as sod and calcareous with medium and shallow soil from weakly to heavily soils on hardly gentle slopes, i.e. 96.61 ha or 79% suitable for cultivation of essential oil crops.

1 The area does not account for the existing fruit-bearing vineyards

According to the degree of erosion, steepness and exposure of slopes it is possible to establish the categories of erosion hazard of the lands of the studied area (Table 3). It can be seen from the Table that half of the land belongs to category IV and 40% belongs to VI, which requires the development of erosion control measures and the establishment of their rational placement carried out in the land management projects during cultivation of essential-oil-bearing plant.

Table 2. Characteristics of land on slope steepness

Indicators	Slope steepness (°)							
	0-2.5	2.5-5	5-7.5	7.5-10	10-15	15-20	>20	Total
Area (ha)	4.76	10.67	39.24	19.75	41.06	5.82	0.23	121.54
Specific weight (%)	3.91	8.78	32.29	16.25	33.79	4.78	0.19	100

Table 3. Characteristics of land on erosion hazard categories

Indicators	Erosion hazard categories						
	III	IV	V	VI	VII	VIII	Total
Area (ha)	3.16	60.88	8.51	47.71	-	1.27	121.54
Specific weight (%)	2.60	50.09	7.00	39.26		1.04	100.00

The characteristics of the studied land, including the establishment of its modern use and the degree of cultural state, are given in Table 4. As can be seen from the Table, most of the land is not currently used, which requires the carrying out of cultural and technical works on the uprooting, cleaning of stones, and primary tillage of soil (at current rates the cost of clearing from wood and grass vegetation will be 20.51 thousand rubles per 1 ha; from stones - 1.56 thousand rubles per 1 ha; primary tillage - 32.76 thousand rubles per 1 ha. The overall price will comprise 54.78 thousand rubles per 1 ha [11]).

Table 4. Characteristics of land

Type of land	Area (ha)	Degree of forest cover	Prevailing soil	Texture	Humus horizon width
Deposit	112.05	Low	Medium sod-calcareous medium stone medium washed	Light clay	Medium
Tilled soil	1.37	-		Light clay	Medium
Forest steppe	8.12	High		-«-	-«-
Total	121.54				

The studies show that the selected land has a complex topography and unfavorable soils for agricultural use due to heavy cobble content and increased dryness, as well as medium washed and hardened soil. However, for essential oil crops that are unpretentious for soils that have a good root system impeding the development of surface runoff, with a rational farming system these lands may be suitable for their cultivation due to favorable climatic conditions. Figure 4 and Table 5 show homogeneous lands suitable for plantations of perennial ether-carrying plants (lavender and rosemary), as well as lands suitable for medicinal crop rotation, where it is possible to cultivate wormwood, as well as other valuable essential oil crops, i.e. chamomile, calendula, etc.

Figure 4. Type design of land massifs for cultivation of essential oil crops

Table 5. Summary of land suitability to be used in crop production

Indicators	Area (ha)	Samples/land use
Essential oil crop rotation	47.89	Wormwood, chamomile, calendula
Perennial plantations	49.60	Lavender, rosemary
Other lands	24.05	Wild plants (hawthorn, dogrose)
Total	121.54	

4. Conclusion

The type design of land massifs for the cultivation of aromatic crops should be based on the analysis of the soil type, its grain sizes, degree of erosion, topography (slope, slope exposure), and cultural state of the land plot taking into account crop requirements for growing conditions, ability to protect the soil from erosion. The type design of land plots for the cultivation of ethers is an integral stage of land management projects (arrangement of territory).

References

[1] Cherkashina E 2014 *Economics and Organization of Rational Use and Protection of Lands of Essential Oil and Drug Industry in the Russian Federation: Doctorate Thesis* (Moscow: State University of Land Management) p 419

[2] *Climate of Anapa* Available at: http://www.anapacity.com/kurort-anapa/klimat-anapa.html

[3] Dzhirgalova E A 2004 *Youth and Science: Third Millennium* Materials of Scientific and Practical Conferences (Elista) pp 84–5

[4] Kudinov M A, Kukhareva L C and Pashina G V 1986 *Essential Oil and Spicy Cultures* (Moscow) p 248

[5] Collective monograph 2017 *Essential Oil Cultures* (Krasnodar: Prosveschenie-Yug) p 295

[6] Pushkina G P, Malankina E L, Tkhaganov R R and Morozov A I 2010 Effective application of growth regulators and micronutrients on essential oil crops *Achievements of Science and Technology of the AIC* [in Russian – Dostizheniya nauki i techniki APK] **7** 17–9
[7] Munnu S 2004 Effects of plant spacing, fertilizer, modified urea material and irrigation regime on herbage, oil yield and oil quality on rosemary in semi-arid tropical climates *Journal of Horticultural Science and Biotechnology* **79(3)** 411–5
[8] Gül M, Kart M C O and Sirikci B S 2016 Determining costs and profitability of lavender farms in Isparta province of Turkey *Journal of Essential Oil Bearing Plants* **19:3** 686-92
[9] Rahmati M, Azizi M, Khayyat M H, Nemati H and Asili J 2011 Yield and oil constituents of chamomile (Matricaria chamomilla l.) flowers depending on nitrogen application, plant density and climate conditions *Journal of Essential Oil Bearing Plants* **14:6** 731-41
[10] Asgharpour N and Honarvar M 2016 Identification and comparison of essential oil composition of Artemisia sieberi and Artemisia aucheri cultivated in the South of Iran *Journal of Essential Oil Bearing Plants* **19:3** 756-61
[11] Ministry of Agriculture of the Russian Federation 2016 *on Approving the Procedure for Determining the Cost of Cultural and Technical Land Reclamation Necessary to Bring the Land Plot from Agricultural Land to Conditions Suitable for Agriculture* (Moscow) 28.12.2016 No. 600

IOP Conf. Series: Earth and Environmental Science **350** (2019) 012056 doi:10.1088/1755-1315/350/1/012056

Information support of management of the land resources of the Russian Federation

A D Demyanova, E A Schastlivetskaya and S A Lipski

State University of Land Use Planning, 15, Kazakova Str., Moscow, 105064, Russia

E-mail: Demyanova_nasty@mail.ru

Abstract. Informational support is an integral part of effective land management. At the present stage, the Unified State Register of Real Estate (hereinafter - EGRN) is an informational database, which management decisions are based on. However, the quantity and quality of the information contained in the EGRN is not enough to pursue an effective management policy related to the country's land resources. This work presents a model of interaction between various informational systems, which allow adjusting and supplementing the EGRN database, thereby ensuring an increase in the quality of management related to the country's land resources.

1. Introduction

The management of the land resources is a global process, its implementation at the present stage involves various components of power from local governments to various ministries and the Government of the Russian Federation.

Thus, each structure deals with a separate aspect in the overall complex of land resources management processes, which use different registers of specific information and special informational systems for the management.

From the point of view of informational support of land resources management processes, there is a number of supporting measures, software packages, various information systems and registries, its usage nowadays has a very narrow direction [1, 2].

2. Materials and methods

The use of the EGRN as an informational database for land management is the most promising from the viewpoint of various aspects. However, the analysis of the current status of the EGRN, its informational content and data quality showed the necessity of supplementing its database with the information contained in other informational systems [3]. The relevance of creating an integrated informational support of land resources management processes was identified based on the interaction of information registers of various ministries and departments.

Management targets set new tasks for informational systems, the solution of which is possible if there is sufficient, objective, reliable information about the objects of management. The analysis of the existing informational systems of various ministries and departments revealed the presence of information relevant for the purposes of carrying out an objective and rational policy in the area of the country's land resources [4, 5]. Thus, the issue of a rational policy in the area of the country's land resources can be considered by the interaction of different ministries and departments with interdepartmental informational support [6].

The analysis of the EGRN database as well as other informational recourses was carried out, such as: Register of Federal Property (The Federal Agency for State Property Management, aka Rosimushchestvo), FGIS "Demetra" (The Federal Service for Veterinary and Phytosanitary Surveillance, aka Rosselkhoznadzor), AIS GRR (The Federal Agency for Fisheries, aka Rosrybolovstvo), FGIS ESIMO (The Federal Service for Hydrometeorology and Environmental Monitoring of Russia, aka Roshydromet), FGIS PTK State Control, PTO UONROS (The Federal Service for Supervision of Natural Resources, aka Rosprirodnadzor), Forestry Register (The Federal Forestry Agency, aka Rosleskhoz), Consolidated State Register of Subsoil and Licenses (The Federal Agency for Subsoil Use, aka Rosnedra), EISGS (The Ministry of Construction and Housing and Communal Services of the Russian Federation, aka Minstroy of Russia), GIS TEK (The Ministry of Energy of the Russian Federation, aka Minenergo of Russia), register of transport, infrastructure and vehicles (The Ministry of Transport of the Russian Federation), AIS "Tax-3" (The Federal Tax Service of Russia), the register of military facilities and territories (The Ministry of Defense of the Russian Federation).

Combining information in a single informational system that meets the functions of rational management of the land resources creates the necessity of implementing integrated information support in this area. To solve these problems, a model of integrated information support was developed for the land resources management in the Russian Federation [7].

During the development of the model, abstract-logical, economic-statistical, historical (in retrospective study of modern problems, for example, systems of land management bodies), graphical (in identifying the dynamics of the processes studied) methods were used which allowed identifying certain patterns and interdependence of some elements of the presented model [8].

3. Study of the integrated informational support model for the land resources management in the Russian Federation

In order to effectively manage the land resources, the application model of integrated informational support is needed, by means of interaction between the Government of the Russian Federation, self-governing authorities, federal ministries and their subordinate organizations, presented in Figure 1.

It should be noted that the Russian State Register, as the department responsible for maintaining the EGRN, is a subordinate institution of the Ministry of Economic Development of the Russian Federation. For clarity, on the scheme the Russian State Register is separated from the Ministry of Economic Development of the Russian Federation.

From the point of the informational correlation, there is a great diversity between the agencies represented in the model (Table 1).

Greater interdependence between registries of information is noticed between the EGRN and the registries of the Ministry of Agriculture of the Russian Federation (Rosselkhoznadzor), the Ministry of Natural Resources and Environment of the Russian Federation (Rosprirodnadzor, Rosleskhoz, Rosnedra), that is to say that for the effective implementation of the powers assigned to these institutions to manage the land resources, the information of the EGRN is important and vital. At the same time, the data transmitted to the EGRN from these organizations will be of great importance in terms of fullness, relevance of the registry of agricultural lands, status and usage of lands of other categories.

The up to date information from the Ministry of Energy of the Russian Federation, the Ministry of Transport of the Russian Federation and the Ministry of Construction, Housing and Utilities of the Russian Federation, is also largely important for filling the EGRN database; to the much less extent these ministries are interested in downloading information from the EGRN.

IOP Conf. Series: Earth and Environmental Science **350** (2019) 012056 doi:10.1088/1755-1315/350/1/012056

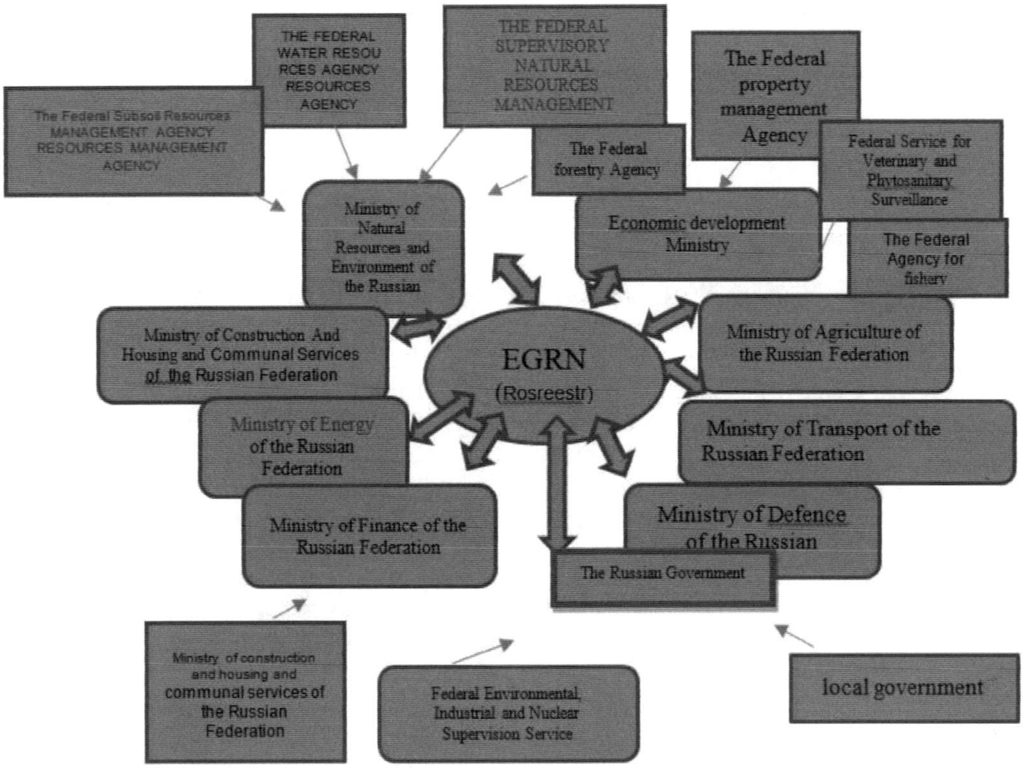

Figure 1. Model of integrated information content of the EGRN: the title, authors and affiliation

Table 1. Level of interrelation of information of the EGRN and registers of various departments

The name of the Department	The importance of the information resource of this Department for the EGRN	The importance of the EGRN information for this Department
Rosimushchestvo	important	important
Rosselkhoznadzor	important	important
Rosrybolovstvo	important	moderately important
Rosgidromed	important	moderately important
Rosprirodnadzor	important	moderately important
Rosleskhoz	important	important
Rosnedra	important	moderately important

Relevant information of the EGRN as means of implementing the most effective management policy is necessary for the Ministry of Defense of the Russian Federation, in terms of ensuring the security and independence of the Russian Federation, as well as the Ministry of Finance of the Russian Federation, in terms of ensuring the collection of more comprehensive land tax

Let us consider the informational systems in details, the interrelation of the FGIS the EGRN which is relevant and necessary for the mutually effective support of the land resources management functions (Figure 2).

The informational systems presented on the scheme already exist and are used in the corresponding departments with the corresponding activity. However, nowadays the complex interaction of these systems does not exist; therefore, the informational support for the land resources management currently has a "lumpy" character. The interaction of the FGIS the EGRN with informational systems presented in the model will contribute to the formation of integrated informational support for the effective management of the land resources [9].

In the interaction of informational systems, the data of the EGRN will be filled with information about all categories of lands, especially the format of information on agricultural land of particular value will be expanded. In terms of data on forestry and water areas, a decrease in disputable situations is expected with apposition of forestry and water lands on lands of other categories.

Filling the Register with information about the subsoil will allow providing a more efficient and rational policy in the process, including distribution and assignation of land.

Information about the facilities of industry, transport, power grid, energy and other objects will allow adjusting the information of the EGRN in terms of the types of permitted territories usage and the presence of restrictions and encumbrances, which should significantly reduce the number of court proceedings provoked by land disputes in this area. This will provide more comfortable conditions for the development of the land market in the country that guarantees the buyer's rights [10].

Reliable information about the lands of the Ministry of Defense of the Russian Federation will allow effectively pursuing the policy of ensuring both the internal and external interests of the state.

Improving the quality of data, including the state of the land, will allow adjusting the cadastral value of land parcels, thereby ensuring the charge of the "fair" land tax, based on unbiased monitoring data, and land supervision [11].

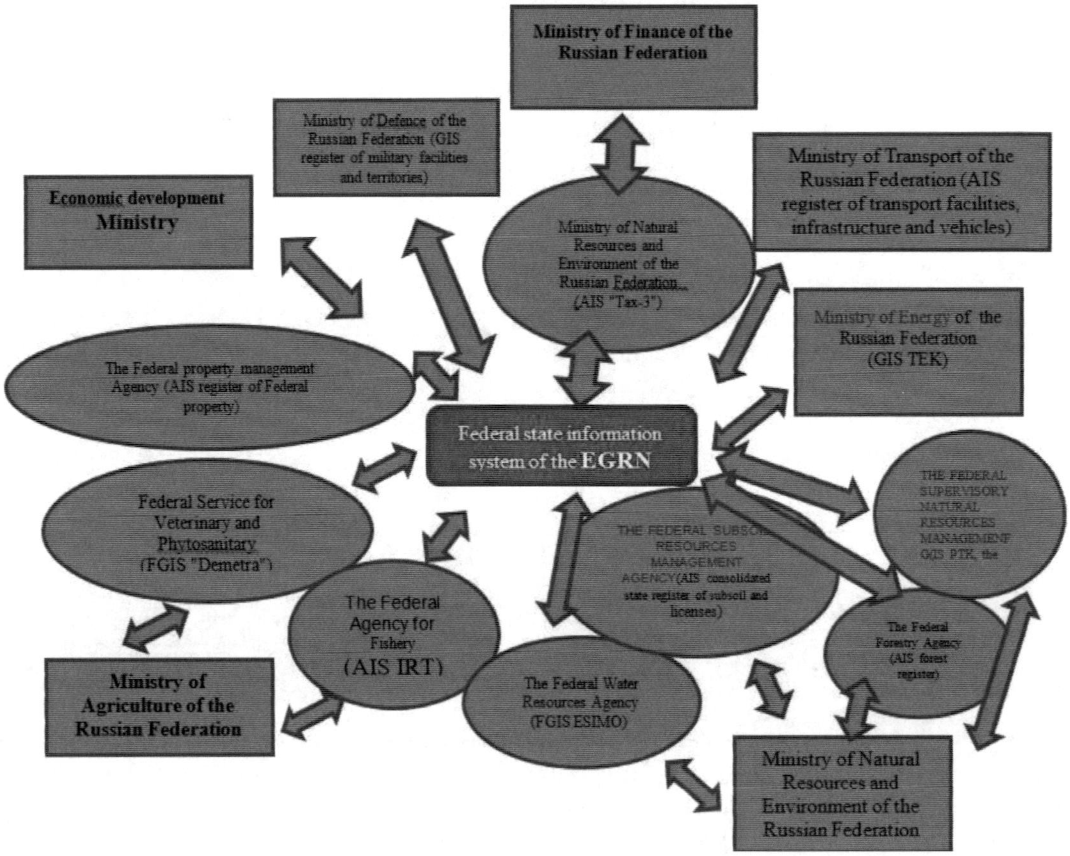

Figure 2. Federal state information system of EGRN integrated content model

IOP Conf. Series: Earth and Environmental Science **350** (2019) 012056 doi:10.1088/1755-1315/350/1/012056

4. Conclusion

Effective and rational management of the land resources of the Russian Federation is only possible in presence of relevant, reliable and sufficient information. Currently the EGRN is a database for these purposes, its information needs to be improved. The presented model of integrated informational interaction will allow improving the quality, quantity and composition of information in this registry.

References

[1] Demyanova A D and Schastlivetskaya E A 2017 *Land Management, Land Monitoring and Cadaster* [in Russian – Zemleustrojstvo, kadastr i monitoring zemel] **11(154)** 55–8

[2] Lipski S A and Demyanova A D 2018 *Property Relations in the Russian Federation* **2(197)** 40–5

[3] Boltanova E S 2016 *Property Relations in the Russian Federation* [in Russian – Imuschestvennye otnoshenija v Rossiyskoy Federacii] **7(178)** 14–23

[4] Petrogradskaya A A 2018 Theoretical and Legal Implications of the Adoption of FL-218 "On State Registration of Real Estate" *Bulletin of Volzhsky University named after V N Tatishchev* **4** 42–50

[5] Popov E V and Semyachkov K A 2018 *Region economy* [in Russian – Economika regiona] **4(14)** 1088–99

[6] Sinenko V A, Shiyapov T I and Parpura D I 2018 *Russian University's of Peoples Friendship J. of Agronomy and Animal Industries* **13(1)** 45–53

[7] Stepanov O A 2018 *Law. Journal of Higher School of Economics* [in Russian – Pravo. Zhournal Vysshey shkoly ekonomiki] **3** 4–23

[8] Shagaida N I and Alakoz V V 2017 *Earth for People* (Moscow: Tsentr strategicheskikh razrabotok. Economic Development)

[9] Motlokhova E A 2015 *Bulletin of Khabarovsk State University of Economics and Law* **6(80)** 166–72

[10] Varlamov A A, Gal'chenko S A, Antropov D V, Komarov S V and Shapovalov D A 2018 *Interexpo Geo-Siberia* **3(2)** 237–51

[11] Sinenko V A 2018 *Bull. of Science and Practice* [in Russian – Bulleten nauki i praktiki] **10(4)** 384–404

Monitoring the process of bioremediation and revegetation of a phosphogypsum waste stack by remote sensing

N R Ermolaev[1], V P Belobrov[1], S A Yudin[1] and E L Torochkov[2]

[1]V.V. Dokuchaev Soil Science Institute, 7, Pyzhyovskiy lane, Moscow, 119017, Russia
[2]"PhosAgro" group, 75, Severnoe highway, Cherepovets, 162622, Russia

E-mail: hukitoc94@gmail.com

Abstract. Production of phosphorus-containing fertilizers is related with the problem of reclamation of phosphogypsum dumps formed during the processing of Apatite-nepheline ore. In this regard, some attention should be paid to biological remediation as the most effective ecological method. Biological reclamation is one of the ways to protect the environment in the formation of phosphogypsum dumps. In this article, we propose a method for assessing the effectiveness of remediation using phyto monitoring using satellite data, which are used as the results of remote sensing. The proposed method allows estimating the volume and state of biomass of plants forming vegetation cover on the surface of man-made massif.

1. Introduction

Production of phosphorus-containing fertilizers at Balakovo branch of "Apatit" is a complex, multifunctional process, as a result of which phosphogypsum is formed and accumulated. Form of phosphogypsum dump is in compliance with the established rules and regulations for the protection of the environment. Biological remediation can be considered as an additional way to protect the environment. Many experts in biological restoration of lands pay much attention to processes of natural overgrowth of the disturbed territories in many respects defining the choice of plants for carrying out a biological stage of the planned reclamation. The study of the potential of natural vegetation restoration involves considering the sequence of successions as stages of development and restoration of disturbed habitats by vegetation [1]. Vegetation and field experiments on biorecultivation on the phosphogypsum dump during 2013-2015 showed that the use of the proposed technology makes it possible in a shorter time (2-3 years) to obtain a stable vegetation cover from the most adapted plants, to which, first of all, the crested wheatgrass (*Agropyron cristatum*). Remote sensing technologies can be used to determine the effectiveness of the biorecultivation process, diagnostics and monitoring of the earth's surface, which are an effective means of collecting spatial and temporal information about vegetation [2, 3]. One of the methods of processing and interpretation of remote sensing data is the calculation of indices of vegetation, soil, hydrological, etc. The most common and convenient for determining the dynamics of phyto productivity of the territory is the vegetation index NDVI (Normalized Difference Vegetation Index) [4].

2. Materials and methods

The research area is located in Balakovo district of Saratov region and is a landfill phosphogypsum storage area of about 150 hectares (Figure 1). The object of study is the surface of the landfill phosphogypsum "Apatite". The assessment of the effectiveness of biological reclamation for the reporting period is based on both the previous basic studies at the experimental sites, and conducted by the staff of the Balakovo branch of "Apatit" in 2016-2018 considering the formation of terraced slopes and the distribution of sludge on them, as the main substrate. To analyze the state of vegetation, a site was selected in the southern part of the phosphogypsum dump, where the formation of terraces for reclamation began from the beginning of 2015 [5].

For a detailed analysis of the restoration of vegetation cover in this area, the boundaries of the two most characteristic and undisturbed areas with vegetation were determined, as well as those of the control area with no vegetation (Table 1).

Table 1. Properties experimental plots on the phosphogypsum waste stack in Balakovo

Plot	Properties	Coordinate	Elevation
Headland	Native vegetation	51°54'49.94"/47°54'48.53"	52
Terrace	Native vegetation	51°54'42.98"/47°54'56.40"	54
Control	freshly poured phosphogypsum	51°54'34.13"/47°55'17.07"	68

Figure 1. Experimental plots on the phosphogypsum waste stack in Balakovo 1. "Headland", 2. "Terrace", 3. "Control"

NDVI index is the best parameter used to characterize fitoproducton in environmental studies. It is a standardized index showing the presence and condition of vegetation (relative biomass). Its value is used as a contrast between the characteristics of two channels from a set of multispectral raster data – chlorophyll pigment absorption in the red channel in the range of photosynthesis 0.55-0.68 μm (RED) and high reflectivity of vegetation in the infrared channel corresponding to the spectrum range 0.73-1.10 μm (NIR). The NDVI process creates a single-band dataset that mainly reflects the density and intensity of vegetation. Different reflection in red and infrared channels allows controlling density and relative intensity of vegetation using spectral reflection of solar radiation. Healthy and dense vegetation usually shows better near-infrared reflection than in the red region of the visible spectrum. If the leaves have a moisture deficit, are withering or dead, they become more yellow and reflect much less in the near infrared. The infrared range is absorbed by clouds, water and snow, and is reflected by rocks and bare soil almost as much as the range of the red region of the visible spectrum [6, 7]. The most suitable for large-scale data mapping and analysis, from our point of view, is the Sentinel-2 system. Sentinel-2 includes multispectral images in 13 channels with visible, infrared and short-wave parts of the spectrum. The images are of high resolution in the necessary NDVI channels to determine the index [8]. Decoding of remote sensing data, mapping and support operations were carried out in Q-gis.

Processing of remote sensing data was carried out in several stages:

1) Selection and loading of necessary images

The first stage of obtaining data of NDVI of the surface is to download remote sensing data. One of the main sources of remote sensing data is the files of the U.S. Geological survey (USGS). The site presents multispectral images of Sentinel, Landsat, MODIS, digital elevation models of different platforms, etc

2) Correction of images

After receiving multispectral images, the data is processed and analyzed in GIS programs. For the correction and processing of satellite images in q-gis there is a specialized plugin called Semi automatic classification.

3) The calculation of the NDVI

The next step in data processing is the calculation of the NDVI vegetation index. For its calculation are used images of RED and NIR channels. Calculations are made by the formula (NIR-RED)/(NIR+RED). When applying the formula to calculate the vegetation index, the raster calculator calculates and builds a raster image reflecting NDVI values on the surface of research area [9].

4) Spatial analysis and statistical analysis of data and presentation of results

The end of data processing is spatial analysis and statistical analysis. This stage consists in obtaining digital data on the state of vegetation at the studied sites

3. Results and Discussion

The statistical analysis of the data allows us to talk about a significant difference between the control area and the areas of “headland” and “control” (Figure 2, table 2). This difference suggests that the terracing of the slopes of the phosphogypsum waste stack and the overlap of their substrate allows us to successfully carry out bioremediation.

Table 2. Results of T-test experimental plots data

	T-test	df	p value
headland-terrace	0.784	24	0.441
headland-control	9.788	24	0.001
terrace- control	4.175	24	0.001

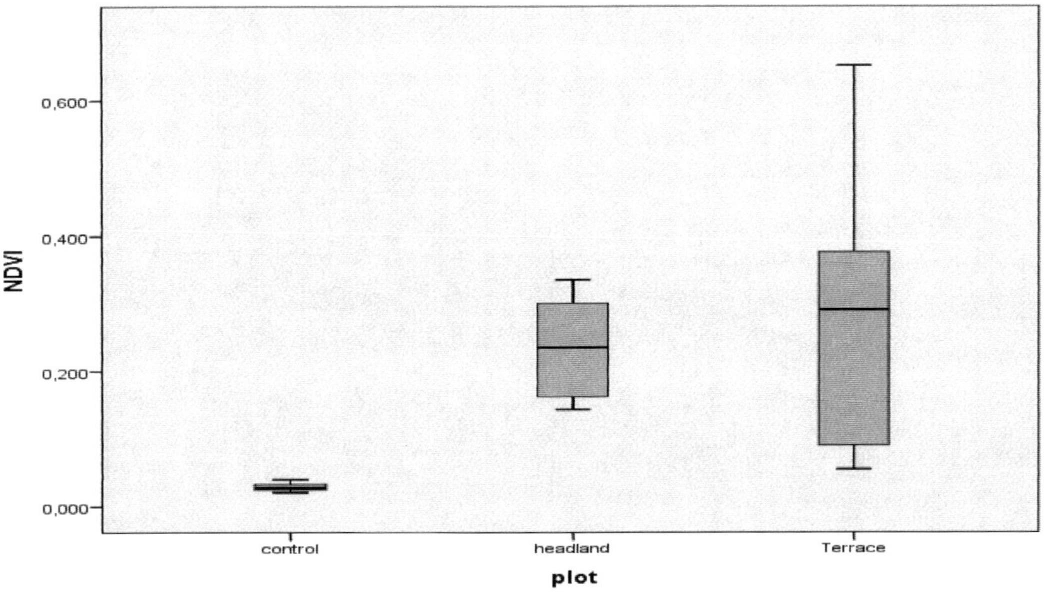

Figure 2. Box-plots of NDVI data

In addition, the obtained NDVI raster images were classified into groups according to the values of NDVI: group 1 – below 0.2 (conditionally without vegetation); group 2 – above 0.2-0.3 (occupied by vegetation in the early stage of development or oppressed vegetation); group 3 – 0.3-0.6 (well-developed vegetation); group 4 – more than 0.6 (highly developed vegetation). Due to the lack of vegetation, the first group was not taken into account. Thus, polygons were built, reflecting the area occupied by vegetation for each selected month. Areas covered by vegetation were calculated using standard vector data processing tools. The analysis of NDVI values made it possible to confidently identify the presence of vegetation on the surface of the phosphogypsum dump. High NDVI values were observed at the sites where the most "lush" vegetation was observed. Based on statistical analysis of the data obtained, graphs were constructed, reflecting the dynamics of vegetation with different values of the vegetation index from 2016 to 2018.

Analysis of changes in the NDVI value in space and time indicates that the plants on the phosphogypsum dumps are under conditions of continuous growth and recovery. The area occupied by the vegetation on the dump increased almost 3.5 times (figure 4), and the NDVI value increased from 0.01 - 0.02 to 0.49-0.56 (figure 3). In addition, the qualitative composition of vegetation has changed significantly.

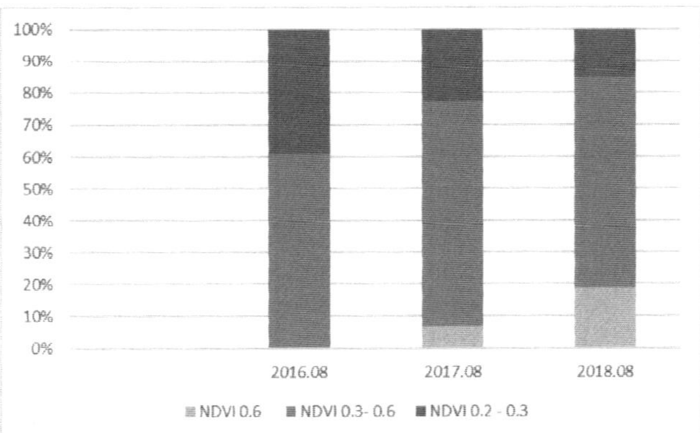

Figure 3. Qualitative change of vegetation on the surface of dumps

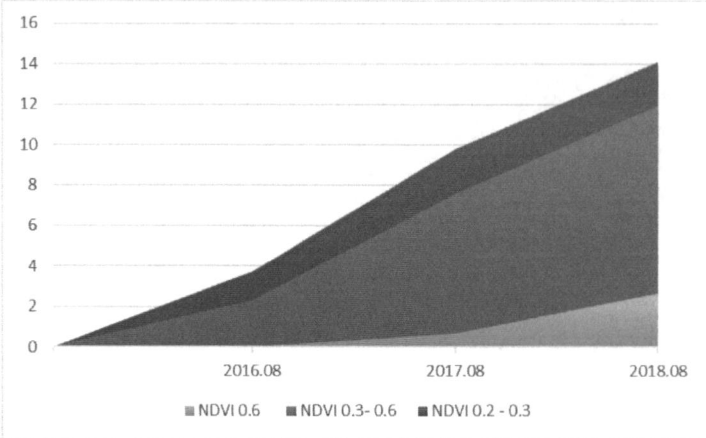

Figure 4. Quantitative change of vegetation on the surface of dumps

4. Conclusion

Remote sensing is an important tool in addressing environmental monitoring issues. This study has shown that satellite data provide information on spatial and temporal trends in the growth and restoration of vegetation on phosphogypsum dumps. This information can be used to assess the parameters of phytomass growth in the process of changing the phenological phases of plant development and climatic conditions of growth. The results of phytomonitoring are necessary for the design, monitoring and evaluation of the effectiveness of technologies for the reclamation of phosphogypsum dumps.

References

[1] Beck P S A, Wang T J, Skidmore A K and Liu X H 2008 *Int. J. of Remote Sensing* 4277–83
[2] Weber D, Schaepman-Strub G and Ecker K 2018 *Ecological Indicators* 447–60
[3] Chang J and Shoshany M 2016 *IGARASS* pp 5300–3
[4] Iftikhar A and Cawkwell F 2016 *J. of plant ecology* 649–71
[5] Belobrov V, Lebedeva M, Abrosimov K, Grebennikov A, Torochkov E and Ryashko 2018 *Spanish J. of Soil Sci.* 183–92
[6] Taleai M and Yameqani A 2018 *Ksce J. of civil engineering* 279–91
[7] Sitokonstantinou V 2018 *Remote sensing* 1–21
[8] Lessio A and Fissore V 2017 *Imaging* 1–16
[9] Silveira O and Martiniano E de Mello 2018 *Int. J. of remote sensing* 2597–619

Development of a statistical forecast model to improve accuracy based on statistical analysis of weather historical data for the Kalmyk region

V M Goryaev[1], G Y Kazakova[1], D B Bembitov[1], E N Dzhakhnaeva[2] and E V Sangadjieva[2]

[1]Gorodovikov Kalmyk State University, 11, Pushkin Ave., Elista, 358000, Russia
[2]Gorodovikov Kalmyk State University, 5th microregion, Elista, 358011, Russia

E-mail: goryaeff@mail.ru

Abstract. The article presents the main assumption that the averaging and smoothing models for the time series of meteorological data are local-stationary with slowly varying values. Accordingly, to estimate the current average, a local average was chosen, which was used to forecast for the near future and can be considered as a compromise between the average model and the random walk model (without drift). The same strategy can be used to evaluate and extrapolate a local trend. Average monthly values of temperature, humidity and pressure for the period from January 1928 to December 2018 were used as initial data at 4 synoptic stations. Comparative statistics on dependencies and forecasts are given. The selected method — a moving average — is often referred to as a smoothed version of the original series, since short-term averaging in a given window has the effect of smoothing irregularities in the desired series. The objective of the study was to adjust the degree of smoothing for the moving average, as an optimal balance between the performance of averaging models and the simplicity of the random walk model or between quality and cost. The task was solved on the basis of the software implementation of the SARIMA model, which required a lengthy adjustment of the initial data and significant manipulations with time series, however, in the end, a successful model was selected.

1. Introduction

Weather forecasts typically use a 12-month weather database and their totality is called the "test reference year" (TRY) or typical meteorological year (TMY). The choice of such typical weather conditions for the given localizations is important in modeling with the weather forecast, thermal characteristics of buildings, structures and other, which ultimately gives the designers the opportunity to observe long periods of the studied data or to select the desired, typical year from the time sequence of TMY. Numerical weather prediction (NWP) data is a form of weather model data. NWP focuses on conducting current weather observations and processing these data using computer models to predict future weather conditions. Knowing the current weather conditions is just as important as numerical computer models processing the data.

Purpose of the Study is to examine the capabilities of the seasonal autoregressive integrated moving average model for weather forecasting and to identify trends in climate change in the region based on the Python 3.6 (Miniconda) code.

IOP Conf. Series: Earth and Environmental Science **350** (2019) 012058 doi:10.1088/1755-1315/350/1/012058

2. Materials and methods of research

One of the most popular and frequently used time series models is the autoregressive integrated moving average model (ARIMA) [1–3]. The main assumption made to implement this model is that the considered time series is linear and follows a certain known statistical distribution, such as the normal distribution. The ARIMA model has subclasses of other models, such as autoregression models (AR) [3], moving average (MA) [4]. For seasonal forecasting of the time series of Box and Jenkins (BJ) [5], they proposed a rather successful version of the ARIMA model and specifically, its seasonal version is SARIMA [6]. But a serious limitation of these models is the assumed linear form of related time series, which becomes inadequate in some practical situations; for this, some authors have proposed various non-linear stochastic models [6], but from the point of view of implementation they are not as simple and convenient as ARIMA models. A review of the literature shows that there is no single model that consistently outperforms other models in all situations; therefore, this article attempts to compare approaches in order to find the best method for forecasting weather in the region.

The article uses data on the monthly weather data at the station Elista No. 1 from 1966 to 2018 and Station No. 2 in the period 1927-2018. For the collection of information, the values of time series from open sources are used: the laboratory of the automated information system of Roshydromet and NOAA.

Weather station Elista:

- Synoptic index: 34861, Height above sea level: 133 m
- Geographic latitude: 46.315488, longitude: 44.279401°

```
nms=['S_IS',1,2,3,'YE_I','MO_I','DA_I','SR','PN','NV','SV','TV','DP','OV','VV','AD','KO','SO','GD','TP','ADR']
test = pd.read_table(r"...\Inp\sr77.csv", sep=';' , engine='python', error_bad_lines= False, parse_dates= {'Datetime':[1,2,3]}, names=nms, header= None ); test_original = test.copy()
```

Initially, partial indexing and cutting of rows, sampling of time series, separation and re-sampling for different months with different aggregates were performed.

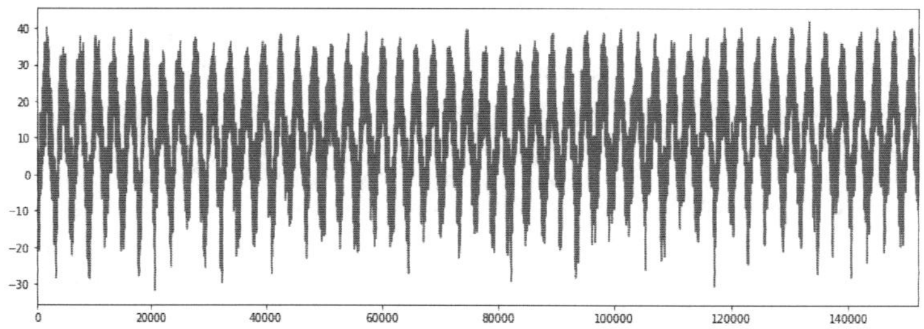

Figure 1. Temperature chart 1967-2018 in Elista

Creating a SARIMA model consists of four systematic steps (identification, evaluation, diagnostic testing and application or prognosis). The components of the series were studied and installed for stl-seasonal and trend decomposition using the LOESS method ("STL).

3. Results and discussion

For the analysis of rainfall data to build a predictive model, we used the approach of Box - Jenkins. The first-order seasonal difference is the difference between the observation and the corresponding observation of the previous year and is calculated as $z_t = y_t - y_{t-s}$ (SARIMA (p, d, q) × (P, D, Q)s). The daily smoothing of the hourly temperature forecast data for June 2019. The confidence intervals for the smoothed values are shown below.

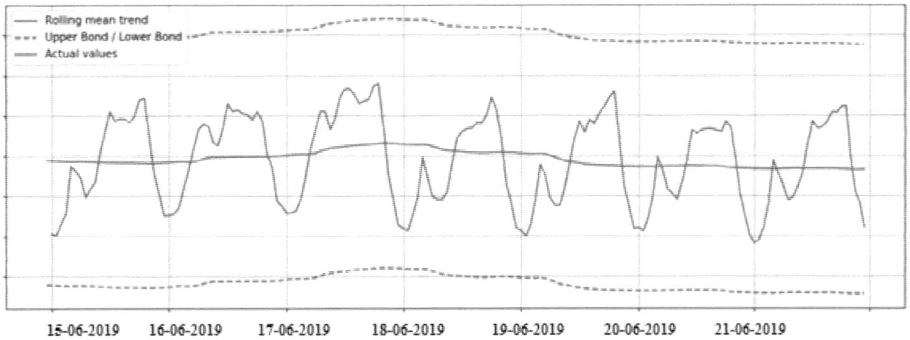

Figure 2. Sliding average with window =7

On the graph, the vertical axis is represented by the following equations:

$$C_n = \sum_{t=1}^{n-h} \frac{(y(t)-\hat{y})(y(t+n)-\hat{y})}{n}, \; C_0 = \sum_{t=1}^{n} \frac{(y(t)-\hat{y})^2}{n}$$

The horizontal axis represents the time delay (previous time steps) h

```
pd.plotting. ac_plot(df['TV'].mean)
pd.plotting.lag_plot(df[df['TV'].mean])
```

The data shows a trend to increase in temperature by 1.1 ° C over the past 50 years, if we take the statistics for a decade (observed from 1927) from 01/01/1928 to 12/31/1937 and from 01/01/1998 to 12/31/2018, the difference is even more striking - 1.79.

```
df[data['dates'] <='1991-01-01']['OV'].mean()
# average humidity for the Republic of Kalmykia - 68.882, c '1991-01-01 = 71.388, (respectively
for average temperatures: 9.363 и 10.211).
```

To determine whether residues are white noise and the type of distribution, it is necessary to construct a normal residual probability graph, with the Akaike information criterion (AIC) and Schwartz Bayes criterion (SBC) being used for the criteria for selecting the model itself.

Seasonal autoregressive integrated moving average with exogenous regressors (SARIMAX) is one of the main assessment classes that can be accessed through statmodels and their result classes.

```
# Imitation of time series …, build a model
class ARe2(...MLEModel):
```

Information output on the resulting model in Table 1.

```
set the parameters that give the lowest AIC
p, q, P, Q = result_table.parameters[0]
```

Table 1. Model Results of SARIMA (2,1,3)x $(1,1,0)_{24}$

Dep. Variable:	Ads №	Synoptic number:	34861
Model: SARIMAX (2, 1, 3)x (1, 1, 0, 24)			
Date:	Sat, 06 Apr 2019		
Q-test Ljung — box:	54.106	Test Jarque-Bera (JB):	47.908
Prob (F- statistics) (Q):	0.070	Prob(JB- statistics):	0.000
heterogeneity (H):	0.590	distortion:	-0.650
Prob(H-statistics):	0.040	Excess:	5.090

The result object has many attributes and methods that can be expected from other Statsmodels results, you can make a prediction and a prediction after the evaluation and the end period can be specified as a data type.

From the graph in figure 3 it follows that the model was able to successfully approximate the initial time series, fixing the daily seasonality, the general downward trend and even some anomalies. On model deviations, it can be fixed that the model reacts rather sharply to changes in the structure of the series, but then quickly returns the deviation to normal values. This feature of the model allows you to quickly create anomaly detection systems, even for noisy data without spending on data preparation and model training.

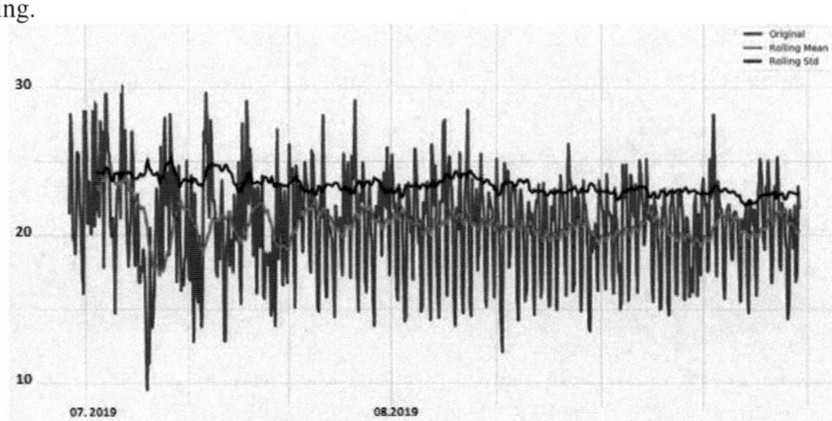

Figure 3. Moving average and standard deviation

The model parameters from table 1 are significant. Graphs of the residuals in figure 4 show that the distribution of residuals of the proposed model is Gaussian (white noise). This is clearly seen in the figure below. Therefore, the proposed model is justified, as it fits well with the test data. With the exception of some minor bursts, which lies outside the 95% confidence interval, all other points are within the confidence interval.

The predicted figures, as a rule are quite close to the actual points. The model performs its predictive function.

Figure 5 shows the results of the study.

```
plotSARIMA(ads, best_model, 50)
```

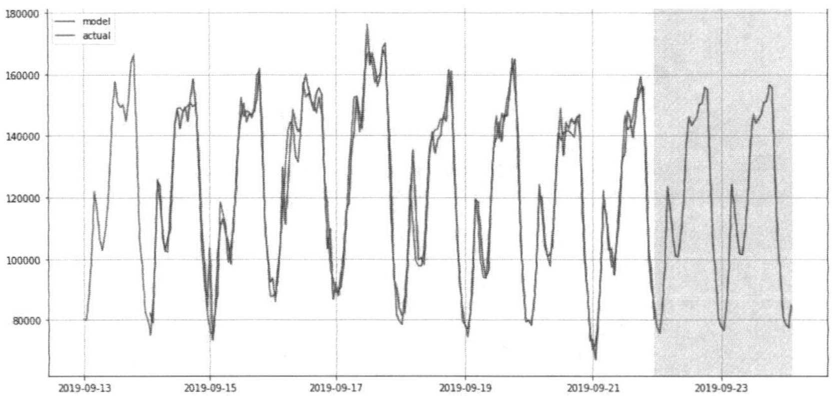

Figure 4. Average absolute percentage error (4.46%)

As a result, adequate forecasts were obtained. This model was mistaken on average by 4.46%, which is generally not bad. However, the total cost of preparing the data, finding the series in a stationary mode and choosing the parameters may not be so costly.

During the initial selection of the delay parameters, it is necessary to find a balance between the optimal prediction quality and the length of the forecast horizon.

```
...
fit1 = sm.tsa.statespace.SARIMAX(Train.Count, order = (2,1,3), seas_order =(0,1,0,24)).fit()
y_hat_avg['SARIMA'] = fit1.predict(start="2019-9-1", end="2019-9-25", dynamic=True)
plt.plot(y_hat_avg['SARIMA'], label ="SARIMA")
```

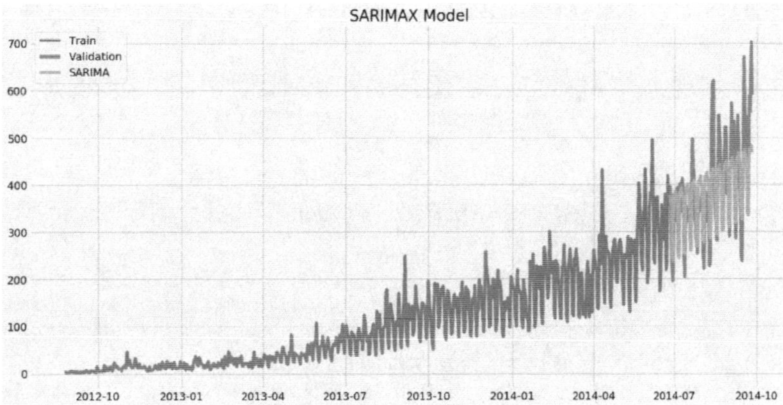

Figure 5. Forecast SARIMA (2,1,3)x (1,1,0)$_{24}$

The forecast is shown by the red line and dots compared to the test data set from January 1967 to December 2018 by the blue line. 95% confidence interval overlapped by yellow lines. The predictive SARIMA model is a valuable tool with potential for early warning, detecting weather changes and can provide reliable information for proactive work, because the prediction values from the SARIMA model (2, 1, 3)x(1, 1, 0)$_{24}$, the most suitable for this model, the forecast values.

4. Conclusion

It was established that the totality of data on average temperatures in the town of Elista from 1966 to 2018 is stationary, which is confirmed by autocorrelation, partial autocorrelation plots and the Dickey Fuller test. To exclude the seasonal component in the time series of average temperatures, it was necessary to carry out one seasonal differentiation. Candidate models were developed as specified in the model building process according to the Jackins approach, and AIC values were obtained for each candidate model. The final model was chosen. SARIMA (2, 1, 3)x(1, 1, 0)$_{24}$.

By subtracting moving averages from baseline observations, specific seasonal values are obtained, what remains in specific seasonal variations usually represents a stationary horizontal row with two effects that cause specific seasonal variations to deviate from an absolutely straight line: seasonal effects and random error in the original observations.

References

[1] Asamoah-Boaheng M 2014 *International Journal of Statistics and Applications* **4** 292-298
[2] Thibault J C and Senocak I 2009 *Journal of Supercomputing* **9(2)** 693-719
[3] Balyani Y, Niya G F and Bayaat A 2014 *Geographic Space* **12(38)** 127-144
[4] Box G, Jenkins G and Reinsel G 2015 *Time series analysis: Forecasting and Control* (John. Wiley & Sons, Hoboken) pp 27-98
[5] Goryaev V M, Dzhahnaeva E N, Lidzhi-Goryaev V V and Al-Kilani V H 2018 *Modern knowledge intensive technologies* [in Russian – Sovremennye naukoemkie tekhnologii] **12(1)** 56-61
[6] Lazareva V G, Bananova V A and Nguen Van Zung 2017 *Advances in modern natural sciences* [in Russian – Uspehi sovremennogo estestvoznaniya] **12** 178-183

Environmental requirements in land management of land use facilities

A Varlamov, O Gvozdeva and R Zhdanova

Department of Land Use and Cadastre, State University for Land Management, 15, Kazakova St., Moscow 105064, Russia

Abstract. The authors of the article studied the role of arranging the territory on an ecological and landscape basis as a set of measures aimed at the use of a certain territory (part) of a land use facility to meet the needs of population and production based on the use of landscape and environmental research and design methods. Based on the example of Kantemirovsky district of the Voronezh Region the system of the ecological framework development of the territory of the region as an integral part of ecological and landscape arrangement of the territory of the region is shown. The work presents the beam-field agricultural landscape "Vysochanovsky" with the introduction of the ecological and landscape system in the agricultural enterprise Druzhba of Kantemirovsky district of the Voronezh Region. The work is based on the study of domestic scientists among which are V.V. Dokuchaeva, D.V. Novikova, A.A. Varlamova, M.I. Lopyreva, A.G. Bogdanov.

1. Introduction

The main condition for the contribution of land assets to the economic growth and sustainable development of the country is the application of land use management methods aimed at the extension of national wealth and well-being of the society while maintaining environmental potential.

The strategic goal of modern land management is the organization of rational use and protection of land resources based on optimizing multi-structured land use, meeting the requirements of its spatial, economic, environmental sustainability, creating safe living conditions for the population and carrying out economic activities, increasing the efficiency of agricultural and forestry land use. The landscape plays an important role in the ecological, natural and social fields. It is considered a basic component of the natural and cultural heritage. Besides, it contributes to the growth of the well-being of the population, serves as the main spatial resource for the development of many economic sectors related to the territory.

2. Materials and methods

Based on the analysis of qualitative state of land resources and environment, the presence of the territorial organization or arrangement of the economic complex it can be concluded that it is necessary to take into account the natural and social and economic component in the process of ecological-landscape arrangement of the territory of land use facilities.

The most important principle of the land use is the obligation to preserve the ability of landscapes to reproduce their useful properties and preserve natural mechanism of reproduction, i.e. social and economic functions - the landscape plays a role aimed at meeting a particular need of society. At the same time, the contradictions between the nature and society may arise and become aggravated, especially since in modern conditions the society's dependence on the state of the natural environment does not decrease, but only increases [1].

The ecological and landscape organization of the territory allows us to take into account and regulate the following functional groups of the needs of society related to landscapes [1]:

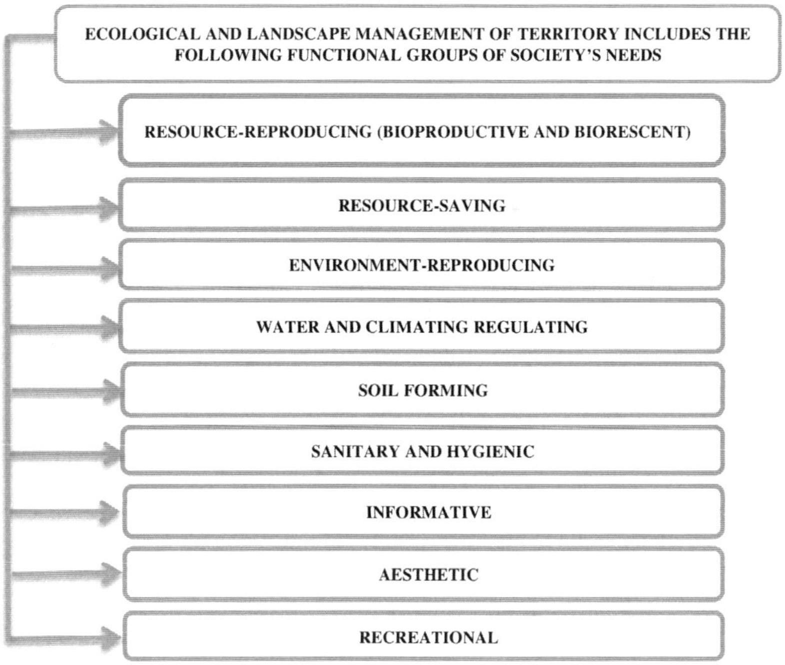

Figure 1. Functional groups of society's needs taking into account the ecological-landscape arrangement of the territory

The results of the impact of human activities on the landscape can be described with the following indicators:

a) changes in the structure, state, functioning of the landscape;
b) changes in the current landscape development dynamics;
c) disruption of the course of natural cycles and trends of natural self-development;
d) changes in response to man-made loads;
e) environmental restrictions (Figure 2).

The origin of the scientific theory of ecological-landscape arrangement of the territory is associated with the name of V.V. Dokuchaev, who proposed a systematic landscape-ecological approach to the use of land and other natural resources.

The experience of ecological-landscape arrangement of rural areas abroad has shown the following:

- the priority is given to measures for the development of the land use system and the ecological-landscape arrangement of rural areas;
- the basis for the development of the land use and the ecological-landscape organization of the territory is a set of land management activities and landscape planning;
- the protection of natural environment and the development of land use are becoming an independent branch of research, design and project development, and even a policy direction;
- the development methods of land use, ecological and landscape arrangement of its territory and organization of these activities can be applied in the development of land management digrams and projects [2–4].

An information and logical model of the sequence of actions and the composition of research and design activities carried out for each stage of work indicated in the model (Figure 2) is appropriate for design on the ecological and landscape basis.

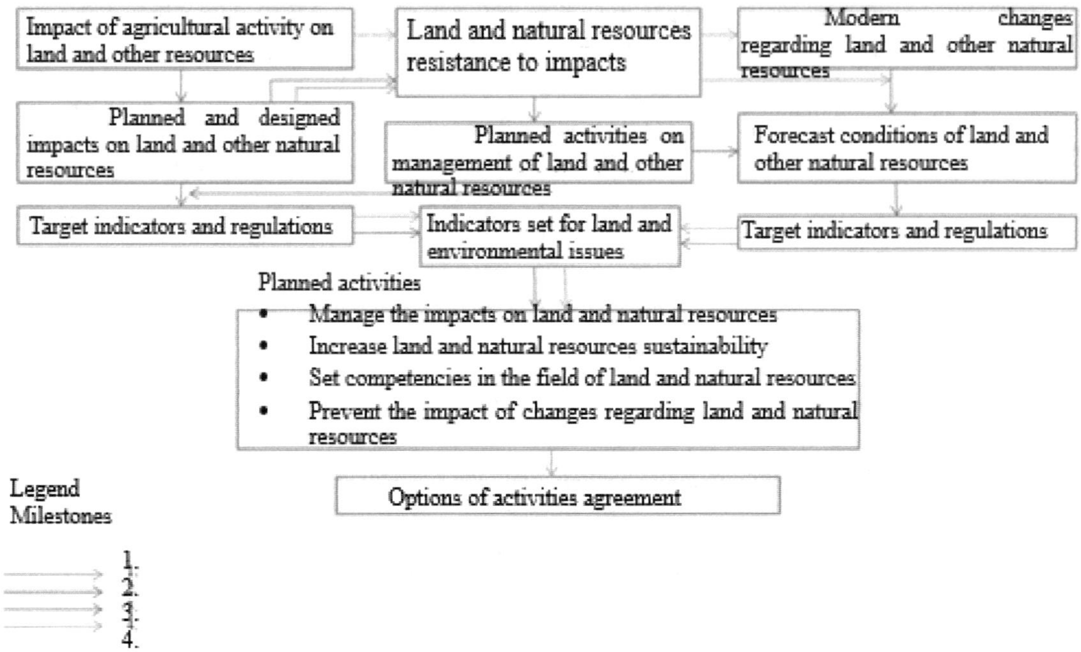

Figure 2. Information and logical model of the sequence of actions in the development of measures for the protection of land-use facilities (by D.V. Novikov). Milestones: 1 – Evaluation of the processes that have caused modern changes of land and other natural resources conditions, 2 – Identification of the problems regarding land and other natural resources management when planning environmental activities, 3 – Development of recommendations regarding strategies of land and environmental management in the region, 4 – Development of several complex strategy aimed at land and environmental management in the region.

Therefore, the main areas of integrated solution of problems of rational use and protection of land resources of the region are the following:

- development of a strategy for the development of the economic complex, the restoration and preservation of natural resource potential, the increase in land productivity and the protection of landscapes in all territories;
- the establishment of prerequisites for sustainable development and ecological rehabilitation of territories, etc.;
- establishment of conditions for the development of ecological balance, development of infrastructure, a combination of various forms of management of land-use facilities;
- development of the territories of the region based on the establishment of rational use and land protection;
- optimization of the existing land use system by improving the structure of land, the rational structure of crops, the abandonment of monoculture;
- the development of ecological landscape organization of the territory [5].

3. Results and discussion

The introduction of sustainable land use mechanism as an integral part of economic mechanism of environmental management includes the following: establishment of differentiated land payments; economic incentives for rational land use; application of economic sanctions for a significant reduction in soil fertility and for violation of land legislation; introduction of monetary compensation for temporary conservation of land, measures of economic incentives for improving the ecological condition

IOP Conf. Series: Earth and Environmental Science **350** (2019) 012059 doi:10.1088/1755-1315/350/1/012059

of land and increasing soil fertility, stimulating the production of environmentally friendly agricultural products [6, 7].

The introduction of adaptive land-use systems of the Voronezh Region based on the ecological and landscape arrangement of the territory will allow annually preventing the environmental losses (55 million tons of soil, 1.5 billion m^3 of water, 7.1 million tons of nutrients) and increase agricultural productivity by 40-50%.

The main blocks of the ecological framework at the level of administrative district are the shelter-protection and water-regulating forest belts. Under the conditions of a dissected relief the linear blocks of the ecological framework should occupy all the areas that are free from the land used, taking into account the location of arranged elements of the territory, i.e. fields, work sites, roads. The system of development of the ecological framework of the territory of the region as an integral part of the ecological and landscape arrangement is shown by the example of Kantemirovsky district of the Voronezh Region. The elements of the ecological framework should be supplemented on the basis of the projects of ecological and landscape arrangement of the territory of specific agricultural enterprises to assess the condition of each land mass and its further use [8].

The ecological frame of the territory of Kantemirovsky District of the Voronezh Region represents a single network with a minimum number of breaks and includes both a natural network and agricultural landscapes. The area of the environment stabilizing lands included in the ecological frame makes up 75.32% of total area of the region, which makes it possible to carry out environmental protection, compensating and sanitizing functions in this area.

Besides, to justify the design decisions when drafting land management diagrams, the value of prevented environmental and economic damage, the productivity index of agricultural landscape, the coefficient of ecological diversity of the territory, the indicators of reduction of soil volume loss, the loss of humus and nutrients, the reduction of proportion of agricultural land subject to various negative effects are calculated.

The production and economic results should be assessed in terms of the amount of agricultural production required by the society and obtained from one unit of a land area while maintaining and increasing the soil fertility, including: an increase in crop yields, reducing the cost of production, increasing the net income and profits; increasing the annual economic effect [9].

When implementing the actions provided by the ecological framework, the agricultural landscape shifts into the state of stability at K = 1.39 from the state of destruction at K = 0.28 (the ratio of the area of stabilizing land to destabilizing) in Kantemirovsky district of the Voronezh Region.

The coefficient of natural protection of the territory, as the ratio of the area of landscapes with environment stabilizing functions to the total area of the region shifts to sustainable (K = 0.65) from the critical level (K = 0.49).

The agricultural enterprise area was 10,000 hectares. The area of arable land was 6,000 hectares. The forest cover of the territory was 17%, including arable land under the forest belts, i.e. 5%. Forest stripes and processing were performed in a horizontal manner. The average area of seedling plot was 20-40 ha. The crop rotations were introduced. The soil fertility amounts to 5.5% of humus. The grain yield during the drought in a family farm of Bogdanov was 32.5 centner/ha in 2010; in the Voronezh Region it was 7 centner/ha. The survey of the territory was made from a satellite in November 2010 (Figure 3).

In 2014, the middle-stabilizing (improving) sites amounted to 75.32% (576.85 ha), whereas in 1975, it was 35.75% over an area of 296.01 ha (Figures 4 and 5). The destabilizing (worsening) lands made up 42% on the area of 347.53 hectares in 2014 and 64.2% on the area of 531.57 hectares in 1975. In 2014, the area within the landscape was 924.38 hectares (2014), 827.58 hectares (1975).

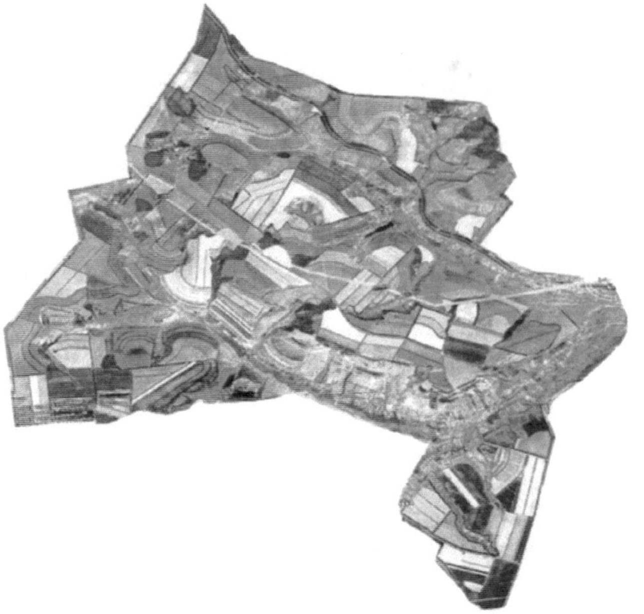

Figure 3. Landscape and ecological (adaptive) arrangement of agricultural system in agricultural enterprise Druzhba, Kantemirovsky district, Voronezh Region

The territory is arranged on an ecological and landscape basis with the purpose of establishment of conditions for mobilization of natural resources and the use of the adaptive potential of crops, conducting economically efficient, socially-oriented and environmentally friendly agriculture, stabilizing nature management and land use by creating environmentally sustainable agricultural landscapes, eliminating degradation and pollution of lands.

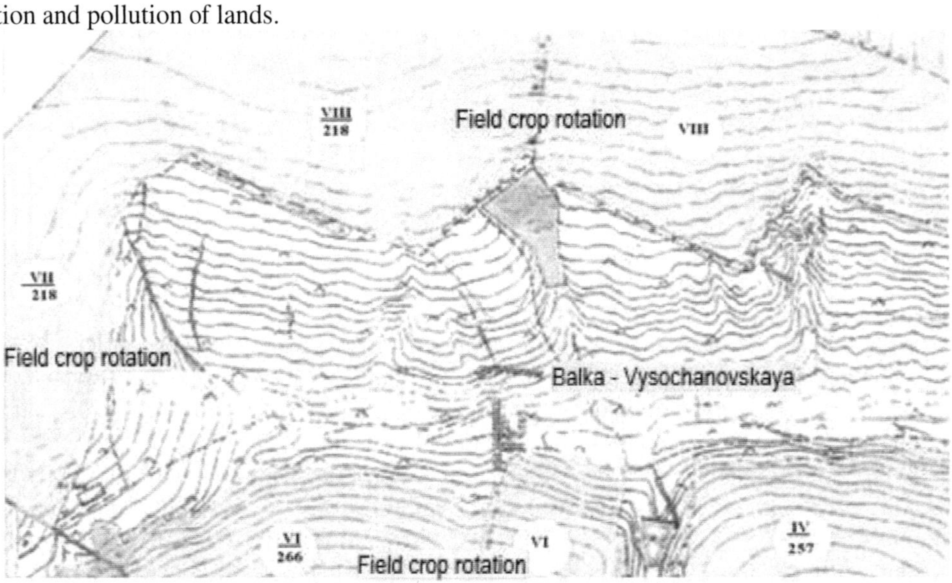

Figure 4. Beam-field agricultural landscape "Vysochanovsky" before introduction of ecological and landscape system into agricultural enterprise "Druzhba", Kantemirovsky district, Voronezh Region (1976)

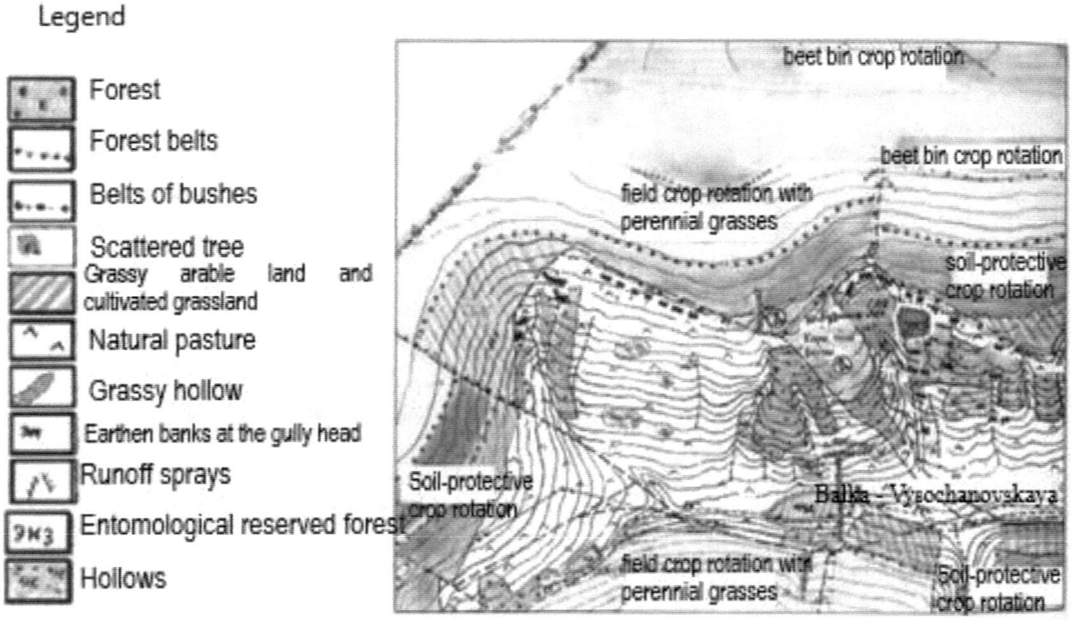

Figure 5. Beam-field agricultural landscape "Vysochanovsky" with the introduction of ecological and landscape system into agricultural enterprise "Druzhba", Kantemirovsky district, Voronezh Region (fragment)

Figure 5 shows the beam-field agricultural landscape "Vysochanovsky" with the introduction of the ecological and landscape system in the agricultural enterprise Druzhba of Kantemirovsky district of the Voronezh Region (actual fragment). Farming has become sustainable.

In the framework of Kantemirovsky district of the Voronezh Region the family farm of Bogdanov corresponds to 50 and 65 points of yield in terms of fertility in the district (Figure 6).

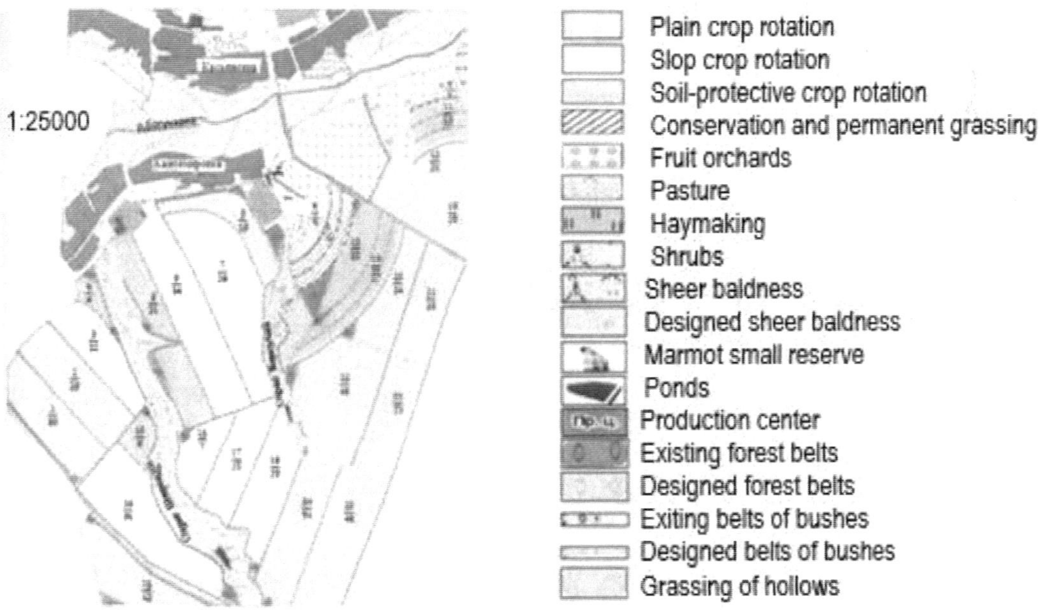

Figure 6. Territorial arrangement of ecological and landscape system of farming by A. Bogdanov

The analysis of data implies that the yields from the family farm of A.F. Bogdanov were 1.8 times higher than the average. As a result of the measures taken in the surveyed Vysochanovskaya section in the agricultural enterprise Druzhba of Kantemirovsky district of the Voronezh Region, the tendency towards stabilization of the humus horizon thickness and its increase has changed. The humus content in the study area increased by 0.24% [9]. The water regime of soils has changed as a result of which the desalinization process of salt marshes to solonetz and in different degrees of saline soils has occurred. The plant cover has undergone significant changes. The water-cut plants were not marked, which indicates the absence of water erosion. The herbage density has changed from 60% to 85% of the projective cover.

4. Conclusion

In the Kantemirovsky district of the Voronezh Region, when implementing measures provided by the ecological framework, the agricultural landscape shifts into the state of stability at K = 1.39 from the state of destruction at K = 0.28 (the ratio of the area of stabilizing land to destabilizing).

As a result of the measures taken in the surveyed Vysochanovskaya section in the agricultural enterprise Druzhba of the Kantemirovsky district of the Voronezh Region, the tendency towards stabilization of humus horizon thickness and its increase has changed.

For more than 40 years of study of landscape farming there has been stored enough information to think that the farming system of agricultural enterprise Druzhba can be taken as a proven ecological example for the Voronezh region. Its general ecological principles can be used throughout the Central Black Soil Region and in some other regions with similar environmental conditions.

References

[1] Galchenko S A, Varlamov A A and Bogdanova O V 2018 Theoretical and methodological foundations for formation of sustainable land management system *IOP Conf. ser. Materials Science and Engineering* 012141
[2] Pakhomova N V and Richter K K 2001 *Economics of nature management and environmental protection* (Moscow: DECA)
[3] Zelepugin A D, ed 2011 *Scientific basis of land use management* (Voronezh)
[4] 1998 *Economics of land use and management* (Voronezh: Voronezh state agrarian University)
[5] Antropov D V, Zhdanova R V and Gvozdeva O V 2016 Influence accounting of zones with special conditions of use of territories in the development of effective system of agricultural land use *Int. Agricultural J.* **4** 15–7
[6] Misra A K, Prakash O and Ramasubramanian V 2004 Forewarning powdery mildew caused by Oidinm mangiferae in mango (Mangifera indica) using logistic regression models *Indian J. of Agricultural Sciences* **74(2)** 84–7
[7] Pankratz A 1983 *Forecasting with univariate Box – Jenkins models: concepts and cases* (New York: John Wiley & Sins)
[8] Girusov E V, ed 2007 *Economy and ecology of nature management* (Moscow: unity-Dana)
[9] Kvasov A Yu, Lopyrev M I, Shevchenko V E, Stekolnikov K E, Pokusaev V V and Bogdanov A G 2014 *What should be a sustainable high-yielding agriculture. Kantemirovsky experiment: experience and opportunities* (Voronezh)

Zoning as a tool of land management

S I Komarov, D V Antropov, A A Varlamov, S A Galchenko and R V Zhdanova

State University of Land Management, 15, Kazakov st., Moscow, 1050641, Russia

E-mail: komarov@zemcad.ru

Abstract: The article describes the zoning of Russia as part of strategic planning. The draft project "Spatial Development Strategy" developed by the Ministry of Economic Development is analyzed. The composition of the zones of this project is compared with authors' zoning results based on land and property development indicators and influencing factors. The conclusion about poor accounting of land resources and land relations in strategic planning was drawn.

1. Introduction

The national land fund is a national wealth whose rational and effective use is impossible without an appropriate management system. In the market economy, it is necessary to develop organizational, legal and economic mechanisms that form an environmentally and economically viable system of objects and subjects of land management, increase budget revenues and implement the most economically and environmentally sustainable forms of state, municipal and private land management [1]. The land management system may federal, regional, municipal and intra-economic depending on the form of ownership of land and a body performing management functions. The main way to achieve strategic goals of sustainable development of Russia and ensure national security is implementation of strategic national priorities, including priorities of socio-economic development of the Russian Federation which is impossible without an effective system of land management.

Land management includes land management, record-keeping (real estate cadaster maintenance), land monitoring, land supervision and control, land use planning and other activities. One of these activities is zoning. The institute of zoning was formed as a result of the need to develop mechanisms for land management, regulation of land management relations caused by various state and public processes (urbanization, scientific and technical progress, expansion of industrial production, pollution of the environment, etc.). Various regulatory documents mention assessment, cadastral, town-planning, functional, territorial, ecological, economic, agricultural and other types of zoning. Regulation of zoning activities is complex and intersectoral; conditions and zoning procedures zoning are regulated by various branches of legislation. Thus, it is necessary to solve theoretical and methodological issues of formation of an effective land management system based on a scientifically justified zoning system which can become a form of management and control over the territory management, a tool of political power aimed at the dynamic and sustainable development of land and natural resources. The authors have already addressed some types of zoning. Contradictions between the results of different types of zoning make it extremely difficult to use them in managing the land and property system of the region, especially in planning and creating a territory development strategy as one of the main management documents according to the law "On Strategic Planning in the Russian Federation". It is necessary to use a complex zoning method accounting for various factors and natural, social, economic and anthropogenic features of regions and municipalities.

The institute of zoning of territories was formed as a result of the need to develop mechanisms for land management, regulation of land management relations due to various public processes. The term "zoning" is often defined as urban planning zoning which is not correct, since urban zoning is a special type of zoning, since "zoning" combines a variety of different types: legal zoning of agricultural land and settlements, restrictive zoning, cadastral, estimated, functional, construction zoning, division of the territory of a national park into functional zones, etc. Many legal scholars addressed the issue of territories. However, most of them dealt with the legal content of the zoning institute, specifics of a legal regime in individual territories, and did not take into account that the content of relations in the area of zoning is wider than the subject of legal regulation of land relations.

Over the past years, there has been a tendency to identify the concept "zoning" with territorial zoning. Many scientific publications analyzing pros and cons of the transition from land categories to the territorial zoning [2-6] appeared. When developing a scientifically justified zoning system for land management purposes, it is necessary to study and take into account the and foreign experience. In other countries, the zoning is an aspect of the territorial planning.

In foreign countries, it is common to distinguish between two groups of zoning systems – American (high autonomy of local governments, declaration of policies in various areas in the form of strategic socio-economic development plans, zoning of the entire territory and its components) and European (a set of national systems that differ from each other). The main goal of zoning is to determine the scope for limiting rights to property in the communal interests.

In the US, zoning is a regulatory tool of the government. Development of territories began with monofunctional zones (tools for urban planning). They are based on the principle of land division into residential, commercial and industrial areas. Monofunctional zones have negative economic consequences and worsen the social situation. The reason for these negative consequences are associated with urbanization, environmental pollution, racial and socio-economic segregation [7]. Four types of zoning are currently used in the USA: Euclidean, functional, stimulating, and territorial. The Euclidean zoning is most common [8]. It is characteristic of areas where the land use is limited to certain activities within the district. The positive aspects of this type are many years of experience, accumulated practice and easy application. The negative sides are as follows: the lack of flexibility and the outdated model of territorial planning [9].

In Delhi, the planning department is responsible for development of the five-year plan and annual plans which include evaluation of financial resources, maintenance, monitoring and evaluation of the five-year and annual plans. For each 15 zones, a zonal development plan should be developed. In addition to zones, there are subzones which are part of a zone with a strictly specified use approved by the model plan [10]. Thus, Western zoning systems are territorial development planning systems (legal zoning) aimed at defining the scope of restrictions on the rights to real estate objects used in the interests of the community. In Russia, the zoning is a complex multidimensional concept: a legal institution, a mechanism for regulating economic and other activities, and a management function.

2. Research methods

The land fund of Russia which is an object of land management is heterogeneous. Land resources have different climatic (the territory of the country stretches from permafrost to subtropics), soil (65 main soil types), areal (the area of the largest Russian region (Sakha Republic) is 3,568 times larger than its smallest subject (Sevastopol), legal (division into 7 categories of land and 13 types of permitted use), economic (the cadastral value of land plots differs 217 times), administrative (the share of regional ownership of land varies from 32 to 100%) characteristics. It is necessary to unite the regions by their characteristics affecting land management.

The Federal Law "On State Strategic Planning" laid a basis for development of strategic planning documents. The main regional document is the strategy of socio-economic development. This document regulates economic aspects of the territory, including land management and economic land and property relations. All the Russian regions adopted strategies for their socio-economic development until 2025 or 2030. According to [11,12], less than a quarter of the Russian regions included the analysis of their land resources and regional land and property relations in these documents. More than 60% of the regions included no more than 2 strategic priorities in their strategies. In 2018, the Ministry of Economic Development developed a draft Spatial Development Strategy implying zoning of the national territory into 14 macroregions. The country was zoned based on natural, climatic, economic, social, development characteristics of the regions. The land and reals estate objects located on were not accounted when determining boundaries and composition of macroregions. The authors zoned the Russian regions based on the land resource and reals estate objects development indicators. 72 characteristics were used divided into several groups were used:

1. **Indicators of cadastral registration and property title registration,** including the number of registered land plots, issued certificates, registered rights, etc.;
2. **2. Indicators characterizing the land fund of the region**, including the structure by categories, areas of land plots, etc.;
3. **Indicators of public land control,** including the number of inspections, violations, fines, etc.;
4. **Indicators reflecting the state of the land market,** including the number of transactions, the area of land plots sold and leased, etc.;
5. **Cost indicators,** including the cadastral value of land, the amount of land tax, rental charges, etc.

3. Strategic clustering results

Using the k-means method and Statistica, the Russian regions were clustered according to the above characteristics into 14 zones (Table 1, Figure 1).

Table 1. Zoning of Russian regions based on land resources and property development

Zone	Subjects of the Russian Federation	Zone	Subjects of the Russian Federation
1`	Leningrad Region, Nizhny Novgorod Region, Perm Territory, Chelyabinsk Region	8	Ryazan Region, Tambov Region, Republic of Dagestan, Omsk Region, Tomsk Region
2	Rostov Region, Republic of Bashkortostan, Republic of Tatarstan, Samara Region, Sverdlovsk Region, Kemerovo Region	9	Lipetsk region, Astrakhan region Republic of North Ossetia Udmurtia Yamalo-Nenets Autonomous District
3	Belgorod Region, Tver Region, Tula Region, Vologda Region, Volgograd Region, Stavropol Territory, Khanty-Mansi Autonomous Area, Altai Krai, Irkutsk Region, Novosibirsk Region, Primorsky Krai	10	Arkhangelsk Region and Nenets Autonomous Area, Zabaykalsky Krai
4	Vladimir Region, Murmansk Region, Saratov Region, Khabarovsk Krai, Magadan Region and Chukotka Autonomous Region	11	Voronezh region, Ivanovo region, Oryol region, Novgorod region, Pskov region
5	Bryansk region, Kaluga region Kursk region, Orenburg region, Kurgan region, Tyumen region, Republic of Buryatia, Amur region	12	Kostroma Region, Smolensk Region, Chechen Republic, Chuvash Republic, Kirov Region, Penza Region, Kamchatka Territory
6	Moscow, Moscow region, Krasnodar region, Krasnoyarsk region	13	Republic of Karelia, Komi Republic, Republic of Adygea, Republic of Kalmykia, Republic of Ingushetia, Republic of Karachay-Cherkessia, Republic of Mordovia, Republic of Khakassia, Sakhalin Region
7	Yaroslavl Region, Kaliningrad Region, St. Petersburg, Ulyanovsk Region, the Republic of Sakha (Yakutia)	14	Republic of Crimea, Sevastopol, Republic of Kabardino-Balkaria, Republic of Mari El, Republic of Altai, Republic of Tyva, Jewish Autonomous Region

The first zone includes Leningrad, Nizhny Novgorod and Chelyabinsk regions and Perm Krai. These regions are characterized by a significant amount of registered land plots and registered rights (the average seasonal values of these characteristics exceed the average Russian by 30-100%); a high degree of urbanization and industrial development (the share of the average seasonal land area of settlements and industrial and other special purpose lands is 155% and 104% of the average Russian indicators, respectively) and low values of forest land areas and specially protected areas (39% and 19%, respectively); a large number of land plot transactions; a significant cadastral value of land plots, and a significant amount of land taxes. Comparing the composition of this zone with the composition of macro-regions in the Spatial Development Strategy, one can see that Nizhny Novgorod Region and Perm Krai are part of one macro-region. Leningrad and Chelyabinsk Regions belong to a different macroregion.

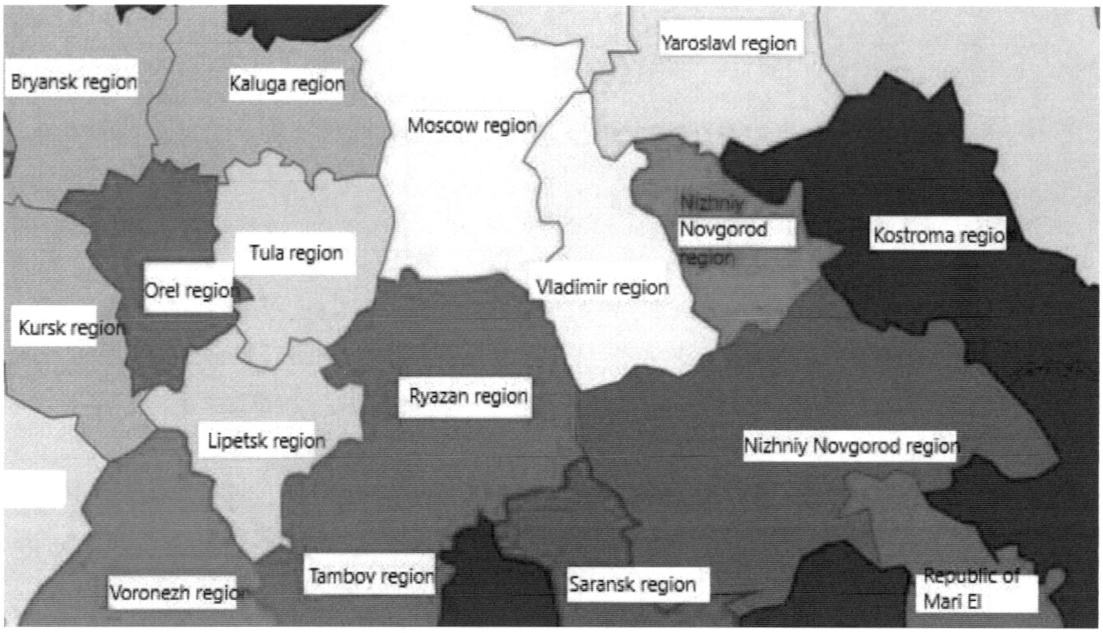

Figure 1. A cartogram of the authors' zoning of the Russian territory

The third zone includes Belgorod, Tver, Tula, Vologda, Volgograd, Irkutsk and Novosibirsk regions, Stavropol, Altai and Primorsky Krais, and the Khanty-Mansi Autonomous District. Among these regions, Tver and Tula ones were referred to macro-region 1, and Novosibirsk region and Altai Krai - to macro-region 11. All other regions belong to different macro-regions, meanwhile the land and property spheres of these regions have much in common: the average Russian amount of land taxes and rental charges; low land market activity; the share of industrial, agricultural and residential land areas is above the average Russian; a high share of state owned land; a high degree of efficiency of the public land control.

None of the zones contains more than two regions falling into the same strategic macro-region; in zones 10 and 14, all regions belong to different macro-regions.

4. Conclusion

The results allow us for conclusions that characteristics of the land and property complex were not taken into account when developing the draft Spatial Development Strategy of Russia and regional strategies. The zoning of Russia carried out by the authors on the basis of land and property characteristics opens up broad opportunities for land management.

5. Acknowledgments

The study was funded by the RFBR as part of the research project № 18-010-01016)

References

[1] Varlamov A, Galchenko S and Bogdanova O 2018 Theoretical and methodological foundations for formation of sustainable land management system *IOP Conf. ser. Mater. Sci. and Engineer.* **451(1)** 012141
[2] Lipsky S A 2016 Land management as a condition for replacing the institute of land categories with territorial zoning *Property relations in the Russian Federation* [in Russian – Imuschestvennye otnosheniya v Rossiyskoy Federatsii] **11** 18–23
[3] Bondar P A and Cooper P A 2017 The transition from land division into categories to territorial zoning: pros and cons *Agrarian and land law* [in Russian – Agrarnoe i zemelnoe pravo] **10** 18–21
[4] Chursin A I, Tikhonov E A and Kosmatova A O 2015 Analysis and prospects of transition to territorial zoning *Int. J. of Applied and Basic Research* **8–1** 75–7
[5] Samofalova L P, Larionov A O, Bocharov M N and Chuprin D A 2016 Territorial zoning of land *Advances in modern science and education* [in Russian – Uspekhi sovremennoy nauki i obrazovaniya] **7(12)** 71-73
[6] Klyushnichenko V N, Ilinykh A L and Kiseleva A O 2017 Consequences of abandonment of targeted land categories *Interexpo Geo-Siberia* **3–2** 152–5
[7] *Urban Stormwater Management in the United States* 2009 (National Academy of Sciences)
[8] Hirt S A 2014 *Zoned in the USA: the origins and implications of American land-use regulation* (Cornell University Press)
[9] Stephani Carl J and Marilyn C 2012 *Zoning 101* 3rd ed (National League of Cities)
[10] Saini Harshita *Zoning of Delhi* Available at: http://www.academia.edu/8812223/Zoning
[11] Antropov D V and Komarov S I 2018 Analysis of the effectiveness of land management in the region based on the method of integrated (cluster) zoning of territories (on the example of agricultural land) *Int. Agricultural J.* **5** 16–9
[12] Komarov S I 2017 Strategic regional planning in the field of land and property relations *Property Relations in the Russian Federation* [in Russian – Imuschestvennye otnosheniya v Rossiyskoy Federatsii] **7** 82–98

IOP Conf. Series: Earth and Environmental Science **350** (2019) 012061 doi:10.1088/1755-1315/350/1/012061

Ecological and biological features of adaptation of deciduous plants to chemical stress

G E Larina[1,2], L G Seraya[1], N N Polyakova[2], E N Baranova[3] and O O Beloshapkina[4]

[1]State University of Land Management, 15, Kazakov st., Moscow, 105064, Russia
[2]All-Russian Research Institute of Phytopathology, 5, Institute st., Bolshie Vyzemy, Moscow Region, 143050, Russia
[3]All-Russian Research Institute of Agricultural Biotechnology, 42, Timiryazevskaya st., Moscow, 127550, Russia
[4]Russian State Agrarian University - Moscow Agricultural Academy named after K.A. Timiryazev, 49, Timiryazevskaya st., Moscow, 127550, Russia

E-mail: galara7@yandex.ru

Abstract. In the urbanized environment, young plants lack moisture due to the limited space. This violates photosynthesis and respiration. To normalize respiration, the plant reduces excessive evaporation and evens out the water balance showing signs of xeromorphy. For the human environment, decorative and appearance of greenery are crucial. The purpose of the article is to study adaptive abilities of deciduous plants when using glyphosate-based herbicidal preparations, 2,4-D, dicamba, triasulfuron, and clopyralid. Anatomical and morphological changes at the cellular and organismic levels in young Tilia cordata Mill plants were identified: the size and density of stomata; frequency of stop cells of stomata; plant height; the number of shoots. The use of glyphosate, 2,4-D, dicamba, triasulfuron, clopyralid stimulated the height of shoots and increased the frequency of occurrence of guard cells with a length “0.15-0.19” mm. When using herbicides as a factor of chemical stress, the number of stomata per 1 mm^2 of the leaf surface increased. Ecological and biological plasticity of *Tilia cordata* Mill under environmental pollution (chemical stress) was identified. It is useful for directed formation of xeromorphic signs in young plants and further planting in urbanized conditions.

1. Introduction

Small-leaved linden young plants are an excellent model of woody plants resistant to urbanized conditions. They have unique gas and smoke resistance, shade tolerance, frost resistance and durability. They are promising plants for biomonitoring and remediation of anthropogenic and man-made territories [1, 2]. Heart-shaped linden, or small-leaved linden (Tilia cordata Mill.) is widespread in Europe and Western Asia.

During a period of leaf growth and development, the stomatal apparatus forms. Its state characterizes the response of the plant to changes in environmental conditions [3]. The size and density of the stomata are associated with the regulation of transpiration which ensures the maintenance of turgor and thermoregulation, the exchange of air masses and carbon dioxide absorption, carbon assimilation during photosynthesis [4]. According to researchers, xeromorphic signs are characterized by small adaxial and abaxial epidermis cells; an increased number and a decreasing size of stomata; a reduced stomatal conduction in the movement of water vapor through the stomata [5]. An increase in the number of stomata is considered as a method of adapting a plant to unfavorable abiotic factors [4]. Therefore, observations of xeromorphic signs are required to regulate the water regime and gas exchange of plants under aggressive chemical stress (use of chemicals).

The purpose of the article is to study adaptive abilities of small-leaved linden plants under chemical stress (use of chemicals).

2. Materials and methods

The experiment was carried out in. Bogorodskoe, Vladimir region. The soil of the experimental plot was medium-loamy, sod-podzolic, well-cultivated, with a pronounced lumpy-grainy structure characterized by a pH value equal to 6.1 (close to neutral); humus content was 2.9%.

The experiments were carried out on two-yearl old linden plants (T.*cordata*) in 2015-2016. In the experiment scheme, herbicidal preparations of different chemical nature were used: glyphosate (1.5 l / ha in ai); 2,4-D (0.9 l / ha for ai); dicamba dimethylamine salt, then dicamba (0.96 l / ha ai); triasulfuron (7.5 g / ha in the ai); clopyralid (0.06 l / ha for ai). Manual weeding followed by cultivation of the rows was used as a control variant. Location of the plots was randomized, the experiment was repeated four times. During the growing season, descriptions, photographic recording of plants and phenological observations were carried out.

Weather conditions were favorable for the growth of plants. The average annual temperature was + 14.1 ° C; the amount of precipitation was 51.6 mm; relative humidity was 71%. Years of observation had different aridity levels; the amount of precipitation was 2-3 times lower than the average annual data: 2015 - 19.2 mm, 2016 - 25.6 mm. In general, the first year (2015) was the warmest year in the history of meteorological observations (+ 2.4 ° C to the average annual rate). The period with the sum of temperatures above +10 ° C was 163 days in comparison with the average annual data (144 days).

Annually, 5–7 pieces of leaves were taken from the middle of the crown from each plant (repetition - 10 plants) which differed in maturity [6]. The anatomical and morphological structure of leaves was studied using imprints made from the surface of uncut leaves. The method for preparing a plant object to study the state of stomata involved production of a thin transparent film of stomata imprints (replicas) using polymer compounds in an organic solvent (nitrocellulose, polymethyl methacrylate in acetone) followed by removal of the imprint and studying it using a light biological microscope [7]. The imprint of the leaf surface was placed in a drop of water on a glass slide; the object was fixed with a cover glass. Microscopic examination was performed using a light biological microscope at a magnification of 40x (a combination of a 4x objective and a 10x eyepiece) and 800x (40x and 20x). The division value for the 4x objective and 10x eyepiece was 0.025 mm. The The number of repetions was 10. At a 80x magnification of the microscope, the field of view was 2.24 mm^2.

In the middle part of the leaf between the veins, the number of stomata was calculated in terms of 1 mm^2 of the surface of the leaf plate, and the length of the stomatal fissure formed by the stomata stop cells was measured. Using a digital camera, photos of the main (colorless) cells of the covering tissue of the leaf or the upper (adaxial) epidermis were produced. Using a mechanical odometer, we measured a contour of the area of a polygon shaped like five main cells of the epidermis lying side by side on the surface [5]. The area was calculated in mm^2 for one cell. The number of repetions was 10.

The experimental data were analyzed in Microsoft Excel (2013). Standard statistics data were calculated for each variational series: ($X \pm \sigma$) arithmetic mean and a standard deviation, a coefficient of variation (Kv). Correlation and regression anaysis was used to interpret the results. To establish the reliability of differences between the compared signs, the Student's t-test was used. When testing statistical hypotheses, a 5% level of significance or $P \leq 0.05$ was used.

3. The study of the structure of the modified lead-tin-base bronze

According to the averaged data for two years, a significant increase in the number of stomata per 1 mm^2 of the surface of the leaf plate was established when processing the studied herbicides from 34% to 73% relative to the control variant.

Adaptation of the stomatal apparatus. The maximum number of cells per unit area was determined for the option with dicamba (402 ± 141 pcs./mm^2), glyphosate (373 ± 24 pcs./mm^2), triasulfuron (365 ± 104 pcs./mm^2), 2,4-D (350 ± 39 pcs./mm^2) clopyralid (310 ± 75 pcs./mm^2).

Dicamba, 2,4-D and triasulfuron are drugs of systemic action against dicotyledons, including perennial plants, which include small-leaved linden. In the options with dicamba and triasulfuron, a sharp increase in dispersion values was observed. It is characterized by an increase in the average number of stomata due to the effective adaptation of linden plants in response to chemical stress.

According to the averaged data for two years, a significant increase in the number of stomata per 1 mm^2 of the leaf plate surface by 34-73% was established when using herbicides.

The comparative analysis of the area of the main cells of the adaxial epidermis identified deviations for all options (1.22-1.34 mm^2) with the exception of glyphosate (a decrease by 1.04 ± 0.8 mm^2). Glyphosate inhibits an enzyme used in the synthesis of aromatic amino acids tryptophan, tyrosine and phenylalanine. It is able to penetrate through into leaves and move throughout the plant to the growth points acting on the whole vegetative plant, including woody plants. Therefore, the inhibitory effect of glyphosate is manifested in a decrease in the area of adaxial side cells of the epidermis (from 20% and higher compared with the control option) which indicates signs of xeromorphism and is aimed at reducing excessive evaporation and loss of plant moisture. When uusing glyphosate, a decrease in the area of adaxial side cells of the epidermis is a reaction (adaptation) of linden plants to chemical stress.

Extensive experimental materials made it possible to analyze the frequency of occurrence of stop cells of stomata of various lengths. The data array was divided into dimension groups by the method of scaling: 0.05-0.09 mm; 0.10-0.14 mm; 0.15-0.19 mm with a measurement accuracy of 0.002 mm. Analysis of a sample of experimental data taking into account the chemical nature of the substance showed that in the options with glyphosate, 2,4-D, dicamba, there was an increase in the frequency of occurrence of stomatal closure cells with a size of "0.15-0.19 mm" by 40% compared to the control option, triasulfuron and clopyralid, where the frequency of occurrence did not exceed 30%.

Stomatal movements and the length of stomata cells are associated with redistribution of potassium ions between trailing and accompanying cells and production of abscisic acid (ABA). Potassium affects the physical state of colloids of plasma and cell walls, metabolism and the flow of water with solutes. But plant protection from drying by stopping stomata occurs at the early stages of water deficiency; this phytohormone, an antagonist of growth stimulants or auxins, enters the plant with water including dicamba and 2,4-D. An increase in the size of cells is due to an increase in evaporation or a decrease in the amount of available water entering from the inner parts of the plant [7, 8]. As an adaptation effect, an increase in the amount of bound water over the amount of free water occurs [8].

Table 1. Observation data on linden plants

Option, parameter	Winning for one plant (piece/plant) after using chemicals	
	single	double
control (without herbicides)	3.4-6.7	4.4-4.9
$X \pm \sigma$	5.0±1.6	4.6±0.3
Kv, %	32.7	5.4
clopyralid	2.6-7.7	5.4-6.4
2,4-D	3.7-5.1	3.8-5.9
triasulfuron	3.8-7.8	4.2-8.8
glyphosate	3.7-5.2	3.5-5.2
dicamba	3.8-8.6	3.8-7.1
$X \pm \sigma$	5.2±1.9	5.4±1.6
Kv, %	38.3	29.7

Phenological development under stress - spawn formation. The highest growth rates were observed in 2-4 year old plants. In the control option, the growth height was 74-89 cm. The stress factor (herbicide) influenced the dynamics of linden growth: regardless of the frequency of use, the height increased by 2-12% compared to the control plants. Analysis of the growth rate identified an increase in the growth rate when using herbicides by 20-51%; in the scheme of twofold application, the growth rate slowed down and amounted to 45-63of the control one.

Among the changes in response to chemical stress, the shoot growth was studied (Table 1). In all options, formation of shoots was 5 ± 1 pieces per plant and did not depend on the frequency of appalications of the preparations.

The maximum shoot formation (7-9 shoots/plant) was observed when using dicamba. The result is related to the chemical nature of the active ingredient. Dicamba stimulates the growth of younger plant tissues. In addition, dimethylamine salt of dicamba is distinguished by its high solubility in water (above 75 g/100 ml); sodium salt of dicamba and other herbicides have worse solubility (38 g/100 ml). Therefore, active formation of shoots is a stress reaction to moisture entering with a dissolved herbicide through the pvascular system stimulating the development of young tissues.

4. Conclusion

In general, the use of herbicides (glyphosate, 2,4-D, dicamba, triasulfuron, clopyralid) caused adaptive changes (reaction) in small-leaved linden plants. An increase in the number of stomata per 1 mm^2 of the leaf plate surface by 34-73% was observed. A significant decrease in the size (area) of the adaxial side of the epidermis when suing glyphosate (more than 20%) was determined. It characterizes the development of xeromorphism signs in young linden plants. It was suggested using the parameter "length of the stomata guard cells" as a response (adaptation) of linden plants to chemical stress. In young deciduous plants, an increase in the frequency of occurrence (40% and above) of guard cells with a length of "0.15–0.19 mm" (with a measurement accuracy of 0.002 mm) can be a signal of prolonged exposure to chemical stress.

In general, adaptive anatomical and morphological changes in young *Tilia cordata* plants can be assessed as an adaptation of linden plants to chemical stress. Signs of xeromorphy aimed at normalizing respiration, reducing excessive evaporation and equalizing water balance are indicators of this process. Thus, planting of linden in the conditions of environmental pollution (chemical stress) are promising in urbanized conditions characterized by the lack of moisture.

References

[1] Mozolevskaya E G, Sokolova E S, Kuzmichev E P, Belova N K and Kulikova E E 1996 *Scientific works of MGUL* **283** 37–64
[2] Gorelova S V and Frontasyeva M V 2017 *The Use of Higher Phytoremediation* (Springer: Cham.) pp 103–155
[3] Sellin A and Kupper P 2007 *European J. of forest research* **126(2)** 241–251
[4] Aasamaa K, Sõber A and Rahi M 2001 *Functional Plant Biology* **28(8)** 765–774
[5] Ziyatdinova K Z and Urazgildin R V 2013 *Bulletin of the Chelyabinsk state university* **7(298)** 181–184
[6] *Anatomical research methods of cultivated plants* 1989 ed G.I. Moskaleva (Leningrad: All-Union Academy of Agricultural Sciences, All-Russian Research Institute of Plants named after N I Vavilov)
[7] Tretyakov N N et al 1990 *A workshop on physiology of plants* (Moscow: Agropromizdat) pp 135–271
[8] Borzenkova R A and Hramtsova E V 2006 *The management to laboratory researches of a big special workshop on physiology and biochemistry of plants* (Yekaterinburg: USU publishing house) pp 14–26

Information support for sustainable development of coastal areas of the Far East

P P Lepekhin[1], V M Stolyarov[1], N A Ivanova[1] and S V Savinova[1]

State University of Land Use Planning, Kazakova str., 15, 105064, Moscow, Russia

E-mail: shampolamo@gmail.com

Abstract. The article considers problems and prospects of reducing the risks of occurrence of emergency situations of natural and technogenic character in the coastal river areas, as well as the usage for operational informing of the population of a comprehensive system that includes Federal, regional and local information centres. It is shown that the solution of the problem is based on the formation of information about emergency situations, their consequences, the state of radiation, chemical, medical, biological, explosive, fire and environmental safety in these areas, taking into account the coastal and transboundary situation to ensure their sustainable development. The authors defined the indicators and presented a strategy for collecting information and proposed a comprehensive system of support for management decisions aimed at preventing natural and man-made emergencies for the river coastal territory (RCT) of the Far East regions.

1. Introduction

In the era of globalization, everyone relies on information on the basis of global networks such as the Internet, the global information space. These trends are fundamentally changing the way in which information is provided to all governing bodies.

The authors believe that the implementation of the concept of sustainable development of coastal areas of the Far East, which required a change of the natural paradigm and reorientation to rapidly renewable resources, should be based on the use of new information technologies in order to develop and make management decisions, providing stakeholders with relevant information in the requested form on-line.

Evaluation of the intensity of the development of the coastal zone of China is given, for example, in the article Wei Xu, Yue-e Dong and others [1]. Problems and prospects for the development of the Far East are presented and analyzed in the article by Michael J. Bradshaw and Nicholas J. Lynn [2]. We rely on these and other sources when conducting our own research

2. Materials and methods

As the object of the study, the regions of the Far East, the situations that require an urgent solution to the problem of Informatization and updating of information flows for all stakeholders are considered. The solution of the problem is based on the use of highly scientific information technologies and expert systems with artificial intelligence to account for and preserve biodiversity and bioresources. The authors give proposals for solving the problems of sustainable nature management in coastal areas on the basis of the coastal-border-basin-territorial principle of information analysis and the creation of various predictive models of management solvers. They believe that the implementation of the concept of sustainable development of coastal areas of the Far East, which requires a change of the natural management paradigm and reorientation to rapidly renewable resources, should be based on the use of new information technologies to develop and make management decisions, provision of relevant information to interested parties in the requested form on-line.

3. Structure of information support of sustainable development of coastal areas taking into account their transboundary character

Information is essential for the optimal management of coastal areas. It is a limited and expensive resource of executive authorities, management bodies of various industrial structures, individuals and legal entities that are somehow connected with the nature of their activities with the territories located in the coastal, transboundary and border areas of water bodies.

Table 1 shows the relationship of information with the objectives and decisions based on the information support for the object of study.

Table 1. Connection of the nature of the information with the completeness of the decision

Information Status	Complete	Incomplete	
Decision	With certainty	With risk	With uncertainty
Objective expectations	Unequivocal	Quasi- Unequivocal	Ambiguous
Subjective sensations	Confidence	Uncertainty	

The use of the basin organization (in contrast to the administrative-territorial principle of information analysis) in information technology is the most promising, however, statistical information (data on physical-geographical, ecological-economic, natural-resource, labor, etc. components), as a rule, is given in administrative districts, which, on the one hand, greatly simplifies the calculations, but, on the other hand, becomes unacceptable for the river-border-transboundary territories.

Increasing demands for information efficiency in the management of coastal areas as economic and natural objects necessitates the creation of network technologies that are developed in accordance with the requirements of modern conditions for the functioning of management bodies. Management of natural processes should be based on reliable information about the past, present and future states of natural and anthropogenic subsystems, such systems are monitoring systems.

Information systems, as well as monitoring information systems, are an integral part of the management system, human interaction with the environment (environmental management system), since information on the current state of the environment and trends in its change should be the basis for the development of measures for the protection of nature and taken into account in planning the development of coastal areas. The results of the assessment of the existing and projected state of the coastal area, in turn, provide an opportunity to clarify the requirements for the subsystem of observations [3].

The problem of pollution of the Amur river appeared in the middle of the twentieth century. The accident at the chemical plant and the emissions of benzene in 2005 in the Songhua river (China) showed the absence of the modern and effective mechanism to address transboundary problems in the basin. It was the basis for the development of cooperation between the two countries in the field of environmental protection with the creation of joint mechanisms for the protection and use of natural resources in the coastal areas and river basin of the Amur river, the largest in the Far East.

The order of the Ministry of natural resources and ecology of the Russian Federation of October 8, 2014 adopted "Guidelines for the implementation of state monitoring of water bodies in terms of observations of the state of the bottom, banks, the state and mode of use of water protection zones and changes in the morphometric characteristics of water bodies or parts thereof". This order provides the composition of observed and measured parameters, frequency of observations and surveys, forms and methods of processing observations, which should be linked to the scale of the water body and the intensity of natural and anthropogenic rearrangements of riverbeds. For individual water bodies, the water situation on which has specific features, combinations of different methods are used in different sequences. The main results of remote sensing are space and aerial photographs of water protection zones, shoreline and water areas of water bodies, as well as materials of remote sensing data processing: transformed images, photographic plans and photographs with special thematic content and other materials.

Table 2. List of observed parameters and frequency of observations

The basis for monitoring the state of the river bottom, the banks, changes in the morphometric characteristics of coastal areas	Type of observation	Observation parameters	Frequency of observations
		River bottom, banks, morphometry	
Planned clearing of channels; landslides, erosion processes, etc.	The condition of the bottom of the water body	1. Characteristic channel forms 2. Change of bottom relief, ΔZ, m 3. Foreign objects at the bottom, location: a) type of object; b) size; c) potential danger; 4. Sediment (degree of sedimentation)	Once a year during the summer-autumn low flow period
	State of the coastline, deformation of the coast	4 Shoreline position 5 Shoreline change over the observation period, ΔX, m 6 The area of flooded areas ΔS, m^2, 7 Area of wetlands S, m^2 8 and its change ΔS, m^2	1 time in 5 years during the summer low water period
The basis for monitoring the state of the river bottom, the banks, changes in the morphometric characteristics of coastal areas	Type of observation	Observation parameters	Frequency of observations
		Sections of rivers where a narrowing of the river bed is observed due to Hydro Technical Facilities, bridge structures	
Threat of flooding	The condition of the bottom and banks of the water body	1. Foreign objects at the bottom, location: a) type of object; b) size; c) potential danger. 2. The presence of islands; 3. The area of plots under shrub and woody vegetation.	Once a year during the fall low water period
		Water protection zone, river coastal territory (authors' proposals)	
The threat of flooding; congestion of trees with roots pulled down by water	State of the river coastal area	a) civil engineering objects b) engineering infrastructure objects c) objects of cultural heritage d) protected areas	Once a year during the summer-autumn low flow period

Information is collected for: changes in the position of the shoreline; dynamics, causes of changes in the shoreline; consequences and potential danger of changes in the shoreline; the presence of foreign objects in the riverbed. The list of necessary parameters and frequency of their determination established by Methodical instructions are given in table 2.

Information gathering must be performed for the coastal basin of the transboundary territories. To do this, it is necessary to develop a strategy for collecting information corresponding to statistical data and innovative technologies, local capabilities, the required accuracy, and the result using the above guidelines.

The authors believe that the method of collecting information, its analysis and processing for the completeness of obtaining reliable and relevant information about the state of the object of study, should include the following actions:

1) analysis of initial (stock) materials on the state and use of land in riverside coastal-transboundary territories;

2) selection of the most informative indicators for monitoring river coastal border-transboundary territories based on a geo-ecological assessment of the territory and danger of natural and man-made regional risks;

3) determining methods for selecting monitoring indicators, developing approaches to collecting information in accordance with local conditions, using Earth remote sensing data (aerospace surveys, data obtained from unmanned aerial vehicles);

4) collecting information on additional indicators using modern innovative technologies for obtaining and processing information on the state of land and water resources; verification and analysis of all selected indicators for coastal area (CA) objects;

5) creation of a database of completed studies by combining and unifying all the materials received so that the results can be obtained promptly at the request of the portal user;

6) creation of a server for the primary processing of information on the basis of incoming requests from ordinary users of the portal, the creation of a CA geoportal and metadata portal, including modules for monitoring transboundary territories, coastal areas, shores, waters, engineering structures, bioresources, etc.

Based on the considered actions, the author proposes a comprehensive system of support for management decisions aimed at preventing natural and man-made emergencies for the river coastal territory (CA) of the Far East regions.

The main methods of CA data collection include: expert opinion, remote data collection, statistical data study, in-situ measurements, sampling of water levels in the rivers of the basin (on the basis of in-situ criteria, expert opinion of nature users and land users), processing, forecasting and modeling.

The development of river coastal areas depends on many factors, among which we can note such as: 1) currently, there is no developed program For the development of river coastal territory; 2) there is great competition between the various owners of natural resources located in this territory; 3) there is ambiguity in the formalization of the definition of the boundaries of the river coastal territory; 4) the discrepancy between the boundaries of administrative territories and natural landscapes, etc..

Based on this, there arises the need to develop, first of all, an information system allowing one to design the use of these natural and man-made coastal territories, taking into account natural and man-made regional risks in the system of sustainable development.

According to the authors, the creation of the coastal territory information system (CA IS) will allow one to ensure, in the necessary volume, all interested users with information on the status and use of these territories, represented by thematic cartographic materials, diagrams, graphs. This will allow information exchange with the Unified State Register of Real Estate, cadastres and registers of natural resources and information systems of other business entities [4].

Thus, CA IS is an integrated information system based on a set of reliable information about the state and dynamics of the use of the coastal area. In this connection, the precise selection and mapping of floodplain areas, the assessment of the main floodplain arrays, the study of the scale and degree of their possible degradation are of particular importance. On the basis of satellite images and topographic maps Egidarev EG studies of the effects of hydroelectric power stations on the adjacent territories in the Amur

River basin were conducted, and a high flood plain of the Amur and its major tributaries, which have a catchment area of over 10 thousand km^2, was identified. The total area of the floodplain-channel complex in the valleys of large watercourses of the Amur basin is 80 341 km^2 [5]. Taking into account the location of existing hydraulic facilities in the basin, it was found that in 2013 the transformation of natural flow to varying degrees occurred in the watercourses, forming half of all floodplain areas with a total area of 42 752 km^2 [6]. Modern methods of obtaining spatial information allow a high degree of accuracy and in a short time to conduct research and perform mapping and modeling on any territory of interest.

The formation of the information base of CA IS is based on materials monitoring the environment and the state of the land; on USRN data; cadastres and registers of natural resources; statistical data. Information exchange of relevant information between the copyright holders of such information on the state of the coastal territory should be carried out through the Unified Information Center, taking into account information on the status of transboundary territories. Therefore, it is necessary to obtain the necessary information to fully assess the condition of the coastal territory and predict its condition. Information in CA IS is formed on thematic blocks: legal, cartographic, economic, environmental, economic use; recreation; special use areas.

The source data in the CA IS are based on basic cartographic material, data on land users, users of natural resources, data on water, land, forest, mineral resources, biological resources, wildlife, engineering structures, specially protected areas, restrictions on use. Each database block (DB) consists of a coordinate grid description, including a complete set of information units. There are 11 such blocks in the IS, among the generally accepted blocks, special attention should be paid to the blocks "CA risks" and "Transboundary territories", which include data on regional natural and technological risks and data on international transboundary territories, playing a special role in the collection, systematization, data processing, assessment and forecasting of the state of man-made objects and natural coastal areas.

The block of risk description of the coastal territory includes data on the boundaries of the CA, taking into account areas with a special mode of land use, as reflected in the Unified State Register of Real Estate. The table "land resources" contains a comprehensive description of the soil cover of the CA in accordance with the accepted classification.

The cartographic basis of CA IS can be represented as a digital spatial model using 3D technologies. As sources of information for creating a digital terrain model, you should use topographic maps, materials of aerospace and photographic surveys, materials of topographic and geodetic works, data from river and sea bottom surveys, remote sensing materials, including those obtained from unmanned aerial vehicles (UAVs), presented in a single coordinate system.

4. Conclusion

The introduction of elements of CA IS at the present stage can be provided under the current state program of the Russian Federation "Information Society (2011 - 2020), approved by the decree of the Government of the Russian Federation of October 20, 2010 N 1815-p", which defines the main priorities of strategic development in the field of information and communication technologies.

CA IS will allow coordinating and directing the efforts of state authorities, regions of the Far East, local governments, economic entities and citizens to improve the informatization of coastal area management taking into account regional natural and man-made risks and the international transboundary characteristics of the Far East regions.

References

[1] Wei Xu, Yue-e Dong, Xin Teng and Pan-pan Zhang 2018 Evaluation of the development intensity of China's coastal area *Ocean & Coastal Management* **157** 124-129 Available at: https://doi.org/10.1016/j.ocecoaman.2018.02.022.

[2] Bradshaw M J and Lynn N J 1998 Resource-based development in the Russian Far East: Problems and prospects *Geoforum* **29(4)** 375-392 Available at: https://doi.org/10.1016/S0016-7185(98)00021-9.

[3] Murasheva A A 2006 *Theory and methods for the formation of the information mechanism of environmental management in the region (by the example of Far Eastern Federal District),* doctoral dissertation (Moscow)
[4] Lepekhin P P and Murasheva A A 2015 Information-modeling system for solving regional environmental problems *Earth Sciences* **1** 24-32
[5] Egidarev E G 2012 Mapping and assessment of floodplain complexes in the Amur River valley *Bulletin of the Far East Branch of the Russian Academy of Sciences* **2** 9–16
[6] Egidarev E G 2014 Evaluation of the environmental impact of placer gold mining in the Amur River basin *Geoecology, engineering geology, hydrogeology, geocryology* [in Russian – Geoecologiya, inzhenernaya geologiya, gidrogeologiya, geokriologiya] **5** 429-441

Sustainable development of the Amur river coastal areas

A A Murasheva[1], V N Khlystun[1], P P Lebedev[1], P P Lepekhin[1] and A V Vdovenko[2]

[1]State University of Land Use Planning, Kazakova str., 15, 105064, Moscow, Russia
[2]Pacific National University, 136, Tikhookeanskaya Str., office 241-center, Khabarovsk, 680035, Russia

E-mail: amur2@nln.ru

Abstract. The sustainable development of the coastal areas of the Amur River is considered as an integrated use of the natural-technical system consisting of a set of interrelated elements aimed at ensuring the state of socio-economic sustainability of the coastal areas of the rivers of the Far Eastern Federal District as a whole. The improvement of the integrated management mechanisms was considered on the basis of a comprehensive system analysis of the coastal areas of the Amur River. This analysis was carried out in several stages, using elements aimed at ensuring the sustainability of the development of the coastal areas of the rivers of the Amur River basin.

1. Introduction

The coastal territory of rivers is considered as a special kind of resources. It combines all types of natural resources, population, production capacity, cultural and intellectual potential.

The coastal territory, which includes natural and man-made resources, is characterized most often by superior man-made impacts on the natural possibilities of the territory in question. Sustainable development of coastal areas of rivers is a complex of measures aimed at organizing economic activities in accordance with the principles of sustainable development and subsequent integrated management. The sustainable development of the coastal areas of the Amur River, the characteristics of the exploitation of such lands, the level of anthropogenic and other resources at different times were considered by both Russian [1] and foreign authors [2]. The article presents a study in the development of criteria and indicators that most effectively monitor the achievement of sustainable development of the coastal area by the example of the Amur River.

2. Materials and methods

The object of the study is the territory of the Amur River and its tributaries. The rivers of the Amur basin are key transportation arteries, including those of international importance. In this regard, the Amur River plays a particularly important role. Medium and small rivers are widely used for timber rafting. The energy resources of the rivers in the region are large and amount to millions of kilowatts, whose energy is used at the largest hydroelectric power stations, electricity is supplied to the neighboring states - China, Mongolia.

Intensive economic development in the Amur river basin, in its coastal territory, both from the Russian Federation and from neighboring countries and, above all, China, require close attention to the environmental situation in the region, assessment, forecasting and development of management decisions that contribute to the sustainable development of the coastal areas of the Amur basin, as the most saturated from an economic point of view, concentration of significant population density and features of international transboundary character and high risk of floods and flooding of adjacent territories.

3. Indicators of sustainable development of coastal areas of rivers

Most countries include their policies for sustainable development of coastal areas in the programs of socio-economic and environmental policies aimed at achieving the sustainable development of these areas. The legislation of many states widely stimulates water conservation and environmental protection activities, taking into account the fact that payments for the destruction of elements of nature are much higher than the costs of preventive measures. The effectiveness of the functioning of the coastal territory of rivers is determined by many factors - socio-economic, technological, environmental and political.

The traditional functions of rivers and their coastal territories in the history and evolution of the development of settlements, as world experience shows, are: strategic (water - way, water - border, water - communication); defensive - natural boundary; communication - the river as a transport and trade artery; economic - the river as a factor in the formation of the economic structure; representative - coastal areas as the basis for the formation of the architectural appearance of cities, settlements; public - coastal areas as zones of economic and other activity.

Improvement and development of transport infrastructure in the coastal zone (construction of roads, bridges, and other communications) has become the dominant method in the transformation of coastal areas of rivers. With the development of economic activities in coastal areas, social and environmental functions have become increasingly important. With the growth of the ecological consciousness of society, the quality of the environment of the coastal territories began to be regarded as one of the most important resources for improving the quality of life.

The basis of a systematic approach to solving problems arising in the coastal area is the potential of coastal areas in aspects, among which an important role is played by: the environmental aspect — restoring and maintaining the natural balance; economic aspect - optimization, increase of efficiency of use of land resources, water area. Comprehensive assessment and implementation of the potential of coastal areas in the relationship functional, social, cultural, aesthetic, environmental, economic aspects should be a necessary part of the strategy for sustainable development of coastal areas.

New approaches to the formation and interaction of communication, public, recreational, cultural, representative, ecological functions of the coastal areas of rivers form the basis of the integrated organization of the infrastructure of coastal areas as a factor of their sustainable development and improvement of habitat quality. However, the unsustainable use of coastal territory resources and the lack of scientifically based management approaches often lead to negative consequences [3]: loss of economically valuable land resources; loss of land resources of high natural and aesthetic value; the disappearance of aquatic and terrestrial species of animals and plants; loss of objects of historical, cultural and archaeological value; air pollution and other.

Sustainable development of the coastal areas of the rivers of the Far Eastern Federal District is a set of measures aimed at organizing economic activities in accordance with the principles of sustainable development and their subsequent integrated management based on the definition of a set of indicators of sustainable development of coastal areas taking into account regional peculiarities.

World experience in the development of sustainable development indicators shows that there are two approaches to their construction: 1) building a system of indicators, each of which reflects specific aspects of sustainable development. Most often, within the framework of the overall system, the following subsystems of indicators are distinguished: environmental; economic; social; institutional 2) the construction of an integral, aggregated indicator, on the basis of which one can judge the degree of sustainability of socio-economic development. Aggregation is usually carried out on the basis of three groups of indicators: environmental-economic; ecological and socio-economic; environmental.

We consider it necessary to use the indicator "capacity of the territory" and "ecological condition" as the main indicator of sustainable development. Capacity ensuring its sustainable development, optimal loads due to natural-ecological, demographic and anthropogenic factors. The proposed indicators require analysis and their selection to adapt to specific conditions and, possible expansion, clarification, which is based on a political decision.

Sustainable development involves targeting indicators to the upper and lower limits. However, this choice is conditional. The loss of controllability of socio-economic and environmental processes, the situation is beyond the lower limit and belongs to the disaster area. Sustainable development of the coastal area is characterized by the efficiency of its use, environmental stability and social security, and demographic integrity, i.e. the ratio of the population of the territory and its possibilities are in balance. Thus, the indicators selected as indicators of sustainable development should be considered in dynamics and have a time reference. The types of indicators and their scope of application are presented in Figure 1 [4].

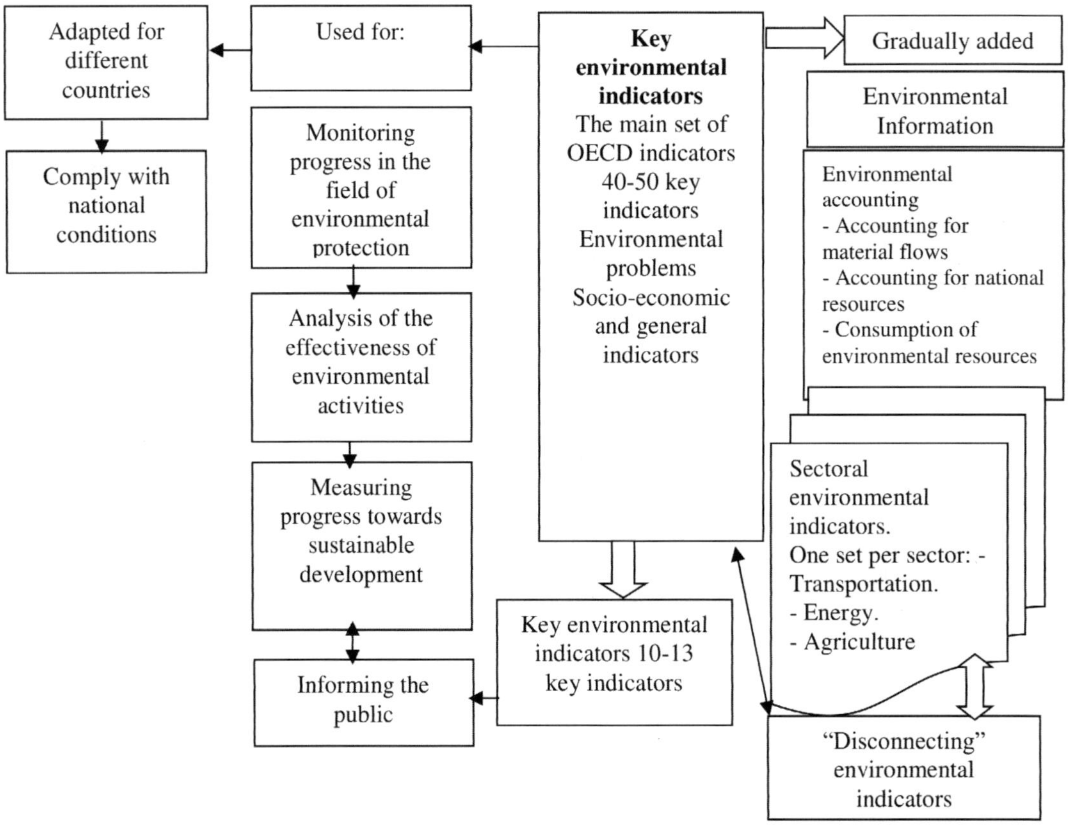

Figure 1. Types and areas of application of indicators concerning the organization of economic cooperation and development (OECD).

Currently, environmental indicators are the dominant indicator in the sustainable development of coastal areas, since the strengthening of human activities on the sustainable development of coastal areas is becoming increasingly important. The main environmental indicators of sustainable development are considered to include: 1. The volume of emissions of harmful substances into the atmosphere, %. 2. The volume of discharges of harmful substances into water bodies, into the urban sewage system, %. 3. The share of formation and removal of toxic waste production activities, %. 4. The volume of discharge without treatment of polluted and toxic water, %. 5. A share (degree) of danger of technological accidents and accidents, %. 6. The volume of costs for the normalization of the ecological situation and the restoration of disturbed natural complexes, %. The World Bank indicators from the short "green" reference book for the Russian Federation include the following indicators (Table 1) [5].

Table 1. Indicators for the Russian Federation from the World Bank's short "green" directory [5]

	Indicator	Subindicator
Atmosphere	Changing of the climate	1. CO_2 emissions from fuel consumption (Roshydromet data) 2. Greenhouse gas emissions 3. Concentration of priority air pollutants in urban areas (Roshydromet data)
	Air quality	4. Emission of harmful substances, total and hazard classes
Land	Agriculture	5. Agricultural land 6. Mineral fertilizer use 7. Pesticide use
	Forest	8. Forest area,% of total land area 9. Forest area by category 10. The intensity of deforestation (using the calculated cutting area)
	Land Desertification	11. Deserted lands (regional estimates, one-time estimates)
	Urbanization	12. Lands of settlements 13. Lands of industry, transport and other non-agricultural purposes
Fresh water	Fishing	14. Annual catch of major species in main basins
	Amount of water	15. Annual intake of groundwater and surface water, in % of the total available water reserves 16. The volume of circulating and consistently used water as a percentage of water withdrawal from water sources
	Water quality	17. Discharge of polluting wastewater to surface water 18. Discharge of pollutants into surface waters
Biodiversity	Ecosystems	19. Lands of specially protected natural territories 20. Protected areas
	Species	21. Availability of main selected species (one-time estimates)
Economic structure	Economic indicators / results	22. GDP per capita 23. Share of investment in GDP 24. The coefficient of renewal of fixed capital 25. Labor productivity 26. «True savings» (assessment)
	Trade	27. Trade balance in goods and services
	Financial position	28. Share of debt in GNP 29. Inflation rate

Table 1. Continuation

	Indicator		Subindicator
Consumption and production patterns	Consumption of materials		30. Intensity of use of materials (material intensity)
	Energy use		31. Annual energy consumption per capita
			32. Share of renewable energy
			33. Energy use intensity (energy intensity)
	Waste generation and waste management		34. Toxic waste generation (by class)
			35. Use and disposal of toxic waste
	Transport		36. Number of passenger cars per 1000 population
	Population welfare	Employment	37. Unemployment rate
			38. Income differentiation ratio
		Revenue sharing	39. Population with money incomes below the subsistence minimum
		Housing	40. Housing security of the population
			41. The proportion of the number of families registered for housing
		Recreation	42. Children's health institutions, rest homes, health resorts

The use of only the indicators listed in Table 1 does not allow one to sufficiently assess the impact of the activities of economic entities on the sustainable development of coastal areas. In this regard, the authors expanded the classification of environmental indicators of sustainable development of coastal transboundary territories, proposed new environmental indicators:

1. The share of pollution of the water environment by economic objects in the total amount of pollution, %.
2. The share of participation of economic entities in environmental measures to reduce the negative impact on the water environment in relation to the total amount of work carried out or projects implemented, %.
3. The share of damage caused by objects of management - pollutants of the water environment, which can be avoided by implementing investment and environmental projects, participating in environmental activities, %.

As a result of the analysis of statistical materials, reporting materials and recommendations of Federal, regional and international environmental organizations, for the rivers of the Amur basin and the Sakhalin rivers, environmental indicators were selected that reflect the pressure, impact, state of natural and man-made subsystems, which include [6]: seizure and violation of land; violation of the hydrological regime of groundwater and surface water; pollution of water resources with various wastewater; pollution of land with mining waste. In addition, the ecosystem of the Amur river basin can be significantly damaged by the installation of flood control facilities, while there may be: land degradation caused by changes in soil processes; reduction of spawning areas, which adversely affects the fish productivity of rivers; salinization of land; lowering water levels in the river after channel regulation leads to a decrease in the level of groundwater.

4. Conclusion

Thus, the considered mechanisms for the development of coastal areas of the Amur river, taking into account the transboundary nature and a set of recommendations for the selection of sustainable development indicators, can become the main basis for the sustainable development of coastal areas on the basis of ecology, protection of shores and water bodies from destruction and pollution, on the basis of innovation, competitiveness, social responsibility.

References

[1] Kim V 2012 Anthropogenic actions to the hydrological regime of the Amur river *Quaternary International* **279–280** 244-245 Available at: https://doi.org/10.1016/j.quaint.2012.08.571.

[2] Lan-lan Yu, Zi-qiang Xia, Jing-ku Li and Tao Cai 2013 Climate change characteristics of Amur River *Water Science and Engineering* **6(2)** 131-144 Available at: https://doi.org/10.3882/j.issn.1674-2370.2013.02.002.

[3] *Distance learning programme for integrated coastal zone management "CoastLearn"* Available at: http://www.biodiversity.ru/coastlearn/iczm_rus/urgencies.html

[4] Tarasova N P and Kruchina E B 2006 Indices and indicators of sustainable development, in *Sustainable development: Nature – Society – Human* [in Russian – Ustoichivoye razvitie: Priroda – Obschestvo – Chelovek] vol 2 (Moscow) Available at: http://www.cawater-info.net/ecoindicators/pdf/tarasova_kruchina.pdf

[5] Bobylev S N and Makeyenko P A, Eds 2001 *Indicators of sustainable development of Russia (Ecological-economic aspects)* (Moscow: TSPRP)

[6] The official website of the Amur Basin Water Management, Available at: http://www.amurbvu.ru/

[7] Murasheva A A, Vdovenko A V, Stolyarov V M and Lepekhin P P 2017 Economic justification of the effectiveness of measures aimed at preventing the negative impact of water in river coastal areas (by the example of Khabarovsk) *Moscow Economic Journal* **4**

Typification of coastal transboundary areas in the context of sustainable land use in the Far Eastern Federal District

A.A. Murasheva, P P Lepekhin, E M Chepurin and E P Ananicheva

State University of Land Use Planning, Kazakova str., 15, 105064, Moscow, Russia

E-mail: amur2@nln.ru

Abstract. The article presents a number of proposals on the typification of the transboundary territories of the Far East of Russia, based on the peculiarities of geographic location and access to natural borders. A specialized designation of such regions has been proposed and a number of indicators have been developed that allow a comprehensive assessment of the ecological status of transboundary territories.

1. Introduction

Land use, forest management, the general economic and environmental situation near the common border between the Russian Federation and the People's Republic of China are characterized by high intensity of use. At the same time, it is necessary to note such an important feature as a general imbalance in the development of the territories of Russia and China. Heilongjiang Province, which borders on the Trans-Baikal Territory, the Jewish Autonomous Region, the Amur Region, the Khabarovsk Territory and the Primorsky Territory, is characterized by a higher degree of urbanization, the level of development of the territory and the man-made load. Based on the above, we note that these features and differences should be considered as specific transboundary gradients and should be taken into account when conducting environmental, economic, social research in the specified region. The article considers some proposals on the typification of transboundary territories for use in conducting a comprehensive assessment of the ecological status of these territories. Article [1] is devoted to the analysis of the development of transport infrastructure in Russia and the Far East, without which we should not talk about the typification of territories in general. An analysis of the patterns of development of transboundary territories, as well as the organizational structure of management of transboundary territories, is presented in article [2].

2. Materials and methods

Immediately before the main part of this article, we will establish the peculiarities of development and cooperation within the border area of the Far Eastern Federal District:

1. The nature of relations between Russia and China is significantly different from the nature of relations between countries in Western Europe, which coordinate joint actions in the framework of such a geopolitical entity as the European Union.

2. When conducting environmental studies, the territory of a single state, either Russia or China, should not be subject to analysis. The environmental situation in the border areas is interrelated and interdependent, and therefore must be viewed through the prism of a single cross-border area.

3. The specificity of the transboundary territories, which we mentioned above, is determined, among

IOP Conf. Series: Earth and Environmental Science **350** (2019) 012064 doi:10.1088/1755-1315/350/1/012064

other things, by a complex of natural and economic factors.

4. Potential environmental or economic problems in the region should be solved by the parties jointly, despite the high difference in the types of economic development and the scale of environmental transformation.

Based on the foregoing, the proposals on the typification of the Far Eastern transboundary territories were developed on the basis of the analysis of economic activity, the degree of natural environment, the degree of man-made and anthropogenic loads.

3. Proposals for the typification of the coastal transboundary territory in the far Eastern Federal District

As noted above, the specific transboundary gradients that characterize environmental, economic, social and other conditions are the first indicators. These include: the area of the region, population density, gross regional product, availability of acreage and others. It should be noted that in Heilongjiang province these indicators are significantly higher than in the southern regions of the far Eastern Federal District. Nevertheless, the area of forests in the FEFD exceeds the same indicator of Heilongjiang province according to expert estimates more than 4 times, but the production of wood per 1 hectare of area in China is about 3.5 times more. One resident of the far Eastern Federal district has 10 times more forests, 25 times more protected areas. However, the main indicators of the advantage on the side of the Chinese province. This indicates a higher intensity of resource use and more significant interference in the geo-and ecosystem of the region and can not affect the state of the Russian territories.

Russian researchers P.Ya. Baklanov and S.S. Gansei believes that the most important methodological issue that is resolved when studying and assessing the state of a territory is the question of the principles and approaches to its "division" [3].

Another approach, developed by B. Van der Velde and R. Martin [4], identifies border regions based on ethno-cultural and economic interaction, for example: alienated border regions - in which there are no links due to military actions, religious, cultural, ethnic differences; coexisting border regions - there is some economic and cultural interaction between them; interdependent border regions - there is significant interaction between them; integrated border regions - characterized by free movement of population, capital, goods. Researcher V.S. Korneevets proposes an approach in which we determined: 1) core regions; 2) transitional regions with subtypes: developing, stable, crisis; 3) peripheral resource regions with subtypes: developing, stable, crisis; 4) development corridors; 5) regions of developed cross-border cooperation [5].

As part of our study, we consider it appropriate to offer a typology of border regions according to the type of borders to which these regions go. Table 1 below shows the typification of the regions and introduced an additional type "T", which indicates access to all types of borders - land, sea, river.

One of the existing classifications of state boundaries divides them into natural and social. The first are including hydrographic. The second are borders that are the result of human activity, in particular the delimitation and demarcation of them on the ground. Thus, the bipro-border and "T"- border regions have a pronounced advantage, which stands out in the form of the presence of natural and artificial borders simultaneously. In view of this, the probability of development of cross-border relations and interactions expressed in cross-border trade and tourism significantly increases. After that, border infrastructure will inevitably be improved.

Table 1. Typification of border regions according to the type of borders to which they reach

Type of border area	Characteristic
Land border	Regions have access only to land state borders.
Marine Border	Regions have access only to maritime state borders.
Bipro-border	*Regions have access to both land and maritime state borders, or land and river state borders.*
"T" border	**Regions have access to land and sea and river state borders**

The border regions of the Amur basin of the Far Eastern Federal District have high potential opportunities for active foreign economic activity. In turn, this causes their high integration and investment potentials. The presence of such a classification sign as bipro-border and "T" border should increase the integration potential of the region. In the Amur basin, the most promising regions are the Amur region, the Khabarovsk Territory, and the Primorsky Territory. The main areas of integration and cooperation in these regions with the neighboring state are represented by such sectors as forestry and energy. Table 2 shows the typification of the regions of the Far East according to the type of borders.

Table 2. Typification of border regions of the Far East

The subject of the Russian Federation	Land border	Marine Border	River border	Bipro-border	"T" border
Amur region	+		+	+	
Jewish Autonomous Region	+		+	+	
Primorsky Krai	+	+	+		+
Sakhalin region		+			
Khabarovsk region	+		+	+	

For a number of previous years and today, cooperation and implementation of programs in the field of vegetable production (Primorsky Territory, Amur Region), housing construction (Primorsky Territory), industrial construction (Khabarovsk Territory), joint fishing and fish processing (Primorsky Territory) continue in border regions. We believe that the development of multidimensional cooperation with the APEC countries can reduce tensions in the field of demography, migration, and ecology in the studied region. At a minimum, infrastructure development as a result of cooperation should become one of the benchmarks, due to which catastrophic population decline (natural and migration) will not be observed, but, on the contrary, will create the necessary socio-economic conditions conducive to the sustainable development of the Russian Far East.

The ecological situation of the region under study is significantly influenced by anthropogenic factors, which include industrial, civil, transport construction, agricultural production, the construction of hydrotechnical and transport facilities, mining, and a number of other.

Let us describe the indicators that allow us to draw conclusions about the typification of transboundary territories and their environmental status today. The analysis of these indicators is made on the example of the coastal areas of the Amur river and, we believe, most fully reflects their current environmental condition.

Indicator 1. *Air emissions from stationary and mobile sources of pollution.* The main sources of air pollution in the river coastal areas of the Jewish Autonomous Region are road transport, stationary sources of the manufacturing industry. A significant proportion of emissions from the production and distribution of gas, electricity and water. Air emissions are due to the development of the Amur mining and smelting cluster. In addition, one can single out a contribution to the pollution of the agro-industrial development zone with a specialization in the cultivation and processing of soybean, the activity of the Teploozersk cement plant, and fish-breeding plants. The mining cluster is concentrated in the economically developed part of the Obluchensky district [6].

The air basin of the Jewish Autonomous Region is experiencing anthropogenic pressures not only from regional sources, but also due to the transboundary transport of pollutants from China. Therefore, an important role in shaping the level of air pollution in the transboundary coastal territory belongs to the processes of transboundary transport of suspended particles and pollutants from adjacent territories. At present, the modern monitoring system for transboundary transport is not developed at the proper level and development of measures for its improvement is required.

Indicator 2. *The volume of wastewater discharged into surface water bodies.* The main polluters of the water basin of the coastal areas of the Jewish Autonomous Region are objects of various types of economic activity in the region, among which industrial and domestic wastewaters of large populated areas, agriculture, forestry and mining have a significant impact. The most important factor affecting the quality of water in the rivers of the region for a long time remains the transboundary transfer of pollutants from the territory of the PRC (the river Sungari) [7].

A significant impact on coastal river ecosystems has their intensive development, which leads to an increase and, consequently, an increase in pollution discharges. The urbanization centers of Birobidzhan, Amurzet, Leninsk and Nizhneleninskoe villages have the most adverse environmental impact on the aquatic ecosystem, since these settlements have the largest amount of raw water discharges per inhabitant.

Indicator 3. *The danger of endogenous and exogenous processes.* As a result of the assessment of the risk of manifestation of endogenous processes (earthquakes), it was established that the territory under consideration is characterized by a very high degree of danger of their manifestations (Table 3).

Table 3. Evaluation of the risk of endogenous processes

Hydrographic unit to which the catchment area belongs	Maximum seismicity of the area (points on the MSK-64 scale)	Degree of danger	Evaluation score
Amur river from the mouth of the r. Bureya to the mouth of the r.	10,0	very high	10
Total			10
Number of indicators			1

The results of determining the degree of danger of exogenous manifestations (EM) in the territory under consideration are summarized in Table 4, from which it can be seen that the most dangerous types of exogenous geological processes here are channel processes (coastal processing) and flooding.

Indicator 4. *The nature of the hydrochemical state of the river.* Within the framework of the joint Russian-Chinese and Russian state monitoring of the quality of transboundary water bodies in the territory of the Jewish Autonomous Region, since 2007, hydrochemical observations have been carried out in the Pashkovo, Amurzet, Nizhneleninskoe, and Nizhnespasskoye settlements. In addition, hydrological observations are conducted on the left tributaries of the Amur: B. Bir, Bijan, Samara, Arkhara and Khingan [8]. Observations of the quality characteristics of bottom sediments are practically not conducted.

Table 4. Estimation of the risk of exogenous geological processes

The main types of EM characteristic of the territory in question	The degree of danger of manifestations of EM	Evaluation score
landslide	Moderate	4
mudslide	Moderate	4
karst	practically absent	1
submergence	Moderate	4
gully erosion	Moderate	4
shore processing (channel processes)	high	7
cryogenic	Moderate	4
flooding	very high	10
Total		38
Number of indicators		8

The qualitative composition of the waters of the Amur River is formed by the influx of substances from the waters of its major tributaries - the Zeya and Bureya rivers in the territory of the Russian Federation and the Sungari river in the territory of the PRC. In this regard, starting from 2007, Russian and Chinese experts have been conducting joint monitoring of the quality of water in the r. Amur. The waters of the Amur River are characterized mainly as "dirty", quality class 4 "a".

The obtained criteria, which characterize the level of pollution of bottom sediments and soils, the danger of manifestation of endogenous and exogenous processes, made it possible to carry out an integrated assessment of the ecological state of the territory under consideration (Table 5).

Table 5. Integral assessment of the ecological state of the basin

Total number of indicators	The total estimated points	Average rating score	Ecological condition corresponding to the	Ecological state based on the results of the assessment of the level of pollution of bottom sediments and soils
13	70	5,38	Unfavorable	Very unfavorable

The average score, calculated taking into account all available indicators, allows one to estimate the ecological state of the research area as "unfavorable". However, the ecological status of this territory should be assessed as "very unfavorable" regardless of the results of the assessment on the average score due to the fact that for certain areas of the territory and the riverbed of the Amur river, in accordance with the information of the Amur basin, an "extremely dangerous" degree of chemical pollution of soils and bottom sediments was revealed [8].

Indicator 5. *State of the nature reserve fund of coastal areas.* According to the State report "On the state and environmental protection of the Russian Federation in 2017" in the Russian Federation there were about 12 thousand protected areas of Federal, regional, local significance with a total area of 232.7 million hectares. the total area of protected areas from 2010 to 2018 increased by 25.4 thousand hectares. the far Eastern Federal District is characterized as the region with the largest share of protected areas in its territory. According to the report, protected areas occupy up to 58 % of the total area of the region, and the total area is equivalent to 123 258.8 thousand hectares [9].

Indicator 6. *Flood risk level of river coastal territory.* Assesses the risk of occurrence of one of the most serious regional hydrological risks in the far Eastern region – flooding, which clearly demonstrated the highest and longest for the last 150 years summer-autumn flood in the Amur river basin in the summer of 2013.

This indicator is one of the most significant, and under certain conditions (the passage of catastrophic floods) can be the main and only indicator characterizing the state of river coastal areas.

Indicator 7. *Frequency of occurrence of tree halls.* This phenomenon in the basin of the river. Cupid is not well understood, there are only brief descriptions of it. One of the leading scientists of the Far Eastern Branch of the Russian Academy of Sciences, AN. Makhinov notes that logging and fires cause an increase in the uneven flow of water in the river, which leads to the intensification of channel re-formations, erosion of the banks and even more intensive influx of trees into the river beds with intensive formation of creases. For the considered section of the river transboundary territory, the occurrence of wood cleats is estimated as average [10].

4. Conclusion

We believe that on the basis of the above proposals on the typification of a coastal transboundary territory, as well as indicators that form the basis for studying the ecological status of territories to a greater extent, a model of an integrated approach to the study of transboundary territories of the Russian Federation can be developed that can ensure their sustainable development condition and respond flexibly to technological, environmental or economic challenges and threats.

References

[1] Pugachev I, Kulikov Y and Yarmolinsky A 2018 Current trends in development of the transport infrastructure of large cities of the Far East, Russia *Transportation Research Procedia* **36** 622-626 Available at: https://doi.org/10.1016/j.trpro.2018.12.144.

[2] Shiuh-Shen Chien and Dong-Li Hong 2018 River leaders in China: Party-state hierarchy and transboundary governance *Political Geography* **62** 58-67 Available at: https://doi.org/10.1016/j.polgeo.2017.10.001.
[3] Baklanov P Ya and Hansei S S 2004 Concepts and types of border and transboundary territories *Izvestiya of Russian Academy of Sciences. Geography Series* **4** 27–34
[4] Van der Velde V and Martin R 1997 So Many Regions, so Many Borders. A Behavioural Approach in the Analysis of Border Effects *37th European Congress of the European Regional Science Association* Rome (Italy), August 26-29, 1997
[5] Korneevets V S 2010 *Transnational and transboundary regions as specific forms of the territorial organization of society,* doctoral dissertation (St Petersburg)
[6] State report "State and environmental protection in the subjects of the Russian Federation," available at: http://ecogosdoklad.ru/
[7] Makhinov A N and Shuguang L 2013 *Formation of the relief of river beds and river banks* (Khabarovsk: FEB RAS)
[8] Vdovenko A V and Kim A V 2015 On the issue of zoning and regulation of economic activities in coastal areas *Bulletin of the Pacific State University* **1** 51–57
[9] Murasheva A A and Vdovenko A V 2016 Optimization of the use of the coastal area taking into account the transboundary features of the basin. Amur *Electronic scientific publication "Uchenye Znakie PNU"* **7(3)** 81-85 No. 383 in the list of HAC (2015-2018)
[10] Bernard F 2013 *Geography and Tourism* (Kazimierz Wielki University) **1** 49–56

Digital land management

T V Papaskiri[1], A E Kasyanov[2], N N Alekseenko[3], V N Semochkin[1], E P Ananicheva[1] and A A Shevchuk[1]

[1]State University of Land Management, 15, Kazakova str., Moscow, 105064, Russia
[2]Moscow Agricultural Academy named after K.A. Timiryazev, 49, Timiryazevskaya st., Moscow, 127550, Russia
[3]Basic Department of Digital Land Management, State University of Land Management, 15, Kazakova str., Moscow, 105064, Russia

E-mail: t_papaskiri@mail.ru

Abstract. The article deals with the issues of modernization of land management as a branch of the economy. Due to the digitalization of the economy, modernization is aimed at creating a digital land management system. Restructuring and digitalization of land management can create annual economic effects in all industries dependent on land management.

1. Introduction

Implementation of the federal target program “Digital Economy of the Russian Federation” approved by Decree of the Government of the Russian Federation No. 1632-r of July 28, 2017 is the main task of all sectors of the national economy. Therefore, a separate program for digitalization of agriculture is one of the priorities in the context of economic sanctions [1]

The Russian Federation needs a new concept of land policy and land management; should improve a paradigm of land and property regulation, update a land management system based on a new land management system supported by the state. This approach will avoid economic losses from irrational use of land resource [2, 6]. This was emphasized in [2–7, 9].

It is necessary to create digital land management capable to update the national land management system and land policy through upgrading the spatial data infrastructure.

2. Materials and methods

The SWOT analysis revealed a direct influence of the land management sector on the national economy through a connecting link - the basic dependence of most economic sectors on the land resource which is their spatial reference.

Results of the poor land policy manifested themselves in land reserves suitable for agricultural purposes (Table 1).

The development of these lands is a major national economic task requiring large financial expenditures. It is necessary to attract specialists in the field of land management; however, there is no full-fledged organizational structure capable of implementing this task. Therefore, it is necessary to create a centralized structure capable of performing the full range of land management tasks.

Table 1. Reserves of lands suitable for agricultural use or more intensive use, million hectares

No	Name	Areas (million hectares)
1.	unused:	
	arable land	20.0
	forage land	34.5
2.	unclaimed plots of arable land	16.6
3.	redistribution fund land	43.7
	including agricultural land	11.9
	arable land	3.5
4.	other categories of agricutural land	24.4
	including arable land	6.2
5.	land in unsatisfactory condition:	
	irrigated	1.15
	drained	2.58
6.	land of bankrupt agricultural organizations	15.9
7.	9.7 million plots	111.9

A significant reduction in financial costs and a relatively rapid restoration of all functions of land management are possible only through its automation, informatization, and digitalization. Priority tasks in the field of land management and land policy whose urgent implementation is possible due to digital land management technologies are as follows:

1. Land inventory;
2. Compilation of the General scheme of land management in Russia, Russian regions and municipalities;
3. Development of lands suitable for agricultural or more intensive use.
4. Development of inter-farm and on-farm land management projects;
5. Development of land redistribution projects;
6. Development of integrated digital land management projects;
7. Establishing boundaries of federal land plots and land management support of measures for the delimitation of federal land ownership;
8. State control and protection of land;
9. Monitoring of crops;
10. Soil mapping and mapping of differential fertilization;
11. Digital relief modeling for land management;
12. Development of agricultural regulations;
13. Evaluation and modeling of crop yield;
14. Identification of valuable productive agricultural land;
15. Development of projects for the improvement of agricultural land, etc.

Digital land management can ensure quality, mass scale and short deadlines of land management works.

3. Results and discussion

Digital agriculture on the basis of smart land management cannot be created without a digital land management system. Digital land management is a basis of the digital economy, digital agriculture based on adaptive-landscape approaches, field monitoring, efficient planning and use of land resources, application of GIS, CAD and information and communication technologies, comprehensive accounting of large factors affecting design and management decisions. [6, 8]

Land management digitalization requires a unified integrated information system, land management forecasting, designing automation based on automated land planning design systems, expert analysis of land use based on neural network analysis, expert and intelligent systems, ICT .... etc.

The land management digitalization project is a complex systemic task affecting the maintenance of land management and the national land management system requiring cardinal structural changes, changes in the relationship of government and departmental structures.

The land management digitalization project should be implemented step-by-step. At the first stage, it is necessary to finalize and put into operation an integrative-integrated automated land resource design system for agriculture and rural development (ICAS-AGRO). Models and software products which carry out land management analysis and land planning, as well as all related electronic services of the agricultural industry can operate ion its basis [6, 9]

Thus, digital technologies of smart agriculture must be integrated on the basis of digital land management. Otherwise, the lack of "land management" will cause huge economic losses (poor use of land resources, land degradation, non-rational land use, etc.).

These approaches can be implemented through the creation of an integrative-integrated computer-aided design of land resources which is a software platform for the integrated design of land management in agriculture and other sectors of the rural economy, in programs for sustainable rural development. The technological environment for interaction of all participants in land management, including land resource planners, land users and landowners, credit and financial organizations, expert and controlling organizations should be created.

The system can be created on the basis of previously developed mathematical models and autonomous functional software modules, including:

- discrete model of crop rotation taking into account the cost of maintaining fertility;
- a discrete model of allocation of land plots at the expense of land shares based on preferences of real estate investors and forming optimal land areas for using modern farming technologies;
- methods for analyzing spatial variability of yield characteristics depending on the conditions of the agrolandscape, including relief (morphometric characteristics);
- a unified systematized database of field experiments of the Agrochemical Service of the Ministry of Agriculture of Russia and the Agrogeos geonetwork;
- a module for automated yield assessment depending on agroclimatic and soil conditions and the agrolandscape (an experimental model was developed for one of 23 large agroclimatic regions);
- a module for automated construction of optimal crop rotations (an experimental model has been developed);
- a module for automated calculation of the cadastral value of agricultural land (an experimental model has been developed; there are results of implementation of government contracts for calculating the cadastral value of agricultural land in 24 regions of the Russian Federation);
- a module of automated operations with cartographic information (an experimental model has been developed).

The general concept of this approach is creation of an automated system of flexible support for multi-step processes of making and implementing long-term collective decisions in the field of land management based on semantic web technologies.

The system will provide a wide range of interested users with an access to tools for the automation of land planning based on information on agricultural lands and rural areas. It must ensure the implementation of a long-term strategy for structuring the land market and technologizing land management services.

Creation and subsequent use of the system will reduce transaction costs of agricultural producers using modern innovative technologies by ensuring the optimal inclusion of these technologies in the production system of agricultural regulations modeled by land management projects.

Development and implementation of the system will provide strategic directions of automation and optimization of land management design processes and implementation of rural development projects.

Implementation of the project will increase the level of annual financial revenues in all related and land-dependent industries and additional annual income by to 40%; the total annual cumulative economic effect of restructuring and digitization of land management will exceed 1 trillion rubles.

IOP Conf. Series: Earth and Environmental Science **350** (2019) 012065 doi:10.1088/1755-1315/350/1/012065

The integrated complex automated system for designing land resources (ICAS-AGRO) is a software package; each module is able to work autonomously and can be integrated into other software packages.

The system is designed to automate the land management design, including implementation of land management projects and their economic justification, bringing them to industrial implementation with online access to the capabilities of the System through a web interface.

The system provides the user with tools of land planning automation, including: coordination of the work of specialists working on a single project in a distributed mode; coordination and examination of projects with a project regulatory status; mass supervision of project implementation; updating projects; monitoring of projects and fixation of monitoring results; archiving and storing projects and generating standard solutions; use of model projects or individual decisions and information in subsequent works on land planning; assessment of the potential of land resources and their use in production projects.

The issue of providing an automatic mode of information exchange with other specialized information resources (distributed databases) and software and hardware complexes will be solved in stages. There will be an information exchange with state and municipal information resources created and maintained by the Ministry of Agriculture of Russia, the Federal State Registration Service, other agencies, services and authorities, and commercial organizations. To this end, when developing the System, the formats and interfaces of the database will be coordinated with the existing government and commercial systems.

Satisfaction of the growing demand for food is an issue of using new fertile land plots whose main source is Russia. With 20% of world land resources, Russia using modern agricultural technologies will be able to provide about 1.5 billion people with agricultural products; together with CES and BRICS countries, it provides three billion people with food. The dynamics of the service market will be closely linked to the dynamics of agricultural products markets, since involvement of land in agricultural production is based on the land management works (land inventory; spatial planning, land management, etc.).

The System with "... a set of program modules will form a comprehensive sound solution of interrelated land management tasks which will make it possible to achieve the maximum economic effect and produce additional products through better land management solutions and use of new information resources.

When designing the System, solutions that are applicable to any national land management system (building an intelligent three-dimensional digital model of the territory based on land clusters and other model objects, providing tools of three-dimensional design, attracting and interpreting data on soil composition, climatic conditions, erosion processes, production-technological and transport-logistic modeling of farms) are used. At the same time, for the adoption and justification of specific design decisions, standards and criteria established by national regulatory acts (developed and approved in the Soviet Union) are used. If we change these standards and criteria, we can develop projects in other national land management systems. It is necessary to separate legal and accounting aspects of land relations from the issues of land design as an economic resource [6]

Thus, the System may be relevant for former Soviet republics (Belarus, Kazakhstan, Moldova, Georgia, Tajikistan, Ukraine), countries developing based on the Soviet land management system (China, Mongolia, Vietnam, etc.), countries that do not have developed systems for designing land resources for economic purposes, monitoring and controlling of land management [6].

Efficiency of land management activities depends on the degree of readiness of land management producers and experts.

For this purpose, the following additional educational training profiles can be used: computer equipment and digital technologies; applied mathematics and programming; applied informatics and programming.

4. Conclusion

In order to improve land management in the Russian Federation, it is necessary

1. To implement land management projects (land inventory, land management forecasting, land management schemes, etc.) in order to carry out a comprehensive reform of the entire industry. To form

a land management system and focus its main functions in the Federal Agency for Land Management subordinate to the President of Russia.

2. To transfer the function of rendering land management services from the Ministry of Economic Development to the Ministry of Agriculture (Department of Land Management subordinate to the Federal Agency for Land Management).

3. To Develop and implement a federal target program (subprogram) “Digital land management and priority types of land management in the Russian Federation (2019-2025)” [8]. To create a research infrastructure (scientific research institutes, laboratories, etc.). To develop and maintain improved computer and information support for the industry.

4. To develop and implement the Federal Target Program "Land Resources and Land Management of Rural Territories of the Russian Federation (2019-2025)". To improve the use of land resources.

Calculation of the economic effect of the program "Digital land management and priority types of land management in the Russian Federation (2020-2025)" and reduction of the time compared with traditional methods showed that the expected economic effect of modern technologies (2 indicators out of more than 20) will be 15.9 billion rubles with a total cost of the program of 2.6 billion rubles. Taking into account that the total estimated number of facilities required for land management works is about 1.8 million, at least one million specialists in land management will be required to ensure stable employment in land management and related branches of the national economy.

6. In general, the expected annual effect of the Program will be more than 100 billion rubles. According to our calculation, under regular and timely implementation of all land management tasks, the specified level of annual revenues should increase in all related and land-dependent industries and increase an annual additional income from 30 to 40% (more than 1 trillion rubles).

References

[1] Decree of the Government of the Russian Federation of 15 April 2014 no. 313 *On approval of the state program of the Russian Federation Information Society* (2011–2020)

[2] Volkov S N and Khlystun V N 2014 Land policy: how to make it more efficient *Int. Agricultural J.* **1–2** 3–6

[3] Volkov S N, Komov N V and Khlystun V N 2015 How to achieve effective land management in Russia *Int. Agricultural J.* **3** 3–7

[4] Volkov S N 2001–2009 *Land management* In 9 volums (Moscow: Kolos)

[5] 2017 *Land management ensuring the implementation of state programs and priority national projects for the development of the agro-industrial complex and other sectors of the economy* ed S N Volkov (Moscow: GUZ)

[6] Papaskiri T V 2018 On the concept of digital land management *Land management, cadastre and monitoring of lands* [in Russian – Zemleustroistvo, kadastr i monitoring zemel] **11** 5–11 (Moscow: Panorama, Athena)

[7] 2018 *Main directions of agricultural land use in the Russian Federation for the future* ed S N Volkov (Moscow: GUZ)

[8] Papaskiri T V 2014 Development of the Federal Target Program On the creation of a computer-aided land management design system and an application package (PPP) for the implementation of priority types of land management and related work in the territory of the Russian Federation *Land management, cadastre and monitoring of land* [in Russian – Zemleustroistvo, kadastr i monitoring zemel] **4** 14–25 (Moscow: Panorama, Athena)

[9] Khlystun V N 2018 Priority research areas in the field of land relations and land management *Land management, cadastre and monitoring of land* [in Russian – Zemleustroistvo, kadastr i monitoring zemel] **8** 5–11 (Moscow: Panorama, Athena)

Modern technologies of digital land management

T V Papaskiri[1], A E Kasyanov[2], N N Alekseenko[3], V N Semochkin[1], E P Ananicheva[1] and I V Volkov[1]

[1]State University of Land Management, 15, Kazakova str., Moscow, 105064, Russia
[2]Moscow Agricultural Academy named after K.A. Timiryazev, 49, Timiryazevskaya st., Moscow, 127550, Russia
[3]Basic Department of Digital Land Management, State University of Land Management, 15, Kazakova str., Moscow, 105064, Russia

E-mail: t_papaskiri@mail.ru

Abstract. The article deals with the issues of digital land management based on the use of modern information technologies. The digital land management system is created by digitalizing the agricultural industry and other economic sectors. Along with a digital land management system, staff training is should be carried out.

1. Introduction

Interaction with the Unified Federal Information System on Agricultural Lands (UFIS AL) created by the Ministry of Agriculture of Russia will improve digitization of agriculture. The solution is application of new technologies [1, 5–8]. Modern land management requires complex mathematical models to form design objects, choose and justify design solutions. Due to the use of spatial data (cartographic materials, elevation models, data on areas, distances, quantitative and qualitative characteristics of soils and landscapes, infrastructures and resources, production and consumption centers, etc.), calculations are required [2-4, 6-8]. To perform calculations, computing resources are required. Thanks to the use of cloud technologies, the digital land management system will reduce the level of technical requirements for the user's when integrating disparate information and computing resources required for project development (BigDATA).

Thus, the relevance of the “Digital Land Management” Concept aimed at solving the above-mentioned problems becomes a priority and is a strategic objective of the entire agrarian sector of the national economy. The development of land management as a science and an industry relies on the NTP and NTR. At The goal of digital land management is as follows:

Creation and implementation of a technological geo-information platform of digital land management as a mechanism for implementation of the national agrarian policy, rational use and protection of agricultural land. Smart land management is designed to provide the spatial basis for digital transformation of the following areas: smart field, smart farm, smart storage, smart greenhouse, smart garden.

2. Materials and methods

A scientific and technological breakthrough in agricultural production is impossible without digital technologies for land planning of rural areas which are a key segment of “smart agriculture”. The lack of systemic generators in agriculture, land management, digitalization of management decisions will cause huge economic losses, irrational use of land resources, land degradation, etc. [5-7]

It is necessary to ensure digital transformation of agriculture by implementing the “Digital agriculture” concept through digital technologies and platform solutions in order to ensure the technological breakthrough in the agro-industrial sector and increase labor productivity in “digital” agricultural enterprises through digital land management [6, 7, 9].

It is necessary to create

- new technologies and software systems for making inventory of agricultural land in order to identify land plots suitable for agricultural development, amelioration and improvement, determine their their boundaries, areas and location using a set of remote sensing methods (aerial photography, laser scanning, UAV and space devices);
- new technologies and software systems for agricultural zoning to identify valuable productive agricultural land, compile a register and indetify types and parameters of agricultural land management;
- an intellectual reference-analytical system of inventory and monitoring of soil and land conditions based on domestic digital, remote, geo-information technologies and computer simulation methods.
- a scientific and technological platform providing a support for planned business decisions, generation of a scientifically grounded spectrum of operational technological capabilities and proposals for precision management of crop production; a modern scientific and technological base for rational planning, forecasting, management and legal support of digital (smart) agricultural land management [5–8].

3. Results and discussion

Thus, digital technologies of smart agriculture in the management system “smart land management” - “smart field” must be integrated on the basis of digital land management. Otherwise, the lack of a systemic generator in the form of “land management” will lead to huge economic losses, irrational use of land resources, land degradation, etc.

There are five stages of goal implementation:

I. Development of a digital platform structure for smart land management

II. Formation of a system for planning and forecasting land management at the federal, regional and municipal levels based on BigData information flows

III. Formation of a system of automated land management, digital agricultural regulations and development of a system of electronic land management documents

IV. Formation of an effective land management system, including land ownership in the national agro-industrial complex, a system for involving unused agricultural land into circulation based on spatial data.

V. Staffing on the basis of a single online platform "Open Agrarian Education" including educational programs in digital land management.

<u>The content </u>of the stages (problems solved at each stage)

I. Development of a digital platform structure for smart land management.

- collection, storage, processing, analysis of information about agricultural land;
- accounting of land, drainage systems and hydraulic structures;
- monitoring of the state of land;
- integration and comprehensive analysis of data on qualitative characteristics of land;
- visualization of monitoring results;
- providing authorized users and interested persons with information on agricultural land.

II. Formation of a system for planning and forecasting land management at the federal, regional and municipal levels based on BigData information flows

- creation of an automated planning system for optimal (adaptive-landscape) land management, including:
- a block for collecting, updating and storing data on the state of land;

- a block for monitoring the state of land;
- a block for multi-purpose assessment of land suitability and crop yield modeling;
- a block for forecasting crop yields;
- a block for planning location of agricultural land and crops, designing adaptive-landscape farming systems and agricultural technologies;
- a block for cadastral valuation of land and their value.

III. Formation of a system of automated land management design, digital agricultural regulations and electronic land management documents

- development of intelligent technologies for land management documentation;
- validation of the system and demonstration of its capabilities on the example of a number of pilot projects of adaptive landscape agriculture;

IV. Formation of an effective land management system, including land ownership, a system for involving unused agricultural land into circulation based on spatial data.

- development of a system for collecting land cadastral information;
- creation of an information processing system;
- development of a system of management decision-making (development of robotic technologies based on artificial intelligence systems);
- development of digital projects for adaptive-landscape land management;
- development of a technological platform for using unused agricultural land;
- allocation of high-value agricultural land and development of regulations for tits use;
- formation of electronic atlases for the land resource management system.

V. Staffing on the basis of a single online platform "Open Agrarian Education" including educational programs for training in digital land management.

- creation of a network of centers for implementing adaptive land management systems;
- creation of a network of centers for advanced training of specialists for centers of competence

The following scientific and technical results ca be achieved:

- a new technology for agricultural zoning of a Russian region to identify valuable productive agricultural land plots, compile a register and establish types and parameters of permitted use of agricultural land plots;
- a software package for agricultural zoning of a Russian region to identify valuable productive agricultural land plots, compile a register and establish types and parameters of permitted use of agricultural land plots;
- instructions for agricultural zoning of a Russian region to identify valuable productive agricultural land plots, compile a register and establish types and parameters of permitted use of agricultural land plots.
- a new technology for making inventory of agricultural land to identify land plots suitable for agricultural development, amelioration and improvement, and determine their boundaries, areas and location using Earth remote sensing methods (aerial photography, laser scanning, UAVs and spacecraft);
- a software package for the inventory of agricultural land in order to identify land plots suitable for agricultural development, amelioration and improvement and determine their boundaries, areas and location using Earth remote sensing methods (aerial photography, laser scanning, UAVs and spacecraft);
- instructions for the inventory of agricultural land in order to identify land plots suitable for agricultural development, amelioration and improvement and determine their boundaries, areas and location using Earth remote sensing methods (aerial photography, laser scanning, UAVs and spacecraft).
- modified versions of the UAV for using in training purposes;
- modified hardware and software products for using in training purposes.

Development of an educational service for new educational programs (using UAVs for training purposes) to study and apply the latest digital technologies;

- the advanced training program "Zoning of agricultural lands of a Russian region";
- the advanced training program "Automated technologies for zoning agricultural land";
- the advanced training program "Modern technologies for studying the state of agricultural lands for the purpose of optimization and regulation of their use";
- the advanced training program "Automated technologies for studying the state of agricultural lands using remote sensing methods";
- the advanced training program "Technologies of zoning, regulation and organization of rational use of land in rural areas of the Russian Federation based on remote sensing methods";
- an educational methodical system and a digital educational resource for the advanced training program "Zoning of agricultural lands of a Russian region";
- an educational methodical system and a digital educational resource for the advanced training program "Automated technologies for zoning agricultural land";
- an educational methodical system and a digital educational resource for the advanced training program "Modern technologies for studying the state of agricultural lands for the purpose of optimization and regulation of their use";
- an educational methodical system and a digital educational resource for the advanced training program "Automated technologies for studying the state of agricultural lands using remote sensing methods";
- an educational methodical system and a digital educational resource for the advanced training program "Zoning technologies, regulation and organization of rational use of land in rural areas of the Russian Federation based on remote sensing methods";
- advanced training of specialists in the program "Zoning of agricultural lands of a Russian region";
- advanced training of specialists in the program "Automated technologies of agricultural land zoning";
- advanced training of specialists in the program "Modern Technologies for Studying the State of Lands";
- advanced training of specialists in the program "Automated technologies for studying the state of agricultural land using complex Earth remote sensing;
- advanced training of specialists in the program "Technologies of zoning, regulation and organization of rational use of land in rural areas thorium of the Russian Federation on the basis of the complex methods of remote sensing of the Earth".

4. Conclusion

In order to improve the land service of the Russian Federation, increase efficiency of land management, it is necessary to implement a number of measures.

1. In order to implement the Order of the Government of the Russian Federation of November 8, 2018 No. 2413-p on measures aimed at improving the legal regulation of land relations, it is necessary to include provisions on new principles and methods of digital land management in the federal law "On Land Management" (new edition).

2. It is necessary to develop and implement a federal target program (subprogram) "Creation of digital land management and implementation of priority types of land management and related works in the territory of the Russian Federation (2019-2025 years)". It is necessary to create a research infrastructure (scientific research institutes, problem laboratories, etc.), develop a computer and information support for the industry.

3. It is necessary to organize centers of digitalization of the agro-industrial complex for solving applied problems and developing software (application packages — modules of the agro-industrial

complex information system) on the basis of federal educational and methodological associations of universities,

4. Leading universities have to implement curricula for bachelor and master degree students in "Land Management and Cadastre".

5. To actualize teaching methods by implementing online courses, digital services of national and global online platforms, expanding the use of interactive methods, individual educational trajectories, project training.

References

[1] Decree of the Government of the Russian Federation of 15 April 2014 no. 313 *On approval of the state program of the Russian Federation Information Society* (2011–2020)
[2] Volkov S N and Khlystun V N 2014 Land policy: how to make it more efficient *Int. Agricultural J.* **1–2** 3–6
[3] Volkov S N, Komov N V and Khlystun V N 2015 How to achieve effective land management in Russia *Int. Agricultural J.* **3** 3–7
[4] Volkov S N 2001–2009 *Land management* In 9 volums (Moscow: Kolos)
[5] Volkova S N 2017 *Land management for implementing federal programs and priority national projects for the development of the agriculture and other sectors of the economy* (Moscow: GUZ)
[6] Papaskiri T V 2018 On the concept of digital land management *Land management, cadastre and monitoring of lands* [in Russian – Zemleustroistvo, kadastr i monitoring zemel] **11** 5–11 (Moscow: Panorama, Athena)
[7] 2016 *Organizational economic mechanism of formation of the computer-aided design system in land management* Thesis (Moscow)
[8] 2018 *Main directions of agricultural land use in the Russian Federation for the future* ed S N Volkov (Moscow: GUZ)
[9] Khlystun V N 2018 Priority research areas in the field of land relations and land management *Land management, cadastre and monitoring of land* [in Russian – Zemleustroistvo, kadastr i monitoring zemel] **8** 5–11 (Moscow: Panorama, Athena)

Methods of land management when locating tea plantations

T V Papaskiri[1], Yu Peng[1], A E Kasyanov[2], V N Semochkin[1], E P Ananicheva[1] and I V Volkov[1]

[1]State University of Land Management, 15, Kazakova str., Moscow, 105064, Russia
[2]Moscow Agricultural Academy named after K.A. Timiryazev, 49, Timiryazevskaya st., Moscow, 127550, Russia
[3]Basic Department of Digital Land Management, State University of Land Management, 15, Kazakova str., Moscow, 105064, Russia

E-mail: t_papaskiri@mail.ru

Abstract. The article deals with the issues of modernization of land management as an economic branch. Due to digitalization of the economy, modernization is aimed at creating a digital land management system. The article analyzes land management methods used for location of tea plantations. The potential of the tea industry in China and other leading tea producing countries was analyzed.

1. Introduction

Identification of new land areas suitable for tea cultivation (selection of a land plot) is an important issue when organizing tea plantations. Design companies often have to design new tea farms in a certain area with established boundaries [1, 5–8, 12]. As a result of the pre-design works, land of such farms is either unsuitable for tea due to its natural conditions, or long-term and costly reclamation measures are required for its development.

Sometimes land allocated for tea farms is of poor quality. This land is allocated because tea production can cover any costs of funds and materials. This approach causes great damage to the agro-industrial sector and cannot be justified [1, 2, 5, 9–12]. Organizations of tea farms should be preceded by detailed surveys and feasibility studies that confirm the possibility of observing optimal conditions for organizing such farms. In all cases, it is necessary to have a sufficient amount of suitable land, close to transport communication systems and processing plants [3, 9, 10].

These requirements can be taken into account only on the basis of a special land management project. It is necessary to establish an exact sequence of all phases of a land management project, including preparatory works which include a detailed survey of the area. [4]

2. Materials and methods

Tea is a subtropical evergreen plant. Some of its varieties (grown in Russia and Georgia) are so hardy that they do not die even at temperatures below –20 ° C. [1, 3, 5, 9, 10]

The tea plant grows in areas located between 45^0 north latitude and 30^0 south latitude at an altitude of 300 to 2000 meters above sea level. Tea plants are usually grown in the jungle or in the places of any old plantings. When preparing a tea plantation, old roots are uprooted, pampas grass is sown to loosen soil. The ground sprouted with grass is plowed and tea bushes are planted [1, 5, 9, 12]

IOP Conf. Series: Earth and Environmental Science **350** (2019) 012067 doi:10.1088/1755-1315/350/1/012067

However, the choice of location for tea plantations and their orientation in the direction of bushes require compliance with certain conditions. The main factor is high temperatures in summer and autumn and relatively low ones in winter. For the active growth of tea plants, long light days are required. Otherwise, tea leaves will lose their aroma. The tea plant is very susceptible to air pollution, the air must be clean and humid. [1, 12]

Tea bushes must be placed on the slopes, so that moisture does not stagnate under the roots. It is a moisture-loving plant that requires constant moisture. It is necessary to locate tea plantations at an altitude of 1.5 km above the sea level.

For the tea plant to produce green leaves, heat and moisture are required. Camellia feels well at a high humidity and heavy precipitation. But tea plants do not stand water stagnation under the roots; they should be washed with water rather than be located in water. [1, 5, 12]

In the provinces of China, India and other countries with tropical and subtropical climatic conditions, peasants grow tea bushes every year. They grow them on mountain slopes with a warm climate and high humidity. Most of the fertile fields can yield up to four or five times a year. The most valuable and elite yield is the first or the second one.

Tea cultivation is an extremely time-consuming process. Labor of farmers is hard, because in a few hours it is possible to collect only a few kilograms of tea leaves, of which only one fourth can be used as a finished product. In addition, during one working day, peasants travel tens of kilometers from one plantation to another one.

Leaves are manually sorted into several varieties: young unblown buds, buds and leaves used to make the best varieties. The leaves from the middle part of the bush are used for coarser varieties.

Manual tea harvesting is necessary, as machine harvesting can damage wet leaves and impair tea taste and aroma. Plantations are mainly located in the countries with cheap labor.

It is necessary to analyze the following issues:

- Production specialization;
- Selection of tea varieties;
- Choosing a site for a tea plantation;
- Natural conditions;
- Reduction of cost and increase in the product quality by improving working conditions.

3. Results and discussion

Selection of a site for a tea plantation. Areas that were under the forest or shrubs, or annual crops are used for this purpose. In all wet areas, a drainage network is arranged.

In order to mitigate harmful effects of dry and cold winds, forest protective stripes are arranged. Flat areas and slopes with a steepness up to 30° are used. Continuous tillage is made to a depth of 45 cm, carefully loosening and cleaning the soil from tree roots, rhizomes, weeds, stones, etc.

Plots with poor and weakly structured soils are sowed with perennial cereal and leguminous grass. [1, 5, 12]

The soil is plowed. For plowing, phosphate fertilizers are applied (200-300 kg per 1 hectare). Rows are created: in flat areas and on slopes with a steepness up to 8° - straight rows with a distance of 1.5 m; on slopes with a steepness of 8° - along the horizontals (across the slope) with a distance of 1.25 m.

To eliminate soil erosion, the following measures are used:

a) deep tillage across the slope;

b) division of rows along horizontals, thickened sowing of tea seeds;

c) a creation of a water regulating network and anti-erosion terraces;

d) sodding of the slopes of terraces and inter-rows of tea plantations with soil-fixing grasses.

Tea plantations are created in spring and autumn. For sowing, seeds with a size of 1.2 cm in diameter are used. They are sown in nests, 5-7 seeds in each nest at a distance of 25-35 cm from each other.

When the belt sowing method is used, seeds are sown in grooves at a distance of 5 cm between them. The seeding depth is 4-5 cm.

In areas exposed to dry winds, 5–7 cm. The average seeding rate, depending on the width of the aisles and the method of sowing, 150 to 250 kg / ha of future tea plants located on the southern slopes and 350 - are used for laying out each hectare of plantations. 400 or more kg of seed on the northern slopes. [1, 5, 12]

In arid districts, to protect the soil from drying out, rows of tea seeds are covered with fern, grass or branches. In May, in order to protect tea seedlings from sun, seedlings are protected with green branches. For shading, annual plants are used (bush beans, soybeans) which are sown in the rows.

The main agricultural measures used at tea plantations are as follows: weeding in the rows of tea seedlings 3-4 times during the growing season and soil loosening; repair of plantations in spring and autumn; on thinned areas of annual plantations, seeds are sowed, on 2-year-old and older plantations, seedlings are planted after soil loosening to a depth of 35 - 40 cm; planting with a clod; cultivation of rows in spring and summer, weeding; soil loosening; autumn-winter digging or plowing between the rows to a depth of 12 - 15 cm since the 3rd year. On well-developed plantations, autumn-winter tillage should be done every 3-4 years.

Forming and cutting of 2-3-year old tea bushes from mid-February to April are required. Bushes are cut in order to cause the branching, create a skeleton which ensures further development of a compact and wide crown. The trellis pruning of full-aged plantations is aimed at leveling all trellis by raising low bushes, maintaining the shape, height and width, and enhancing the emergence and reduction of fruiting. For pruning, a tea-cutting machine has been designed [1, 5, 12]

The correct choice of sites determines productivity and durability of future plantings. At the same time, tea variety requirements to the environmental conditions, soil and climatic features and possibilities of mechanization of tea cultivation must be taken into account (Figure 1). In Russian conditions, for example, in Sochi, the vertical limit of possible land development for tea plantations is an altitude of 500-600 m above the sea level, although the potential has not been fully studied. Slope with a steepness of 22-25° or less can be used for tea plantations. On steeper slopes, it is advisable to leave natural vegetation.

The following t issues are addressed when selecting sites and locating tea plantations:

1. Identification of new lands suitable for tea cultivation.
2. Determination of the size of land areas and rational arrangement of territories suitable for tea cultivation.
3. Expansion of areas.
4. Selection of tea varieties.
5. Improvement of productivity by improving seed and planting material quality, using fertilizers, choosing the best sites.
6. Removal of old plantations, preparation and planting of new ones.
7. Irrigation.
8. Application of fertilizers during the growing season based on data obtained from UAV monitoring.
9. The use of technical equipment during harvest.
10. Monitoring of the content of tea plantations.

Identification of new areas of land suitable for tea cultivation of tea. Design organizations have to design new tea farms within established boundaries.

Lands of some farms are either unsuitable for tea cultivation due to their natural conditions, or long-term and costly reclamation measures are required for their development.

Lands allocated for tea farms can have poor quality. They are selected due to the fact that tea production can justify any costs of funds and materials. It can cause damage.

Organizations of tea farms should be preceded by detailed surveys and feasibility studies that confirm the possibility of observing optimal conditions for organizing such farms. [3–5, 12]

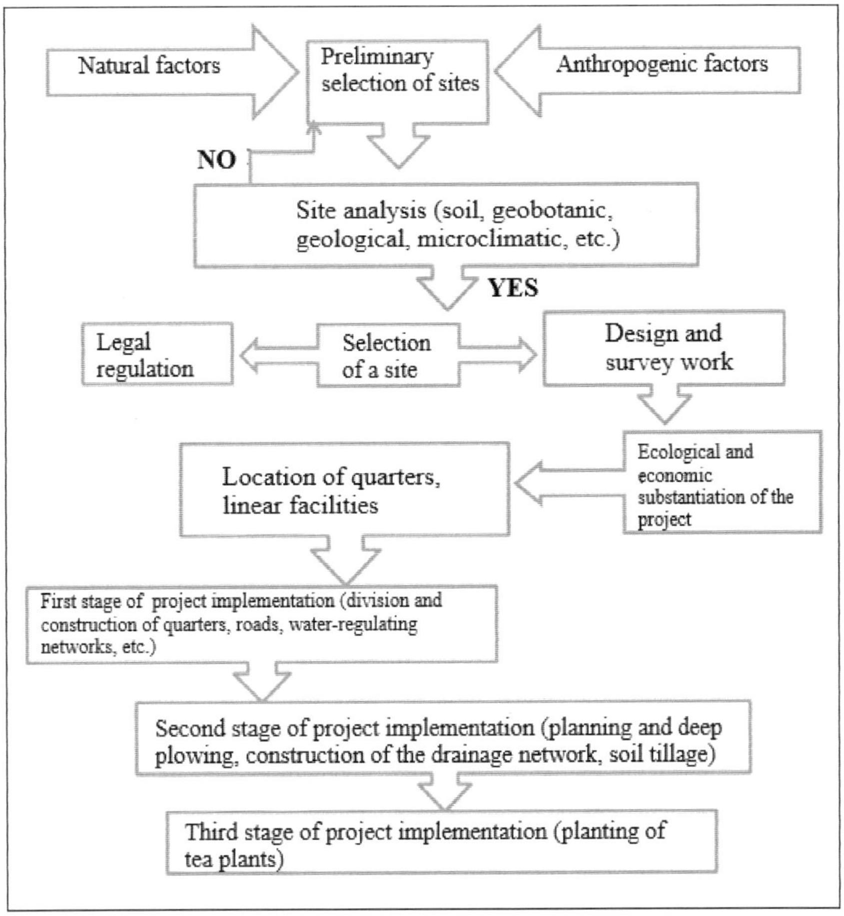

Figure 1. Infological scheme of site selection and tea plantation location [12].

Sites are selected taking into account plant requirements for soil and climatic conditions, on the basis of farm management projects, soil and other surveys. Design and exploration determine practical suitability and economic feasibility of various plots used for tea plantations. Roads, water regulation networks, and protective forest belts should be created. A drainage network should be created [1, 5, 12].

Tea plants are responsive to mineral fertilizers; they are applied according to agrochemical cartograms reflecting the actual content of mineral substances in the soil.

Today, various organizations have developed indices of plant availability with nitrogen, phosphorus and potassium NDVI. These indices are used to determine the degree of nutrients. Using unmanned aerial vehicles (UAVs) equipped with special spectral chambers, it is possible to determine the degree of nutrients for tea plants and identify their optimal rates.

One of the reserves to increase the yield of tea plantations is alignment of trellises. Until now, in some farms, tea lanes are wavy curves. This surface reduces the leaf area and makes it difficult for workers to collect tea leaves. Differentiated trimming of plants can level these trellises.

4. Conclusion

1. When reviving the tea industry in Russia, it is necessary to use the experience of the Chinese agro-industrial complex, including mechanisms of state regulation, taxation and integrated development of rural areas based on food, social, environmental and other types of security.

2. When choosing a tea plantation site, slopes with a warm climate and high humidity are optimal. Most plantations can give yield 4-5 times a year. The most valuable and elite tea is of the first and second yields.
3. Application of these results will provide an additional annual income from tea production.

References

[1] Daraselia M K, Vorontsov V V, Gvasalia V P and Tsanava V P 1989 *Tea culture in the USSR* ed R D Pantshava (Tbilisi: Metsniereba)
[2] Volkov S N and Khlystun V N 2014 Land policy: how to make it more efficient *Int. Agricultural J.* **1–2** 3–6
[3] Volkov S N 2018 *Land Policy and Land Management in China* (Moscow: GUZ)
[4] Volkov S N 2001–2009 *Land management* In 9 volumes (Moscow: Kolos)
[5] Gochalashvili M M 1983 *Biological basis of tea leaf culture* (Tbilisi: Tsonda)
[6] 2017 *Land management ensuring the implementation of state programs and priority national projects for the development of the agro-industrial complex and other sectors of the economy* ed S N Volkov (Moscow: GUZ)
[7] Papaskiri T V and Yunlong P 2017 *Land management atlas of China* (Moscow: GUZ)
[8] Karkashadze N I and Gvelesiani R K 1988 *Economy and organization of tea production* (Tbilisi: Ganatlebi)
[9] 2018 *The main directions of agricultural land use in the Russian Federation for the future* ed S N Volkov (Moscow: GUZ)
[10] Khlystun V N 2018 Priority areas of research in the field of land relations and land management *Land management, cadastre and monitoring of land* [in Russian – Zemleustroistvo, kadastr i monitoring zemel] **8** 5–11 (Moscow: Panorama, Athena)
[11] Cherkashin E V 2013 *Economy and organization of rational use and protection of land in the essential oil and drug industry* (Moscow: GUZ)
[12] Papaskiri T V and Yunlong Peng 2018 *Economics and land management of the tea industry on the example of southern China* (Moscow: GUZ)

Prediction of agricultural land use

A Rasskazova and Yu Sinits

State University for Land Management, 15, Kazakova St., Moscow 105064, Russia

Abstract. In the article the authors examined the role of forecasting in the system of agricultural land use management. Particular attention was paid to the choice of a method for predicting agricultural land use. The classification of formalized methods of agricultural land use forecasting was highlighted, which include extrapolation and modeling methods. The authors examined the features of each forecasting method and the possibility of its use for predicting agricultural land use. The experience of forecasting agricultural land use in India is considered. The study was based on the work of Russian and foreign scientists like A.A. Varalamov, S.A. Galchenko, A. Pankrats, S. Makridakis and others.

1. Introduction

The economic development of any agricultural region directly depends on its land and resource potential. The management of agricultural land use is a complex system that includes the forecasting and planning of land use, land supervision, land monitoring and so on.

The forecasting with agricultural land use is the most important pre-project activity and initial stage of the agricultural land use management process the main goal of which is rational use and protection of land.

To achieve goals regarding the prediction of agricultural land use, it is necessary that the chosen method of forecasting should correspond to the maximum extent.

2. Materials and methods

Among the forecasting methods used in the agricultural land use management system, the extrapolation and modeling methods should be highlighted.

The idea of the extrapolation methods is expressed in the analysis of changes in the objects of research in time and the distribution of the revealed regularities into the future. Inside the extrapolation methods, there are as follows: simple extrapolation methods, least squares, exponential smoothing, moving averages and others [1].

The method of simple extrapolation involves calculation of the average value of indicator, which forms the basis of the short-term forecast. So, for example, it is necessary to substantiate a short-term forecast of products of a particular agricultural land use, i.e. it is necessary to determine the arithmetic mean value:

$$X_n = \sum \frac{Xi}{n}, \quad (1)$$

where X_n is the predicted value; Xi is the volume of agricultural products; n is the number of years under consideration.

Analytical alignment in forecasting is a finding of mathematical function that most accurately describes the change trend.

IOP Conf. Series: Earth and Environmental Science **350** (2019) 012068 doi:10.1088/1755-1315/350/1/012068

Among the modeling methods aimed at forecasting the agricultural land use the most commonly used methods are the methods of network, structural and statistical modeling and economic and mathematical methods [2].

With the help of structural models, it is possible to distinguish qualitatively homogeneous groups based on a large number of features. For example, land use in similar conditions.

The network modeling is based on the construction of the activity network. Each type of work in the activity network can be represented by an arrow that connects the initial and final event (Figure 1).

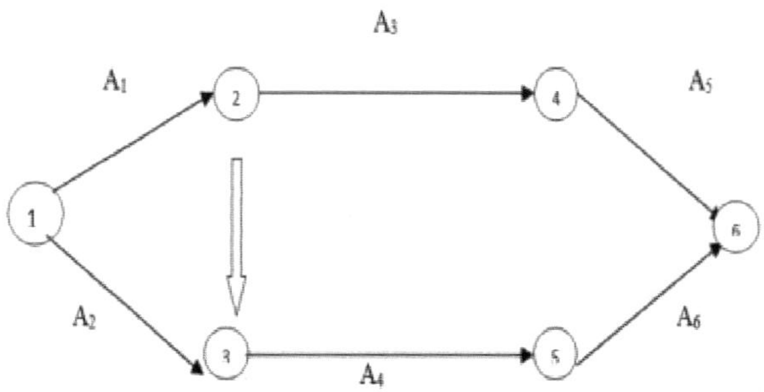

Figure 1. Activity network

Each type of work is shown by two numbers on the graph, corresponding to the initial and final events. In this example, the type of work (1.2) is indicated by A1; the type of work (1.3) is A2. Each type of work is characterized by two parameters: the duration of its implementation and the number of resources necessary for its implementation.

Neural network modeling is used along with other methods of forecasting of agricultural land use.

A neural network is a set of special mathematical functions with a variety of parameters that are configured in the process of learning from past data. Then, the neural network processes the initial real data and gives its forecast of future behavior of the system under study.

When applying this method for the purpose of predicting the use of land resources, the layer of input signals will be the values of the factors that have the greatest impact on the indicator being studied, i.e. values of independent variables. The values of the predicted indicators (dependent variables) corresponding to the existing set of values of the input variables are used as the output layer.

Let's consider the methods for predicting agricultural land use abroad, in particular, the experience of India. The following indicators are considered as predictable, i.e. the output of products from the area of agricultural crops, the spread of pests of cultivated plants and the diseases, as well as the dynamics of the volumes of agricultural crops.

For agricultural forecasting in India, the following statistical forecast models are applied: regressive, time series, and probabilistic.

The regressive models are multiple linear regression (for predicting output per unit area), meteorological indices (for predicting output per unit area, and the number of crop pests), as well as a regressive logic model (for predicting low or high yields, the presence or absence of pests / diseases, etc.) [3].

The regressive model which is very close to the model of multiple linear regression has the following form:

$$Y = \beta_0 + \beta_1 X_1 + \beta_2 X_2 + \ldots + \beta_p X_p + \varepsilon\,, \quad (2)$$

where β_0 and β_i (i = 1,2,... p) are evaluated variables and ε is random errors of representativeness; Y is the predicted variable; X_1, X_2, ..., X_p are forecast variable; p - shows the number of variable forecast.

The obtained assumptions (limitations, conditions) of the inverse reaction of Y to the predicted X_1, X_2, ..., X_p are linear regression β_0, $\beta_1, \ldots, \beta_p$. The drop-down errors are independent of each other and equally distributed (idd) random variables with an average zero value can fluctuate within σ^2. The errors are independent from each other (pairwise mixed moment of the second order is zero), according to which the predictor variable X_1, X_2, ..., X_p is not accidental and error-free [4-6].

In practical application, the indicator of output per unit area is designated as Y, and biometric indicators of crops as X (i = 1,2, ..., p).

Depending on the crop, the biometric indicators of seedlings are plant growth, the number of spike-heads of panicles, etc. In addition to the "growth index" the indicators of a certain period of time are taken during the harvest period.

To build weather indices, the parameters of changing weather forecast, using the following formulas are taken [6]:

$$Y = a_0 + \sum_{i=1}^{p} a_i Z_i + \sum_{i \neq j}^{p} a_{ij} Z_{ij} + e, \quad (3)$$

$$Z_i = \sum_{w=1}^{m} r_{iw} X_{iw}, \quad (4)$$

$$Z_{ij} = \sum_{w=1}^{m} r_{ijw} X_{iw} X_{jw}, \quad (5)$$

where Y is the dependent variable which shows the output per unit area / pest; e is a random error; r_{iw} is the correlation coefficient; Y c I denotes the variable value of weather parameter; r_{ijw} shows the resulting volume of production; i^{th} and j^{th} are variables of the weather parameter during the week w^{th}; m indicates the weekly forecast; p indicates several weather parameters; Z_i , Z_{ij} the predictors that depend on the basic weather indicators of maximum and minimum temperatures, precipitation in the form of rain, relative humidity, etc.

These models use the weather parameters for a week for a year or several years.

In cases when the predictor is a qualitative indicator, a regressive logical model is used. In the absence of data on the number of pests of agricultural crops / diseases of agricultural crops, such qualitative indicators as incidence / non-occurrence, low / high morbidity, etc. are taken. The preferred statistical model for analyzing such dichotomous responses is the binary logistic regression model. It is used to describe the relationship of several independent variables to a binary dependent variable. The logistic regression is used to obtain the probabilities of occurrence of various classifications [7, 8].

$$P(E = 1) = \frac{1}{1+\exp(-z)}, \quad (6)$$

Such model has the following form (5):

$$P(E = 1) = 1 / (1 + \exp(-z)), \quad (7)$$

where z is the weather parameter function.

If $P(E = 1)$ is greater than 0.5, then the probability of the disease is higher, and if it is less, then the probability of occurrence of the disease is minimal. For more accurate forecast, the value of up to 0.7 is taken.

Time series models reflect the analysis of data for a specific time interval. The basic principle of the model is to predict what will happen after certain period of time, without determining the factors that have an impact on this.

Time series models are used in short term forecasting. For this type of forecasting it is enough to have from 20 to 50 indicators. The more indicators we have, the more accurate the forecast is.

Time series models include exponential smoothing and an autoregressive integrated moving average.

A probabilistic model is built on a Markov chain to predict the yield of agricultural products per unit area.

The above mentioned models are effectively applied in agricultural forecasting. Depending on the results obtained, the course of the direction of the state policy in the field of agriculture is determined.

3. Results and Discussion

As studies show the application of the method of analytical alignment for the purpose of agricultural land use, the analytical functions align well the initial dynamic range (the correlation coefficient exceeds 0.7). This significantly reduces the autocorrelation (the autocorrelation coefficient decreases four times or more), its actual value is much less than the critical table value, which at 5% probability level of significance and the length of the original time series (16 years) is equal to 0.323.

All selected functions are suitable regarding the forecast, i.e. with their help one can extend the current trend for the future and find the most likely limits of deviations of the predicted values, for example, the yield of winter wheat [9].

Besides, the method of exponential smoothing is used to determine the effectiveness of agricultural land use. This alignment is particularly strongly fluctuating time series for subsequent prediction. This method allows you to make reasonable predictions based on the time series, having a moderate relationship in time, and provides a greater account of the indicators achieved in recent years. Let's consider the application of this method and define the predicted value of crop yields. We presume that the calculation period will be qualitatively similar to the last segment of the period under study. The coefficients of equation (estimates of the coefficients), found with the exponential smoothing of the last value of the index in the original time series, are used for the subsequent prediction.

Statistical modeling for the purpose of determining the effectiveness of land use is performed on the basis of a single regression equation and on the basis of a system of regression equations. The forecast options can be viewed as both one equation and a set of them [10].

The studies conducted using the regressive analysis methods at agricultural enterprises of the Ryazan region allowed determining the list of main independent variables for the inclusion into the statistical models for such important indicators as crop yield Y_1, productivity of farm animals Y_2, labor productivity Y_3, and production cost Y_4 (Table 1).

Table 1. Statistical model data

Indicators	Independent variables
Crop yield Y1,	Evaluation points, differential income per 1 ha, cost recovery, the level of production mechanization, capital supply, capital intensity, irrigation level, forms of labor and production organization, the amount of mineral and organic fertilizers applied; coefficients characterizing the degree of development of the management system
Productivity of farm animals Y2	Feed area, quality of feed, security of the main ingredients of nutrition; breed composition of animals, labor costs, coefficients of application of progressive zootechnical measures, costs of veterinary measures
Labor productivity Y3	The level of mechanization of labor and production, indicators of the level of specialization of production, indicators of social infrastructure, wages, working conditions, provision of social infrastructure
Production costs Y4	Crop yields, productivity of farm animals, the share of individual cost items, product quality indicators, the share of mechanized labor and production

In order to effectively manage agricultural land use, the cluster analysis can be applied. This method of forecasting involves the selection of compact and distant groups of objects, looking for a "natural" partition of the aggregate into the areas of clusters of objects. The clustering process is a system consisting of a series of interconnected elements. An example of the use of the clustering method in forecasting the agricultural land use can be the clustering of municipalities in the region according to the following indicators: the efficiency ratio of agricultural land use, the level of profitability of all activities,%; net income of agricultural products to the total cost, rub./rub.; profits of agricultural enterprises from all activities, rub/ha ..

4. Conclusion

To sum up, it should be noted that when choosing a method for predicting agricultural land use, the following factors are of particular importance, i.e. the purpose of the forecast, the characteristics of the available data, the forecast period, the permissible probability of errors, the cost of the forecast.

For effective management of agricultural land use at various administrative and territorial levels, it is necessary to create a reliable and effective system of forecasting and planning indicators, from various aspects, characterizing the state of land resources and the effectiveness of their management. The requirements for the composition and structure of indicators are formed in accordance with the types of the subject and the tasks set.

References

[1] Komarov S I and Rasskazova A A 2018 *Forecasting and planning the use of land resources and real estate* (Moscow)
[2] Galchenko S, Varlamov A and Bogdanova O 2018 Theoretical and methodological foundations for formation of sustainable land management system *IOP Conf. ser. Materials Science and Engineering* 012141
[3] Sinitsa Yu S 2016 *Economic efficiency of land cadastral systems of the Russian Federation,* candidate thesis, State University of Land Management (Moscow)
[4] Agrawal R, Mehta S C, Bhar L M and Kumar A 2004 *Development of weather based forewarning system for crop pests and diseases – Mission mode project (NATP)* report IASRI publication
[5] Jain R C, Jha M P and Agrawal R 1985 Use of growth indices in yield forecast *Biometrical J.* **27(4)** 435–39
[6] Makridakis S, Wheelwright S C and Hyndman R J 1998 *Forecasting: Methods and Applications* (New York: John Wiley)
[7] Misra A K, Prakash O and Ramasubramanian V 2004 Forewarning powdery mildew caused by Oidinm mangiferae in mango (Mangifera indica) using logistic regression models *Indian J. of Agricultural Sci.* **74(2)** 84–7
[8] Pankratz A 1983 Forecasting with univariate Box Jenkins models: concepts and cases (New York: John Wiley&Sins)
[9] Askolskaya E A 2010 Some aspects of methodology for the selection of experts based on assessing the degree of confidence in the level of professional judgment *News of the Orenburg State Agrarian University* **4(28–1)** 170–2
[10] *Electronic textbook on statistics StatSoft* 2001 (Moscow: StatSoft) Available at: http://www.statsoft.ru/home/textbook/default.htm

Analysis of the water bodies zones influence on the cadastral value of garden and horticultural land plots

E N Bykowa, A A Bogolyubova and D D Simonova

Saint-Petersburg Mining University, 2, 21st line of Vasilevskiy Island, Saint-Petersburg, 199105, Russia

E-mail: Bykova_EN@pers.spmi.ru

Abstract. The paper proposes new methodological provisions for determining the cadastral value of garden and horticultural lands with water bodies zones. A method was developed for calculating the cadastral value differentiation coefficient, which takes into account the presence of a water protection zone on the land plot. A hypothesis was formulated about the influence of water bodies zones on the cadastral value of these lands and it was confirmed for the water protection zone. The analysis of statistical data was made, which was the basis for selecting analogous objects for calculations. Two stages of the cadastral value modeling were carried out, the results were compared to obtain the desired coefficient. The analysis of the obtained models quality confirmed the validity and adequacy of the cadastral value model.

1. Introduction and problem statement

Currently, the most acute problem in obtaining reliable data on the cadastral value of land plots is in incomplete consideration of all influencing price-forming factors. One of the topical issues of the cadastral valuation process is to determine the dependence of value on various kinds of constraints (encumbrances), significantly influencing the implementation of rational and full use of land in accordance with their intended purpose. In this paper, we consider water protection zone and a coastal protective strip of a water body. According to the Russian legislation they are refered to the zones with special conditions of use of the territories.

One of the segments of real estate, where there are significant restrictions on the implementation of economic activities in the areas of water bodies zones, is garden and horticultural plots. The assessment of such lands is carried out according to the approved in 2017 methodology, where a list of recommended influence factors is proposed. In the cadastral valuation of the lands of Saint-Petersburg in 2018, factors that did not correspond to the specified list were used for this segment, but, according to the experts, they were more important and fully justified for inclusion in the valuation model. The list of price-forming factors methodological guidelines for the cadastral assessment already contains the factor "The presence of encumbrances (restrictions) of the land plot", but there are no methodologies for that assessment in mass form. This once again confirms the relevance of the topic and the necessity to develop proposals for more accurate and fair cadastral valuation.

According to the last cadastral assessment of the land plots in Saint-Petersburg, which was conducted in 2018, factors for the presence of water bodies zones in garden and horticultural plots were not taken into account. Consequently, the land plots, partially or fully located in the specified zones and with restrictions on use, were assessed as free from these protection zones, which indicates incorrectness of the obtained results of the cadastral assessment, and, therefore, the unfairness of the the tax burden distribution between the subjects of this land market segment. In view of the foregoing, the purpose of the study was to improve the methodology for determining the cadastral value of garden and horticultural lands by calculating the coefficient of accounting for water bodies zones.

The practical significance lies in the possibility of using the results to modernize the existing methodology for determining the cadastral value of garden and horticultural lands located in settlements in order to ensure social justice in the implementation of tax policy.

2. Review of literature and the prerequisites for the modernization of the existing cadastral valuation methodology

The issues and ways of modernization as a methodology for cadastral valuation of horticultural lands and technical guidelines are described in the scientific works of many researchers. Separate developments that take into account the influence of zones with special conditions of use of the territories (ZWSCUT) were proposed by several Russian scientists. A.A. Varlamov gave the formula for calculating the value of land, taking into account the coefficient ZWSCUT, but the method of obtaining coefficients was not presented [1]. Kretinin K.V. and Kuznetsov N.A. proposed the formula for calculating the amount of damage caused by overhead transmission lines [2]. Sutyagin V.Yu. offered the calculation of the coefficient, which takes into account the degree of restriction of rights due to the establishment of a protected zone, by an expert method [3]. Kovyazin V.F. and Romanchikov A.Yu. presented a method for calculating the cadastral value of forest lands, which includes special forest factors [4]. The beginning of all these studies in the field of the influence of ZWSCUT on the value of land was laid in the USA by several researchers in 1979. Gustafson R.J., Grumstnup P.D., Herdrickson E.R. and Meyer M.P. They studied the negative impact of airborne power lines on economic activities in the agricultural lands. It was revealed that power lines affect the efficiency of air scattering of agrochemicals, machine tillage, and their electric field reduces crop yields [5].

One of the authors of this article Bykowa E.N. considered the cadastral valuation of horticultural and gardening lands associations in detail in one of her earlier work [6]. The work is based on the determination of the influence of ZWSCUT on the cadastral value of land by an expert, expert-analytical and statistical methods. These zones are considered as a price-forming factor in determining the cadastral value of land, and their accounting is associated with socially fair taxation.

Although the Federal Standard for Assessment No. 4 establishes the cadastral valuation taking into account the encumbrances (restrictions), but as a price-forming factor the availability of the ZWSCUT is not included. Justification of the influence of water bodies on the activities was carried out in weight terms by an expert method in the article [6]. According to the work in water protection zones activity limits are 14 and 16% for garden and horticultural plots and in coastal protective zones 18 and 20%. Spatial analysis of the territory showed that about 56% of garden and horticultural area associations of Saint-Petersburg are covered by ZWSCUT. By this fact the importance of the factor «presence of ZWSCUT on a land plot» in determining the market value of the consideration land is proven [6].

3. Gathering and analysis of information

To calculate the cadastral value of garden and horticultural lands, the State Cadastral Valuation Method is used, approved by order of the Ministry of Economic Development of Russia on 12.05.2017 No. 226. According to it, the cadastral value of land plots of horticultural and gardening associations within the segment «Gardening and horticulture, low-rise residential buildings» is determined on the basis of a statistical analysis of market prices and other information about real estate, as well as other methods of mass real estate valuation.

The land plots of garden non-profit associations (GNA) «Slavyanochka-3» (217 land plots) and of GNA «Slavyanochka-2» (121 land plots) located in the borders of Saint-Petersburg were used to study the considered influence.

The cadastral value of land plots that are not in the areas of water bodies was calculated on the basis of price-forming factors used in the cadastral assessment of 2018. The list of such factors is formed by a specially created group of experts based on a typical list of factors of a particular segment.

Initially, based on the analysis of the market of the land plots in question, a list of all price-forming factors was formed that could influence their market value. Then, the type of factor was determined for each factor - qualitative or quantitative. For qualitative, a list of factor gradations (rank value), algorithms and description of these algorithms for calculating the values of factors were developed. Independent and importantly influencing factors were identified in the course of mathematical modeling of the significance price-forming factors of the cadastral assessment of 2018. These factors include: the area of the land plot; location (influence of highways and influence of local centers); engineering support (availability of electricity, water, heat, sewage and gas supply at the plot); air pollution level; «greening» of the plot; provision of the plot area location with social infrastructure objects; availability of access roads. Also two new factors were included: «availability of water protection zone on the plot» and «presence of the coastal protective strip on the plot». It will allow to estimate the difference in the cadastral value calculated in 2018 and the value determined with the inclusion of new factors. To obtain the values of price-forming factors and market data were used legitimate databases of state information resources «Regional geoinformation system of Saint-Petersburg» (RGIS), Public cadastral map and market data base of the City Department of Real Estate Inventory and Valuation (CDREIV).

After agreeing on the list of segment factors, there is a necessity in the transition to the definition of a reference land plot for each association. Therefore, a plot with an average GNA area and an incomplete set of utilities (power supply and water supply), which is typical of these GNAs, was selected. The use of a comparative approach in determining the value involves the search for reliable information about transactions or offers on the market of the segment to which the object of valuation belongs. Due to the CDREIV database, statistical information on objects-analogues was selected, which was necessary for calculating the value of the reference land plot. The required information includes individual characteristics of each analogue, its market value at the date of the offer in the market or purchase and sale transactions.

For the presentation of statistical market information, estimated objects and water bodies zones was used the geoinformation system (GIS) «MapInfo Professional 11.5». Figure 1 shows a part of the land plots partially or fully located within the boundaries of a water protection zone and a coastal protective zone and for which the cadastral value will be calculated.

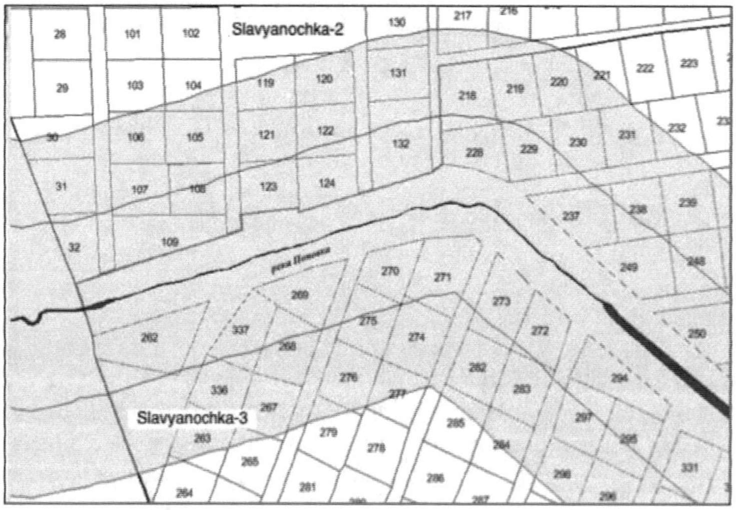

Figure 1. Objects of assessment of partially or fully located within the boundaries of water areas

4. Modeling the cadastral value of garden and horticultural lands

To accomplish the research goal, the modeling was carried out under the conditions of the presence and absence of a water protection zone and a coastal protective strip on the land plots. The first step in building models was to make adjustments for the elements of comparison: terms of sale (discount for

IOP Conf. Series: Earth and Environmental Science **350** (2019) 012069 doi:10.1088/1755-1315/350/1/012069

the offer was taken into account according to CDREIV data) and market conditions (the dates of sale of objects were taken into account according to the method specified in [7]). Then, we made a qualitative analysis of the characteristics and prices of analogous objects, based on determining the rank values of the quality characteristics. The next step was calculated weighted quality indicators, but it was necessary to establish the weighting factor of each price factor for this [8]. In order to fully assess the situation on the market, weights of factors were calculated by a mathematical method for calculating weights using the feature «Search for Solution» of MS Excel [9, 10]. The results of calculating the weights for each of the models are presented in Table 1. The weighted relative quality indicators of the models are determined by multiplying the obtained weights by the values of the relative coefficients.

Table 1. The weighting factors of price-forming factors for 1 and 2 models

Model / Factor	Plot area	Location	Engineering support	Level of air pollution	Greening area of the plot	Provision of the plot location district to social infrastructure	Presence of access roads	Share of water protection zone on the land	Share of coastal protective strip on the	Sum
Weight for 1 model, %	7.8	21.3	25.9	17.9	6.3	10.2	10.6	-	-	100
Weight for 2 model, %	1.4	11.7	28.6	14.0	12.2	12.2	5.0	18.6	2.2	100

At the next stage, the coefficients of factors were calculated, the value of which determined the feasibility of introducing them into the value model (the factor was excluded with a coefficient of significance less than 0.3 according to the Cheddock scale). The test for multicollinearity revealed the interdependence between the factors «the share of the water protection zone on the land plot» and «the share of the coastal strip on the land plot» (coefficient of inter-factor correlation is more than 0.75). As a result, the second of the indicated factors was excluded from the calculations, since its coefficient of partial correlation is lower, and therefore weights and weighted quality indicators are recalculated.

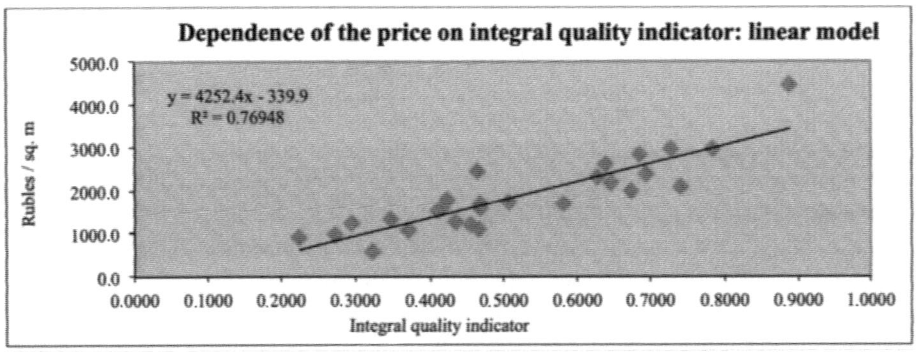

Figure 2. Dependence of the price on integral quality indicator: linear model (taking into account the water protection zone)

The construction of the regression equations was made on the basis of the calculated integral quality factor (IQF) and the price of land. 4 types of the equations were built for each of the 2 models (linear, exponential, logarithmic and power) to determine which of them best describes the statistical information. According to the results of the analysis, the most reliable was a linear model (Figure 2) for both models without taking into account the water protection zone. Coefficients of determination are 0.64 and 0.77, which indicates that the model is above average.

Determination of the cadastral value of the land burdened with ZWSCUT is proposed to be calculated using the formula presented in the monograph [5]. It contains the ZWSCUT accounting coefficient for the ith burdened part of the land plot. This coefficient can be determined by different methods depending on the development of the land market, but in this study it is calculated for the water protection zone based on the obtained cadastral value models. It includes the ratio of the received specific cadastral value (SCV) without considering ZWSCUT ($Y_{without\ ZWSCUT}$) and SCV taking into account ZWSCUT ($Y_{with\ ZWSCUT}$) according to the formula:

$$C_i = \frac{Y_{\text{without ZWSCUT}}}{Y_{\text{with ZWSCUT}}} = \frac{y = 4049.1x - 27.959}{y = 4252.4x - 339.9} \quad (1)$$

The calculation of the water protection zone coefficient is presented in Table 2.

Table 2. Calculation of the water protection zone coefficient for cadastral valuation of garden and horticultural lands

Model 1 (without water protection zone)		Model 2 (with water protection zone)	
R^2	0.64	R^2	0.77
$F_{calc.}$	45.19	$F_{calc.}$	83.45
Level of significance F	4.80E-07	Level of significance F	1.92E-09
$F_{crit.}$	4.24	$F_{crit.}$	4.24
Value, rub/sq.m	1346.59	Value, rub/sq.m	1233.07
Area, sq.m	1034.8	Area, sq.m	1034.8
Value, rub	1393454.73	Value, rub	1275977.09
Water protection zone coefficient (C_{WPZ})			0.92

5. Analysis of the results

The results obtained in the research process allow us to draw a number of conclusions concerning both the improvement of the methodology for determining the cadastral value of land in general and the results of calculating the coefficient of water protection zone accounting directly. In general, the proposed changes in the methodology for cadastral valuation of land burdened by ZWSCUT allow for ensuring socially fair land taxation.

Modification of the method in terms of accounting for the factor «availability of ZWSCUT» allows one to use as its parameters the share of the area occupied by ZWSCUT, as well as the rules for using the territory within its limits. Using these optios will help to calculate the coefficient of accounting for a specific type of ZWSCUT. In this case, the method of calculating these coefficients depends on the development of the land market and the availability of information to the population about the presence of restrictions on the use of the territory.

In general, it can be said that there is a market reaction to the presence of a water protection zone, which is confirmed by the significance of this factor in modeling. The influence of the water protection zone is logical (a «-» sign with regression equation coefficients) and corresponds to the hypothesis of a decrease due to its presence in the price of land.

IOP Conf. Series: Earth and Environmental Science **350** (2019) 012069 doi:10.1088/1755-1315/350/1/012069

6. Conclusion

Thus, it should be noted that the use of the calculated coefficient of accounting the water protection zone in determining the cadastral value of land plots of garden and horticultural associations by the example of Saint-Petersburg showed the following.

Firstly, when covering a part of land plots of equal size with a water protection zone, the cadastral value on average decreases by 65 000 rubles (5% of the value), which, respectively, reduces the land tax for right holders by 195 rubles.

Secondly, the greatest change in the cadastral value concerned the land plots completely located within the boundaries of the water protected zone, the value of which on average decreased by 120 000 rubles (9% of the value; land tax decreased by 360 rubles).

References

[1] Varlamov A A and Gal'chenko S A 2000 *Land cadaster: study guide* (Moscow: Russian state agrarian university MSXA named after K.A. Timiryazev) p 288

[2] Kretinin K V and Kuznetsov N A 2003 Compensation payments to agricultural enterprises from users of engineering networks and communications *Monitoring of lands in the resource management system of the regions of Russia* 110-113

[3] Sutyagin V Yu 2016 Caution, protected zone! Or what you need to know when challenging the cadastral value *Scientific notes of the Tambov Regional Branch of the Free Economic Society of Russia* (Tambov: Tambov state university named after G.R. Derzhavin) 54-68

[4] Kovyazin V F and Romanchikov A Yu 2018 Problems of determining the cadastral value of forest land *Journal of Mining Institute* **229** 98-104

[5] Gustafson R J, Grumstnup P D, Herdrickson E R and Meyer M P 1979 Land lost production under and around electrical transmission line structures *Transaction of the ASAE and CSAE* 16

[6] Bykowa E N 2018 *Assessment of land with burdens in use. Theory and methods* p 240

[7] Vol'nova V A 2015 *Review of options for calculating adjustments in real estate valuation by a comparative approach* p 56

[8] Gribovskii S V, Ivanova E N, L'vov D S and Medvedeva O E 2003 *Property valuation* p 704

[9] Gavrilova A S 2015 Using a qualimetric model in a comparative approach *Evaluation questions* **02(80)** 24-32

[10] Gavrilova A S 2016 Using a qualimetric model in applying a comparative approach *Property Relations in the Russian Federation* **7(178)** 46-574

Protection of coastal territories of Russian water reservoirs using land management methods

O A Sorokina, L E Petrova, A V Dontsov, A V Fedorinov and D V Antropov

State University of Land Use Planning, 15, Kazakova ave., Moscow, 105064, Russia

E-mail: sorokinaoa81@gmail.com

Abstract. The article discusses protection methods for coastal areas of large Russian reservoirs from flooding. The object is the Saratov water reservoir which has a significant destructive effect on the adjacent territories. The methods of land monitoring and land management are applied to protect real estate objects from flooding. Recommendations for improving territorial planning schemes of municipalities adjacent to reservoirs are suggested.

1. Introduction

In Russia, the total flood-prone territories is 400 thousand km^2 of which up to 50 thousand km^2 are annually flooded. Thousands of settlements with a population of about 4.6 million people, economic facilities and more than 7 million hectares of agricultural land are flooded.

Historically, water reservoirs have been located upstream of rivers or in large settlements. As a result of their destruction, settlements are flooded and a significant damage is caused to life and health of residents and economies of regions. In the Far East, Volgograd, Kemerovo Region, the Altai Republic, Khakassia, Tyva, Altai and Krasnoyarsk Krais, large-scale floods caused a multimillion-dollar damage to real estate objects.

In most cases, dam accidents occur during their construction or initial period of operation. During this period, construction defects manifest themselves, the filtration regime and deformations of the structure stabilize. During the next 40–50 years, the state of the structure remains stable, and accidents are not not probable. After that, the risk of accidents increases as a result of deformation of building materials.

In Russia, the average age of hydraulic structures is 50 years; analysis of emergencies indicates the need for continuous monitoring and preventing floods. Land management activities play a significant role.

The operation of reservoirs is regulated by two legal acts: the rules for using water resources of reservoirs and the rules for technical operation and improvement of reservoirs.

These documents [1] compiled on the basis of hydrological calculations cannot be used by land surveyors and urban planners, since information on the zones of influence of reservoirs is presented schematically on small-scale maps; there are no data describing boundaries of these zones; there is no information on degradading lands. The lack of areal data on the impact of reservoirs on coastal areas made it impossible to take into account flooding and land degradation in urban planning and land management documents. It is necessary to take into account the influence of zones with special conditions while determining the cadastral value of land. The causes of various damages are the lack of information about restrictions on economic activities (and, accordingly, water protection zones and coastal strips) and elaboration of a mechanism to compensate land owners for the damage caused by such restrictions [2, 3]. It has been established that the presence of a water protection zone has an effect on the economic nature of real estate objects located in this zone [4–6].

This article desccribes methods used for land monitoring and management aimed at planning the use of real estate objects in areas subject to flooding. Recommendations for improving territorial planning schemes of municipalities adjacent to reservoirs are suggested.

2. Materials and methods

The theoretical basis of the study is works of Russian and foreign scientists [7–10], official reports, resolutions, methodological manuals, and federal and regional legal acts.

The research object is the Saratov water reservoir built in 1967–1968 on the Volga river. Its physical parameters are as follows: the length is 357 km, the average width is 5.1 km, the area of the mirror at the normal retaining level is 1,831 km^2, the length of the coastline at the NLW water is 865 km.

The reservoir was built for hydropower, water supply, fisheries, irrigation, recreation, timber rafting and agriculture. It is located in Ulyanovsk, Samara and Saratov regions.

The Saratov reservoir has a significant destructive effect on the adjacent territories:

- the length of landslide slopes is 570 km, including dangerous areas - 85 km (15% of the coastal perimeter),
- the total bank processing area is about 50 km^2, the maximum width of the intensive processing zone at different coastal areas is 60–100 m,
- the total length of the coastline is 65 km, the length of the fortified bank is 25 km.
- the area of flooded coastal areas is 180 km^2 (11% of the mirror area).

On the basis of the comparative statistical analysis of open data, measures of land management aimed at protecting the banks of the reservoir have been ranked by their efficiency.

More than 15 municipal districts adjoin the reservoir. They have general plans for the development of the territory. Measures to combat land degradation are declarative: specific locations of planned erosion control measures in the coastal areas have not been established, and their volumes have not been determined.

The assessment of territorial planning schemes of municipal districts made it possible to supplement sections on the protection of coastal areas.

3. Results and discussion

The construction of shore protection structures has the most significant effect on bank destruction.

On the basis of the collected Internet data, a database on the implemented analogue objects, the volume of prevented bank processing, m/year and the cost of one meter of bank protection measures has been created. The linear functional dependence of the cost of bank protection (y) on the bank processing (x) has the form $y = 79.784 + 51.285 * x$. On its basis, you can build a financial forecast.

On the basis of these calculations, the feasibility of bank protection works in 12 settlements of Samara Region was determined (Figure 1). The total length of bank protection is 39 kilometers, of which 22 kilometers is a new construction object.

Taking into account the ongoing processes in the territories adjacent to the Saratov reservoir, the types of coastal fortifications were ranked according to their effectiveness, taking into account the volume of capital investment: relatively cheap sloping coastal fortifications, stripes, the most expensive resistant belt and retaining walls. Resistant belts are expedient for protection of settlements, while sloping coastal fortifications are used on the inter-settlement territories.

Additional measures to strengthen the slope were recommended: replanting slopes, seeding slopes with special plants or surface trimming with natural turf; rock dumping and erection of additional mounts at its junction with e coastal shoals; alluvium of gentle sandy beaches; maintenance and protection of existing forest plantations and shrubs on the slopes and in adjacent reservoir areas; special water-intercepting upland shafts and channels.

Figure 1. The scheme of the bank protection facilities of the Saratov water reservoir in Samara region

In case of flooding of adjacent territories, it is recommended to develop working projects for land drainage taking into account two main areas of agricultural application of the zones:

1) radical or superficial improvement of forage lands planting appropriate herbs;

2) drainage followed by creation of meadow and arable land plots (this direction is beneficial for large and compact areas over 100 hectares).

The agricultural use of shallow waters makes it possible to develop agriculture and animal husbandry in the areas adjacent to the reservoir (cheap fodder production due to natural aquatic vegetation and cultivation of moisture-loving crops - wild rice, canary). These lands can be used to collect fodder hydrophytes (reed, water manna, aloe telorez, pierced and brilliant and rutus spiky). These plants are highly productive, occupy large areas and have excellent fodder qualities.

For the further rational use of adjacent lands, it was proposed to develop land management measures:

1. Special survey, identification and assessment of the size of flooded territories, bank processing, compilation of thematic maps using GIS systems;

2. Ecological and economic, agro-ecological and ecological-landscape zoning (zoning);

3. Inter-farm land management, rational reformation of the municipal land fund;

4. Development of measures for improving natural landscapes, restoring and improving soil fertility, eliminating waterlogging.

The construction of bank protection structures is carried out within the boundaries of the water protection zone. To construct them, a draft water protection zone and a coastal protective zone of the reservoir are being developed. The width of the water protection zone and coastal protective strip is determined in accordance with Article 65 of the Water Code of the Russian Federation. The boundaries of the water protection zone and coastal protective strip of the reservoir, their coordinates and reference points are described. The lack of information about the boundaries of zones causes various damages, because the lack of information on the exact location of land (the real estate cadastre) and information on restricted application of lands, boundaries of special zones (the border register) have a direct impact on the value of economic regulators of land management.

Thus, the structure and composition of the cadastral value are influenced by various pricing factors (restrictions on the application of the object located in the water protection zone). The results of studies have shown that there is a negative relationship between the cost and the presence of water protection zones. It is obvious that using the mechanism of redistribution of land payments at the stage of cadastral value formation, it is possible to compensate for losses arising from restrictions on the use of land in special areas.

The availability of this information will be useful for those who carry out assessment activities taking into account the influence of special zones on the total value of real estate objects. It will help reduce damages caused by the lack of information on water protection zones in the unified state register of real estate objects when calculating the cadastral value of special land plots. It can also be used when making decisions on the withdrawal or repurchase of these objects.

4. Conclusion

The aspects of land management activities carried out in areas affected by reservoirs showed the need for their integrated application in land management schemes taking into environmental and economic damage. These materials will reduce the damage caused by the lack of information in the state information registers, have a direct impact on the establishment of various types of land payments and contribute to the sustainable land use.

It was proved that bank destruction can be stopped by building bank protection facilities. Protective belts can be built for protecting settlements while sloping coastal fortifications can protect inter-settlement territories. The recommendations for improving the territorial planning of municipal areas by adding sections on the protection of coastal areas were provided. These measures will contribute to the rational socio-economic development of territories, the effective use of flood-prone areas and reduce possible damage to real estate objects.

References

[1] Lidin K L, Meerovich M G, Bulgakova E A, Vershinin V V and Papaskiri T V 2018 Applying the theory of informational flows in urbanism for a practical experiment in architecture and land use *ESPACIOS* **39(1)** 12

[2] Varlamov A, Galchenko S and Bogdanova O 2018 Theoretical and methodological foundations for formation of sustainable land management system *IOP Conf. ser. Materials Sci. and Engineer.* **451(1)** 012141

[3] Tsarenko A A, Schmidt I V and Tarbaev V A 2017 Actual aspects of cadaster relations in the monitoring system of land and rural areas *J. of Environmental Management and Tourism* **8 1(17)** 116–22

[4] Voronkova O Y, Akhmetshin E M, Sycheva I N, Shpakova R N, Pashkova E Y and Poltarykhin A L 2018 Economic mechanism of regulating land relations in the agricultural sector of Russia *Europ. Res. Studies J.* **21(4)** 280–91

[5] Amirova E F, Voronkova O Yu, Pyurveeva K A, Shatalov M A, Panteleeva T A and Sorokina O A 2018 Functioning of agroindustrial complex in the conditions of digital economy *Int. J. of Mechanical Engineer. and Technol.* **9(13)** 586–94

[6] Vershinin V V, Murasheva A A, Shirokova V A, Khutorova A O, Shapovalov D A and Tarbaev V A 2016 The solutions of the agricultural land use monitoring problems *Int. J. of Environmental and Science Education* **11(12)** 5058–69

[7] Karpenko L V 2009 Forecasting ecologically hazardous phenomena resulting from the impounding of swamps by the Boguchany reservoir *Geography and Natural Resources* **30(2)** 126–30

[8] Shirokova V A, Khutorova A O, Gurov A F, Shirokov R S and Yurova Y D 2017 Environmental monitoring of small rivers of the Moscow region (on the example of the river Osetr) *GeoConf. SGEM* **18** 355–62

[9] Ozerova N A, Shirokova V A, Krivosheina M G and Petrosyan V G 2017 The spatial distribution of Sosnowsky's hogweed (Heracleum sosnowskyi) in the valleys of big and medium rivers of the East European Plain (on materials of field studies 2008–2016) *Russ. J. of Biological Invasions* **8(4)** 327–34

[10] Klyushin P V, Lukyanova T S, Shirokova V A, Khutorova A O and Savinova S V 2018 The current state of boggy territories in the Meshchyora national park *18th Int. Multidisciplinary ScientificGeoConf. SGEM–2018* **18(2.1)** 775–82

Rural development in the Far Eastern Federal District

V N Khlystun[1], A V Vdovenko[2], V M Stolyarov[1] and E M Chepurin[1]

[1]State University of Land Use Planning, 15, Kazakova Str., Moscow, 105064, Russia
[2]Pacific National University, 136, Tikhookeanskaya Str., office 241-center, Khabarovsk, 680035, Russia

E-mail: vms88@inbox.ru

Abstract. The article presents the material relating to the current state and prospects of development of rural areas in the Far Eastern Federal district of Russia. It is shown which factors are restraining and which factors are stimulating, as well as the degree of influence of each group on the development of rural areas of the Far East of Russia. The practical significance lies in the fact that the results of the study can be useful in the development and justification of rural development programs and regional agricultural policy.

1. Introduction

Currently, in the light of the latest trends in the development of the Russian national economy, as well as global economic relations, there is a need to address issues related to the efficiency of agricultural land use, increase the profitability of agricultural production, ensuring sustainable development of rural areas.

A considerable number of works are devoted to the issues of increasing the efficiency and sustainable development of rural areas in the Far East and the country as a whole [1–3].

One of the objectives of this work is to develop proposals for sustainable development of rural areas of the far Eastern Federal district of the Russian Federation on the basis of increasing the profitability and efficiency of regional agricultural production. The development of proposals was carried out within the framework of the adopted State program for the development of agriculture in the Far Eastern Federal district.

2. Materials and methods

In the development of proposals, the analysis and forecast of the rational use of agricultural land was carried out, the methods of correlation-regression and trend analysis were used, the balance method was used on the materials obtained from public sources (reports, statistical material). As a result of the application of these methods, the main problems in the agro-industrial complex of the far Eastern Federal district are identified, proposals for the development of rural areas are given.

3. Study of the development of the agricultural sector and rural areas in the far Eastern Federal district

The agricultural sector of the Far East operates in difficult climatic and socio-economic conditions. The main reasons for this are the geographical location, climate, remoteness from major industrial centers and rural areas. More than 80 % of the region's territory belongs to the far North and equated areas.

The economic reforms of the early 1990s produced different results, including both positive and negative ones. The negative results, for example, include a long disparity in prices for agricultural and industrial products, which in turn led to a shortage of working capital and fixed assets, as a result – many agricultural enterprises became unprofitable, production and sales fell to extremely low levels, there was an outflow of the rural population.

Severe natural and climatic conditions in the far East have strengthened these trends and, together with the aggressive economic environment, have created a threatening situation for agriculture in the Far East.

The positive results of agrarian and land reforms in the far East (as well as throughout Russia) include, of course, the emergence of private agricultural organizations, the abolition of the state monopoly on agricultural land, the emergence of incentives for the rural population to process their own plots, to remain in rural areas, increasing the potential of rural areas. Positive results cannot be achieved without comprehensive state support.

At the moment, the Federal law "On the development of agriculture" of 29.12.2006 N 264-FZ, as amended in December 2018, defines the main provisions and vectors of the development of the state agrarian policy [4–6].

In order that the negative phenomena were as much as possible weakened or could be more systematically controlled, in the process of implementing the state agrarian policy it is necessary to take into account the regional features of the Far East, including its economic sectoral structure, the features of the social sphere and so on [7].

Therefore, the study of trends, factors and conditions that determine the results of the agricultural sector in a particular region is not only scientific, but also practical, and thus economic entities will be able to more reasonably approach the improvement of business and increase the production of competitive products.

We believe that the effectiveness of rural development is consistent with such criteria of efficiency of agricultural and land policy as: increasing the number of working population and fertility in rural areas; increasing the area of rural areas; increasing the area of effectively used productive land; the formation of effective mechanisms of state influence on the sustainability and balance of production and economic indicators of agricultural and industrial activity; ensuring the protection of land fertility; increasing the share of environmentally friendly products; increase in exports of agricultural raw materials and products of food and processing industry with simultaneous growth of the share of deep processing products in total exports.

The agricultural sector of the Far East is characterized by uneven provision of land resources and especially arable land. Rural areas exist and operate in difficult conditions. It is worth noting that the agricultural lands of the Far East are characterized by a sufficiently low natural fertility, and the organizational scheme of agriculture is complicated by the remoteness of production and sales centers of both production resources and finished agricultural products.

An important aspect that also has an impact is the state of the material and technical base of enterprises and the labor potential of the rural population. Nevertheless, taking into account all the above, we can say that all activities that contribute to a more complete and effective use of land can be combined into the following groups:

1. Inclusion in the production use of each hectare of land assigned to the economy: it is unacceptable that they fall out of economic turnover.

2. Increase of economic soil fertility is primarily irrigation and drainage, chemical reclamation, application of fertilizers, development of crop rotations, surface and radical improvement of meadows and pastures.

3. Preservation of fertility and conservation of soil: field-protective afforestation, soil conservation technologies and crop rotations, the system of measures on combating water and wind erosion.

4. Rational use of economic soil fertility: the use of the most productive varieties, improved seed production, improvement of plant placement schemes, compliance with the optimal timing of agricultural work, the fight against plant diseases, pests and weeds.

5. Organizational and economic activities: improving the structure of acreage taking into account

market conditions, deepening specialization, the use of progressive forms of organization and remuneration, improvement of forms of management, etc.

Organization of effective use of agricultural land is possible only on the basis of rational organization of the territory of specific agricultural activities. To do this, it is necessary to develop land management projects with a detailed study of organizational,economic and agro-economic justification.

The far Eastern Federal district is the largest district of Russia, its area is 6,217 thousand km^2 or 36.4% of the country. More than 80% of the territory of the Far East are legally assigned to the regions of the Far North and equated to them areas, 90% of them are covered by permafrost, in the remaining territories there is a long seasonal freezing of soils, thus, in the land Fund of the region, agricultural lands occupy 1.1% of the territory, and arable land — only 0.4% [8].

The main agricultural territories of the far Eastern Federal district are the Amur region, Primorsky and Khabarovsk territories, as well as the Jewish Autonomous region, they are called the southern territories of the far Eastern Federal district, which are part of the Amur river basin. In these regions concentrated 4,244 thousand hectares of agricultural land or 78.7% of all farmland in the Far East.

In Primorye and in the South of the Amur region, many crops are cultivated, including heat-loving ones: soybeans, rice, corn, tomatoes, grapes. The area of the southern territories is 1.5 million square kilometers, which is 24% of the entire territory of the Far East.

On the territory of the South of the Far East, the largest areas of land are occupied by forests. The potential of soils in the region is small, but sufficient for the formation of average yields in conditions of limited use of fertilizers.

In order to increase the productivity of agricultural land and obtain sustainable crops, a regular set of measures for soil treatment is necessary: land reclamation and irrigation, the use of innovative technologies to restore soil fertility.

During the period of agrarian reforms in the far East, there was a deterioration in the use of agricultural land. If in 1990 – 6,649 thousand hectares of land were used in agricultural production, including 3,191 thousand hectares of arable land, in 2014 – 5,405.1 thousand hectares of land and 2,565.4 thousand hectares of arable land, and by 2019, according to various estimates, the area of arable land has decreased to 2,500 thousand hectares.

The structure of agricultural production is dominated by soybean. The main regions in which soybean sowing is carried out are the Amur region (up to 70% of all crops are concentrated), the Primorsky region (18.3 % of crops), the Jewish Autonomous region (9.1 % of crops), the Khabarovsk region (2 % of crops).

It is necessary to say about the investment potential of the region in question. It should be noted that in the region as a whole, there is a tendency for a decline in investment in the agricultural sector. The material and technical base of the manufacturers is deteriorating, there are no comprehensive programs for updating equipment and machines. Unfortunately, for most manufacturers, access to subsidized loans is denied, in view of this, the renewal of fixed capital is not possible.

Note also the disappointing trends in the state of the park of agricultural machines. The park of combine harvesters in the region decreased by 8 times, forage harvesters - by 16 times, and potato harvesters by 28. Disposal of other types of equipment also exceeds its intake. Plows, seeders, cultivators, sowing complexes, sprayers, equipment for fodder among the scarce machines.

Due to the reduction, and in some cases, the cessation of the production of machines in Russia, agricultural producers were forced to buy imported equipment until 2014, but after the start of the international competition struggle for many people, the purchase of foreign equipment was not available.

Due to the prevailing economic conditions, work on land reclamation has practically ceased. A significant part of irrigated and drained land is taken out of circulation and without reconstruction of the amelioration systems it is impossible to put them into economic circulation. More than half of the irrigated and drained lands are in unsatisfactory condition.

Under current conditions, it is unrealistic to count on a massive influx of private investment in the agricultural sector of the Far East in the coming years.

The widening of the gap between the living standards in the city and in the countryside increases the social unattractiveness of the rural areas, stimulates the migratory moods of the rural population, threatens the disruption of the development programs not only of the agricultural sector, but also of the entire economy of the region.

The rural population of both Northern and southern border areas is declining. In the southern territories, the rural population decreased by 20% by 2019. Local residents continue to leave the far Eastern Federal district.

Since 2018, the outflow of population has increased and for the first 4 months only due to the outflow of local residents, the number has decreased by 20 thousand people. Unfortunately, there are no actual statistics on the outflow of the rural population for 2019, but table 1 contains data on the migration outflow for the period from 2000 to 2015, which generally allows us to understand the processes of the outflow of the rural population [9].

Young people and people of working age leave the village most actively, the population is aging, and the gender balance is violated. Similar processes are taking place in rural areas. According to the forecast, by 2020, with an increase in the total load by 25.5 %, 49.6 % will fall to children and 50.4 % - to persons older than the working age.

Table 1. Migration growth rates (per 10 000 people)

Territory	2000	2005	2010	2012	2013	2014	2015
Russian Federation	24.7	20.0	19.0	20.6	20.6	18.8	16.8
Eastern federal district	-83.4	-80.5	-49.4	-31.8	-53.0	-39.8	-39.0
Primorsky territory	-44.3	-50.6	-34.8	-5.7	-36.7	-20.4	-14.5
Khabarovsk territory	-38.3	-92.9	-30.8	-4.0	-22.2	-18.9	-36.9
Amur region	-91.0	-99.9	-59.9	-52.8	-70.9	-15.6	-46.7
The Jewish Autonomous region	-51.7	-158.8	-49.1	-89.2	-124.8	-108.2	-120.4

Attracting foreign labor is an objective process, but it should be clearly regulated by the state. It is advisable to attract labor from the CIS countries, to simplify the procedure for obtaining Russian citizenship to persons who have moved to permanent residence in the far East.

The involvement of citizens and companies of China, the Republic of Korea and North Korea creates conditions favorable not only for foreign companies, but for all parties of economic cooperation

In recent years, more than 90% of agricultural enterprises do not conduct training, which leads to the degradation of the labor force. Low qualification of personnel leads to inefficient use of all resources and stimulates production decline.

We believe that the basis for the development of proposals for sustainable development of the region should be applied methods of mathematical modeling, including the blocks presented in table 2 [10, 11].

When developing the forecast of development of agriculture of territories of the far Eastern Federal district till 2030 two enlarged scenarios are developed:

The first – inertial, which reflects the actual trends in the development of agriculture. It is assumed that in the future they will remain unchanged. It is expected that the improvement of certain elements of the economy of the agricultural sector will be gradual, evolutionary in nature, and this will cause a low rate of growth of agricultural production.

The second – an optimistic scenario - is based on the assumption of the normative provision of agriculture with resources and its management in accordance with the zonal systems of agriculture. This scenario is based on optimistic trends and can be used as an indicator of the development opportunities of agricultural production sectors, which should be sought.

IOP Conf. Series: Earth and Environmental Science **350** (2019) 012071 doi:10.1088/1755-1315/350/1/012071

Table 2. The structure of the economic and mathematical model of forecasting the production of agricultural products and the need for material and technical resources

A	**Box 1. Forecast of production volumes of agricultural products**
1.	Econometric model
1.1	Trend models (dependences linear, exponential, power, logarithmic),
1.2	Correlation and regression models $У = a + b_1x_1 + b_2x_2 + \ldots + b_nx_n$ – multiple correlation
2.	Simulation modeling
B	**Box 2. Forecast of material and technical resource requirements**
1.	Balance model
1.1	$Q_{ftlz} = \sum[(У_i \cdot N_i \cdot K_x) \cdot S_i]$ - where Q_{ftlz} – fertilizer demand, total; $У_i$ – I-th crop yield; N_i – fertilizer application rate per 1 centner of the i-th type of product; $К_x$ – correction factor for soil nutrient content; S_i – sown area of the i-th crop $Q_{fualu} = \sum(N_{iy} \cdot S_i)$ - where Q_{fualu} – demand for fuels; N_{iy} – fuel rate at a certain yield of the i-th crop; S_i – sown area of i-th culture $Q_{pocc} = \sum(N_i \cdot S_i)$ - where Q_{pocc} – electricity demand; N_i – the rate of electricity consumption per unit of the i-th type of product at the projected yield; S_i – sown area of i-th culture
1.2	$Q_{ik} = \sum[(N_{.} \cdot U_i) \cdot n] : K_{ik}$ - where Q_{ik} – volume of i-th type of feed; $N_{.}$ - feed requirements per 1 head of livestock; U_i – the proportion of the i-th type of feed in the structure of feed; $n_{.}$ – livestock population; $К_{ik}$ – coefficient of nutritional value of the i-th type of feed
3	$V_i = Q_{ir} + \sum_{i=1}^{m} Q_{ip} - \sum_{i=1}^{n} Q_{iv}$ - where V_i – the projected volume of the region's demand for i-th kind of product; Q_{ip} – the volume of receipt of the i-th product in the region by m - channels; Q_{iv} – the volume of export of the i-th type of product by n - channels and its consumption for further processing in the region

4. Conclusion

According to the developed proposals for the development of agricultural production in the Amur region, there is a steady growth in the acreage of grain crops. In agricultural organizations, the growth will be 15.7% by 2030 in relation to 2015, in the Peasant Farm – by 22.2%. As noted earlier, in the Amur region, most of the arable land has been developed and used by various agricultural producers. Accordingly, the increase in gross revenues is achieved through the intensification of production. For all crops in the forecast period projected yield growth of 10-18%.

In the Khabarovsk territory, the implementation of proposals for the first option for the development of grain production will increase the acreage in agricultural Enterprises in 2025 in relation to 2011 by 0.5 thousand hectares, the second option — by 1.3 and the third option — 2.6 thousand hectares.

In the Primorsky Territory, in the first variant of the development of the agricultural sector, there will be a slight increase in acreage in all categories of farms and will amount to 334.4 thousand hectares in 2030. Potato and vegetable growing in the region should be carried out on the basis of a rational combination of potato and vegetable production for local consumption with the formation of specialized enterprises focused on the cultivation of seed products and focused on export to the northern regions of the Far-Eastern Federal District.

Agri-food industry can be one of the main sectors in the economy of the Far East. For the Russian leadership, the need for self-reliance has become apparent since the beginning of the country's economic and political isolation. After the establishment of sanctions, import substitution is used both as a response mechanism to Western sanctions for illustrative purposes, and as a timid attempt to sovereign the Russian economy. The development of rural areas of the Far East is impossible without an integrated approach to the development of agricultural production, ensuring access of producers to credit resources, popularization of the rural lifestyle.

References

[1] Uzun V, Shagaida N and Lerman Z 2019 Russian agriculture: Growth and institutional challenges *Land Use Policy* **83** 475-487
[2] Griewald Y, Clemens G, Kamp J, Gladun E, Hölzel N and Dressler H 2017 Developing land use scenarios for stakeholder participation in Russia *Land Use Policy* **68** 264-276
[3] Smirnov S 2015 Economic Fluctuations in Russia (from the late 1920s to 2015) *Russian Journal of Economics* **2** 130-153
[4] Boyko A, Bogomaz E, Glaz N, Emel'yanova E, Esinovskiy A et al 2013 *Agrarian sector of the Far East: problems and development prospects* ed Shelepa A (GNU DVNIIEP AIC of the Russian Agricultural Academy)
[5] 2012 *The State Program for the Development of Agriculture and Regulation of Agricultural Products, Raw Materials and Food Markets for 2013–2020*
[6] *On approval of the state program of the Russian Federation "Socio-economic development of the Far East and the Baikal region" Resolution of the Government of the Russian Federation of 15.04.2014 No. 308* Legal system "Consultant Plus", available at: http://www.consultant.ru/document/cons_doc_LAW_162190/
[7] Khlystun V and Alakoz V 2016 Mechanisms for the inclusion of unused land in agricultural turnover *Economics of agricultural and processing enterprises* [in Russian – Ekonomika sel'skokhozyastvennykh i pererabatyvajuschikh predpriyatiy] **11** 38-42
[8] *Official website of the Ministry of the Russian Federation for the Development of the Far East,* available at: https://minvr.ru/
[9] *Official website of the Agency for the Development of Human Capital in the Far East,* available at: https://hcfe.ru
[10] Vdovenko A and Kim L 2013 *Handbook on the Features of land management in the Far Eastern Federal District (*Khabarovsk: Pacific State University) p 117
[11] Murasheva A 2005 The development of the territories of the Far Eastern region and ways to solve the environmental problem *Cadasturu, organizerateritoriului, ingineriamediului* **13** 80-85

IOP Conf. Series: Earth and Environmental Science **350** (2019) 012072 doi:10.1088/1755-1315/350/1/012072

Application of index of general toxic effects of biologically available polycyclic aromatic hydrocarbons as leading indicator in assessing the quality of industrial-urbanized territories

V V Vershinin[1] and A S Nartov[1,2]

[1]State University of Land Use Planning, 15, Kazakova str., Moscow, 105064, Russia
[2]Institute for Chemical Reagents and High Purity Chemical Research Center of Kurchatov Institute, 3, Bogorodsky Val str., Moscow, 107076, Russia

E-mail: v.vershinin.v@mail.ru , nartovalexander@gmail.com

Abstract. The authors consider the important issue relating to the assessment towards the terrotory exposed to some toxic effects of man-made polyaromatic hydrocarbons (PAHs). As the main criterion applied in referring a territory to a particular level, it is proposed to use indices of the total toxic effects of bioavailable PAHs. Depending on the values of these indicators, we point out seven levels and propose for each one some evidence-based recommendations on the most rational use of industrialized urban territories.

1. Introduction

The level of technogenic pressure on all components of the environment is growing rapidly, which is the inevitable consequence of the technical development of society. The global scale of environmental challenges and their consequences began to be realized relatively recently, about 60-70 years ago, and in some countries even later. Owing to a gradual understanding of the seriousness of environmental issues, over the past few decades, there have been identified some technogenic factors that have the most critical impact on certain components of the environment. Including identified groups of pollutants, the content of which must be controlled in the soil, water, atmosphere and in objects of wild nature. The list of such compounds includes polyaromatic hydrocarbons, related to persistent organic pollutants and having a carcinogenic and mutagenic toxic effect [1].

Today, new methods of recultivating land contaminated by various ecotoxicants, including PAHs, continue to be actively introduced in the world [2]. However, the limited land area suitable for living and conducting human activities, with an increasing population in the world and its needs, necessitate development of a scientific approach to management applied towards polluted territories, not only in terms of improving their quality. In our view, it is required also the substantiated distribution of the land resources between the tenant farmers, based on urgent data of geo-ecological monitoring.

2. Methods and materials

Recommendations towards the lands use exposed to man-made and technogenic impacts were given on the basis of the levels of toxic effects of bioavailable PAHs in the designated territory.

The indices of the total toxic effects of bioavailable PAHs, proposed by the authors, were used as an indicator determining these levels: $I_{2\text{-}3}$ (for light polyaromatic hydrocarbons, containing less than 4 aromatic rings in the structure) and $I_{4\text{-}6}$ (for heavy ones, containing 4-6 aromatic rings), each ring was calculated using the formula (1).

$$I = \sum_{i=1}^{n} c_i \cdot f_i \cdot \frac{1}{lgK_{oc}} \quad (1)$$

where c_i is the concentration of i PAH in the soil sample; f_i - a toxic equivalence factor $i$of that PAH [3]; K_{oc} is the binding constant of i PAH of this soil [4].

A feature of these indices is that, unlike most of these indicators, they take into account not only the total concentration of polyaromatic compounds in the soil, but also the biological effects of each of them.

The division of the index into two indicators $I_{2\text{-}3}$ and $I_{4\text{-}6}$ is due to the fact that the pressure of saturated vapors significantly decreases when a fourth aromatic ring appears in the PAH structure. This is the reason why heavy polyaromatic hydrocarbons are more characteristic of being in the form of solid particles adsorbed by soil than of light PAHs, which are largely present in the atmosphere as vapors [1].

The terrotories allocation within the level classification was conducted in line with the scale, based on not mathematical laws but on qualitative changes which were observed in the environment when transferring from one level to another. The assignment of a territory to one or another level was made by comparing the current values $I_{2\text{-}3}$ and $I_{4\text{-}6}$ for this territory with their maximum possible values and between each other.

3. Results and Discussion

Proposals for the most rational use of lands (in the authors' opinion) exposed to man-made impact, depending on the level of some toxic effects of bioavailable polyaromatic hydrocarbons are presented in Table 1. The indicator $I_{2\text{-}3}$ allows to indirectly judge about the state of atmospheric air in a certain territory, and the indicator $I_{4\text{-}6}$ - mainly about the quality of the soil, for the reasons mentioned above. Thus, the recommendations are given on the basis of the synergy of these two indicators.

The advanced analytical equipment, used to measure the concentration of poly-aromatic hydrocarbons in the soil, possesses the completely low limit of the detection [5]. Thus, in the case of complete absence of analytical signals corresponding to compounds of the PAH class, it is possible to speak of the most favorable ecological condition of the territory (level 1), which is best suited to accommodate recreational zones, children's and medical institutions, as well as agricultural lands, on which food products are produced (especially those which are not subjected to further heat treatment).

If heavy PAHs begin to be detected in the soil, while volatile substances (level 2) are also completely absent, then there is a low-hazardous toxic effect of polyaromatic hydrocarbons. Most likely, light PAHs are predominantly present in the atmosphere, but they are not highly toxic. Thus, the territory is suitable for housing residential areas.

The territory should be attributed to level 3, if low molecular weight and heavy PAHs are present in the soil, with the additional condition that the index $I_{2\text{-}3}$, attributed to the maximum value of this indicator exceeds the similar ratio for the index $I_{4\text{-}6}$. This indicates to the prevailing effect of light PAHs on the territory. It is quite acceptable to place agricultural lands on which technical and other crops are growing, largely being subject to further processing. Lipophilic compounds like polyaromatic hydrocarbons are likely to penetrate plant cells, mostly settling on the leaves from the atmosphere, rather than with nutrient juices from the soil (although, the question about what effects more PAHs sorbing by plants is still open - their content in the soil or in the atmosphere) [6, 7]. However, as mentioned above, volatile PAHs are low carcinogenic and the low value of the indicator $I_{4\text{-}6}$ for the soil indicates low concentrations of highly toxic polyaromatic hydrocarbons in the atmosphere. Thus, a territory with a relatively high level of PAH in the atmosphere, but low in the soil should have a weak effect on the quality of agricultural products under production.

The territories assigned to level 4 differ from the previous group because here the ratio of the index $I_{2\text{-}3}$ to its maximum value is less than the same ratio for the index $I_{4\text{-}6}$. This indicates that the soil contains mostly polyaromatic hydrocarbons with a high toxic equivalence factor close to 1. At the same time, the probability of a rather high content of light PAHs in the air (at the level of values close to the MPC value) is high. Conducting agricultural activities on these lands is inappropriate because of rather high concentrations of PAHs in the soil and in the atmosphere. The continuous presence of people in the territory is also undesirable, however, non-permanent exposure to low carcinogenic compounds in quantities below MPC is acceptable. Therefore, in the territories of level 4, it is advisable to place the working areas of enterprises (with a working day of not more than 8 hours) and transport roads.

Table 1. Recommendations for lands use to be subject to human impact

Level of carcinogenic effects on territory	$I_{2\text{-}3}$	$I_{4\text{-}6}$	Objects whose accommodation or location is most rational in given territory
(1)	Below the detection limit	Below the detection limit	Recreations, children's and medical institutions, agricultural lands for food production
(2)	Below the detection limit	$I_{4\text{-}6} < I_{max}$	Residential dwellings
(3)	$I_{2\text{-}3} < I_{max}$; $\frac{I_{2-3}}{I_{max}} > \frac{I_{4-6}}{I_{max}}$	$I_{4\text{-}6} < I_{max}$	Agricultural lands for industrial and other crops undergoing to further processing/treatment
(4)	$I_{2\text{-}3} < I_{max}$; $\frac{I_{2-3}}{I_{max}} < \frac{I_{4-6}}{I_{max}}$	$I_{4\text{-}6} < I_{max}$	Work areas of industrial enterprises with the possibility of permanent presence of employees in the territory without the use of personal protective equipment, transport and other communications.
(5)	$I_{2\text{-}3} \geq I_{max}$	$I_{4\text{-}6} < I_{max}$	Technological buildings, a long stay in the territory of which requires the use of personal protective equipment (warehouses, remote controlled industrial installations, etc.)
(6)	$I_{2\text{-}3} < I_{max}$	$I_{4\text{-}6} \geq I_{max}$	Technological buildings of I and II class of danger, even a short stay in the territory of which requires the use of personal protective equipment
(7)	$I_{2\text{-}3} \geq I_{max}$	$I_{4\text{-}6} \geq I_{max}$	The territory is not suitable for living and carrying out any types of economic activity. Land reclamation is required.

Exceeding the maximum index $I_{2\text{-}3}$ with a relatively low value of the indicator $I_{4\text{-}6}$ gives evidence to refer the territory to level 5, from which the term "pollution" can be used. In the air of these areas, obviously, an excess of the MPC of light PAHs should be observed, with a relatively low content of the most toxic high-molecular compounds. It is allowed to place household buildings (warehouses, industrial facilities that do not require the constant presence of personnel nearby, etc.). If necessary, a long stay in the area should be accompanied with the use of some personal protective equipment, primarily respirators.

In case of exceeding the maximum value $I_{4\text{-}6}$, but with the value $I_{2\text{-}3}$ below the maximum, the territory can be attributed to 6 level of toxic effects. Such a territory is contaminated with PAH with a high toxic equivalence factor. The probability of exceeding MPC of light polyaromatic hydrocarbons in the atmosphere is high. These lands are suitable only for the placement of technological facilities, for which I and II class of danger is obviously established and even a short-term stay within their limits requires the use of some personal protective equipment.

Finally, exceeding the maximum values for both indicators shows the critical polution of the territory with polyaromatic hydrocarbons. No economic activity here should be carried out, reclamation is required.

4. Conclusion

It is obvious that timely and well-planned measures for the rehabilitation and improvement of the quality of the territory should be aimed at reducing the toxic effect of bioavailable polyaromatic hydrocarbons. However, conducting these events are not always possible due to financial, technical or other reasons. In this case, the recommendations given in the article can be used as a basis in land management of unclaimed lands. Knowing the current level of toxic effects of PAHs on the territory, it is possible to limit the choice of infrastructure facilities located on it.

Also, with the help of the proposed recommendations, the inverse challenges can be solved, namely, the selection of the most optimal place for a new infrastructure object, from the point of ecology location, taking into account its future toxic effects on some neighbouring territories and the nature of their building.

References

[1] Hussein I A-Sh and Mona S M M 2016 *Egyptian J. of Petroleum* **25** 107–23
[2] Kadri T, Rouissi T, Kaur Brar S, Cledon M, Sarma S and Verma M 2017 *J. of Environ. Sci.* **51** 52–74
[3] Nisbet I C and LaGoy P K 1992 *Regulatory Toxicology and Pharmacology* **16** 290–300
[4] Perminova I V, Grechishcheva N Yu and Petrosyan V S 1999 *Environ. Sci. Technol.* **33** 3781–87
[5] Dong C-D, Chen C-F and Chen C-W 2012 *Int. J. Environ. Res. Public Health* **9** 2175–88
[6] Bakker M I, Casado B, Koerselman J W, Tolls J and Kollöffel C 2000 *The Sci. of Total Environ.* **263** 91–100
[7] Li J, Shang X, Zhao Z, Tanguay R L, Dong Q and Huang C 2010 *J. of Hazardous Materials* **173** 75–81

The use of covariance matrices in laboratory processing of geo-environmental data of field survey for development projects of land use planning

V V Vershinin, A V Fedorinov and A V Dontsov

State University of land use planning, 15, Kazakova str., Moscow, 105064, Russia

E-mail: v.vershinin.v@mail.ru

Abstract. The paper introduces the possibility of using covariance matrices in the factor analysis. The factors affect the effectiveness of issues' solution, concerning the involvement of unused agricultural lands in the active economic turnover in land use planning projects. The usage of the analysis tool, proposed by the authors, helps to improve the efficiency and objectivity of project decision substantiation in land use planning.

1. Introduction

One of the most important problems of current land management in Russia is the involvement of unused agricultural lands in the active economic turnover.

According to various data [1-3], in Russia such areas of unused land shares are currently about 20 million hectares.

The All-Russian Agricultural Census of 2016 revealed the results, showing the reduction in the total land area belonging to agricultural commodity producers. It is more than 100 million hectares compared to the results of 2006. Moreover, there were only 125 million hectares used by rural producers in 2016 despite the remaining amount of 142,7 million hectares [3].

As our experience shows, there is the most effective way to solve this problem. It is used to identify the unused land plots and involve them into the turnover, based on the special project development of land use planning [4]. The role and importance of land use planning on the whole, and particularly its projects, have been repeatedly emphasized by such specialists in land management as S. N. Volkov, V. N. Khlystun, N. V. Komov and others [5-7].

The development of such projects should be based on the use of an optimal economic and mathematical apparatus.

2. Models and methods

The principal component analysis (PCA) is used for maximum factor consideration which provides the offered measures efficiency for the development of unused land plots. It is one of the main methods allowing the reduction of data dimension with the least information loss.

The principal component calculation can be performed in two ways. The first one is the calculation of singular matrix data decomposition. The second one is the calculation of eigenvectors and eigenvalues of the original data covariance matrix. The usage of the covariance matrix has significantly allowed simplifying and, thus, accelerating the consideration process of factors, affecting the project decision choice.

The objective of principal component analysis has at least four basic versions. We recommend using the data approximation, applying linear manifolds in these works.

3. Results and discussion

The assigned task is to minimize the involvement cost of unused land plots among agricultural lands in active economic turnover. According to the work results, obtained in the Yaroslavl region, the task solution is determined by the influence of 13 key factors (table 1).

Table 1. Analyzed factors and their weight characteristics

Factor number	Factor name	Factor weight (from 1.1 to 2.1)
1	The distance from a settlement	1.3
2	The distance from an economic center	1.5
3	The distance from paved roads	1.6
4	The degree of land plot stocking occupation in the period of 1 – 5 years	1.2
5	The degree of land plot stocking occupation in the period of 5 -15 years	1.4
6	The degree of land plot stocking occupation in the period of 15 -25 years	1.6
7	Land plot area	2.0
8	Land plot form (compactness factor)	1.8
9	The degree of land plot peat formation	1.5
10	The degree of brushwood of the land plot	1.5
11	The species composition of land plot soil	1.6
12	The necessity of roads' repair and (or) their reconstruction	1.8
13	The previous use of the land plot	1.1

To evaluate each factor, its significance indicator ("weight" factor) is used (table 1). The weight value is set on the basis of a pair statistical sampling using the covariance matrix. It is a factor cumulative indicator which influences the costs for land plot involvement in economic turnover and the possibility of land plot further usage.

In order to improve the accuracy (objectivity) of territory exploration analysis for each of the listed 13 factors, the factor "internal" weight is set. The process of its setting depends on the characteristic change of the considered factor. The calculation example is shown in table 2.

Table 2. The value of "internal balance" for the analysis of the factor "the necessity of road construction and their repair"

Position number	Construction and repair of roads	Value of factor "internal weight"
1	The minor (small) repair of road surface	1.1
2	The repair of road surface up to 500 m	1.2
3	The repair of road surface from 0.5 to 2 km	1.3
4	The repair of road surface of more than 2 km	1.4
5	The construction of new roads up to 500 m	1.5
6	The repair of the road surface part and new road construction up to 500 m	1.6
7	The construction of new roads from 0.5 to 1 km	1.7
8	The construction of new roads from 1 to 1.5 km	1.8
9	The construction of new roads from 1.5 to 2 km	1.9
10	The construction of new roads of more than 2 km	2.0

Table 3. The results of factor analysis (fragment)

The land plot number (according to inventory records)			1	2	3	…	37	38	The number of participants
Factors	the distance	from settlement (1.30)* — the value	1.30	1.26	1.58	…	1.62	1.49	38
		from settlement (1.30)* — the value considering the weight	1.69	1.64	2.05	…	2.11	1.94	
		from economic center (1.50)* — the value	1.00	1.00	1.02	…	1.33	1.33	38
		from economic center (1.50)* — the value considering the weight	1.50	1.50	1.53	…	2.00	2.00	
		from paved roads (1.60)* — the value	1.22	1.23	1.24	…	1.51	1.68	38
		from paved roads (1.60)* — the value considering the weight	1.95	1.97	1.98	…	2.42	2.69	
	the forestry area	trees from 1 to 5 years (1.20)* — the value	0.00	0.00	0.00	…	0.00	1.30	8
		trees from 1 to 5 years (1.20)* — the value considering the weight	0.00	0.00	0.00	…	0.00	1.56	
		trees from 5 to 15 years (1.40)* — the value	1.30	2.00	1.20	…	2.00	0.00	11
		trees from 5 to 15 years (1.40)* — the value considering the weight	1.82	2.80	1.68	…	2.80	0.00	
		trees from 15 to 25 years (1.60)* — the value	0.00	0.00	0.00	…	0.00	0.00	5
		trees from 15 to 25 years (1.60)* — the value considering the weight	0.00	0.00	0.00	…	0.00	0.00	
	Land plot area (2.00)*	the value	1.60	2.00	1.40	…	1.20	1.60	38
		the value considering the weight	3.20	4.00	2.80	…	2.40	3.20	
	Land plot form (1.80)*	the value	1.80	1.80	1.80	…	1.80	1.80	38
		the value considering the weight	3.24	3.24	3.24	…	3.24	3.24	
	the degree of peat formation (1.50)*	the value	0.00	0.00	0.00	…	1.80	0.00	6
		the value considering the weight	0.00	0.00	0.00	…	2.70	0.00	
	brushwood of the land plot (1.50)*	the value	0.00	0.00	0.00	…	0.00	0.00	6
		the value considering the weight	0.00	0.00	0.00	…	0.00	0.00	
	soils (1.60)	the value	1.40	1.40	1.40	…	1.40	1.40	38
		the value considering the weight	2.24	2.24	2.24	…	2.24	2.24	
	The necessity of roads' repair and (or) their reconstruction (1.80)*	the value	1.20	0.00	0.00	…	1.30	0.00	14
		the value considering the weight	2.16	0.00	0.00	…	2.34	0.00	
	The previous use of land plot (1.10)*	the value	1.70	1.80	1.70	…	1.70	1.80	38
		the value considering the weight	1.87	1.98	1.87	…	1.87	1.98	
The sum of land plot factor values			19.7	19.4	17.4	…	21.4	21.6	
Land plot location (according to the total factor influence)			21	18	10	…	29	31	
The grouping of land plots			2	2	2	…	4	4	

*Factor weight (from table 1)

It should be noted that appraisers, as a rule, resort to expert weight distribution, giving the reasons of any kind.

The research procedure is replaced with the use of the mathematical method of weigh coefficient calculation. For this purpose, we have calculated the parameter which is opposite to the specific weight of the corrective sum for each analog in the total sum of analog correctives. The greater the specific weight, the smaller the weight coefficient, and vice versa. The calculation is conducted according to the following formula (figure 1):

$$C = \frac{(|S_A|+1)/(|S_{1\ldots n}|+1)}{(|S_A|+1)/(|S_1|+1)+(|S_A|+1)/(|S_2|+1)+\cdots+(|S_A|+1)/(|S_n|+1)} \quad (1)$$

where C – the target coefficient; n – analog number; Sa – corrective sum for all analogs; S1…n – the sum of analog correctives for which the calculation is conducted; S1; S2; Sn – the sum of the 1st, 2nd and n- analog correctives.

It is obvious that the dependence is not linear. Otherwise, the weight distribution would be much easier (in direct proportion) [8].

There is the example of analysis results, received by the authors (table 3). The analysis is conducted to evaluate the possibility of involving unused agricultural lands in the active economic turnover using the covariance matrix.

4. Conclusion

Summation of the weight characteristics of each of the considered factors with their subsequent correction, allows grouping the land plots and ranking them according to the conditions, capabilities and efficiency of their development (table 4).

The solution of such task enables using new indicators, necessary to implement the concept of monitoring development in the Russian Federation, which was repeatedly mentioned in scientific literature [9].

Table 4. The land plot distribution according to development stages with the consideration of cultivated land maximum productivity

№	Stage number	The number of participants	The land plot number in inventory records	The total area, ha
1	The first stage	12	1, 2, 3, 4, 5, 6, 8, 9, 18, 22, 31, 32	305.44
2	The second stage	7	24, 25, 26, 27, 33, 35, 37	278.54
3	The third stage	19	10, 11, 12, 13, 14, 15, 16, 17, 19, 20, 21, 23, 28, 29, 30, 34, 36, 38	281.7
	Results:	38		865.68

Table 5. The costs for development of unused land plots from among agricultural lands

Land plot grouping	The land plot number in inventory records	The land plot area, ha	The average development costs for 1 ha, thous. RUB	The total development costs, thous. RUB
1	5, 8, 25, 27, 30, 31, 34, 36	244.39	(24) 25	6110
2	1, 2, 3, 7, 9, 13, 14, 15, 16	170.29	(31) 30	5109
3	4, 11, 17, 18, 22, 23, 24, 26, 28, 29, 32, 33	365.98	(34) 35	12809
4	6, 10, 12, 20, 21, 37, 38	81.44	(39) 40	3258
5	19, 35	3.58	more than 45	161
Total amount:	-	865.68	-	27447

Each of 13 factors are considered when analyzing the cost value data for the involvement (development) of unused land plot in the turnover. The data usage helps to obtain the average cost values for all participants, belonging to each of 5 selected groups.

The comparison of these data shows the presence of almost identical intervals between the selected groups (figures in brackets; table 5). It indicates the effectiveness and reliability of authors' method, used to assess the factors of land plot development with the help of the covariance matrix.

References

[1] Volkov S N 2017 The contemporary state of land relations, land management and land use planning in the Russian Federation and the scientific substantiation of their basic regulation directions in agro-industrial complex *Materials for the report at the meeting of the Presidium of the Russian Academy of Sciences 28 March 2017* (Moscow: State university of land use planning)

[2] Lipski S A 2017 Integrated approach to the involvement of unused land in agricultural production *Nikon readings* **3** 71–5

[3] Laikam K E and Fomin A A 2019 About some problems of agricultural land inventory in the Russian Federation *Int. Agricultural J.* **2(369)** 7–12

[4] Vershinin V V and Fedorinov A V 2018 Some systematization aspects of modern land management plans *Materials for the Parliamentary hearings of the Council of Federation of the Federal Assembly of the Russian Federation for Stolypin readings at the State university of land use planning of 19 April 2018* eds S N Volkov and A A Fomin (Moscow: State university of land use planning) 54–8

[5] Khlystun V N 2018 Modern land policy and directions of its improvement *Economy of agricultural and processing enterprises* [in Russian – Economika selskohoziaystvennyh i pererabatyvajuschih predprijatij] **8** 8–11

[6] Papaskiri T V, Cherkashina E V, Chepurin E M, Petrova L E and Semochkin V N 2019 The land policy of the Russian Federation under the threat of a new version of the law On Land Management (opinion of experts) *Land management, land monitoring and cadaster* [in Russian – Zemleustrojstvo, kadastr i monitoring zemel] **4(171)** 11–35

[7] Lipski S A 2015 Private ownership for agricultural lands: advantages and disadvantages Experience of two decades *Studies on Russian Economic Development* **26**(1) 63–6

[8] Sleptsov S V 2018 *The calculation of weight coefficients in comparative approach* Available at: https://www.top-ocenka.com/ocenka-blog24.html

[9] Volkov S N, Shapovalov D A, Klyushin P V, Shirokova V A and Khutorova A O 2017 Solutions of problems in defining indicators of agricultural land within the framework of activities for the implementation of the concept of development monitoring in the Russian Federation *Int. Multidisciplinary Scientific GeoConf. SGEM* **17**(52) 819–28

IOP Conf. Series: Earth and Environmental Science **350** (2019) 012074 doi:10.1088/1755-1315/350/1/012074

Digital land management: new approaches and technologies

S N Volkov, E V Cherkashina and D A Shapovalov

State University of Land Use Planning, 15, Kazakova str., Moscow, 105064, Russia

E-mail: shapoval_ecology@mail.ru

Abstract. The article analyzes problems of implementation of the "digital agriculture" program. The role of modern land management in implementing tasks of effective land management is described. Opportunities of modern departmental information systems when addressing issues of geographic information support of the agro-industrial complex are considered. New approaches and digital land management technologies providing a significant (up to 30%) increase in the efficiency of land use were suggested. The need for training the staff under digitalization of the industry was emphasized.

1. Introduction

In the agricultural industry, land is a spatial basis, a natural resource amd a production tool whose rational use determines the efficiency of the industry and the state of the natural environment. Therefore, "smart land management" using "smart land management" methods should be at the core of digitalization of agriculture.

This indisputable fact is due to the following reasons:

1. The main issues of development of the industry (smart field, smart farm, smart garden, etc.) should not be considered in the overall system of f the agricultural organization. The farming system is an element of an integrated economic system. It is determined by the need for the development of field crop amd animal husbandry, fodder production, etc.

2. All the main agricultural branches are united by a single territory within its borders and connected by elements of the production and social infrastructures (roads, cattle routes, ameliorative networks) which requires the interconnected solution of the following issues:

- streamlining of land and property relations (only 30% of land plots in the village are registered) which does not allow for the planned use and circulation of agricultural land;
- accounting and assessing the quality of land in order to link the location of agricultural sectors to the territory by taking into account production and territorial properties of land (soil fertility, location of plots, land-reclamation, etc.);
- designing the single interconnected system of land reclamation, anti-erosion and environmental protection measures whose boundaries coincide with catchment areas, basins of small rivers, areas of wind erosion, etc.

At the same time, smart land management should be both a means of obtaining and processing information, as well as a mechanism for making management decisions on land management regulation, fertility enhancement and soil protection.

The World Land Management Review prepared by the United Nations Economic Commission for Europe states that "... the level of civilization of society is determined by the level of land management and land use". "... Due to the lack of land management in Russia, millions of hectares of previously cultivated agricultural land are empty." The data of the National Union of Land Surveyors of Russia shows that more than 60 million hectares of land consisting of 6.2 million plots may be involved in agricultural circulation.

It is clear that for agriculture of Russia without using this reserve, assessing the land and resource potential, formalizing the ownership of rural land, it is impossible to achieve economic success.

Therefore, land management should be the main mechanism for implementing the agrarian policy, the main factor in improving the competitiveness of agriculture. It should be integrated into the overall system of economic management.

2. Materials and methods

Large arrays of processed information, complex and multivariant processes of land use, various branches of agriculture and the stochastic nature of production require the use of digital land management technologies, modern multi-level databases, variable software solutions, intelligent systems for solving production and design tasks, changing the land management system.

Digital land management is a system of geo-information support for agricultural territories, including the on-line processing of significant geospatial information flows (big geo data) and the issues of land ownership structure and systemic territorial development of farms.

Currently, the main information resource of digital land management is the Unified Federal Information System on Agricultural Lands.

This system provides the Ministry of Agriculture and subordinate organizations with up-to-date, relevant and reliable information on agricultural lands. The data are obtained during the state monitoring of these lands. The system collects, stores, processes and analyzes information about agricultural land, land accounting, land-improvement systems and hydraulic structures, systematic monitoring of the state and use of land, providing stakeholders with information about agricultural land.

The system contains information

- about the boundaries and areas of agricultural land and crops;
- data on land users;
- indicators of soil fertility;
- information on negative processes, land reclamation facilities and other information.

3. Information resource of digital land management

To create a modern, efficient land management system, this information base is not always sufficient. The data updating system is not operational, not relevant and reliable; it does not contain decision-making blocks. In particular, according to the Analytical Center, out of 83 agricultural regions, 22 ones did not report data to the UFIS. Out of the regions that reported the information, more than 70% do not have reliable information on crops in the sown areas (database completenes is less than 50%). The situation is similar with information about land users. In 42 regions of the Federation, the completeness of the land user database is less than 50%. There are significant errors in the cartographic provision of the UFIS AL. It is based on data from Landsat satellites with a 30 m spatial resolution which makes it impossible to construct accurate contours of lands and identify their purpose and actual application. Remote sensing materials from RapidEye satellites (6.5 m resolution) and ALOS / AVNIR-2 (10 m resolution) were applied on the territory of the current agrochemical survey cycle (580 thousand km^2). However, specification and correction of the system data were fragmentary and of poor quality. As a result, 60-70% of the data do not correspond to the real boundaries of the fields and contain erroneous information about the areas and location of agricultural land plots in Russia. This is confirmed by selective comparisons with the data of current UAV flights made by university specialists in Tambov, Yaroslavl, and Smolensk regions (Figure 1) [1, 2].

Figure 1. Comparative analysis of remote sensing data of the UFIS AL and UAV survey data

We have suggested adaptive technologies for the development of integrated land management projects, allowing to study the state of lands, predict individual indicators and production results, and manage land resources solving important agricultural problems. The main ones are as follows:

- assessment of agricultural lands;
- identification and involvement of unused lands into circulation;
- registering valuable productive agricultural lands;
- development of zone-based agricultural regulations;

- drafting of adaptive landscape farming and land management projects.

Special software and unique equipment, including robotic geodetic total stations, ground-based and airborne laser scanners, unmanned aerial vehicles with multispectral imaging equipment in the visible and infrared ranges, space monitoring data with a spatial resolution of up to 10 cm, cloud geo-information systems processing a large amount of geospatial information, automated land management on-line design systems are required. The use of digital land management systems will allow to

1. create an information computer-based system for assessing the quality and location of agricultural land plots based on the assessment of their productive and territorial properties (analogues of the American LESA system - Land Evaluation and Site Assessment System and the Soviet system of intra-farm land assessment) [3].

2. form land ownership in the agricultural industry, carry out these processes ensuring the delimitation of all forms of ownership, register all agricultural land plots. This will attract additional financial resources to the agricultural industry against security of land plots and increase the taxable base.

3. plan the land management at the federal, regional and municipal levels as in the European Union, the USA and China [4], which will link the development of agriculture with socio-economic and natural conditions of the country and regions and business opportunities.

4. develop agricultural (land management) regulations, measures against land degradation and soil fertility reproduction, enhance the efficiency of land use.

5. develop regional business-oriented projects of adaptive-landscape land management in order to link the farming system (crop cultivation technologies: a processing system, fertilizers, soil reclamation, plant protection, seed production, a system of machines) to the land, and improve the efficiency of its use.

Evaluation and planning (programming) of crop yields based on the intra-field organization of the territory is the most important component of these projects.

Currently, the methods for evaluating and planning crop yields are based on the analysis of data on obtained as a result from aerospace observations, taking into account weather conditions possibilities for various types and volumes of field work on crop processing, fertilization, harvesting, etc.

They are rather accurate, but take into account the existing situation which characterizes optimal location of crops and their structure regarding the soil quality based on the suitability of crops and the location of economic centers. This makes it impossible to obtain large yields focused on the maximum possible and rational use of arable land resources (soil fertility) and does not allow for proper management of production of high stable yields.

To solve this problem, we have developed and tested methods for estimating and planning crop yields based on the multivariate analysis of geospatial information in the context of crop rotation fields, taking into account their field organization and separation into homogeneous areas.

This will solve the following issues:

- improvement of the accuracy of estimation and prediction of crop yields with a probability of up to 95%;
- an increase in the yield by 25-30% by optimizing the location of crops on the optimal land plots;
- reduction of production costs for the cultivation of crops by 15-20% by taking into account technological properties and location of land plots;
- coordination of cultivation technologies and specific areas of arable land;
- development of a system of anti-erosion and environmental protection measures within the boundaries of fields and working plots (agrotechnical, forest reclamation, hydrotechnical, organizational and economic) [5, 6] (Figure 2).

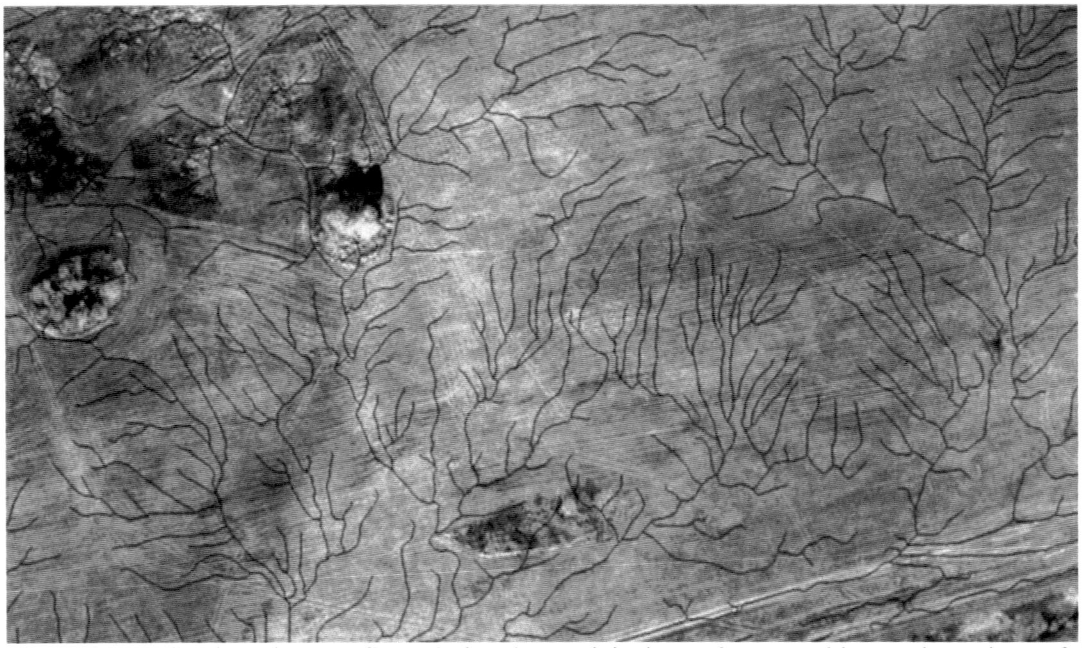

Figure 2. Localization of stream flows during the precipitation and snow melting on the territory of the Company n.a. Michurin (Smolensk region) for agricultural landscape-based territory management

4. Conclusion

Thus, using the example of pilot projects in several regions, it was shown that:

1. In the agricultural o industry, the technological breakthrough based on digital transformation is possible only by creating optimal soil-agrotechnical and organizational-territorial conditions that ensure a significant increase in crop yields on the basis of planning and programming, production output per one worker, a decrease in material costs of fuel, electricity, soil protection means, plants, labor, etc..

2. Modern digital technologies can optimize key parameters affecting crop yields: soil fertility; technological land properties; location and structure of land.

3 For the widespread use of digital technologies in the agro-industrial sector, it is necessary to train staff. A single online platform "Open Agrarian Education" can be created. This educational platform will be part of a unified national system of online education linked to information systems of the Ministry of Agriculture of Russia: "Electronic Atlas of Agricultural Lands", the System for Monitoring and Forecasting Food Security, the Federal System of Information on Agricultural Lands.

4. It is necessary to make proposals to the draft Federal Law "On Land Planning" in order to create a digital system of land use and land management.

5. The most significant scientific areas of digital land use and land management are development of projects aimed at improving technologies, methods and algorithms for collecting, processing and managing information support for the most important digital technologies, including;

- Development of an information hardware-software system for involving agricultural land plots into active economic circulation, including in connection with their use not for the intended purpose or in violation of the legislation of the Russian Federation based on the developed digital criteria and modeling of agricultural production efficiency taking into account long-term development of the agro-industrial sector and ensuring food security of the Russian Federation;
- Development of a system for automating zoning of agricultural territories under the influence of technological development of the agricultural industry based on integrated aerospace information in order to update information state resources for the sustainable development of agricultural lands.

References

[1] Rukhovich D I 2016 Use of remote sensing data for the organization of retrospective monitoring of land cover *Earth from space: the most effective solutions* [in Russian – Zemlia iz Kosmosa – Naibolee effektivnye reshenija] **5(21)** 41–9

[2] Kalinina N V, Rukhovich D I, Pankova E I and Chernoussenko G I 2019 Quantitative calculation of parameters for the allocation of zones and facies of the areas of distribution of chestnut soils of Russia based on climate-soil-particle size index *Soil sci.* **3** 304–16

[3] *US Law on Federal Land Policy and Land Management* 2016 Comp.: The Bureau of Land Management of the Department of the Interior and the Legal Department of the Department of Labor of the United States (Washington; Moscow: GUZ) pp 24–5

[4] Volkov S N 2005 *Land Management. Land management abroad* vol **7** pp 71–8 (Moscow: KolosS)

[5] Khitrov N B, Vilchevskaya E V, Rukhovich D I and Rukhovich A D 2018 Retrospective monitoring of soil and land cover of closed depression on the Azov plan *Int. Multidisciplinary Scientific Geoconf. SGEM* **18(3-2)** 443–8

[6] Loshakov A V, Klyushin P V, Shirokova V A, Khutorova A O and Savinova S V 2019 Environmental problems of treatment of saline lands for agricultural needs in the first agro-climatic zone of the Stavropol territory *South of Russia: Ecology, Development* [in Russian – Yug Rossii: Ekologiya, Razvitie] **14(1)** 105–16

Current geoecological status of the specially protected natural area "Crane Homeland" (based on materials proposed by the Ministry of Ecology and Environment of the Moscow Region)

T S Lukianova[1], V A Shirokova[1,2], A O Khutorova[1] and Yu D Yurova[1]

[1]State University of Land Use Planning, Kazakova str., 15, Moscow, 105064, Russia
[2]S.I. Vavilov Institute for the History of Science and Technology of the Russian Academy of Science, Baltiyskaya str., 14, Moscow, 125315, Russia

E-mail: Yuliya.yurova.1996@mail.ru

Abstract. The article describes the current state of the geoecosystem of the reserve "Crane Homeland". The authors take into consideration the consequences of improper drainage land-reclamation carried out in past decades in Apsarevskoe tract and Dubna water-swamp complex. Cutting down and uprooting the forest strip along the river Dubna in the vicinity of the Island village have led to the destruction of bird (Cyanistes cyanus) nests. In addition, the article points to poaching and spring fires occurred in a number of other places.

1. Introduction

Consideration of the territory of the reserve "Crane Homeland" and its study is due to its constant monitoring to save undisturbed geosystems and their components in the natural state, as well as the maintenance of ecological balance.

For the scientific study of the region in connection with changes in the study area due to the impact of drainage, violations of the natural complex by local farmers and visiting tourists, as well as residents of the private sector, it is necessary to conduct environmental monitoring of the territory of the reserve "Crane Homeland".

Land protected geoecosystems include transitional and upland swamps, wetlands, small-leaved and spruce-small-leaved forests with oaks, hazelnut and broad-leaved ferns, gray-grass and birches moist. The areas surrounding the studied region have been also exposed within many decades to the impact of human activity and the considerable loading from the neighboring areas. As a result, there is a need for a comprehensive study of the degradation of protected area [1, 2].

2. Methods and materials

The assessment of the geosystem of the reserve "Crane Homeland" was carried out on the basis of the long-term studies of protected species of plants and animals. The studies included data on the key ornithological areas, forest genetic reserves and valuable soil objects. In addition, there were data of the Ministry of ecology and nature management of the Moscow region, data of the Union of bird protection of Russia and data of the Ministry of forestry included in the Red book of the Russian Federation.

IOP Conf. Series: Earth and Environmental Science **350** (2019) 012075 doi:10.1088/1755-1315/350/1/012075

3. Results and discussion

The purpose of this kind of research is the maintenance of preservation and restoration of natural systems, maintenance and preservation of habitats of rare species of plants and animals; monitoring of plant and animal species listed in the Red Book of Russia, as well as conducting research work of the specially protected natural area "Crane Homeland".

The species listed in the Red book of Russia and the book of the Moscow region include: Cypripedium calceolus, Rubus chamaemorus, Goodyéra répens, Empetrum nigrum, Dactylorhiza, Birch humifusa, Rubus arcticus, Nammarbya Marsh, Mjakotnica univalent or Stagachka, Shejhcerija palustris. There are vulnerable species that need monitoring and observation: mezereum or Daphne, Lycopodium complanatum, Platanthera bifolia.

The Red book of Russia includes rare and vulnerable species of animals: Ciconia nigra, Anser erythropus, Circus macrourus, Circaetus ferox, Aquila danga, Aquila pomarina, Aquila chrysaetus, Haliaeetus albicilla, Falco peregrinus, Lagopus, Numerius arquata, Lanius excubitor, Cyanists cyanus.

The red book of the Moscow region presents the following species: Ciconia ciconia, Grus grus, Anas Acuta, Anas strepera, Tringa totanus, Tringa stagnatilis, Philomachus pugnax, Gallinago media, Limosa, Xenus cinereus, Pernis apivorus, Milvus migrans, Circus cyaneus, Falco columbarius, Strix uralensis, Surnia ulula, Picus, Dendrocopos leucotos, Picoides tridactylus, Columba Oenas, Sylvia nisoria, Vipera berus, Natrix, Lacerta Agilis, Triturus cristatus, bufotes viridis, Colias palaeno, Coenonympha hero, coenonympha tullia,agriades optilete, Arichanna melanaria, Papilio machaon.

Rare species include Asio flammeus, Caprimulgus europaeus, Eurasian pygmy owl, Anthus pratensis, Falco, Turdus viscivorus [1–5, 7].

The total area of especially protected natural territories "Crane Homeland" is 11974.5 hectares, and it necessary to note the area of land included in the boundaries of the reserve without exemption from commercial use [1–8]. The considered reserve is located in the North-East of the Moscow region of the upper Volga lowland in the zone of distribution of flat and slightly wavy water-glacial, moraine-water-glacial and ancient alluvial-water-glacial slowly drained plains. The absolute heights of the natural-territorial complex vary from a minimum of 119 meters to 151.4 meters above the sea level [9–13]. Dubna bog massif is located on the right bank of the Dubna River, lying along its straightened bed. Within the considered area, the height differences are small absolute marks that vary from 126 m to 133 m.

The top layer of the pre-Quaternary Foundation of the area is the Dolomites of Perm, as well as upper Jurassic clay, siltstone and sand. There are the Dolomites of the Carboniferous in the West. On the study area there are flat water-glacial plains with swamps of all kinds, as well as the right Bank of the valley of the Dubna river with floodplain territory and floodplain terrace. The surfaces of flat areas are composed of sandy loam, water-glacial deposits and loam. Plains have slopes of 1-2°. Partially on smooth surfaces, slightly pronounced lower pits appeared.

On the Western part there are elongated hills ranging from 200-300 m to 1-2.5 km with a height of 2-3 m. The surface of the floodplain of the Dubna river formed at an altitude of about 2 meters above the riverbed, folded by alluvial sediment power loamy sand and sandy loam. The first floodplain terrace with an average height of about 5 meters is composed of ancient alluvial deposits. On the territory of the reserve, in places where waterlogged peat or humus poles are, there are old slopes with different lengths, a maximum of 5 km. They are deposited on water-glacial, ancient alluvial or alluvial deposits. They include the peat and the forming one opposed to phytogenic forms of bumps, tree trunks [9–14].

The common pool area has a slope in the Dubna River, the tributary of the Volga. On the plot there is a network of drainage canals, which are practically in parallel or perpendicularly positioned relative to each other. Melioration systems pass through a quarter of the area of the site. Their length is from 300-600 m to 3-6 km. Overgrown channels at their intersections in the north-west area represent rectangular fireproof quarries (20-30 m - length, 10-15 m - width). In this area there are old ponds (10-20 m wide) in the South-West section oxbow formed up to 60 meters [11, 12, 14].

Soils are represented by sod-podzolic on the sandy sediments. Such soils were formed on the hills.

These soils are formed at higher elevations. Drainage that slowed down in small depressions formed sod-podzol clay soil. In the hollows, on the bottom of the humus, there are gley soils under black alder - humic - gley soils. Peat oligotrophic and peat eutrophic soils are mainly located on bogs. At the same time, alluvial humus - gley soils are widespread on the river floodplain; in ancient riverbeds -alluvial peat - gley soils [2, 3, 7, 15, 16].

Apsarevskoe tract is a moraine-glacial plain with moraine hills that formed surrounded by water-glacial Plains and lowlands. It also includes a small portion of Hotcha River (upstream). Absolute marks vary from 119 m (in the North at the edge of the Hotcha River) to 151.4 m (at the highest point of the hill). The surface of the pre-Quaternary Foundation consists of clays and silt upper Jurassic. The plain is composed of cover loam on water-glacial sediments and moraine. The hills formed lower (height 5-10 m). Their slopes have a maximum steepness of 3-5°. At the sites, which folded seats and hollows, valleys have a drain in the reclamation of swamps [17].

The riverbed of the Hotcha river is straightened. The width of the river bed is 5.5 km, depth - 0.2 - 0.5 m. The bottom of the Hotcha river is sandy, it is found on the bottom of silt. The flow rate is 0.1 m/sec, sometimes it is simply absent. Water consumption in the upper reaches (in the East) 0.03 m^3/sec. In the North, the section of the Hotcha River receives the water flow on the Kostinka River. The width of the left tributary is 1.5-7 m. The Site is transformed by drainage reclamation in the form of channels, which are located on 20% of the territory. The channels have a width of 2 to 8 m, sometimes up to 10 m. There are overgrown ponds of small depth. In the South a transitional swamp has formed, which is about 6 hectares. In the lowlands ponds of anthropogenic origin have formed, they have a rectangular shape with a length of 40-100 m, a width of 30 to 50 meters.

Soils are characterized as agrogenic-transformed. On high plains and hills agro-sod-podzolic soils have formed, which are confined to the loam. At the same time, depressions and lowlands are mostly found in sod-podzol clay. Wetlands are represented by peat oligotrophic peat soils and eutrophic. Sometimes in the area for hollows there are humic-gley soils or humus-gley soils. In the floodplain, the rivers are characterized by the light-humus and alluvial soil.

The reserve of the first has two large arrays of forest. On the Eastern part of the forest there are small-leaved forests with floodplain black-alder trees. On the Western part there are swampy pine forests, on transitional and upland swamps – with pine and birch. There are also spruce, pine, small-leaved and mixed forests.

The Western part of the forest has a greater variety of trees. In the forest waters of the West spruce, pine, birch, aspen, black alder, linden and maple grow. In the southern part of the forests as part of the stand there is spruce, pine, birch, aspen and oak. Sometimes arrays of spruce undergrowth, even oak and undergrowth can be found.

The lower tiers are poor, there are even dead-blood parts of the cover. From the sparse grass-shrub, a tier can be found: cranberries (Vaccínium vítis-idaéa), blueberries (Vaccínium myrtíllus), Lúzula, Solidágo virgáurea, Rúbus saxátilis, Gymnocárpium dryópteris, Stellária holóstea, Lysimáchia, Dáphne mezéreum.

To the North of the site there are well developed small-leaved forests: alder, birch, willow nettle, meadow-nettle with weed. In the undergrowth there is black currant. The grass tier is represented by: Epilóbium, Aegopódium podagrária, geranium. Around the marsh land and pine-birch forests, stripes with Lycopódium annotínum grow.

On the Eastern part of the reserve, black and birch (nettle) forests grow. In their undergrowth there is: black currant, cherry and raspberry. The following herb layer is marked: Urtíca dióica, filipéndula, lycopus europaeus, Chamaenérion angustifolium, Impátiens Nóli-tángere, Deschampsia cespitosa, Geránium robertiánum, Stáchys palústris, dryópteris carthusiána Dry Dryópteris fílix-mas, cómarum palústre, Athýrium fílix-fémina, Ranunculus repens, Thelýpteris palústris, glechoma hederacea, thysselinum palustre, Fallópia convólvulus. On the second section there are separate island forest lands, previously cultivated fields. At present, on the farmland in most of the abandoned meadows and former fields, there are small-leaved species with growing pines. In the depressions, the willows from wetland areas grew. On the Western part and in the center of the site there are swampy areas, pines and birches. Spruce declines, the circle has overgrown with alder forests.

The undergrowth is dominated by: Urtíca dióica and Aegopódium podagrária. One can meet untamed Filipéndula ulmária, Equisétum sylváticum, riv́ále Géum, celandine, Lysimáchia, Páris. Moss grows on the fallen trees. Copious amounts of: bilberry, blueberry, cloudberry, cranberry can be found, including Empetrum nigrum, Lycopódium. In the Eastern part of the site on the outskirts of the small-leaved forest there is a gray alder with a birch and aspen with a thick teenager cherry and alder gray. In the grass tier there are wide Júncus effúsus, Ranunculus repens, Páris, Deschampsia cespitosa, Géum riv́ále. In the North-West the melioration overgrown swamp with a forest of birch is drained.

The areas of the reserve are included in the specially protected natural areas of the Moscow region. Many birds, including rare and protected birds, as well as those listed in the Red book of Russia, nest on its territory, as well as during the flight. The areas of the preserve: Dubna array and Apsarevskoe tract, are important in maintaining populations of rare bird species, rare and protected species of animals. For example, the first site is the habitat of at least 102 species of terrestrial vertebrates (4 species of amphibians, reptiles, 80 species of birds and 14 species of mammals). They belong to the meadow-edge, wetland and forest species of the Moscow region. Within the Dubna wetland there are three main zoocomplex: zoopharmacy mixed forests, zoopharmacy pine forests, pine - sphagnum mires and peat, zoopharmacy open wetland habitats and agricultural lands.

The zooformations of mixed forests include species such as: Erinaceus europaeus, Sciurus, Castor, Lepus timidus, Lepus europaeus, Sus scrofa, Cervus elaphus, Capreólus capreólus, Vulpes, Canis lupus, Lynx lynx, Ursus arctos, Meles meles, Ciconia nigra, Accipiter gentilis, lyurus tetrix, Glaucidium passerinum, Surnia ulula, Caprimulgus europaeus, cuculidae, Picus canus, Dendrocopos leucotos, Dendrocopos major, Picoides tridactylus, Anthus trivialis, oriolus oriolus, garrulus glandarius, troglodytes troglodytes, Phylloscopus collybita, fringílla coél, carduelis spinus, Erithacus rubecula, Sitta europaea, Parus major, Cyanistes caeruleus, Turdus pilaris, Turdus philomelos, Turdus iliacus, Turdus merula, Aegithalos caudatus, Muscicapa striata, Zootoca vivipara, Lacerta agilis, Bufo bufo, Bufotes viridis,Rana temporaria.

To zoopharmacy pine forests, pine-sphagnum bogs and peatery include: Numenius arquata, Xenus cinereus, Tetrao urogallus, Falco columbarius, Dryocopus, Turdus viscivorus, Poecile montanus, Certhia familiaris, Periparus ater, Cyanistes cyanus, Regulus regulus, Rana arvalis. There are rare species of butterflies: Colias palaeno, Coenonympha hero, Coenonympha tullia, Agriades optilete, Carsia sororiata. On semi-aquatic territories the following can be highlighted: Ciconia Ciconia, Anas acuta, Anas strepera, Limosa Limosa, Gallinago media, Tringa totanus, Tringa stagnatilis, Milvus migrans, Circus cyaneus and Circus pygargus, Sylvia nisoria. There are birds listed in the Red book of the Moscow region such as: Circus macrourus, Aquila, Aquila chrysaetos, Circaetus gallicus, Haliaeetus albicilla, Falco peregrinus, Haematopus ostralegus, Lanius excubitor. On the territory there are: Asio flammeus, Falco, Perdix perdix, Tringa glareola.

Among the migratory there are: Anser albifrons, Anser erythropus, Philomachus pugnax, Golden rye, Curlew. Their number sometimes reaches 1500-3000 individuals.

In another area there are about 81 species of vertebrates. There are two kinds of amphibians, one kind of reptile, 60 species of birds and 15 species of mammals. Perch and pike inhabit the rivers. The faunistic complex of terrestrial animals of the Apsarevskoe tract mainly consists of species of meadow-glade, a lesser role is played by forest and wetland. In the seasonal period, thousands of flocks of the grey crane (Grus grus), Anser fabalis and Anser albifrons can accumulate. They fatten up before a long flight.

In mixed forests there are: common hedgehogs (Erinaceus europaeus), protein (Sciurus), beaver (Castor fiber), Lepus timidus and Lepus europaeus, Sus scrofa, Cervus elaphus, Capreólus capreólus, Vulpes vulpes, Accipiter gentilis, Lyrurus, Columba palumbus, Columba oenas, Cuculidae, Picus canus, Dendrocopos leucotos, Dendrocopos major, Anthus trivialis, oriolus oriolus, Corvus Corax, garrulus glandarius, troglodytes troglodytes, Phylloscopus Trochilus, fringílla coélebs, carduelis spinus,

Erithacus rubecula, Sitta europaea, Parus major, cyanists Caerule, Turdus pilaris, Turdus philomelos, Turdus iliacus Linnaeus, Turdus merula, Aegithalos caudatus, Muscicapa striata, Zootoca vivipara, lynx (Lynx Lynx), wolf (Canis lupus), badger (Meles Meles), brown bear (Ursus arctos).

4. Conclusion

On the described territory of the reserve "Crane Homeland" there is now extremely negative pressure exerted by humans. Their actions are the logging, felling and stubbing of forests (strip of at least 50 meters) on a plot near the Dubna river. Essentially, this is the destruction of nesting white blue Tits (Cyanistes cyanus). In spring and autumn, the burning of grass and debris leads to frequent fires.

In summer, during the July anticyclone there is spontaneous combustion of dried peat bogs, which are the areas of high fire hazard. Drained marshes, in general, are the result of improper drainage reclamation in the twentieth century. As a result, geoecosystems suffer, where a huge number of birds and animals live among the vegetation cover of the unique natural complexes of the reserve "Crane Homeland" [18].

References

[1] Belaya N I, Dubinin E P and Ushakov C A 2001 *The geological structure of the Moscow region* (Moscow: Moscow state University press)

[2] Wagner B B and Manucharyan B O 2003 *Geology, relief and minerals of the Moscow region. Textbook for the course Geography and ecology of the Moscow region* (Moscow: Moscow state pedagogical University) p 92

[3] Ogureeva G N and Buldakova E V 2006 Diversity of forests of the Klin-Dmitrov ridge in connection with the landscape structure of the territory *Forest science* **1** 58-69

[4] Wagner B B and Dmitrieva V T 2006 Lakes and reservoirs of the Moscow region *Geography and ecology of the Moscow region (*Moscow: Moscow state pedagogical University) p 76

[5] Spiridonov A I 1974 The Main features of the morphostructure of the Russian plain *Geography* **10** 95-115

[6] Vasilyeva I V 1961 The Border of the Moscow glaciation and its landscape value *Bulletin of Moscow University: Series geography* **3** 62-66

[7] Machkova N A and Sorokin V N 1991 *Climate of Moscow region* (Moscow: Moscow state University press)

[8] Belyaeva N G 2012 Analysis of the structure of forest cover of the South-Western part of the Klin-Dmitrov ridge on the basis of landscape approach *News of Samara scientific center of RAS* **1-6** 1434-37

[9] Volkov S N and Samodurov A I 2012 Features of recreational potential of forest biogeocenoses of the Klin-Dmitrov ridge *Bulletin of the Peoples' friendship University of Russia. Series: agronomy and animal husbandry* **5** 62-65

[10] Kurnaev S F 1982 *Fractional forest-growing zoning of non-Chernozem center* (Moscow: Science) p 118

[11] Melekhov I S 2004 *Essay on the development of forest science in Russia* (Moscow: Moscow state forest University) p 209

[12] Nikitin F A and Nikitin V F 2005 History of development, current state and ways of increasing the productivity of forests of the Klin-Dmitrov ridge *Forestry Bulletin* [in Russian – Lesnoj vestnik] **5** 107-113

[13] Karpachevsky M L et al 2009 *Nature of Moscow region: the loss of the last two decades* (Moscow: Publishing Center for wildlife conservation) p 92

[14] Mironov V V 1977 *Ecology of coniferous species in artificial reforestation* (Moscow: Forest industry) p 232

[15] Ivanenko B I 1962 Forest Zoning of the Moscow region *Collection of works on forestry* Issue 45 (Moscow: Federal budgetary institution all-Russian Research Institute of forestry and

mechanization of forestry) pp 48-53
[16] Rules of care of woods 2007 *Order of the Ministry of nature of Russia* No. 185 67
[17] Smirnova E D 1958 *Rivers and lakes of the Moscow region* (Moscow) p 96
[18] Kurnaev S F 1982 *Fractional forest-growing zoning of non-Chernozem center* (Moscow: Science) p 118

IOP Conf. Series: Earth and Environmental Science **350** (2019) 012076 doi:10.1088/1755-1315/350/1/012076

Current state of the specially protected natural area "Novosurinsky" in Mozhaisk District of Moscow Region

T S Lukianova [1], V A Shirokova[1,2], A O Khutorova[1] and Yu D Yurova[1]

[1]State University of Land Use Planning, Kazakova str., 15, Moscow, 105064, Russia
[2]S.I. Vavilov Institute for the History of Science and Technology of the Russian Academy of Science, Baltiyskaya str., 14, Moscow, 125315, Russia

E-mail: Yuliya.yurova.1996@mail.ru

Abstract. This article is devoted to the description of the status of the specially protected natural area "Novosurinsky" in Mozhaisk district of Moscow region. The study of the reserve requires constant monitoring of its territory. Despite the strict requirements for the protection of the ecosystem, its preservation and maintenance of the ecological balance, there are negative impacts in the form of littering in the studied reserve. The arrival of vehicles and their unhindered movement lead to violations of soil and vegetation. In connection with careless handling of fire, arson of dry grass and forest litter, fires occur.

1. Introduction
As it is specified in the decision of the Executive authority of Moscow region "On the territory of the state wildlife area, plowing of lands, cutting woods of the main use, haying, grazing of cattle, collecting nuts, mushrooms, berries, seeds, medicinal and other plants are forbidden". In addition, the law prohibits fishing and sport hunting, fishing, catching non-commercial animals, collecting botanical and zoological collections. This also includes the provision of land for development, collective gardening, suburban areas, geological and other research, mining, roads, construction of buildings and structures, pipelines, power lines, the use of pesticides, plant protection chemicals, mineral fertilizers, timber rafting, blasting. There are also the device halts, travel and parking of vehicles, tourist parking, recreation and other types of nature that harm the environment.

The reserve "Novosurinsky" is regarded as a complex of biological reserves. Such reserves are designed to protect and restore the number of rare and endangered species of plants, animals, phytocenoses and zoocenoses. According to encyclopedias, the complex of biological reserves can include areas with valuable rare species of trees, shrubs, forest grass plants, forest associations - botanical reserves, as well as areas including rich meadows that have preserved relict and endemic plant species - forest reserves. The same group includes zoological reserves with rare or very rare and valuable faunistic representatives of endangered species of animals. Such types of animals have survived only due to the special natural conditions of the reserve.

The reserve "Novosurinsky" has a total area of 717.1 hectares. The tasks include the preservation of the territory of a specially protected area. The preservation of the habitat of rare plant covers species and wildlife. Geographically the study area is delineated within the urban settlement Mozhaisk and rural settlement Borikovskiy, south of the highway (A-100) to the west and south-west of the village of Kukarino. The reserve includes 2 sections, which are separated by a strip of the Moscow railway (Smolensk direction). Clustering the territory includes area №1 (South) with a total area of 496 hectares and area №2 (North) with a total area of 221.1 hectares [1, 2].

2. Methods and materials

The assessment of the geosystem of the reserve "Novosurinsky" was carried out on the basis of the long-term studies of protected species of plants and animals. The studies included data of the key ornithological areas, forest genetic reserves and valuable soil objects. There are also data of the Ministry of ecology and nature management of Moscow region, data of the Union of bird protection of Russia and data of the Ministry of forestry included in the Red book of the Russian Federation.

3. Results and discussion

The reserve "Novosurinsky" is located on an elevated plot moraine-glacial plain. Through the reserve there is the watershed of the rivers Moscow and Protva. The area is composed of boulder loam moraine of Moscow glaciations. Moraine deposits are directly on the Carboniferous limestone. Everywhere the plain is composed of heavy cover loam. The nature of the relief of the territory characterizes the alternation of flat-topped hills, as well as wetlands with a slight elevation difference. The absolute height of the highest point of site № 1 is on a flat top of the hill. This is a plot of Mozhaisk forestry, which is 225 m above sea level. The height of the lowest point is located in the southeast of the same area №1. It is 195 m above sea level.

The highest point on site №2 is located on the southern border of the slope of the hill from the moraine and is 216 m above sea level. The lowest point of its Northern border on the bottom of the erosion valley equals 196 m above sea level. The total height difference does not exceed thirty meters. The slopes are on average from 1 to 30 degrees.

The erosion network on the territory of the reserve is poorly developed. In relief there is one expressed erosion valley, crossing the second section from South to North. Other depressions in the relief have the form of shallow ravines with damp beams, sometimes swampy. Temporary watercourses form erosion incisions with a width of up to 2 meters and a depth of up to 1 meter exclusively along the bottom of the beams [3, 4].

The surface runoff from plot №1 is aimed into the Mgut river, whereas the surface runoff from plot № 2 of the left tributary of the Protva river drains to the catchment basin of the Moscow river. Seasonal waterlogging with waterlogging is observed in the hollows between the hills. On the territory of the reserve, you can find low-lying swamps. Sod - podzolic - gley soils are most widely represented on the territory. Sometimes on the largest slopes there are simply sod-podzolic soils, and peat and eutrophic soils are formed on swamp sites [3, 4].

Among the vegetation covers there is an old age, oak-aspen, aspen - oak hazel forests with birches. Forests include birches and grass with participation in the grass-shrub layer of species of different ecological-coenotic groups (taiga, weed-forest, and meadow-forest). In addition, forests predominate in the Northern and North-Eastern parts of site №1; they are widely represented in the eastern half of site №2.

Birches and aspens in such forests characterize an average age (up to 80 - 100 years). Trunks with a diameter of that of the oak tree can be equal to 60-70 cm. Some oak trees are with large trunk diameters, which reach 70-80 cm. Tier shrubs are closed, represented by hazel and honeysuckle. The herbal layer is associated with the types of forest herbs: Mercuriális perénnis, Galeobdolon luteum, Pulmonaria obscura, Ásarum europaéum, Víola mirábilis, Ranunculus cassubicus, Stellária holóstea, Aegopódium podagrária, Campánula latifólia. This is a rare, vulnerable species (not included in the Red Book of Moscow region, but they need control and surveillance).

In the depressions and lower parts of the slopes, oak-aspen hazel and horsetail-greenfinch communities with ferns (Polypodióphyta) are widespread. This motley grass presents Fragária moscháta and Tróllius europaéus - rare and vulnerable species (not included in the Red Book of Moscow region, but in need of control and supervision). In addition to the above, there is Crepis paludosa, Geum rivale, Stáchys sylvática, Equisétum sylváticum, Polypodióphyta and Carex sylvatica. Bromopsis benekenii, Brachypodium silvaticum, Chaerophyllum aromaticum, Páris, Ranunculus cassubicus. Deschampsia cespitosa and Filipéndula ulmária are widespread in the territory and grow in the lowlands and in the depressions.

In aspen forests of old age and sparse hazel forests with motley grasses the following can be found: Filipéndula ulmária, Tróllius europaéus, Calamagrostis canescens, Vukvitsa medical, Chaerophýllum aromáticum, Geranium Marsh, giant Fescue, Valeriana officinális, Bistorta officinalis, lysimachia Vulgaris.

Rare vulnerable species are Aconitum lasiostomum, Polemonium caeruleum, Orchid-dactylorhiza fuchsii and Dáphne mezéreum. These species are not included in the Red Book of Moscow region, but need control and monitoring. On the western half of site № 2, among the old-growth forests of aspen, you can find separate areas of aspen-oak and spruce-oak forests with hazel and grasses. There are plants such as Polypodiophyta and Óxalis, and in the groups of spruce forests, you can find taiga species: Pýrola, Trientális europaéa, Orthília secúnda and Vaccínium myrtíllus.

In the depressions between the hills, as well as in the erosion valleys, aspen - oak and honeysuckle - hazel forests with grasses are widely represented. Among these forests, maples and elm bare sometimes grow, interspersed with spruce and birch. The grass layer of this forest is dominated by species typical of forest herbs: Galeóbdolon lúteum, Carex pilosa, Mercurialis perennis, Anemóne ranunculoídes. The following dominate in the lower areas of the slopes: Lamium maculatum, Milium effusum, Stáchys sylvatica, Mercurialis perennis, ferns (Polypodióphyta) and Campánula latifólia (rare and vulnerable) (not included in the Red Book of Moscow region, but in need of control and surveillance). High Aconítum, Actaéa spicáta and Mercuriális (species listed in the Red Book of Moscow region and in need of control and supervision) can be sometimes found. They grow in areas №2 and №1.

On the large acreage parcel there is the wetland fuzzy - birch forest with grey alder, with an undergrowth of spruce and shrubby willows. For such forests, the pronounced representatives are Equisetum sylvaticum, Juncus effusus, Calamagrostis canescens, Lysimachia vulgaris, Carex vesicaria and Carex canéscens, Pyrola rotundifólia and Viola. These communities are bound by the terms of drained forests in the past. In this regard, small drainage ditches were preserved.

In the marshy lowlands there are aspens, oaks and birches. Ferns, nettles and meadowsweet flourish in the lower tier. The usual representatives in the area are Myosóton, Ficaria verna, Geum rivale, Polypodiophyta, Cirsium oleraceum, Aegopódium podagraria, Chrysosplenium ovalifolium.

On the territory of site №1 there are mostly meadows. They are located on the wet forest glades and lawn. It is noted here that: Filipéndula ulmária, Lysimáchia vulgáris, Lysimachia nummularia, Crépis paludósa, Ranunculus repens, Angélica sylveśtris, Deschampsia cespitosa, Anthoxánthum, Festuca pratensis, Círsium palústre, Juncus tenuis and Júncus filifórmis, otentílla erécta, Géum rivále, Prunélla vulgáris, Carex lasiocarpa, Lysimáchia vulgáris, Betonica officinalis, Succisa Pratensis, Coccyganthe, Geránium palústre.

In the lowlands of wetlands in the meadows, the following plants are added to the above species: Alopecurus pratensis, Juncus conglomeratus and Juncus effus, Carex pallescens, Carex flava, Carex leporina, Galium uliginósum, Cirsium heterophyllum, Equisetum sylvaticum L.

A series of large meadows among oak-aspen forests on site №1 occupies meadows grass with the highest floristic wealth. The herbs can usually be found: Calamagróstis epigéjos, Agróstis capilláris, Dáctylis glomeráta, Deschampsia cespitosa, Geranium magnificum, Betonica officinális, Melampýrum nemorósum, Hypericum maculatum, Lysimáchia vulgáris, Crépis paludósa, Succisa pratensis, Succisa pratensis, Equisétum sylváticum, Thalíctrum lúcidum, centaur é phrýgia, Selinum Carvifolia, Anthríscus sylvéstris, Potentílla erécta.

Rare and protected plant species also grow: Gladiolus imbricatus (plant species listed in the Red Book of Moscow region). In addition to the above, there are Trollius europaeus, Dactylorhiza fuchsii and Platanthéra bifólia (not included in the Red Book of Moscow region, but they need control and supervision). The plant "Gladíolus" is widely developed in these areas of site №1, blooms and bears fruit [5, 6].

On a large grassy, swampy glade with shrubby willows and birch undergrowth, besides the "Gladíolus", there are such plants as: Iris or Iris sibirica, Dactylorhíza, Dactylorhíza incarnáta, Listéra ováta (not included in the Red Book of Moscow region, but they need control and surveillance), Dactylorhīza fūchsii, Tróllius europaéus, Polemónium caerúleum. In such glades there are: Deschampsia cespitosa, Alopecurus pratensis, Poa trivialis, Agrostis capillaris, Briza média. Growing sedge: Carex pallescens, panicea, flava, vulpina, pilosa. Grow: Júncus, Succisa pratensis, Geránium palústre, centaur é phrýgia, Filipéndula ulmária, Melampýrum nemorósum, Potentilla, Lysimachia nummularia, Coccyganthe, Thalíctrum lúcidum, Valeriána officinális, Géum riväle, Círsium palústre, Círsium oleráceum, Selínum carvifólia, Hypericum maculatum, Equisétum palústre, Angélica sylveśtris, ranú nculus ácris, Rúmex acetósa, Myosótis scorpioídes, Bistorta officinalis [2, 6, 7].

The fauna on the sites of the "Novosurinsky" is well preserved and represented on the territories of other communities in Moscow region. The fauna of terrestrial vertebrates is based on species that are inherent in deciduous and mixed forests of other regions of the Russian Federation.

The territory is dominated by species that are environmentally related to tree and shrub vegetation. However, the proportion of inhabitants of meadow-field areas among vertebrates is not numerous. Within the reserve there are three toocomplexes or zoopharmacy: zoopharmacy deciduous forests, moist forests zoopharmacy, zoopharmacy open habitats [8, 9].

Mixed and deciduous forests of different types in areas №1, №2 have retained their inherent complex of animal species. Among them are: Lepus timidus, Cuculus canorus, Dendrocopos leucotos (listed in the Red Book of Moscow region), Turdus pilaris, Turdus merula, Turdus philomelos, Fringílla coélebs, Erithacus rubecula, Sylvia atricapilla, Phylloscopus sibilatrix and Phylloscopus trochilus, Aegithalos caudatus, Ficedula hypoleuca, Parus major, here live Rana temporaria and Rana arvalis.

The edges of forests and grasslands on site № 1 are widely presented by: Microtus agrestis, Mustela black, Talpa europaea, Buteo buteo, Falco is a rare and vulnerable species (not included in the Red Book of Moscow region, but they need control and surveillance), Anthus trivialis, Motacilla alba, Apus Apus, Saxicola rubetra, Hirundo rustica, Emberiza citrinella, Carduelis cannabina, Coloeus monedula, Pica pica, Carduelis, Sylvia communis.

Individual wetland depressions and swales, ditches and wet areas in the forest are habitat types that are related to wetlands. This information can be found on sites №1 and №2. It includes Sus scrofa. Here it reaches a high number [10, 11].

Acrocephalus palustris, Locustella fluviatilis, Emberiza schoeniclus nest in biotopes of the plots. You can find three types of frogs: Rana arvalis, Rana temporaria and Pelophylax lessonae. Throughout the reserve, hedgehog, Fox, Sparrowhawk, Goshawk, great spotted woodpeckers are widespread.

In areas №1 and №2 the following vertebrates live: Columba oenas, Pernis apivorus, Nucifraga, Muscardinus avellanarius, which are listed in the Red Book of Moscow region. On the territory of the site you can find bats. They are rare and vulnerable, but not included in the Red Book of Moscow region, their 2 species, and they need control and monitoring. Here you can meet Nyctalus noctula and Pipistrellus nathusii.

Due to the diversity and richness of grasses and grasses forest glades, among oak and aspen forests abundance of flowering plants, there are favorable conditions for the formation of a rich fauna of insects. Neptis sappho, sappho are species of butterflies listed in the Red Book of Moscow region; you can find Colias palaeno, Euphydryas aurinia, Euphydryas maturna, Melitaea diamina, Melitaea phoebe, Argynnis laodice, Lycaena helle, Agriades optilete, Saturnia pavonia, Hyphoraia aulica, Callimorpha dominula, smerinthus Caecus, laothoe amurensis.

8 rare vulnerable Lepidoptera species, not included in the Red Book of Moscow region, but which need to be controlled and monitored, were noted on the territory of plots №1 and №2. These include: Thymelicus sylvestris or Ochlodes sylvanus, Aphantopus hyperantus, Lopinga achine, Erebia ligea, Brenthis ino, Maniola jurtina, Euchalcia variabilis and Pygaera timon [10, 11].

4. Conclusion

Currently on the territory of the "Novosurinsky" reserve there are some negative impacts in the form of litter, intensive recreation, with the consequent disturbance of soil, vegetation and the destruction of the soil. In addition, due to careless handling of fire, arson of dry grass and forest litter, fires occur.

The arrival of vehicles and their unhindered movement lead to violations of soil and vegetation. In addition, the obvious threats to the studied area should include any construction, laying of roads and other communications, increased recreational impact and self-production of peat.

References

[1] *State natural reserve "Novosurinsky" of the Moscow region* (Moscow: Arctic and Antarctic research Institute Laboratory of geoinformation technologies)
[2] Kalikhman T P 2017 The system of specially protected natural areas of the Far Eastern Federal district in atlas mapping *Geography And Natural Resources* **38(3)** 226-237
[3] Kotlyakov V M and Komarov A I 2007 Geography (concepts and terms) *Five-language academic dictionary* (Moscow: Nauka)
[4] Ecology of Moscow region 2001 *Encyclopedia* (Moscow: Modern notebooks)
[5] Grebenshchikov O S 1956 *Geobotanical dictionary* (Moscow: USSR Academy of Sciences; Nauka)
[6] Vishnyakov N V and Zelenskaya O Y 2018 Monitoring of the tourist-recreational activity of specially protected natural territories as an effective indicator of predicted development of tourist territories *South of Russia: Ecology, Development* [in Russian – Yug Rossii: Ekologija, razvitie] **13(4)** 119-128
[7] Bedova P, Bogdanov G, Anichkin A, Vorobeva I and Turmuhametova N 2018 Experience in establishing specially protected natural areas in case of small scale landscapes – objective criteria and arguments to justify *Int. Multidisciplinary Scientific Geoconf. Surveying Geology And Mining Ecology Management Sgem* **5.2** pp 429-436
[8] Salikhov T K, Karagoishin Z M, Svanbayeva Z S, Inkarova Z I, Dukenbayeva A D, Sagatbayev E N and Rakisheva A K 2016 Geoecological assessment of the projected state nature reserve 'Bokeyorda' in West Kazakhstan region *Oxidation Communications* **39(4-II)** 3579-90
[9] Tilekova Z T, Oshakbaev M T and Khaustov A P 2016 Assessing the geoecological state of ecosystems in the Balkhash region *Geography And Natural Resources* 79-86
[10] Shirokova V A, Khutorova A O, Gurov A F, Shirokov R S and Yurova Y D 2018 Environmental Assessment of Anthropogenic Effects on Recreation Zones: Osetr River in Moscow Region *Baltic Geodetic Congress (BGC Geomatics)* pp 181-185
[11] Klyushin P V, Lukyanova T S, Shirokova V A, Khutorova A O and Savinova S V 2018 The current state of boggy territories in the Meshchyora national park *18th Int. Multidisciplinary Scientific GeoConf. SGEM* **2.1** pp 775-782

AUTHOR INDEX